Foundational Theories and Practical Applications of Qualitative Research Methodology

Hesham Mohamed Elsherif
Queens Public Library, USA

A volume in the Advances in Library and Information Science (ALIS) Book Series

Published in the United States of America by
IGI Global
Information Science Reference (an imprint of IGI Global)
701 E. Chocolate Avenue
Hershey PA, USA 17033
Tel: 717-533-8845
Fax: 717-533-8661
E-mail: cust@igi-global.com
Web site: http://www.igi-global.com

Library of Congress Cataloging-in-Publication Data

CIP Pending

Foundational Theories and Practical Applications of Qualitative Research Methodology
Hesham Mohamed
2024 Information Science Reference

ISBN: 979-8-3693-2414-1
eISBN: 979-8-3693-2415-8

This book is published in the IGI Global book series Advances in Library and Information Science (ALIS) (ISSN: 2326-4136; eISSN: 2326-4144)

British Cataloguing in Publication Data
A Cataloguing in Publication record for this book is available from the British Library.

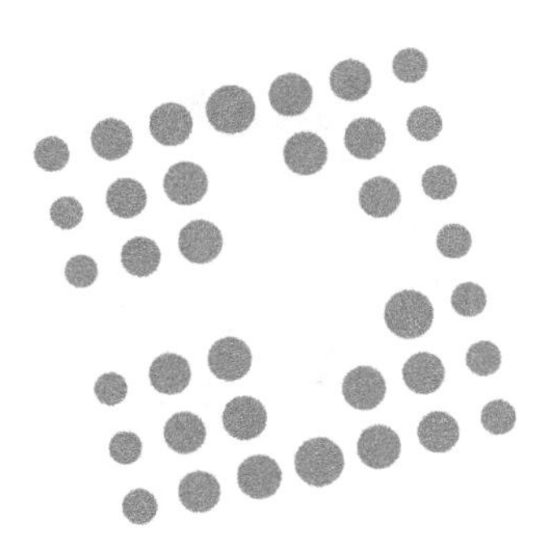

Advances in Library and Information Science (ALIS) Book Series

Alfonso Ippolito
Sapienza University-Rome, Italy
Carlo Inglese
Sapienza University-Rome, Italy

ISSN:2326-4136
EISSN:2326-4144

Mission

The **Advances in Library and Information Science (ALIS) Book Series** is comprised of high quality, research-oriented publications on the continuing developments and trends affecting the public, school, and academic fields, as well as specialized libraries and librarians globally. These discussions on professional and organizational considerations in library and information resource development and management assist in showcasing the latest methodologies and tools in the field.

The **ALIS Book Series** aims to expand the body of library science literature by covering a wide range of topics affecting the profession and field at large. The series also seeks to provide readers with an essential resource for uncovering the latest research in library and information science management, development, and technologies.

Coverage

- University Libraries in Developing Countries
- Librarianship and Human Rights
- Research Habits of Students and Faculty
- State Library Agencies
- Evidence-Based Librarianship
- Academic Libraries in the Digital Age
- Library 2.0 Tools
- Joint-Use Libraries
- Electronic Resources
- Mobile Library Services

IGI Global is currently accepting manuscripts for publication within this series. To submit a proposal for a volume in this series, please contact our Acquisition Editors at Acquisitions@igi-global.com or visit: http://www.igi-global.com/publish/.

The Advances in Library and Information Science (ALIS) Book Series (ISSN 2326-4136) is published by IGI Global, 701 E. Chocolate Avenue, Hershey, PA 17033-1240, USA, www.igi-global.com. This series is composed of titles available for purchase individually; each title is edited to be contextually exclusive from any other title within the series. For pricing and ordering information please visit http://www.igi-global.com/book-series/advances-library-information-science/73002. Postmaster: Send all address changes to above address.

Titles in this Series

For a list of additional titles in this series, please visit: http://www.igi-global.com/book-series/advances-library-information-science/73002

Examining Information Literacy in Academic Libraries
Sabelo Chizwina (North-West University, South Africa) and Mathew Moyo (North-West University, South Africa)
Information Science Reference • copyright 2024 • 313pp • H/C (ISBN: 9798369311431) • US $230.00 (our price)

AI-Assisted Library Reconstruction
K.R. Senthilkumar (Sri Krishna Arts and Science College, India)
Information Science Reference • copyright 2024 • 363pp • H/C (ISBN: 9798369327821) • US $235.00 (our price)

Challenges of Globalization and Inclusivity in Academic Research
Swati Chakraborty (GLA University, India & Concordia University, Canada)
Information Science Reference • copyright 2024 • 301pp • H/C (ISBN: 9798369313718) • US $225.00 (our price)

Multidisciplinary Approach to Information Technology in Library and Information Science
Barbara Holland (Brooklyn Public Library, USA (Retired)) and Keshav Sinha (University of Petroleum and Energy Studies, India)
Information Science Reference • copyright 2024 • 345pp • H/C (ISBN: 9798369328415) • US $245.00 (our price)

Handbook of Research on Innovative Approaches to Information Technology in Library and Information Science
Barbara Holland (Brooklyn Public Library, USA (Retired)) and Keshav Sinha (University of Petroleum and Energy Studies, India)
Information Science Reference • copyright 2024 • 427pp • H/C (ISBN: 9798369308073) • US $285.00 (our price)

Illuminating and Advancing the Path for Mathematical Writing Research
Madelyn W. Colonnese (Reading and Elementary Education Department, Cato College of Education, University of North Carolina at Charlotte, USA) Tutita M. Casa (Department of Curriculum and Instruction, Neag School of Education, University of Connecticut, USA) and Fabiana Cardetti (Department of Mathematics, College of Liberal Arts and Sciences, University of Connecticut, USA)
Information Science Reference • copyright 2024 • 389pp • H/C (ISBN: 9781668465387) • US $215.00 (our price)

Emerging Technology-Based Services and Systems in Libraries, Educational Institutions, and Non-Profit Organizations
Dickson K. W. Chiu (The University of Hong Kong, Hong Kong) and Kevin K. W. Ho (University of Tsukuba, Japan)
Information Science Reference • copyright 2023 • 353pp • H/C (ISBN: 9781668486719) • US $225.00 (our price)

701 East Chocolate Avenue, Hershey, PA 17033, USA
Tel: 717-533-8845 x100 • Fax: 717-533-8661
E-Mail: cust@igi-global.com • www.igi-global.com

Table of Contents

Detailed Table of Contents

This chapter delves into the foundational aspects of qualitative inquiry, contrasting it with quantitative methods by emphasizing its focus on understanding human experiences and social realities through words, narratives, and symbols. It highlights the interpretive and naturalistic approach of qualitative research, emphasizing the study of phenomena within their natural settings to uncover deeper meanings behind human behaviors, attitudes, and interactions. The chapter outlines key characteristics of qualitative research, including its exploratory nature, reliance on descriptive data, and iterative process, while also stressing the importance of context, holistic perspectives, and the researcher's subjective involvement. Additionally, it traces the historical evolution of qualitative research, drawing from its roots in anthropology, sociology, and psychology, and discusses contemporary developments that include digital methodologies and interdisciplinary approaches.

This chapter examines the essential role of theoretical frameworks in research, highlighting their significance as the structural backbone that guides the inquiry process, from conception through analysis. It discusses how these frameworks serve not only as support but also as detailed blueprints that outline the arrangement and interrelations within a study. By positioning research within a broader academic dialogue, theoretical frameworks enhance its credibility and facilitate understanding, linking the study to established paradigms.

In this chapter, the foundational blueprint of qualitative research is articulated, emphasizing the systematic strategies that guide researchers through the investigative journey. The chapter begins by elucidating the concept of research design, which encompasses the selection of appropriate methods of data collection and analysis techniques, and the structuring of the research timeframe and process. It delineates various types of research designs, including descriptive, experimental, observational, comparative, and longitudinal designs, highlighting their distinct objectives and applications.

This chapter delves into the intricate processes of data collection in qualitative research, highlighting the foundational importance and diverse methods used to capture the nuances and complexities of human behavior and societal trends. The core focus lies on how different data collection techniques—ranging from interviews and observations to document analysis and visual methods—can effectively address varied research questions within the qualitative paradigm. The alignment of specific methods with research questions is underscored as pivotal for ensuring the validity of the research outcomes.

This chapter explores the methodologies of fieldwork and immersion, rooted in anthropological and sociological traditions, as essential for gaining profound insights into diverse disciplines. By situating researchers directly within the study environments, these methods offer a unique vantage point to observe and interpret complex social dynamics and cultural interactions. The historical evolution of fieldwork is highlighted, showcasing its transformational impact on research methodologies and its adoption across various fields. The benefits of these approaches include rich contextual insights, enhanced rapport with participants, and the ability to observe dynamic interactions.

This chapter examines the critical stages of data analysis and interpretation in qualitative research, which are essential for transforming collected data into meaningful insights. Qualitative data, characterized by its volume and complexity, requires a methodical approach to uncover depth and context that quantitative data might overlook. The analysis process includes data preparation, coding, identifying themes, and visual representation, each playing a pivotal role in deriving nuanced understandings from textual and visual data. Theoretical frameworks guide the interpretation, ensuring analytical depth.

This chapter explores the distinctive nature and challenges of ensuring quality in qualitative research, contrasting it with the quantitative paradigm's focus on numerical analysis and generalizability. Qualitative research, centered on understanding complex human experiences, debates traditional metrics of quality such as validity and reliability, proposing alternatives like credibility, transferability, and confirmability tailored to its subjective and interpretive methodology. The discourse extends to embracing subjectivity as a strength, employing reflexivity and leveraging methodological strategies like triangulation and member checking to enhance rigor.

Preface

INTRODUCTION

In the evolving landscape of research, there's an unmistakable gravitation towards understanding the 'why' behind phenomena, alongside the 'what'. While quantitative research methods provide statistical insights and generalizability, qualitative methods offer depth, context, and a closer look at the nuances of human experiences. This book delves into the intricate and enriching world of qualitative inquiry, guiding readers through its myriad facets, from foundational theories to hands-on applications.

Purpose and Audience

The primary objective of this volume is to serve as a comprehensive guide to both newcomers and seasoned researchers navigating the qualitative realm. We aim to:

1. Equip scholars, students, and professionals with the knowledge and tools they need to design, conduct, and interpret qualitative studies.
2. Foster a deeper appreciation for the value of qualitative insights, emphasizing the depth and richness they bring to scholarly endeavors.
3. Bridge the gap between theory and practice by providing real-world examples, case studies, and practical tips.

This book caters to a broad spectrum of readers. Whether you're an undergraduate just introduced to qualitative methods, a graduate student working on a thesis, a doctoral researcher deep-diving into complex ethnographic studies, or a professional wanting to glean nuanced insights into your field of work, this book is crafted with you in mind.

Structure of the Book

Our journey begins with an exploration of what defines qualitative research, tracing its historical roots and key characteristics (Chapter 1). We then delve into the core theoretical frameworks that anchor various qualitative approaches—ranging from phenomenology to case study methodologies (Chapter 2).

Once grounded in these foundational concepts, we'll guide you through the nuts and bolts of research design and planning, including the formulation of compelling research questions, ethical considerations, and sampling strategies (Chapter 3).

With the blueprint ready, Chapters 4 and 5 introduce a suite of data collection methods and insights into the immersive experience of fieldwork. These chapters aim to equip researchers with practical knowledge on techniques like interviewing, observation, and focus groups.

Data, once gathered, requires meticulous analysis and interpretation. Chapter 6 unravels the intricacies of this phase, supplemented by digital tools and software that can streamline the process.

Ensuring the quality and rigor of qualitative research is paramount, which Chapter 7 addresses in depth. The art of presenting findings is then elaborated upon in Chapter 8, preparing researchers for both academic and non-academic audiences.

We recognize that the realm of qualitative research is not devoid of challenges, especially in terms of ethics. Chapter 9 delves deep into these concerns, providing guidance on navigating sensitive areas with grace and responsibility.

As research increasingly leans toward mixed methods, Chapter 10 elucidates how qualitative methods can be harmoniously integrated with quantitative ones.

Finally, in Chapter 11, we gaze into the future, highlighting emerging trends and directions in qualitative research.

We invite you to embark on this enlightening journey through the world of qualitative research with us. Let the pages that follow spark your curiosity, refine your skills, and kindle a passion for understanding the depth and intricacies of the human experience.

Hesham Mohamed Elsherif
Queens Public Library, USA

Chapter 1
Introduction to Qualitative Research

ABSTRACT

This chapter delves into the foundational aspects of qualitative inquiry, contrasting it with quantitative methods by emphasizing its focus on understanding human experiences and social realities through words, narratives, and symbols. It highlights the interpretive and naturalistic approach of qualitative research, emphasizing the study of phenomena within their natural settings to uncover deeper meanings behind human behaviors, attitudes, and interactions. The chapter outlines key characteristics of qualitative research, including its exploratory nature, reliance on descriptive data, and iterative process, while also stressing the importance of context, holistic perspectives, and the researcher's subjective involvement. Additionally, it traces the historical evolution of qualitative research, drawing from its roots in anthropology, sociology, and psychology, and discusses contemporary developments that include digital methodologies and interdisciplinary approaches.

DEFINING QUALITATIVE RESEARCH

At its core, qualitative research is interpretive and naturalistic in approach. This means that qualitative researchers study things in their natural settings, attempting to interpret phenomena based on the meanings people assign to them (Geertz, 1973). Such research does not just present facts but instead offers a detailed view of the topic at hand, seeking to uncover the 'why' behind specific behaviors, attitudes, and perspectives (Merriam & Tisdell, 2015).

Qualitative research, distinct in its approach and methodology, primarily seeks to understand the complex interplay of human experiences, beliefs, motivations, and behaviors in their natural settings (Denzin & Lincoln, 2017). Instead of relying heavily on quantifiable metrics and generalizable statistical results, qualitative research delves into the intricate textures and nuances that frame the human experience.

DOI: 10.4018/979-8-3693-2414-1.ch001

1. **Nature of Inquiry**: Qualitative research is inherently exploratory. While quantitative methodologies might ask 'how much?' or 'how many?', qualitative research is more interested in 'why?' and 'how?' It seeks to unravel the underlying motivations, reasons, and contexts of a phenomenon (Creswell, 2017).
2. **Data Types**: Instead of numerical datasets, qualitative research often produces descriptive data, expressed in natural language. This can be drawn from various sources such as interviews, observations, narratives, diaries, or any other form of textual, audio, or visual information (Silverman, 2016).
3. **Constructivist Approach**: Qualitative research often operates under a constructivist paradigm, assuming that reality is socially constructed. This means that there isn't a single, objective reality; instead, there are multiple realities shaped by individual experiences and social contexts (Guba & Lincoln, 1994).
4. **Emphasis on Context**: Context is not merely a backdrop for qualitative research; it is central to it. The environment, socio-cultural setting, historical timeframe, and even the immediate physical setting can significantly influence participants' responses and behaviors. The qualitative researcher takes these factors into account, providing a rich description of the setting to allow for a deeper understanding of the data (Patton, 2015).
5. **Iterative Process**: The design of a qualitative study is frequently non-linear. This means researchers might revisit data collection methods, refine their research questions, or even change their focus based on the emerging insights from their data (Maxwell, 2012).
6. **Holistic Perspective**: Rather than dissecting a phenomenon into smaller, isolated units for analysis, qualitative research aims to capture the entirety of the experience. It seeks to provide a holistic view, considering interrelations, systems, and comprehensive narratives (Miles, Matthis, & Huberman, 2014).
7. **Subjective Involvement**: While objectivity is championed in many research paradigms, qualitative research acknowledges and even embraces the researcher's subjective involvement. Recognizing that pure objectivity is challenging, if not impossible, researchers reflect on their own influences, biases, and roles in the interpretation of data (Peshkin, 1988).

In essence, qualitative research is not about counting or measuring but about interpreting and understanding. It thrives on the richness of human experience, seeking to elucidate meanings, understand complexities, and contribute to academic and practical knowledge in a manner that is intricate, profound, and deeply contextual.

History and Evolution of Qualitative Research

Historically, qualitative methodologies have roots in disciplines such as anthropology, sociology, and psychology. Anthropologists like Malinowski and Boas emphasized participant observation as a means of understanding cultures from an emic, or insider's perspective (Boas, 1922; Malinowski, 1922). Sociologists like Weber argued for the 'Verstehen' approach, focusing on understanding social actions through interpretative means (Weber, 1949).

The roots of qualitative research are as old as human curiosity itself. Historically embedded in disciplines like anthropology, sociology, and psychology, its evolution has been shaped by shifts in philosophical thought, disciplinary advances, and methodological innovations.

Ancient Beginnings: Qualitative modes of inquiry can be traced back to ancient civilizations, where oral narratives, storytelling, and firsthand observations were the principal means of understanding the world (Denzin & Lincoln, 2017).

The precursors to what we understand as qualitative research today were seeded in ancient civilizations, where the intricacies of human behavior, societal norms, and shared beliefs were captured through observation, narrative, and dialogue.

Oral Traditions: Ancient societies were predominantly oral cultures. Narratives, myths, and legends passed down through generations were more than mere stories; they were repositories of collective wisdom, histories, moral codes, and cosmological beliefs. These oral narratives offered deep insights into the social, moral, and spiritual worlds of these cultures (Ong, 1982).

Philosophical Dialogues: The classical Greek tradition of philosophical dialogue, as embodied by Socrates, Plato, and Aristotle, can be seen as a form of qualitative inquiry. Socratic questioning, in particular, was a method of eliciting truth through continuous, probing dialogue, focusing on understanding the 'essence' of concepts such as justice, love, and virtue (Jarratt, 1991).

Ancient Ethnographies: Herodotus, often referred to as the 'Father of History,' engaged in what can be considered early ethnographic work. His 'Histories' were detailed accounts of various cultures, rituals, and traditions of peoples across the then-known world, providing rich qualitative descriptions based on both firsthand observations and reports from others (Herodotus, 425 BCE/2003).

Ancient India's Narrative Texts: The ancient Indian epics, the Mahabharata and the Ramayana, are replete with intricate character developments, dialogues, and moral dilemmas. These texts, while steeped in mythology and legend, provide profound insights into the societal norms, human psyche, and philosophical underpinnings of the time (Doniger, 2010).

Confucian Texts: In ancient China, the Analects of Confucius offer detailed narratives of conversations, thoughts, and teachings of Confucius. These dialogues provide rich qualitative data on moral values, societal norms, and the human condition in ancient Chinese society (Confucius & Lau, 1979).

The ancient world, through its narratives, dialogues, and descriptive accounts, laid the groundwork for the qualitative exploration of human experiences. These early efforts underscore the longstanding human desire to understand, interpret, and make sense of our collective and individual journeys in intricate, nuanced ways.

Anthropological Foundations: The early 20th century saw the rise of anthropological legends like Franz Boas and Bronisław Malinowski. Boas introduced the idea of cultural relativism, emphasizing the importance of understanding a culture from its own perspective (Boas, 1911). Malinowski, in his seminal work in the Trobriand Islands, laid the foundations of ethnography with his approach of participant observation, arguing for the importance of immersing oneself in the culture being studied (Malinowski, 1922).

Anthropology, as a discipline dedicated to the study of human societies and cultures, has significantly influenced the conceptual underpinnings and methodologies of qualitative research. Some of the most profound contributions to the field of qualitative inquiry can be traced back to the pioneering works of anthropologists in the late 19th and early 20th centuries.

Franz Boas and Cultural Relativism: Franz Boas is often recognized as the father of modern anthropology, challenging the prevailing ethnocentric views and racial hierarchies of his time. He introduced the concept of cultural relativism, asserting that every culture should be understood and judged on its own terms, rather than by the standards of another culture. This perspective shifted the focus from evaluating cultures on a perceived evolutionary scale to understanding each culture as a unique entity shaped by its history, environment, and internal logic (Boas, 1887).

Bronisław Malinowski and Participant Observation: Malinowski's work in the Trobriand Islands laid the foundations for modern ethnographic fieldwork. He championed the method of participant observation, where the researcher lives within the community they are studying for an extended period, actively participating in daily activities while observing societal dynamics. Malinowski argued that to understand a culture fully, one must grasp the "native's point of view" and the intricacies of their everyday life (Malinowski, 1922).

Margaret Mead and Cultural Patterns: Margaret Mead's work, especially in Samoa, emphasized the role of culture in shaping individual behavior and personality. Through her detailed ethnographies, Mead demonstrated that many behaviors and roles considered "natural" or "innate" in Western societies were, in fact, products of cultural conditioning. Her work underscored the power of cultural forces in shaping human experiences (Mead, 1928).

Ethnography and Thick Description: Clifford Geertz expanded on traditional anthropological methods by introducing the concept of "thick description." Instead of just documenting observable behaviors, Geertz proposed that anthropologists should provide dense, detailed descriptions, capturing not only the action but also the context, meaning, and symbolism behind it. This approach sought to understand cultures in all their complexity, moving beyond surface observations to deeper interpretations (Geertz, 1973).

Reflexivity in Anthropology: As anthropology evolved, there was a growing recognition of the researcher's role in shaping the research process and its outcomes. Scholars like Lila Abu-Lughod emphasized reflexivity, encouraging researchers to critically reflect on their own positions, biases, and relationships with study participants. This reflective stance added depth and nuance to ethnographic accounts (Abu-Lughod, 1988).

Through these foundational contributions, anthropology has provided qualitative research with methodological rigor, a commitment to deep, immersive understanding, and a respect for the diversity of human experiences.

Sociological Influences: The emergence of qualitative methods in sociology was marked by seminal figures like Max Weber and Erving Goffman. Weber's concept of 'Verstehen,' or interpretive understanding, emphasized the necessity of grasping the subjective meanings individuals attribute to their actions (Weber, 1949). Goffman's symbolic interactionism and dramaturgical analysis illustrated how individuals present themselves in everyday life (Goffman, 1959).

The discipline of sociology, with its focus on human society, interactions, and structures, has made seminal contributions to the development and refinement of qualitative methodologies. These sociological influences have expanded the depth and range of qualitative research, emphasizing the importance of understanding social phenomena within their broader societal context.

Max Weber and Verstehen: A foundational figure in sociology, Max Weber, introduced the concept of *Verstehen*—a German term often translated as "interpretive understanding" or "empathetic understanding." Weber asserted that to truly understand social actions, one must grasp the subjective meanings and motivations individuals attribute to their behaviors. This interpretive approach emphasized the importance of seeing the world through the eyes of those being studied, highlighting the role of individual agency within broader societal structures (Weber, 1949).

The Chicago School and Urban **Ethnography**: The Chicago School, a group of sociologists based at the University of Chicago in the early 20th century, played a pivotal role in shaping

American sociological thought. Researchers like Robert Park, Ernest Burgess, and later, Everett Hughes, employed ethnographic methods to study urban life, human ecology, and social worlds. Their immersive studies in diverse urban settings provided detailed insights into the experiences of immigrants, marginalized groups, and urban transformations, laying the groundwork for modern urban ethnography (Bulmer, 1984).

Symbolic Interactionism: Grounded in the works of George Herbert Mead, Herbert Blumer, and later Erving Goffman, symbolic interactionism posits that individuals construct their realities based on interactions and symbolic meanings. This perspective emphasizes the dynamic nature of social life, where meanings are continuously negotiated, redefined, and interpreted. Goffman's dramaturgical approach, for instance, studied social life as a theatrical performance, where individuals play roles, follow scripts, and use symbols to convey meaning (Goffman, 1959).

Grounded Theory: Developed by sociologists Barney Glaser and Anselm Strauss in the 1960s, grounded theory is a systematic methodology that involves constructing theory from data. Instead of beginning with a predefined theory, researchers collect and analyze data simultaneously, allowing patterns, themes, and theories to emerge inductively. Grounded theory has since become a foundational qualitative methodology, widely adopted across disciplines (Glaser & Strauss, 1967).

Critical Theory and the Frankfurt School: Originating with scholars like Max Horkheimer, Theodor Adorno, and later, Jürgen Habermas, the Frankfurt School critiqued traditional scientific and sociological approaches. They emphasized the role of ideology, power structures, and capitalist systems in shaping social realities. This critical perspective added depth to qualitative inquiry, pushing researchers to question dominant narratives and uncover underlying power dynamics (Horkheimer, 1982).

These sociological influences underscore the discipline's commitment to understanding the nuances of social life. They have enriched qualitative research methodologies, emphasizing the importance of context, interpretation, and critical inquiry in studying the multifaceted tapestry of human society.

Rise of Phenomenology and Existentialism: Philosophers such as Edmund Husserl and Jean-Paul Sartre influenced qualitative thought with their focus on individual consciousness, lived experiences, and existential realities. Their ideas encouraged researchers to delve deep into individual perceptions and experiences to extract meaning (Husserl, 1970; Sartre, 1943).

Phenomenology and existentialism, with their profound focus on individual consciousness, lived experience, and existential realities, introduced novel philosophical dimensions to qualitative research. These paradigms shifted the focus from external behaviors to internal perceptions and meanings, fundamentally shaping the way human experience is studied and understood.

Edmund Husserl and Phenomenology: Often considered the founder of phenomenology, Edmund Husserl introduced a methodology aimed at exploring the essential structures of consciousness. Husserl's phenomenology emphasizes the "intentionality" of consciousness, suggesting that consciousness is always consciousness *of* something. Through 'phenomenological reduction,' Husserl proposed bracketing out one's preconceptions to achieve a pure, unbiased description of the phenomenon, thus capturing its very essence (Husserl, 1913/2012).

Martin Heidegger and Hermeneutic Phenomenology: Building on Husserl's foundation, Martin Heidegger introduced a hermeneutic turn to phenomenology, emphasizing the interpretative nature of understanding. He posited that individuals are always embedded within a historical, cultural, and linguistic world, and thus, understanding any phenomenon always involves interpretation. For Heidegger, the essence of being was intricately linked with temporality and one's relationship to the world (Heidegger, 1927/1962).

Jean-Paul Sartre and Existentialism: Sartre's existentialism brought forth the notions of freedom, responsibility, and existential angst. He proposed that existence precedes essence, suggesting that individuals first exist and then define their essence or nature through their actions and choices. For Sartre, the individual is condemned to be free and must bear the weight of this freedom in the face of life's inherent meaninglessness (Sartre, 1943/1956).

Maurice Merleau-Ponty and the Embodied Experience: Merleau-Ponty emphasized the importance of embodiment in understanding human experience. He proposed that our body is not just an object in the world but the very medium through which the world is experienced. This phenomenological stance has greatly influenced studies focusing on perception, embodiment, and the interrelationship between the self and the world (Merleau-Ponty, 1945/2012).

Existential Phenomenology in Qualitative Research: These philosophical currents have deeply influenced qualitative methodologies. Existential and phenomenological inquiries focus on capturing the lived experiences of individuals, delving into the meanings, perceptions, and essences of their experiences. Such studies aim not just to describe but to interpret, seeking a profound understanding of human existence in all its richness and complexity (van Manen, 1990).

Through the contributions of phenomenology and existentialism, qualitative research has been equipped with philosophical and methodological tools to explore the depths of human experience, moving beyond mere surface descriptions to delve into the very essence of existence and meaning.

Postmodernism and Beyond: The latter half of the 20th century witnessed the rise of postmodernism and poststructuralism. Thinkers like Michel Foucault and Jacques Derrida critiqued traditional notions of truth and reality, emphasizing the constructed nature of knowledge and the role of power in its formation (Foucault, 1977; Derrida, 1978). These perspectives enriched qualitative methodologies, introducing concepts like deconstruction and discourse analysis.

The rise of postmodernist and poststructuralist thought in the latter half of the 20th century profoundly shaped and expanded the horizons of qualitative research. These paradigms offered a critique of conventional narratives, methodologies, and epistemologies, advocating for a more pluralistic and deconstructive approach to understanding reality.

The Postmodern Turn: At its core, postmodernism rejects grand narratives and singular truths, positing that reality is fragmented, multiple, and continuously constructed. This perspective critiques foundational assumptions, questioning the very possibility of objective knowledge. As a result, postmodernist thought encouraged researchers to embrace complexity, uncertainty, and multiplicity (Lyotard, 1979/1984).

Michel Foucault and Power/Knowledge: Foucault's work was pivotal in highlighting the intricate relationship between power and knowledge. He contended that knowledge is never neutral but is invariably entwined with power dynamics, thereby shaping societal structures, discourses, and realities. His concepts of discourse analysis and the archaeology of knowledge have become central in many qualitative studies, focusing on the construction and representation of knowledge within specific historical and social contexts (Foucault, 1969/1972).

Jacques Derrida and Deconstruction: Derrida's approach to deconstruction dissected and challenged conventional linguistic structures and meanings. By demonstrating the instability of language and highlighting the play of differences (*differance*), Derrida emphasized the impossibility of a fixed, stable meaning. This notion has been influential in qualitative studies that aim to deconstruct dominant narratives and explore the fluidity and ambiguity of texts and meanings (Derrida, 1967/1978).

Postcolonial Critiques: Postcolonial theorists, such as Edward Said and Gayatri Spivak, have critiqued the Western-centric viewpoints that often dominate academic discourses. Their works emphasize the importance of acknowledging the perspectives of the 'other,' those historically marginalized or silenced by colonial and imperialist narratives. This critique has been foundational in qualitative research that seeks to amplify subaltern voices and counter-narratives (Said, 1978; Spivak, 1988).

Digital and Multimodal Turn: With the rise of digital technology and the internet, qualitative research has seen a shift towards digital and multimodal methodologies. Researchers now engage in online ethnographies, digital storytelling, and utilize multimedia platforms to capture and represent the nuanced realities of the digital age (Pink et al., 2016).

Postmodernism and its subsequent developments have deeply enriched qualitative research, urging scholars to challenge the taken-for-granted, to embrace the multiplicity of voices, and to continuously reflect upon the dynamic interplay of power, representation, and knowledge.

Contemporary Developments: Today, qualitative research is dynamic and multifaceted, embracing diverse traditions like narrative research, grounded theory, case studies, and more. With the advent of technology, digital qualitative methods, including online ethnographies and virtual focus groups, have also gained prominence (Markham, 2017).

As qualitative research continues to evolve in response to societal, technological, and academic changes, contemporary developments have further broadened the scope and depth of methodologies, tools, and epistemologies in the field. Here, we delve into some of these current trends that are shaping the future trajectory of qualitative inquiry.

Interdisciplinary Approaches: Contemporary qualitative research often defies traditional disciplinary boundaries. There's an increasing fusion of methodologies and theoretical frameworks from anthropology, sociology, psychology, and even areas like digital humanities. This interdisciplinary approach enriches qualitative studies, fostering a more holistic understanding of complex phenomena (Nowotny, 2003).

Narrative Inquiry: Rooted in the belief that storytelling is fundamental to the human experience, narrative inquiry focuses on personal stories and lived experiences. By understanding individuals' narratives, researchers can delve into the intricacies of identity, experience, and societal structures (Clandinin & Connelly, 2000).

Visual and Sensory Ethnography: As researchers seek richer, more multifaceted data, there's a growing emphasis on visual and sensory methodologies. This involves capturing data through photographs, videos, sound recordings, or even taste and touch, providing a more immersive account of participants' realities (Pink, 2015).

Digital and Virtual Methods: With the digital revolution and the ubiquity of online spaces, qualitative researchers are navigating virtual communities, online interactions, and digital artifacts. Techniques such as virtual ethnography, online interviews, and digital content analysis have gained prominence in studying the ever-evolving digital landscape (Hine, 2015).

Ethical Considerations in a Globalized World: As research increasingly spans diverse, global contexts, there's heightened attention to ethical considerations. Issues related to consent, representation, cultural sensitivity, and digital privacy are continually being reevaluated to ensure research integrity (Guillemin & Gillam, 2004).

Post-qualitative Turn: Challenging traditional qualitative norms, the post-qualitative movement, influenced by posthumanism and new materialism, questions the very foundations of representation, interpretation, and human-centric inquiry. Researchers in this vein are exploring more experimental, non-linear, and non-representational methodologies (St. Pierre, 2018).

Embracing Complexity with Mixed Methods: Recognizing that singular methodologies might not capture the complexity of certain phenomena, there's a growing trend toward mixed methods research, which combines qualitative and quantitative approaches. This triangulation offers a more comprehensive view, leveraging the strengths of both paradigms (Creswell & Plano Clark, 2017).

Contemporary developments in qualitative research underscore the field's dynamism and adaptability. As societal landscapes shift and new challenges emerge, qualitative inquiry continues to evolve, reflecting the ever-changing tapestry of human experiences, meanings, and realities.

Key Characteristics and Assumptions:

There are several defining characteristics and assumptions that underlie qualitative research:

1. **Holistic View**: Qualitative research aims for a comprehensive understanding of phenomena. It often considers context, environment, interrelationships, and individual interpretations as central to understanding the entirety of the subject (Patton, 2002).

A holistic view in qualitative research underscores the importance of understanding phenomena in their entirety, emphasizing the interconnectedness and interrelations of all its components. Instead of fragmenting a phenomenon into discrete variables or components, as is common in quantitative research, a holistic approach seeks to grasp the totality of experiences, settings, and contexts.

Characteristics of a Holistic View

a) **Complexity and Interconnectedness**: Holistic research recognizes that social phenomena are complex, shaped by myriad factors that are interrelated. It does not merely focus on individual elements but also explores the relationships, dynamics, and synergies between them (Moran, 2017).

b) **Contextual Understanding**: A holistic approach values the significance of context, acknowledging that phenomena are deeply embedded in their socio-cultural, historical, and environmental settings. The nuances of these settings play a crucial role in shaping experiences and cannot be divorced from the phenomena being studied (Creswell & Poth, 2017).

c) **Emphasis on Systems**: Systems thinking is integral to holistic research. Instead of viewing phenomena as isolated units, they are seen as parts of broader systems where each element affects and is affected by the other (Checkland, 1981).

d) **Dynamic and Evolving Nature**: Recognizing that social realities are constantly shifting and evolving, the holistic approach seeks to capture these dynamics, understanding that phenomena are not static but change in relation to various internal and external influences (Lincoln, Lynham, & Guba, 2011).

Assumptions Underlying a Holistic View

i. **Interdependence**: At the heart of the holistic view is the assumption that all elements within a system or context are interdependent. Changes or influences on one component can have cascading effects on others (Bertalanffy, 1968).
ii. **Non-reductionism**: Contrary to reductionist approaches that break down phenomena into isolated variables, the holistic view assumes that the whole is more than the sum of its parts. This means that understanding each component separately will not necessarily lead to a comprehension of the entire phenomenon (Moustakas, 1994).
iii. **Organic and Non-linear Relationships**: Holistic research operates on the assumption that relationships within systems are organic and non-linear. This means that they do not always follow predictable patterns but can have emergent properties that are not evident when components are studied in isolation (Capra, 1996).
iv. **Subjectivity and Multiple Realities**: Accepting the subjective nature of human experiences, the holistic approach assumes that there can be multiple, valid interpretations of a single phenomenon, all contributing to a richer, more comprehensive understanding (Denzin & Lincoln, 2005).

The holistic perspective in qualitative research, with its focus on totality, interrelations, and complexity, offers a profound and intricate understanding of human experiences and social phenomena. It champions depth, context, and interconnectedness, ensuring that research captures the richness and multifaceted nature of the lived experience.

2. **Inductive Reasoning**: Rather than starting with a hypothesis, qualitative research often begins with an open question or a phenomenon of interest. As data is collected and analyzed, themes and patterns emerge, guiding subsequent data collection and analysis (Strauss & Corbin, 1998).

Inductive reasoning, often juxtaposed with deductive approaches, holds a foundational position in qualitative research. This method of reasoning involves deriving general principles or theories from specific observations or cases, enabling a more open, exploratory, and emergent approach to understanding complex phenomena.

Characteristics of Inductive Reasoning

a) **Data-driven Inquiry**: Unlike deductive reasoning, which starts with a hypothesis that is then tested, inductive reasoning begins with observations, patterns, or specific instances. The research process is fundamentally data-driven, with themes, patterns, or theories emerging from the data itself (Thomas, 2006).
b) **Grounded Theory Formation**: As general patterns or theories are formulated based on specific observations, the resulting theories are said to be 'grounded' in the data. This is the essence of grounded theory methodology, which epitomizes the inductive approach in qualitative research (Glaser & Strauss, 1967).
c) **Open-ended Exploration**: Inductive reasoning encourages a more open and flexible research design. Researchers enter the field with open-ended questions, allowing the research direction to be shaped by participants' perspectives and emerging insights (Charmaz, 2006).

d) **Iterative Process**: The inductive approach is iterative, requiring researchers to move back and forth between data collection and analysis. As patterns emerge, researchers might refine their data collection techniques or pursue new lines of inquiry (Coffey & Atkinson, 1996).

Assumptions Underlying Inductive Reasoning

I. **Emergent Nature of Research**: Inductive reasoning operates on the assumption that one cannot know everything about the phenomenon at the outset. The research process, therefore, is emergent, adapting and evolving based on insights gained during the study (Hammersley, 1992).

II. **Subjectivity of Interpretation**: Inductive approaches acknowledge the interpretative role of the researcher. It is assumed that the researcher's background, experiences, and perspectives can influence data interpretation, and this subjectivity is integral to the research process (Denzin, 1989).

III. **Contextual Understanding**: Inductive reasoning emphasizes the significance of context. It assumes that meanings, behaviors, and experiences are deeply embedded within specific socio-cultural and historical contexts, and these cannot be divorced from the phenomena under study (Geertz, 1973).

IV. **Multiple Realities**: In line with the qualitative paradigm, inductive reasoning rests on the assumption that there can be multiple, often subjective, interpretations or realities of a single phenomenon. These multiple perspectives contribute to a richer, more nuanced understanding (Lincoln & Guba, 1985).

Inductive reasoning, with its emergent, grounded, and context-specific nature, is fundamental to qualitative research's endeavor to capture the richness, depth, and complexity of human experiences. It allows researchers to remain open and receptive, letting the data guide the journey of discovery and understanding.

3. **Flexibility**: Qualitative methodologies are dynamic. Researchers can adapt their strategies based on the data they gather, ensuring a richer and more nuanced understanding of their subject (Creswell & Poth, 2017).

Flexibility stands as one of the hallmark features of qualitative research. Unlike quantitative approaches that adhere more rigidly to predetermined structures, qualitative studies often embrace an adaptable, evolving design, allowing researchers to deeply engage with the nuances, complexities, and unforeseen facets of human experience.

Characteristics of Flexibility in Qualitative Research

a) **Emergent Design**: Qualitative research often begins with general research questions instead of tightly formulated hypotheses. As data collection progresses, the research design, including data collection methods and research questions, may evolve based on what is learned (Merriam, 2009).

b) **Adaptable Data Collection Techniques**: Depending on the emerging insights, researchers may adapt their data collection techniques. For instance, an interview guide might be modified to explore unexpected themes or to delve deeper into particular issues that arise during early interviews (Patton, 2015).

c) **Iterative Analysis**: Qualitative analysis is an iterative process. Researchers might go back to the field for more data or refine their analytic strategies based on initial findings, ensuring that the emerging themes or theories are robustly grounded in the data (Braun & Clarke, 2006).

d) **Diverse Methodological Approaches**: Flexibility in qualitative research is also evident in the plethora of methods and approaches available to researchers, from grounded theory to ethnography, from narrative analysis to phenomenology. Researchers often draw from multiple methodologies or adapt them to suit the specific demands of their study (Denzin & Lincoln, 2011).

Assumptions Underlying Flexibility in Qualitative Research

I. **Complexity of Human Experience**: One foundational assumption is that human experiences and social realities are complex and multifaceted. A rigid, predetermined research design might not capture this complexity, and thus flexibility is essential (Smith, 2015).

II. **Subjectivity and Co-construction**: Qualitative research often acknowledges the co-constructive nature of knowledge, wherein both the participant and researcher play active roles in the research process. This mutual engagement necessitates flexibility as both parties influence the direction and outcomes of the research (Morrow, 2005).

III. **Valuing the Unexpected**: The qualitative research paradigm values unexpected findings or divergent patterns as they can offer profound insights. Flexibility ensures that such unexpected avenues can be pursued, enriching the research (Charmaz, 2014).

IV. **Acknowledgment of Context**: Flexibility is rooted in the understanding that context matters. Since research takes place in dynamic settings with myriad influences, the ability to adapt to these changing contexts ensures that the research remains relevant and grounded (Flyvbjerg, 2006).

In essence, flexibility in qualitative research underscores the field's commitment to understanding human experiences in all their richness and depth. It allows for an authentic, responsive, and nuanced exploration, ensuring that the research remains deeply connected to the lived realities it seeks to understand.

4. **Emphasis on Process**: Instead of just focusing on outcomes or products, qualitative research is equally, if not more, interested in processes. It seeks to understand how individuals and groups interpret experiences, construct their realities, and assign meanings (Charmaz, 2006).

One distinguishing feature of qualitative research is its emphasis on process. Instead of merely focusing on outcomes or end products, qualitative inquiry is deeply invested in understanding the dynamics, evolutions, and sequences through which social phenomena unfold.

Characteristics of Emphasis on Process

a) **Temporal Dynamics**: Qualitative studies often explore the temporal aspects of experiences, tracing the chronological unfolding of events, perceptions, and changes over time (Langley, 1999).
b) **Journey, not just Destination**: The emphasis on process means that the trajectory—how participants arrive at certain viewpoints, decisions, or feelings—is as important, if not more so, than the end state itself (Maxwell, 2012).
c) **Understanding Change and Stability**: Through a process-oriented lens, researchers examine the dynamics of change, transitions, and evolutions. Equally, understanding what remains stable or resistant to change provides rich insights into the underlying structures and constants in a phenomenon (Pettigrew, 1997).
d) **Sequential Analysis**: In emphasizing process, qualitative researchers might employ sequential or narrative analyses, tracing events, decisions, or experiences in the order they occur, seeking to understand causality, influence, and interplay (Riessman, 2008).

Assumptions Underlying Emphasis on Process

I. **Social Reality is Dynamic**: At the heart of this emphasis is the assumption that social realities are not static but continuously evolving. Human experiences, meanings, and perceptions shift in response to various influences and over time (Van de Ven, 1992).
II. **Contextual Interplay**: Process-oriented qualitative research assumes that phenomena are deeply entwined with their context. This means that as contexts evolve, the phenomena themselves might undergo changes (Bhaskar, 2013).
III. **Holistic Understanding**: By emphasizing process, qualitative inquiry operates on the assumption that a holistic understanding of phenomena requires grasping not only their current state but also their origin, evolution, and trajectory (Saldaña, 2015).
IV. **Meanings Evolve**: Central to qualitative epistemology is the belief that meanings are not fixed. Emphasizing process acknowledges that individuals might reinterpret, renegotiate, or redefine their experiences and understandings as they encounter new information, challenges, or contexts (Bruner, 1990).

In essence, the emphasis on process in qualitative research seeks to capture the fluidity, dynamism, and intricacies of human experiences. By tracing the journey and not just documenting the destination, researchers can offer a more comprehensive, nuanced, and authentic account of the phenomena under study.

5. **Subjectivity and Reflexivity**: Qualitative researchers acknowledge their own biases, values, and roles in the research process. Reflexivity involves self-awareness and critical self-reflection on how these biases and values may influence the research (Finlay, 2002).

Within qualitative research, the intertwined concepts of subjectivity and reflexivity stand as core tenets. These concepts challenge the notion of detached, objective knowledge, emphasizing instead the deeply personal, interpretative, and contextual nature of research.

Characteristics of Subjectivity and Reflexivity

a) **Personal Lens**: Subjectivity acknowledges that researchers bring their own backgrounds, experiences, biases, and perspectives to the research process. Rather than being passive vessels, researchers are actively engaged, interpreting data through their unique lenses (Peshkin, 1988).
b) **Ongoing Self-awareness**: Reflexivity involves a continuous process of self-awareness and self-examination. Qualitative researchers are encouraged to constantly reflect upon their roles, influences, and interactions in the research process (Finlay, 2002).
c) **Interpretative Nature of Research**: Given the inherent subjectivity, qualitative research is often deemed interpretative. Findings are not seen as absolute truths but as interpretations shaped by the researcher's and participants' perspectives (Geertz, 1973).
d) **Researcher's Journal or Memo**: Many qualitative researchers maintain reflexive journals or memos, where they document their thoughts, feelings, biases, and the decisions they make throughout the research process. This serves both as a tool for introspection and as a transparent record of the research journey (Ortlipp, 2008).

Assumptions Underlying Subjectivity and Reflexivity

I. **Knowledge is Co-constructed**: Central to the qualitative paradigm is the belief that knowledge is not merely discovered but co-constructed. Both the researcher and the participant contribute to the creation and interpretation of meanings (Guba & Lincoln, 1994).
II. **No Neutral Ground**: Qualitative research operates on the assumption that pure objectivity or neutrality is unattainable. Every research decision, from question formulation to data interpretation, is influenced by the researcher's subjectivity (Harding, 1993).
III. **Reflexivity Enhances Credibility**: While subjectivity might seem to challenge the validity of research findings, reflexivity is seen as enhancing credibility. By being transparent about biases, influences, and decisions, researchers provide a clear trail that others can follow, evaluate, and understand (Lincoln & Guba, 1985).
IV. **Embrace of Multiple Realities**: Subjectivity and reflexivity lead to the acceptance that there can be multiple, equally valid interpretations or realities of a phenomenon. This plurality is not seen as a limitation but as a strength, capturing the richness and diversity of human experiences (Denzin & Lincoln, 2005).

Embracing subjectivity and reflexivity, qualitative research acknowledges the complexities and intricacies of human understanding. By recognizing and reflecting upon their roles, biases, and influences, researchers can offer deep, transparent, and authentic insights into the phenomena they study.

To conclude, qualitative research offers an intricate, detailed, and profound understanding of human experiences and the broader social world. Its emphasis on depth, context, and meaning sets it apart, providing researchers with a robust toolkit to explore the complexities of the human condition.

REFERENCES

Abu-Lughod, L. (1988). Writing Against Culture. In R. G. Fox (Ed.), *Recapturing Anthropology: Working in the Present* (pp. 137–162). School of American Research Press.

Bertalanffy, L. V. (1968). *General System theory: Foundations, Development, Applications*. George Braziller.

Bhaskar, R. (2013). *A realist theory of science*. Routledge. doi:10.4324/9780203090732

Boas, F. (1887). Museums of Ethnology and their Classification. *Science*, *9*(228), 587–589. doi:10.1126/science.ns-9.228.587.b PMID:17779724

Boas, F. (1911). *The mind of primitive man*. MacMillan.

Boas, F. (1922). *Ethnology of the Kwakiutl*. American Museum of Natural History.

Braun, V., & Clarke, V. (2012). Thematic analysis. In H. Cooper (Ed.), *APA handbook of research methods in psychology*. APA. doi:10.1037/13620-004

Bruner, J. (1990). *Acts of Meaning*. Harvard University Press.

Bulmer, M. (1984). *The Chicago School of Sociology: Institutionalization, Diversity, and the Rise of Sociological Research*. University of Chicago Press.

Capra, F. (1996). *The Web of Life: A New Scientific Understanding of Living Systems*. Anchor Books.

Charmaz, K. (2014). *Constructing grounded theory* (2nd ed.). Sage.

Checkland, P. (1981). *Systems Thinking, Systems Practice*. John Wiley & Sons.

Clandinin, D. J., & Connelly, F. M. (2000). *Narrative inquiry: Experience and story in qualitative research*. Jossey-Bass.

Coffey, A., & Atkinson, P. (1996). *Making Sense of Qualitative Data: Complementary Research Strategies*. Sage Publications.

Confucius, & Lau, D. C. (1979). *Confucius: The Analects*. Penguin UK.

Creswell, J. W., & Creswell, J. D. (2017). *Research design: Qualitative, quantitative, and mixed methods approaches*. Sage Publications.

Creswell, J. W., & Plano Clark, V. L. (2017). *Designing and conducting mixed methods research*. Sage publications.

Creswell, J. W., & Poth, C. N. (2017). *Qualitative inquiry and research design: Choosing among five approaches* (3rd ed.). Sage.

Denzin, N. K. (1989). *The research act: A theoretical introduction to sociological methods*. Prentice Hall.

Denzin, N. K., & Lincoln, Y. S. (2018). *The Sage handbook of qualitative research*. Sage publications.

Derrida, J. (1967/1978). *Writing and Difference*. University of Chicago Press.

Doniger, W. (2010). *The Hindus: An Alternative History*. Oxford University Press.

Finlay, L. (2002). "Outing" the Researcher: The Provenance, Process, and Practice of Reflexivity. *Qualitative Health Research, 12*(4), 531–545. doi:10.1177/104973202129120052 PMID:11939252

Flyvbjerg, B. (2006). Five misunderstandings about case-study research. *Qualitative Inquiry, 12*(2), 219–245. doi:10.1177/1077800405284363

Foucault, M. (1969/1972). *The Archaeology of Knowledge*. Pantheon.

Foucault, M. (1977). *Discipline and Punish: The Birth of the Prison*. Vintage.

Geertz, C. (1973). *The interpretation of cultures: Selected essays*. Basic Books.

Glaser, B. G., & Strauss, A. L. (1967). *The Discovery of Grounded Theory: Strategies for Qualitative Research*. Aldine Transaction.

Goffman, E. (1959). *The Presentation of Self in Everyday Life*. Anchor Books.

Guba, E. G., & Lincoln, Y. S. (1994). Competing paradigms in qualitative research. In N. K. Denzin & Y. S. Lincoln (Eds.), *Handbook of qualitative research* (pp. 105–117). Sage.

Guillemin, M., & Gillam, L. (2004). Ethics, reflexivity, and "ethically important moments" in research. *Qualitative Inquiry, 10*(2), 261–280. doi:10.1177/1077800403262360

Hammersley, M. (1992). *What's Wrong with Ethnography? Methodological Explorations*. Routledge.

Harding, S. (1993). Rethinking standpoint epistemology: What is "strong objectivity?". In L. Alcoff & E. Potter (Eds.), *Feminist Epistemologies* (pp. 49–82). Routledge.

Heidegger, M. (1927/1962). *Being and Time*. Harper & Row.

Herodotus. (2003). The Histories. Trans. A. de Sélincourt. Penguin.

Hine, C. (2015). *Ethnography for the Internet: Embedded, embodied and everyday*. Bloomsbury Publishing.

Horkheimer, M. (1982). *Critical theory*. Seabury Press.

Husserl, E. (1970). *The Crisis of European Sciences and Transcendental Phenomenology*. Northwestern University Press.

Husserl, E. (2012). *Ideas: General Introduction to Pure Phenomenology*. Routledge. (Original work published 1913) doi:10.4324/9780203120330

Jarratt, S. C. (1991). *Rereading the Sophists: Classical Rhetoric Refigured*. Southern Illinois University Press.

Langley, A. (1999). Strategies for theorizing from process data. *Academy of Management Review, 24*(4), 691–710. doi:10.2307/259349

Lincoln, Y. S., & Guba, E. A. (1985). *Naturalistic inquiry*. Sage. doi:10.1016/0147-1767(85)90062-8

Lincoln, Y. S., Lynham, S. A., & Guba, E. G. (2011). Paradigmatic controversies, contradictions, and emerging confluences, revisited. In N. K. Denzin & Y. S. Lincoln (Eds.), *The Sage Handbook of Qualitative Research* (4th ed., pp. 97–128). Sage.

Lyotard, J. F. (1979). *The postmodern condition: A report on knowledge*. Manchester University Press.

Lyotard, J. F. (1984). *The postmodern condition: A report on knowledge*. University of Minnesota Press.

Malinowski, B. (1922). *Argonauts of the Western Pacific*. Routledge & Kegan Paul.

Markham, A. (2017). Ethnography in the Digital Internet Era: From fields to flows, descriptions to interventions. In N. Denzin & Y. Lincoln (Eds.), The Sage handbook of qualitative research (5th ed.). Sage.

Maxwell, J. A. (2012). *Qualitative research design: An interactive approach* (3rd ed.). Sage.

Mead, M. (1928). *Coming of Age in Samoa*. William Morrow & Company.

Merleau-Ponty, M. (2012). *Phenomenology of Perception*. Routledge. (Original work published 1945)

Merriam, S. B., & Tisdell, E. J. (2015). *Qualitative research: A guide to design and implementation*. Jossey-Bass.

Miles, M. B., Huberman, A. M., & Saldaña, J. (2014). *Qualitative data analysis: A methods sourcebook* (3rd ed.). Sage.

Moran, D. (2017). *Husserl's Crisis of the European Sciences and Transcendental Phenomenology: An introduction*. Cambridge University Press.

Morrow, S. L. (2005). Quality and trustworthiness in qualitative research in counseling psychology. *Journal of Counseling Psychology*, *52*(2), 250–260. doi:10.1037/0022-0167.52.2.250

Moustakas, C. (1994). Phenomenological research methods. *Sage (Atlanta, Ga.)*.

Ong, W. J. (1982). *Orality and Literacy: The Technologizing of the Word*. Methuen. doi:10.4324/9780203328064

Ortlipp, M. (2008). Keeping and using reflective journals in the qualitative research process. *The Qualitative Report*, *13*(4), 695–705.

Patton, M. Q. (2002). *Qualitative research and evaluation methods* (3rd ed.). Sage Publications.

Patton, M. Q. (2015). *Qualitative research & evaluation methods: Integrating theory and practice*. Sage publications.

Peshkin, A. (1988). In search of subjectivity—One's own. *Educational Researcher*, *17*(7), 17–21.

Pettigrew, A. M. (1997). What is a processual analysis? *Scandinavian Journal of Management*, *13*(4), 337–348. doi:10.1016/S0956-5221(97)00020-1

Pink, S. (2015). *Riessman, C. K. (2008). Narrative methods for the human sciences*. Sage Publications.

Pink, S., Horst, H., Postill, J., Hjorth, L., Lewis, T., & Tacchi, J. (2016). Digital Ethnography: Principles and Practice. *Sage (Atlanta, Ga.)*.

Saldana, J. (2015). *The coding manual for qualitative researchers* (3rd ed.). Sage.

Sartre, J.-P. (1943/1956). *Being and Nothingness: An Essay on Phenomenological Ontology*. Philosophical Library.

Sartre, J.-P. (1946). *Existentialism is a Humanism*. Yale University Press.

Silverman, D. (2016). *Qualitative research* (4th ed.). Sage.

Smith, J. A. (2015). Qualitative psychology: A practical guide to research methods. *Sage (Atlanta, Ga.)*.

Spivak, G. C. (1988). Can the subaltern speak? In C. Nelson & L. Grossberg (Eds.), Marxism and the interpretation of culture (pp. 271-313). Macmillan.

St. Pierre, E. A. (2018). Post-Qualitative Inquiry: From Representation to Rhizome. In *Handbook of Posthumanism in Film and Television* (pp. 1–12). Springer.

Strauss, A., & Corbin, J. (1998). Basics of qualitative research: Procedures and techniques for developing grounded theory. *Sage (Atlanta, Ga.)*.

Thomas, D. R. (2006). A general inductive approach for analyzing qualitative evaluation data. *The American Journal of Evaluation*, *27*(2), 237–246. doi:10.1177/1098214005283748

Van de Ven, A. H. (1992). Suggestions for studying strategy process: A research note. *Strategic Management Journal, 13*(Summer Special Issue), 169-188.

van Manen, M. (1990). *Researching Lived Experience: Human Science for an Action Sensitive Pedagogy*. State University of New York Press.

Weber, M. (1949). Objectivity in Social Science and Social Policy. In E. A. Shils & H. A. Finch (Eds.), *The Methodology of the Social Sciences*. Free Press.

Chapter 2
Theoretical Frameworks

ABSTRACT

This chapter examines the essential role of theoretical frameworks in research, highlighting their significance as the structural backbone that guides the inquiry process, from conception through analysis. It discusses how these frameworks serve not only as support but also as detailed blueprints that outline the arrangement and interrelations within a study. By positioning research within a broader academic dialogue, theoretical frameworks enhance its credibility and facilitate understanding, linking the study to established paradigms.

INTRODUCTION

Theoretical frameworks serve as foundational pillars in research, shaping the lens through which we view, understand, and interpret phenomena. They act as guiding blueprints, providing structure and direction, underpinning every research decision, and molding the very essence of inquiry. This chapter delves into the intricate tapestry of theoretical frameworks, exploring their significance, typologies, and application in qualitative research.

Understanding the theoretical framework of a study is akin to understanding the skeletal structure of a building. It provides not just support but a comprehensive outline detailing the arrangement and interrelation of parts within the research. Researchers use these frameworks to position their study within a broader academic conversation, demonstrating how their inquiries are grounded in, extend, challenge, or deviate from established theories and perspectives (Maxwell, 2013).

Significance of Theoretical Frameworks

1. **Guiding Inquiry**: Theoretical frameworks provide a roadmap for research, directing the formulation of research questions, data collection methods, and analysis strategies. They help researchers make sense of their findings, offering a lens to interpret and contextualize results (Creswell, 2013).

DOI: 10.4018/979-8-3693-2414-1.ch002

2. **Positioning Research in Academic Discourse**: By anchoring research in an established framework, scholars situate their work within a broader academic dialogue. This not only adds credibility but also facilitates readers' understanding by connecting the study to known paradigms (Grant & Osanloo, 2014).
3. **Enhancing Coherence and Consistency**: A well-articulated theoretical framework ensures coherence throughout the research process. It acts as a consistent reference point, ensuring alignment between the research questions, methodology, and conclusions (Ravitch & Riggan, 2017).

Typologies of Theoretical Frameworks

The landscape of theoretical frameworks is vast and varied, encompassing a multitude of perspectives, paradigms, and orientations. Some notable frameworks include:

1. **Constructivism**: Rooted in the belief that knowledge is socially constructed, this framework emphasizes subjective interpretations, co-creation of meaning, and the influence of socio-cultural contexts (Vygotsky, 1978).
2. **Critical Theory**: This paradigm seeks to identify and challenge power structures, emphasizing societal critique, empowerment, and transformative action (Horkheimer, 1982).
3. **Feminist Theory**: Centering on issues of gender inequality, this framework interrogates patriarchal structures, advocating for gender justice, and highlighting women's experiences (Butler, 1990).
4. **Phenomenology**: Focusing on the lived experiences of individuals, phenomenology seeks to understand the essence of phenomena, exploring perceptions, feelings, and interpretations (Husserl, 1913/2012).
5. **Postcolonialism**: This theoretical stance critiques colonial legacies, exploring issues of power, representation, identity, and cultural dynamics in postcolonial societies (Said, 1978).

Application in Qualitative Research

In qualitative research, theoretical frameworks are not just retrospective explanations but proactive guides. They shape the research design, influence methodological choices, and inform data interpretation. Researchers often engage in a dialogic process with their chosen framework, allowing it to shape the research while also adapting it to the nuances of their study. This recursive engagement ensures that the research remains grounded while also contributing to the evolution of the framework itself (Charmaz, 2014).

In conclusion, theoretical frameworks are pivotal in shaping the trajectory, depth, and rigor of research. They offer scholars a structured lens, grounding inquiries in established paradigms while also providing the flexibility to explore, challenge, and expand upon them. As we delve deeper into individual frameworks in the subsequent sections, it becomes evident that these are not mere academic constructs but vital tools that breathe life, depth, and direction into research.

Phenomenology

Phenomenology, as both a philosophical movement and a research methodology, seeks to explore and understand human experiences in their lived richness and depth. Originating from the Greek words "phainómenon" (that which appears) and "lógos" (study), phenomenology is essentially a study of phenomena as they present themselves to consciousness.

Historical Roots

Phenomenology as a philosophical doctrine can be traced back to the works of Edmund Husserl, a German philosopher in the early 20th century. Husserl posited that for true understanding, one must "go back to the things themselves" and capture their pure essence devoid of presuppositions (Husserl, 1913/2012). His work laid the foundation, which was later built upon by philosophers like Heidegger, Merleau-Ponty, and Sartre, each adding nuances and expanding the scope of phenomenological inquiry.

Phenomenology, as both a philosophical tradition and a research methodology, occupies a central position in the human sciences. Its historical trajectory provides a vivid illustration of intellectual evolution, beginning with the foundational inquiries of Edmund Husserl and branching out through the diverse contributions of subsequent thinkers.

Edmund Husserl: The Founding Father

Edmund Husserl (1859-1938) is universally recognized as the founder of phenomenology. Trained initially in mathematics, his transition into philosophy was marked by his interest in understanding the structures of consciousness. Husserl's principal concern was to establish a rigorous science of consciousness that would explore phenomena as they present themselves to the human mind, without preconceptions or biases.

- **"Phenomenological Reduction"**: Husserl introduced the notion of "phenomenological reduction," which means bracketing or setting aside the natural attitude to focus solely on the ways phenomena appear to consciousness (Husserl, 1913/2012).

Phenomenological reduction, often referred to as the "epoche" or "bracketing," is a foundational concept in Husserlian phenomenology. This transformative process seeks to shed the layers of preconceived beliefs, biases, and judgments, allowing phenomena to be perceived and understood in their pure, undistorted essence.

Origins and Context

Edmund Husserl introduced the notion of phenomenological reduction as a means to ground the sciences in a more firm epistemological foundation. He contended that to truly understand the essence of any phenomenon, one must suspend or "bracket" their natural attitude and engage with the world in a radically different manner (Husserl, 1913/2012).

Key Elements of Phenomenological Reduction

a) *Bracketing the Natural Attitude:* The primary step in phenomenological reduction involves suspending the "natural attitude." This means setting aside one's everyday beliefs, judgments, and assumptions about the world, stepping back from the taken-for-granted understanding of reality (Moran, 2000).

b) *Revelation of Phenomena:* By bracketing preconceptions, researchers can encounter phenomena in their pure form, free from cultural, societal, or personal biases. This process reveals the phenomena as they present themselves to consciousness, in their unadulterated form (Husserl, 1970).
c) *Intentionality and Consciousness*: Intertwined with phenomenological reduction is Husserl's concept of intentionality—the idea that consciousness is always directed towards something. The reduction allows researchers to explore this intentionality more profoundly, engaging deeply with the ways in which consciousness interacts with phenomena (Smith, 2003).

Implications for Research

Phenomenological reduction has profound implications for qualitative research:

a) *Depth and Authenticity*: Bracketing allows researchers to delve deep into the phenomena under study, ensuring that their interpretations are grounded in participants' actual experiences rather than external biases or preconceptions (Moustakas, 1994).
b) *Reflexivity:* Phenomenological reduction promotes reflexivity. Researchers continuously reflect upon their biases, ensuring they don't unduly influence the research process or findings (Finlay, 2002).
c) *Challenges in Application:* While phenomenological reduction is a powerful tool, it's not without challenges. Achieving complete bracketing is difficult, and researchers must be vigilant to ensure their preconceptions don't creep back into their interpretations (Giorgi, 2009).

Phenomenological reduction is more than a methodological tool; it's a transformative process that changes the very way researchers engage with the world. It encapsulates the essence of Husserlian phenomenology, underscoring the deep commitment to understanding human experiences in their richest and most genuine form.

- **Intentionality**: A cornerstone of Husserl's phenomenology is the concept of intentionality—the idea that consciousness is always directed toward something, be it an object, an idea, or a phenomenon (Husserl, 1900/2001).

Intentionality is a cornerstone concept in phenomenological thought, particularly within the framework established by Edmund Husserl. At its core, intentionality is about directionality and relatedness; it's the fundamental characteristic of consciousness that it is always directed at or about something.

Origins and Philosophical Underpinnings

The term "intentionality" was revitalized by Husserl, though its roots can be traced back to medieval scholastic philosophy. Husserl's exploration of intentionality sought to delve into the very structures of consciousness and understand how we relate to the world around us (Husserl, 1900/2001).

Key Facets of Phenomenological Intentionality

1. *Consciousness and its Directedness*: Husserl posited that all conscious experiences are intentional; they always have an object. Whether one is perceiving a tree, imagining a unicorn, or recalling a past event, consciousness is invariably about something (Smith, 2003).
2. *Noema and Noesis:* Within the intentional act, Husserl distinguishes between the act of consciousness (noesis) and the object as intended (noema). While noesis pertains to the act of perceiving, believing, or imagining, noema relates to the content or object of that act (Husserl, 1913/2012).
3. *Transcendental Ego*: For Husserl, the transcendental ego is the pure subject of intentional acts. It's the unchanging point of reference, the "I" that persists amidst the fluctuating stream of consciousness (Zahavi, 2005).
4. *Horizons of Intentionality*: Intentionality isn't limited to the immediate object of consciousness. Every intentional act opens up a horizon of further experiences, meanings, and possible explorations, creating a rich tapestry of interrelated perceptions and understandings (Husserl, 1936/1970).

Implications for Phenomenological Research

a) *Depth of Exploration*: Intentionality drives researchers to delve beyond surface perceptions, exploring the depth and breadth of participants' experiences and the ways they relate to their world (Moustakas, 1994).
b) *Relational Understanding:* Acknowledging the inherent intentionality of consciousness prompts phenomenologists to consider the relational dynamics between participants and their environments, experiences, and understandings (van Manen, 1990).
c) *Challenges of Capturing Essence*: The multifaceted nature of intentionality, with its interwoven horizons and layers, poses challenges for researchers striving to capture the essence of lived experiences. It demands meticulous attention, reflection, and description (Giorgi, 2009).

Intentionality, as conceptualized in phenomenological thought, underscores the profound interconnectedness of the self, consciousness, and the world. It compels researchers to approach phenomena with a deep sense of curiosity, exploring not just what is immediately perceived but the myriad of relations, horizons, and layers that constitute human experience.

Martin Heidegger: Existential Turn

Heidegger, once a student of Husserl, took phenomenology in a new direction. While he built upon Husserl's foundational ideas, he infused them with existential concerns, emphasizing the situatedness of human beings in the world.

- **"Dasein"**: Central to Heidegger's philosophy is the concept of "Dasein," often translated as "being-there." Dasein refers to the unique way humans exist and relate to the world. He contended that understanding human existence requires an examination of its total context—historical, social, and existential (Heidegger, 1927/1962).

Martin Heidegger, a seminal figure in 20th-century philosophy, introduced profound shifts in phenomenological and existential thought. His concept of "Dasein," central to his magnum opus "Being and Time" (*Sein und Zeit*), encapsulates his distinctive existential and ontological inquiries.

Genesis of "Dasein"

The term "Dasein" originates from the German "Da" (there) and "Sein" (being). At its most basic, Dasein can be translated as "being-there" or "there-being." However, its implications in Heideggerian thought are profound and multi-faceted.

Key Characteristics of Dasein

a) **Primacy of Being-in-the-World**: For Heidegger, Dasein isn't merely in the world as water is in a glass. Instead, Dasein's existence is fundamentally entangled with the world. This "being-in-the-world" underscores an integral, indivisible relationship between Dasein and its environment (Heidegger, 1927/1962).
b) **Temporality and Historicity**: Dasein's existence is temporal—it has a past (what Heidegger terms "thrownness"), a present ("facticity"), and a future ("projection"). This temporal nature is not just chronological but existential, shaping Dasein's understanding of itself and its possibilities (Heidegger, 1927/1962).
c) **Care Structure**: At the heart of Dasein's existence is the structure of care. Dasein is always already involved with the world in modes of care—whether that involves concern for entities within the world or care for its own being (Heidegger, 1927/1962).
d) **Being-towards-Death**: One of the most discussed aspects of Dasein is its "being-towards-death." Heidegger posits that Dasein has an anticipatory relationship with its own mortality. This awareness of death influences Dasein's choices, actions, and understanding of its own finitude (Blattner, 2006).

Implications for Phenomenological and Existential Thought

I. **Shift from Consciousness to Being**: Heidegger's focus on Dasein marked a pivot from Husserl's emphasis on consciousness. For Heidegger, the question wasn't about how consciousness relates to the world but about the nature of Being itself (Dreyfus, 1991).
II. **Ontological Inquiry**: Heidegger's exploration of Dasein was not just a phenomenological description but an ontological inquiry. He was concerned with the question of Being – what does it mean to be? Dasein became the lens through which he explored this primordial question (Mulhall, 2005).
III. **Critique of Traditional Metaphysics**: Dasein's analysis laid the groundwork for Heidegger's critique of traditional Western metaphysics. He argued that prior philosophical traditions overlooked the fundamental question of Being, getting entangled in ontic (concerning entities) rather than ontological (concerning the nature of Being) inquiries (Heidegger, 1927/1962).

Heidegger's introduction of Dasein revolutionized existential and phenomenological landscapes. By focusing on the existential intricacies of human existence, Heidegger provided a rich, profound framework for understanding our relationship with the world, our temporality, and the very nature of Being.

Maurice Merleau-Ponty: Embodied Perception

Merleau-Ponty brought the body into the phenomenological conversation. He argued against the Cartesian dualism of mind and body, emphasizing instead the embodiment of perception and understanding. Maurice Merleau-Ponty, a central figure in 20th-century existential phenomenology, profoundly influenced discussions surrounding the nature of perception and embodiment. His insistence on the primary role of perception and the intrinsic connection between the body and consciousness offers a paradigm shift from traditional dualistic perspectives.

The Centrality of Perception in Merleau-Ponty's Thought

For Merleau-Ponty, perception is not a mere passive reception of stimuli or a straightforward translation of the external world into internal representations. Instead, it is the foundational act through which the world is revealed to us and through which we engage with the world (Merleau-Ponty, 1945/2012).

- **Perception as Primary**: Merleau-Ponty posited that perception is the foundation of human knowledge and that our bodily interactions with the world shape our understanding of it (Merleau-Ponty, 1945/2012).

Key Aspects of "Perception as Primary"

a) **Inseparability of Seer and Seen**: Central to Merleau-Ponty's thesis is the intertwined nature of the perceiving subject and the perceived object. The two are in a reciprocal relationship, continuously shaping and informing each other (Merleau-Ponty, 1945/2012).
b) **The Body as Perceptive Subject**: Breaking away from Cartesian dualism, Merleau-Ponty posited the body not as an object in the world but as our primary means of engaging with the world. The body isn't just a thing; it is our means of having a world (Merleau-Ponty, 1964).
c) **The Lived Body (Le corps vécu)**: Merleau-Ponty distinguished between the objective body (the body as an object of science) and the lived body (the body as we experience it). The lived body is at the nexus of our perceptual experiences, being both subject and object, both perceiver and perceived (Merleau-Ponty, 1945/2012).
d) **Perception as Pre-Reflective and Pre-Objective**: Perception, in Merleau-Ponty's view, is pre-reflective and pre-objective. Before we think, rationalize, or categorize, we perceive. This raw, immediate engagement with the world is primary, laying the groundwork for further cognitive and reflective operations (Merleau-Ponty, 1964).

Implications for Phenomenology and Beyond

I. **Challenge to Dualistic Paradigms**: Merleau-Ponty's emphasis on embodied perception challenges the Cartesian mind-body dualism, offering a more holistic understanding of human existence that integrates mind, body, and world (Romdenh-Romluc, 2010).

II. **Grounding in Concrete Experience**: By placing perception at the forefront, Merleau-Ponty shifts philosophical focus from abstract theorization to concrete lived experiences, grounding philosophical inquiries in the tangible, everyday encounters of human beings (Carman, 2008).

III. **Influence on Embodied Cognition**: Merleau-Ponty's ideas have resonance beyond phenomenology, particularly influencing contemporary discussions in cognitive science and psychology regarding embodied cognition, which posits cognition as deeply rooted in the body's interactions with the world (Varela, Thompson, & Rosch, 1991).

Merleau-Ponty's assertion of the primacy of perception reconfigures our understanding of consciousness, body, and world. By recognizing the body as our anchor in the world, and perception as our primary mode of engagement, he offers a rich, integrative framework that celebrates the embodied, intertwined nature of human existence.

Jean-Paul Sartre: Freedom and Existence

Sartre's existential phenomenology focused on the themes of freedom, responsibility, and human existence's inherent absurdity. He explored the tensions between individuals' subjective experiences and the external world's objective realities (Sartre, 1943/1956).

As one of the paramount figures of existentialist philosophy, Jean-Paul Sartre embarked on an exploration of human existence that was both profound and radical. At the heart of Sartre's existentialism are the intertwined themes of freedom and existence, which serve to elucidate the nature of human beings and their place in the world.

Fundamental Tenets of Sartre's Philosophy

For Sartre, human beings exist first and only subsequently create their essence through their actions and choices, encapsulated in his famous assertion, "Existence precedes essence" (Sartre, 1946). This existential priority places freedom at the core of human beingness.

Delineating Freedom in Sartre's Thought

a) **Radical Freedom**: Sartre posited that every individual possesses an innate and absolute freedom. This freedom isn't just the capacity to choose between alternatives but represents a deeper, ontological freedom that defines our very being (Sartre, 1943/1956).

b) **Freedom as Responsibility**: This radical freedom comes with immense responsibility. Since there's no predetermined essence or divine plan directing our actions, every choice we make is our own, and we must bear full responsibility for those choices (Sartre, 1946).

c) **Bad Faith (Mauvaise foi)**: Given the weight of such freedom and responsibility, individuals often seek to escape it, a phenomenon Sartre terms "bad faith." This involves self-deception, where individuals deny their freedom and pretend they are bound by external determinants, effectively abdicating responsibility (Sartre, 1943/1956).

d) **Freedom in Situations**: While Sartre emphasizes absolute freedom, he acknowledges that this freedom is exercised within specific situations. While situations might condition our choices, they don't determine them. We always have the freedom to choose how to respond (Sartre, 1943/1956).

The Existential Implications of Freedom

I. **Existential Anguish**: Recognizing one's radical freedom can lead to existential anguish. This isn't merely anxiety about specific outcomes but a deeper anguish that arises from grasping the full weight of our freedom and the responsibility it entails (Sartre, 1946).
II. **Authenticity**: The antidote to bad faith is authenticity—an honest recognition and acceptance of one's freedom and responsibility. An authentic individual embraces their freedom, makes choices deliberately, and accepts responsibility for their actions (Sartre, 1946).

Jean-Paul Sartre's exploration of freedom and existence challenges traditional notions of predetermined human nature or divine destiny. By positing an existence that precedes essence and emphasizing the primacy of freedom, Sartre not only reshapes our understanding of human nature but also underscores the weight of responsibility that accompanies such freedom.

Hannah Arendt: Action and the Public Sphere

Drawing from Heideggerian phenomenology, Arendt explored the nature of power, politics, and action in the public sphere. She emphasized the collective aspect of human experiences and the transformative potential of collective action (Arendt, 1958).

The historical development of phenomenology is marked by rich intellectual dialogues, divergent paths, and profound insights into the nature of human existence. While the roots of phenomenology are anchored in Husserl's groundbreaking work, its branches reach out in myriad directions, reflecting the diverse concerns, interpretations, and innovations of subsequent thinkers.

Hannah Arendt, a pivotal thinker in political philosophy, offered an insightful examination of human action and its intrinsic connection to the public sphere. Drawing upon her experiences as a political theorist and witnessing some of the most tumultuous events of the 20th century, Arendt delved into the nature of power, action, and the realm of human interactions that constitute the public domain.

Arendt's Conceptual Framework

Arendt's magnum opus, "The Human Condition," presented a tripartite division of human activities: labor, work, and action. While labor and work are crucial for human survival and world-building respectively, it's action that Arendt sees as vital for political life (Arendt, 1958).

The Nature and Significance of Action

a) **Distinctiveness of Action**: For Arendt, action stands apart due to its inherent unpredictability and its rootedness in plurality. Action arises when individuals come together, acknowledging and celebrating their distinctiveness (Arendt, 1958).
b) **Power and Action**: Contrary to conventional interpretations of power as something that individuals possess or wield, Arendt posits power as something that emerges when people act together. It's both generated by and sustains collective action (Arendt, 1970).

c) **Narrativity and Immortality**: Actions are not isolated events; they weave narratives. By participating in actions, individuals insert themselves into the human story, ensuring a form of worldly continuity and immortality (Arendt, 1958).

Public Sphere and the Space of Appearance

I. **The Importance of Visibility**: Arendt introduces the idea of the "space of appearance" – a space where individuals appear before one another as distinct entities. This space, constituted whenever individuals gather for political action, is transient and needs continuous action to be sustained (Arendt, 1958).

II. **The Public Realm**: For Arendt, the public sphere is where freedom manifests, where individuals come together to deliberate, discuss, and act. It's in stark contrast to the private realm, which is associated with necessity and survival (Arendt, 1958).

III. **Endangered Public Sphere**: Arendt was wary of the dangers posed by modernity to the public sphere. The rise of the social—a domain that blurs the lines between public and private—threatens to eradicate the space of appearance and the very possibility of political action (Arendt, 1958).

Hannah Arendt's exploration of action and the public sphere provides a profound understanding of the conditions and spaces where political life unfolds. Her insights, rooted in a deep commitment to the values of democracy and human dignity, serve as a beacon for navigating the complexities of political engagement and collective existence.

Key Concepts

1. **Intentionality**: Central to phenomenology is the concept of intentionality. It suggests that consciousness is always about something, meaning our thoughts, feelings, and perceptions are directed towards objects, events, or experiences (Smith, 2003).
2. **Epoche or Bracketing**: Husserl introduced the idea of "epoche", a process where researchers set aside their biases, beliefs, and preconceptions to encounter the phenomenon in its pure form (Husserl, 1913/2012).
3. **Lifeworld (Lebenswelt)**: Heidegger expanded on Husserl's ideas by emphasizing "lifeworld" or the everyday world of human experiences. It's the world as lived, not just as conceptualized or theorized (Heidegger, 1927/1962).
4. **Embodiment**: Merleau-Ponty emphasized the embodied nature of human experience, suggesting that our body is not just an object in the world but a medium through which we experience the world (Merleau-Ponty, 1945/2012).

Intentionality

Intentionality is a foundational concept in both philosophy of mind and phenomenology. Historically rooted in scholastic philosophy and revitalized in modern times primarily by Franz Brentano and later Edmund Husserl, intentionality refers to the capacity of mental states to be about, directed at, or represent something beyond themselves.

Historical Background

The term 'intentionality' derives from the Latin word 'intentio,' which medieval philosophers used to discuss concepts and abstractions. However, its contemporary philosophical significance was largely shaped by Franz Brentano, who highlighted it as the defining characteristic of the mental (Brentano, 1874).

Key Aspects of Intentionality

a) **Directionality of Consciousness**: Intentionality is often described as the mind's inherent ability to "point at" or "be about" something. Every mental act has an object – whether it's a thought about a mountain, a perception of a tree, or a desire for coffee (Husserl, 1900/2001).
b) **Noesis and Noema**: Husserl's phenomenology furthered our understanding of intentionality by distinguishing between the act of consciousness (noesis) and the object as intended (noema). This distinction underscores the relational aspect of conscious experience (Husserl, 1913/2012).
c) **Intentional Inexistence**: Brentano observed that the intentional objects of our thoughts and desires might not exist in reality. For instance, one can think of a golden mountain or a unicorn, even if such entities don't exist in the real world (Brentano, 1874).
d) **Intentionality and Representation**: Later philosophers, notably in the analytic tradition, have linked intentionality to the representational content of mental states. On this view, mental states possess content that represents particular aspects of the world (Crane, 2001).

Implications and Significance

I. **Cognitive Science and Philosophy of Mind**: Understanding intentionality is crucial for cognitive science and artificial intelligence. It raises questions about the nature of representation, the possibility of machine consciousness, and the fundamentals of cognition (Dennett, 1987).
II. **Phenomenological Investigations**: In phenomenology, intentionality serves as a tool to investigate the structures and modalities of conscious experience, driving inquiries into perception, imagination, memory, and other mental phenomena (Husserl, 1900/2001).
III. **Challenges to Physicalism**: Intentionality poses challenges to strictly physicalist accounts of the mind. The "aboutness" of mental states isn't straightforwardly reducible to physical states, leading to debates on the nature and ontology of mental phenomena (Chalmers, 1996).

Intentionality stands as a cornerstone concept in the exploration of consciousness, bridging historical and contemporary philosophical discussions. It underscores the inherent outward-looking nature of conscious states, prompting inquiries into the intricate relationship between mind and world.

Epoche or Bracketing

Epoche, commonly known as "bracketing," is a foundational concept in phenomenological research. Stemming from the Greek word "ἐποχή," which signifies "suspension" or "cessation," this methodological tool was introduced by Edmund Husserl to facilitate an unbiased, fresh exploration of phenomena in their purest form.

Origins and Development

Edmund Husserl, the founder of phenomenological philosophy, posited the method of epoche as a means to circumvent presuppositions and biases, thereby accessing a phenomenon's pure essence (Husserl, 1913/2012).

Key Facets of Epoche

a) **Suspending Judgments**: At its core, epoche entails suspending or "bracketing out" one's preconceived beliefs, judgments, and notions about the phenomenon under investigation. This allows researchers to approach their subject with a "beginner's mind" or fresh perspective (Husserl, 1970).
b) **Revealing Phenomena as They Present Themselves**: By employing epoche, phenomenologists can encounter phenomena in their raw, unaltered form—free from the layers of cultural, societal, or personal interpretations. This method thus seeks to reveal the phenomena as they appear to consciousness, devoid of any adulterations (Moustakas, 1994).
c) **Transcendental Reduction**: Epoche is often linked with the concept of "transcendental reduction." Once biases are bracketed, researchers can focus on the phenomena themselves, thus accessing a transcendental ego that observes without interference (Husserl, 1913/2012).

Implications for Phenomenological Research

I. **Enhancing Authenticity**: The process of bracketing ensures that research findings emerge from the authentic experiences of participants rather than the biases or preconceptions of the researcher (Giorgi, 2009).
II. **Promotion of Reflexivity**: Epoche also fosters reflexivity, prompting researchers to continually reflect on and be aware of their biases, ensuring they don't unduly influence the research process or the emergent findings (Finlay, 2002).
III. **Challenges in Implementation**: While epoche is a valuable tool, its application is not without challenges. Attaining complete bracketing is challenging, and researchers must consistently remain vigilant to ensure their preconceptions don't infiltrate their interpretations (Creswell, 2013).

The method of epoche or bracketing is more than a mere procedural step in phenomenological research—it represents a philosophical commitment to apprehend the world afresh, ensuring that phenomena are explored and understood in their pure, undistorted essence.

Lifeworld (Lebenswelt)

The concept of the lifeworld, or "Lebenswelt" in German, is an indispensable notion within phenomenological philosophy and social theory. Introduced by Edmund Husserl and further developed by later thinkers such as Alfred Schütz and Jürgen Habermas, the lifeworld represents the immediate and pre-reflective world of everyday lived experience.

Origins and Development

Husserl introduced the term "Lebenswelt" to counter the growing scientism and objectivism of his time, emphasizing the importance of our immediate, lived experience as the foundation for all forms of knowledge (Husserl, 1936/1970).

Key Characteristics of the Lifeworld

a) **Pre-Theoretical Domain**: The lifeworld is the world as we immediately live it, not as we theorize or conceptualize it. It is the taken-for-granted backdrop against which all our activities, both mundane and exceptional, take place (Husserl, 1936/1970).
b) **Intersubjectivity**: Central to the lifeworld is the realm of intersubjectivity. It's a shared space where meanings are constituted through social interactions and where individuals engage in mutual recognition and understanding (Schütz, 1967).
c) **Horizons of Meaning**: The lifeworld is replete with horizons—backgrounds and contexts that give meaning to our experiences. These horizons are continuously shifting, molded by past experiences and anticipations of the future (Husserl, 1936/1970).
d) **Crisis of the Lifeworld**: For Husserl, the lifeworld was under threat from the encroachments of scientific rationality. He believed that the objectifying gaze of science, although valuable in its domain, could obscure the richness and depth of lived experience if misapplied (Husserl, 1936/1970).

Implications for Social Theory and Research

I. **Phenomenological Investigations**: Recognizing the primacy of the lifeworld has been foundational for phenomenological researchers, grounding their inquiries in the concrete, lived experiences of individuals (Merleau-Ponty, 1945/2012).
II. **Critical Social Theory**: Jürgen Habermas utilized the concept of the lifeworld in his critical social theory. For Habermas, modernity saw the colonization of the lifeworld by systemic mechanisms (like the market and bureaucracy), leading to a loss of meaning and freedom in everyday life (Habermas, 1987).
III. **Foundation for Ethnographic and Qualitative Research**: Understanding the lifeworld is central to ethnographic and qualitative research methods. Researchers immerse themselves in the lifeworlds of their subjects to gain a deep understanding of their practices, meanings, and perspectives (Geertz, 1973).

The lifeworld, as a foundational concept in phenomenological and social thought, underscores the importance of the immediate, lived world of everyday experience. It reminds scholars and researchers to remain rooted in the tangible, intersubjective realm of human existence, celebrating its depth, diversity, and inherent meaning.

Embodiment

Embodiment is a foundational and multifaceted concept that cuts across various disciplines, from philosophy and cognitive science to anthropology and psychology. Central to this idea is the understanding that the body isn't just a passive vessel but actively shapes our perception, cognition, and experience of the world.

Historical Background

While elements of embodiment can be traced back to ancient philosophies, its modern formulations have roots in the works of existential and phenomenological thinkers, especially in the writings of Maurice Merleau-Ponty, who posited the body as an experiential medium (Merleau-Ponty, 1945/2012).

Key Dimensions of Embodiment

a) **Body as Subject**: Embodiment challenges traditional dualisms that separate the mind from the body. Instead of viewing the body merely as an object, embodiment underscores its role as a subject—a central player in our conscious experience (Merleau-Ponty, 1945/2012).
b) **Perception and Action**: The body isn't just a receiver of sensory inputs but is intertwined with our perceptual processes. Our bodily actions, postures, and movements actively shape the way we perceive and interpret the world (Gibson, 1979).
c) **Cognitive Processes**: Beyond perception, embodiment has significant implications for cognition. Theories of embodied cognition suggest that cognitive processes are deeply rooted in the body's interactions with its environment (Lakoff & Johnson, 1999).
d) **Social and Cultural Dimensions**: Embodiment is also socially and culturally situated. The ways we inhabit our bodies, experience them, and attribute meanings to them are shaped by sociocultural norms, values, and narratives (Csordas, 1994).

Implications for Research and Theory

i. **Reconfiguring Cognitive Science**: The recognition of embodiment has catalyzed shifts in cognitive science, prompting researchers to examine how bodily interactions with the environment shape cognitive structures (Varela, Thompson, & Rosch, 1991).
ii. **Therapeutic and Clinical Domains**: Understanding embodiment has therapeutic ramifications. It informs approaches in psychotherapy, somatic practices, and body-based interventions, acknowledging the interplay between bodily experiences and psychological well-being (Fuchs & Schlimme, 2009).
iii. **Anthropological and Sociological Inquiries**: The embodied nature of human existence offers anthropologists and sociologists a rich terrain to explore how cultural, societal, and historical contexts shape bodily experiences and vice versa (Csordas, 1994).

Embodiment underscores the profound entanglement of our bodies with our conscious experiences, perceptions, and cognitive processes. By highlighting the centrality of the body in shaping our engagement with the world, the concept of embodiment challenges and expands traditional paradigms, offering a more holistic understanding of human existence.

Phenomenology in Research

When applied as a research methodology, phenomenology aims to:

1. **Capture the Essence**: The goal is to describe the essence or core meaning of lived experiences around a phenomenon (Moustakas, 1994).

Phenomenology, with its deep philosophical underpinnings, is distinguished by its earnest pursuit to capture the essence of lived experiences. The endeavor to discern this 'essence' is not merely an academic exercise but a profound commitment to understanding the core of human experiences in all their richness and complexity.

Historical Context

Edmund Husserl, the founder of phenomenological philosophy, championed the idea of returning "to the things themselves." This entailed a rigorous examination of lived experiences to uncover their essential structures and meanings, setting the foundation for phenomenological inquiry (Husserl, 1913/2012).

Understanding the Notion of 'Essence'

a) **Defining Essence**: In the phenomenological context, 'essence' refers to the invariant nature or core characteristics of a phenomenon, revealing what makes it distinct and universally identifiable across varied instances (Moustakas, 1994).
b) **Eidetic Reduction**: Central to discerning essence is the process of eidetic reduction, a method where researchers seek to determine the universal qualities of an experience while setting aside its particular manifestations (Husserl, 1913/2012).
c) **Beyond Empirical Descriptions**: Capturing the essence moves researchers beyond mere empirical or factual accounts. It propels them into the realm of deeper significances, capturing the underlying and often intangible meanings that constitute an experience (Merleau-Ponty, 1945/2012).

The Significance of Capturing Essence in Phenomenological Research

i. **Revealing Deep Insights**: By focusing on essence, phenomenology offers profound insights into the human condition, uncovering layers of meaning often overlooked in other research paradigms (van Manen, 1990).
ii. **Universal Resonance**: While phenomenological studies are rooted in particular lived experiences, the extraction of essence allows for the findings to resonate on a more universal scale, revealing shared human truths and connections (Giorgi, 2009).

iii. **Authentic Representation**: The commitment to capturing essence ensures that participants' experiences are represented authentically, honoring their depth and intricacy (Smith, Flowers, & Larkin, 2009).
iv. **Guiding Reflective Practices**: Grasping essence often demands a reflexive stance from researchers, prompting them to continually revisit, reflect upon, and refine their understandings in the light of emerging insights (Finlay, 2008).

Phenomenology's distinctive aim to capture the essence of lived experiences underscores its dedication to exploring the profound depths of human existence. This endeavor is not just about understanding a phenomenon but about honoring and illuminating the intricate tapestry of meanings that constitute human life.

2. **Prioritize Participants' Voices**: Phenomenological research prioritizes participants' perceptions, feelings, and narratives, seeking to understand their world from their viewpoint (van Manen, 1990).

At the core of phenomenological research lies a profound respect for and commitment to the authentic voices of participants. This methodological stance, rooted in the philosophical foundations of phenomenology, seeks to bring forth the rich tapestry of lived experiences, ensuring that research remains anchored in the genuine narratives of those it studies.

Historical and Philosophical Foundations

Edmund Husserl's call to return "to the things themselves" serves as the bedrock of phenomenological inquiry. This return prioritizes the immediate and direct descriptions of experiences as they are lived and narrated by individuals, paving the way for their voices to take center stage (Husserl, 1913/2012).

Centrality of Participants' Voices in Phenomenological Research

a) **Authentic Narratives**: Phenomenological research seeks to elicit and explore the raw, unfiltered accounts of participants. These narratives, free from external impositions or theoretical constructs, offer genuine windows into the lived worlds of individuals (van Manen, 1990).
b) **Co-Construction of Meaning**: The phenomenologist does not position themselves as an outside expert but rather as a collaborator in the process of meaning-making. Through iterative dialogues, participants and researchers co-construct the essence of the phenomenon under study (Smith, Flowers, & Larkin, 2009).
c) **Honoring Subjectivity**: Recognizing that lived experiences are deeply subjective and personal, phenomenological research not only accepts but celebrates this subjectivity, ensuring that participants' unique perspectives and interpretations are preserved and highlighted (Merleau-Ponty, 1945/2012).

Implications for Research Practices

I. **Deep Listening**: Phenomenological researchers engage in deep, empathetic listening, immersing themselves in participants' narratives and seeking to understand their experiences from within their lifeworlds (Moustakas, 1994).

II. **Reflexivity**: Recognizing the potential influence of their own preconceptions, phenomenologists engage in continual reflexive practices, ensuring that their interpretations remain true to participants' voices (Finlay, 2008).

III. **Ethical Considerations**: Prioritizing participants' voices also involves ethical commitments. Researchers must ensure the safety, dignity, and respect of participants, acknowledging the vulnerability and courage involved in sharing deeply personal experiences (Bevan, 2014).

Phenomenology's commitment to prioritizing participants' voices is more than a methodological decision—it's an ethical and philosophical stance that acknowledges the intrinsic value of individual lived experiences. By elevating these voices, phenomenological research contributes to a richer, more nuanced understanding of the human condition, anchored in the authentic narratives of those it seeks to understand.

3. **Holistic Understanding**: Phenomenologists look for holistic understanding, seeing experiences as interconnected wholes rather than fragmented parts (Giorgi, 2009).

Central to phenomenological research is the aspiration towards a holistic understanding of human experience. Rooted in its philosophical foundations, phenomenology seeks not merely to dissect or analyze but to grasp experiences in their fullness, embracing their intricacies, contexts, and interconnectedness.

Philosophical Underpinnings

Edmund Husserl's phenomenology, with its emphasis on the lived world ("Lebenswelt"), laid the groundwork for a holistic exploration of human experiences. This perspective champions the idea that individuals don't just exist within a vacuum but are intricately embedded within a web of relationships, contexts, and meanings (Husserl, 1913/2012).

Dimensions of Holistic Understanding in Phenomenological Research

a) **Contextual Embeddedness**: Phenomenological research recognizes that lived experiences are deeply embedded within specific socio-cultural, historical, and interpersonal contexts. These contexts are not mere backdrops but actively shape and are shaped by individual experiences (Heidegger, 1927/1962).

b) **Interconnectedness of Phenomena**: Rather than isolating specific aspects of an experience, phenomenology aims to understand how various facets are interconnected, creating a cohesive and integrated whole (Merleau-Ponty, 1945/2012).

c) **Embracing Ambiguity and Complexity**: A holistic approach acknowledges the inherent ambiguities, contradictions, and complexities in lived experiences. It refrains from oversimplifying, instead celebrating the multi-layered nature of human existence (van Manen, 1990).

d) **Integrative Exploration**: Phenomenology integrates various dimensions of human experience—cognitive, emotional, sensory, and existential—offering a comprehensive view of how individuals perceive, feel, think, and exist within the world (Moustakas, 1994).

Implications for Research Practices

I. **Rich Data Collection**: Seeking a holistic understanding necessitates collecting rich, detailed, and comprehensive data. This often involves in-depth interviews, prolonged engagement, and, at times, multiple modes of data collection, such as combining interviews with observations or artifact analyses (Smith, Flowers, & Larkin, 2009).
II. **Iterative Analysis**: Phenomenological data analysis is often iterative, where researchers continually move between parts and the whole, ensuring that detailed insights are integrated into a holistic understanding of the phenomenon (Giorgi, 2009).
III. **Transcending Reductionism**: Phenomenologists resist reductionist tendencies. They aim to capture the essence of experiences without reducing them to mere components or isolated themes, ensuring the totality of the experience is preserved (Finlay, 2008).

CONCLUSION

Phenomenology's commitment to a holistic understanding represents a profound respect for the intricacy and richness of human life. In its pursuit to grasp the totality of lived experiences, phenomenological research offers invaluable insights that capture the fullness, depth, and interconnectedness of human existence.

Phenomenological Approaches

There are several approaches within phenomenological research, including:

1. **Descriptive Phenomenology (Husserlian)**: Rooted in Husserl's work, this approach aims to describe phenomena as they appear, emphasizing epoche and pure description (Husserl, 1913/2012).

Phenomenology, initiated by Edmund Husserl in the early 20th century, seeks to shed light on the structures of consciousness and the nature of human experience. The Husserlian approach, often termed "Descriptive Phenomenology," centers on describing experiences just as they are lived, without any theoretical or interpretative overlay.

Philosophical Foundations

Edmund Husserl's work marked a radical departure from prevailing philosophical trends by placing consciousness at the forefront of inquiry. For Husserl, phenomenology was about a return "to the things themselves," emphasizing the descriptive exploration of phenomena as they present themselves to consciousness (Husserl, 1913/2012).

Key Principles of Descriptive Phenomenology

a) **Epoché (Bracketing)**: Central to Husserlian phenomenology is the idea of epoché, where researchers "bracket" or set aside their preconceptions and judgments about the phenomenon under investigation. This process ensures a fresh engagement with the experience, untainted by biases or presuppositions (Husserl, 1970).

b) **Intentionality**: Husserl posited that all consciousness is consciousness of something. This principle of intentionality highlights the always directed nature of consciousness, implying a perpetual relationship between the conscious subject and the world (Husserl, 1931).

c) **Essential Structures**: Descriptive phenomenology aims to unearth the essential structures or invariant features of an experience. It seeks to discern what makes a particular experience distinctively what it is (Moustakas, 1994).

d) **Constitutive Acts of Consciousness**: Husserl's phenomenology explores the ways in which consciousness constitutes phenomena. This involves understanding the acts of perceiving, imagining, remembering, and so forth, through which phenomena become present to us (Husserl, 1983).

Implications for Research Practices

i. **Rich Descriptions**: In line with its descriptive nature, Husserlian phenomenological research prioritizes rich, detailed, and layered descriptions of the lived experience, emphasizing participants' direct accounts (Spiegelberg, 1982).

ii. **Reductive Analysis**: Researchers engage in a process of phenomenological reduction, analyzing the data iteratively to discern the essence of the experience while continually bracketing their own interpretations (Giorgi, 2009).

iii. **Empathetic Engagement**: Genuine understanding in descriptive phenomenology necessitates an empathetic engagement with participants, allowing researchers to immerse themselves fully in the world of the other (Moran, 2000).

Descriptive phenomenology, rooted in Husserl's philosophical innovations, offers researchers a methodical and rigorous approach to explore human experiences in their pure, undiluted form. By highlighting the structures of consciousness and the essence of lived experiences, this approach contributes rich, profound insights into the tapestry of human existence.

2. **Interpretative or Hermeneutic Phenomenology (Heideggerian)**: Drawing from Heidegger, this approach believes understanding is inherently interpretative. It seeks not just to describe but to interpret the deeper meanings of experiences (Heidegger, 1927/1962).

Originating from the philosophical work of Martin Heidegger, interpretative or hermeneutic phenomenology diverges from Husserl's descriptive phenomenology. While both approaches concern themselves with lived experience, Heideggerian phenomenology places a significant emphasis on the interpretative nature of understanding and the historical, societal, and existential contexts that shape human existence.

Philosophical Foundations

Martin Heidegger's magnum opus, *Being and Time* (1927), laid the groundwork for hermeneutic phenomenology. He moved away from Husserl's emphasis on pure description to explore the interpretative processes inherent in human understanding and the existential question of Being.

Key Tenets of Hermeneutic Phenomenology

a) **The Hermeneutic Circle**: Central to Heidegger's approach is the idea of the hermeneutic circle. It suggests that understanding is achieved through a circular movement between the parts of an experience and the whole, between pre-understanding and reflective interpretation (Heidegger, 1927/1962).
b) **Historicity**: Heidegger emphasized the significance of history in shaping human understanding. Every interpretation is embedded within a historical context, and understanding is a fusion of these historical horizons (Gadamer, 1960/2004).
c) **Dasein (Being-there)**: One of Heidegger's major contributions is the concept of Dasein. It refers to human existence in its unique mode of Being, characterized by its self-awareness, temporality, and care for its own existence (Heidegger, 1927/1962).
d) **Existential Inquiry**: Heideggerian phenomenology, while remaining a study of lived experiences, delves into existential themes such as authenticity, mortality, and freedom, seeking to illuminate the fundamental structures of human existence (Heidegger, 1927/1962).

Implications for Research Practices

i. **Interpretative Analysis**: In line with its hermeneutic nature, Heideggerian phenomenological research places a significant emphasis on the interpretative analysis of data, seeking to unravel deeper meanings and existential significances (van Manen, 1990).
ii. **Acknowledging Pre-understanding**: Researchers recognize and reflect upon their pre-understanding, realizing that every interpretation is shaped by prior experiences, biases, and societal narratives (Laverty, 2003).
iii. **Engagement with Texts**: Hermeneutic phenomenology often involves a deep engagement with texts, where researchers revisit and reflect upon data iteratively, uncovering layers of meaning through this cyclical process (Smith, 2007).

Interpretative or hermeneutic phenomenology offers a profound exploration into the interpretative nature of human understanding and the existential aspects of Being. By emphasizing interpretation, historicality, and existential themes, Heideggerian phenomenology enriches our comprehension of the profound depths of human existence.

3. **Existential Phenomenology**: Drawing insights from existential philosophers like Sartre and Merleau-Ponty, this approach emphasizes issues like freedom, alienation, and embodiment (Merleau-Ponty, 1945/2012).

Existential phenomenology merges the intricate insights of existential philosophy with the methodological rigor of phenomenological research. Pivoting around themes of freedom, authenticity, anxiety, and death, existential phenomenology seeks to illuminate the fundamental questions of human existence and the nature of Being.

Historical and Philosophical Roots:

Existential phenomenology emerged as thinkers like Jean-Paul Sartre, Simone de Beauvoir, and Maurice Merleau-Ponty began to weave existential themes into the fabric of phenomenological inquiry. By doing so, they expanded phenomenology's horizons to encompass deep existential concerns (Merleau-Ponty, 1945/2012; Sartre, 1943/1956).

Principal Themes of Existential Phenomenology

a) **Freedom and Authenticity**: At the heart of existential phenomenology lies the notion of human freedom. It underscores the capacity of individuals to shape their destinies, make choices, and live authentically, steering away from the societal impositions of predetermined roles (Sartre, 1943/1956).
b) **Being-towards-Death**: Drawing from Heidegger's concept of "Being-towards-death", existential phenomenology explores the inevitability of death as a foundational aspect of human existence, emphasizing its role in influencing human choices and the quest for meaning (Heidegger, 1927/1962).
c) **The Embodied Self**: Influenced by Merleau-Ponty's work, existential phenomenology places significant emphasis on the body as the locus of experience. It asserts that human existence is invariably embodied, and this embodiment shapes perception, interaction, and identity (Merleau-Ponty, 1945/2012).
d) **Alienation and Absurdity**: Inspired in part by Sartre and Camus, existential phenomenology delves into feelings of alienation, absurdity, and existential angst. It explores the confrontations individuals face in a seemingly indifferent or absurd universe (Sartre, 1943/1956; Camus, 1942/1955).

Implications for Research and Therapy

i. **Exploration of Lived Experience**: Existential phenomenological research prioritizes the exploration of lived experiences, especially as they relate to fundamental existential concerns such as the search for meaning, freedom, isolation, and finitude (van Manen, 1990).
ii. **Therapeutic Approaches**: Existential themes, when integrated into therapeutic contexts, allow for a deeper understanding of human anxieties, desires, and struggles. Existential psychotherapy, influenced by these themes, aids individuals in navigating life's challenges and in crafting authentic narratives (Yalom, 1980).

Existential phenomenology offers a profound lens through which to view the human condition. By marrying the existential concerns of freedom, death, alienation, and authenticity with phenomenological inquiry, it illuminates the deepest layers of human existence, enriching both research and therapeutic landscapes.

Critiques and Challenges

Phenomenological research, though rich and insightful, is not without its challenges:

1. **Subjectivity**: The deeply personal and subjective nature of phenomenological research can be both a strength and a limitation. Achieving true bracketing is challenging, leading to potential biases (Creswell & Poth, 2017).

Phenomenological research, with its dedication to elucidating the intricacies of human experience, inherently grapples with the concept of subjectivity. While this intimate engagement with subjective realities offers rich insights, it also comes with its own set of critiques and challenges.

Subjectivity in Phenomenological Research

Subjectivity, in the context of phenomenological research, refers to the deeply personal, interpretative, and individual nature of experiences. Phenomenology acknowledges and even celebrates this subjectivity as it seeks to understand phenomena from the vantage point of those who experience them (Moustakas, 1994).

Critiques and Challenges Concerning Subjectivity

a) **Risk of Over-Reliance on Personal Interpretation**: One critique posits that the subjective nature of phenomenological research may lead to an over-reliance on the personal interpretations of both participants and researchers, potentially eclipsing more objective or shared realities (Denzin & Lincoln, 2005).
b) **Issues of Validity and Generalizability**: Given its subjective focus, phenomenological studies often face questions regarding the validity of their findings and the extent to which these findings can be generalized. Some critics argue that the deep dive into individual experiences might not necessarily represent broader patterns or truths (Creswell, 2013).
c) **Potential Bias and Lack of Neutrality**: The intimate involvement of the researcher in phenomenological studies, coupled with their personal biases and pre-understandings, can potentially influence the research process and outcomes (Finlay, 2008).
d) **Difficulty in Achieving Epoché**: Husserl's concept of epoché or "bracketing" entails setting aside personal biases and preconceptions. However, achieving true bracketing is challenging, and critics argue that complete neutrality might be an unattainable ideal (Giorgi, 2009).
e) **Navigating Dual Roles**: Phenomenologists often walk a fine line between being empathetic listeners and critical analysts. Striking a balance between immersing oneself in participants' narratives and maintaining a critical, reflexive stance can be challenging (van Manen, 1990).

Addressing the Challenges

i. **Reflexivity**: Reflexivity involves ongoing self-awareness and self-examination throughout the research process. Through reflexivity, researchers can identify, acknowledge, and address their biases and influences on the research (Finlay, 2002).

ii. **Triangulation**: Using multiple methods, data sources, or researchers can help in validating the findings of a phenomenological study, providing a more comprehensive understanding of the phenomenon (Creswell & Miller, 2000).
iii. **Rich, Thick Descriptions**: Offering detailed and layered descriptions of experiences can provide readers with a deeper understanding, allowing them to determine the applicability of findings to other contexts or settings (Geertz, 1973).

While subjectivity in phenomenological research poses challenges, it's also its strength. By diving deep into the subjective realms of human experience, phenomenology uncovers layers of meaning and understanding often missed in more objective research paradigms. Addressing the critiques and challenges related to subjectivity ensures the continued richness and relevance of phenomenological inquiries.

2. **Time-Consuming**: Given its depth, phenomenological research often requires extensive time for data collection, analysis, and reflection (Laverty, 2003).

Phenomenological research, lauded for its depth and rich exploration of human experiences, often faces critiques concerning the significant amount of time it necessitates. The meticulous nature of this qualitative approach, while providing invaluable insights, poses challenges in terms of the time commitment required at various stages of the research process.

The Time-Consuming Nature of Phenomenological Research

The inherent depth and complexity of phenomenological research demand extended periods of data collection, analysis, and interpretation. Given its commitment to a thorough exploration of lived experiences, the research process can become elongated (van Manen, 1990).

Critiques and Challenges Concerning Time Consumption

a) **Extended Data Collection**: Phenomenological studies often involve long, in-depth interviews or multiple interviews with participants to capture the richness of their experiences. This approach, while exhaustive, is time-intensive (Moustakas, 1994).
b) **Iterative Analysis**: The data analysis process in phenomenological research is iterative. Researchers continuously move between the data and emerging themes, refining their understanding. This cyclical nature, while ensuring depth, extends the duration of analysis (Giorgi, 2009).
c) **Researcher Fatigue**: The prolonged and intensive engagement with data can lead to researcher fatigue. This fatigue can, in turn, impact the quality of analysis and interpretation, especially in large-scale studies (Smith, Flowers, & Larkin, 2009).
d) **Report Writing**: Given the richness of the data, phenomenological research often results in lengthy reports. Crafting these detailed and layered accounts, ensuring they reflect the depth and nuance of participants' experiences, is time-consuming (Langdridge, 2007).
e) **Challenges in Funding and Publication**: Due to the time-consuming nature of this approach, securing funding can be challenging. Additionally, the extensive results might exceed the word limits of many academic journals, posing challenges for publication (Creswell, 2013).

Addressing the Challenges

i. **Clear Research Boundaries**: Researchers can establish clear boundaries in terms of research questions, participant numbers, and data collection methods to ensure the study remains manageable within time constraints (van Manen, 1990).
ii. **Utilizing Technology**: Technological tools, like transcription software or qualitative data analysis software, can assist in streamlining certain aspects of the research process, saving time without compromising depth (Bazeley & Jackson, 2013).
iii. **Collaborative Efforts**: Working in research teams or collaboratively can help distribute the workload, especially during intensive stages like data analysis (Malterud, 2001).

While the time-consuming nature of phenomenological research poses challenges, it's essential to recognize that this investment of time underpins the depth and richness for which the approach is celebrated. By strategically navigating and addressing the challenges associated with time, researchers can continue to harness the strengths of phenomenological inquiry while ensuring feasibility and efficiency.

3. **Generalizability**: Findings from phenomenological studies are deeply contextual, making generalization to broader populations challenging (Polkinghorne, 1989).

At the nexus of phenomenological research lies the pursuit of understanding human experiences in their rich, intricate depth. However, this very depth and specificity have sparked critiques concerning the generalizability of phenomenological findings. Let's delve deeper into the challenges of generalizability within the realm of phenomenological inquiry.

Generalizability in the Context of Phenomenological Research

Generalizability, often associated with the transferability of research findings to broader contexts, holds distinct connotations in phenomenological research. Phenomenological studies prioritize depth over breadth, focusing on the unique lived experiences of individuals or smaller groups rather than larger populations (Moustakas, 1994).

Critiques and Challenges Concerning Generalizability

a) **Limited Sample Sizes**: Phenomenological research typically involves smaller, purposefully-selected samples to capture the essence of lived experiences. Critics argue that the findings derived from such limited samples may not be applicable to broader populations (Polkinghorne, 1989).
b) **Depth at the Expense of Breadth**: The in-depth nature of phenomenological studies, while yielding rich data, often focuses on very specific experiences or contexts. This specificity can make it challenging to draw broader conclusions or apply findings to diverse settings (Creswell, 2013).
c) **Contextual Sensitivity**: Phenomenological findings are deeply embedded in the contexts from which they arise. The socio-cultural, historical, and personal backgrounds of participants play pivotal roles, making it challenging to determine if similar results would emerge in different contexts (van Manen, 1990).

d) **Varied Interpretations**: The interpretative nature of phenomenological research, shaped by both participants' and researchers' perspectives, can lead to multiple understandings of the same phenomenon. This multiplicity poses challenges for deriving universally applicable insights (Smith, Flowers, & Larkin, 2009).

Addressing the Challenges of Generalizability

i. **Reframing Generalizability**: In phenomenology, the aim is not always to generalize in the traditional sense. Instead, researchers often seek "transferability," where the depth and richness of findings allow readers to determine their applicability to other contexts (Lincoln & Guba, 1985).
ii. **Rich, Thick Descriptions**: Providing detailed, layered accounts of participants' experiences can enable readers and other researchers to understand the context fully and determine the potential relevance of findings to other settings (Geertz, 1973).
iii. **Explicit Reflexivity**: By clearly articulating their biases, backgrounds, and interpretative processes, researchers can offer insights into the potential scope and limitations of their findings, helping readers gauge generalizability (Finlay, 2002).
iv. **Collaborative and Cross-Cultural Studies**: Engaging in collaborative research or conducting studies across diverse cultural contexts can provide comparative insights, enhancing the broader applicability of phenomenological findings (LeVasseur, 2003).

While generalizability remains a point of critique for phenomenological research, it's crucial to recognize that the value of this approach lies in its depth and authenticity, not necessarily its broad applicability. By reframing generalizability, embracing rich descriptions, and practicing reflexivity, phenomenologists can address challenges while honoring the unique strengths of their approach.

In conclusion, phenomenology, with its emphasis on lived experiences and deep exploration of human consciousness, offers a profound lens for qualitative research. It captures the richness, intricacy, and nuance of human experiences, providing insights that are both deep and transformative.

Grounded Theory

Grounded Theory (GT) is a research methodology that aims to derive theories from systematic analysis of data. Originally developed by sociologists Barney Glaser and Anselm Strauss in the 1960s, GT emphasizes the emergence of concepts and theories from within the data itself, rather than testing preexisting hypotheses (Glaser & Strauss, 1967).

Historical Context

The birth of Grounded Theory was a response to the prevailing research practices of the 1960s, which heavily leaned towards the verification of theories. Glaser and Strauss sought to flip the paradigm, aiming instead to *discover* theory grounded in empirical data, thus the name "Grounded Theory" (Glaser & Strauss, 1967).

The 1960s marked a transformative period in the field of social sciences. Research methodologies were predominantly dominated by positivist approaches, often emphasizing hypothesis testing against empirical data. Amidst this backdrop, Grounded Theory (GT) emerged, offering a fresh, inductive approach to theory development. Conceived by sociologists Barney Glaser and Anselm Strauss, Grounded Theory sought to move away from mere hypothesis verification to the generation of theories directly rooted in data.

A Response to the Prevailing Research Landscape

The mid-20th century research paradigm predominantly valued quantitative methodologies, often sidelining qualitative inquiries (Kelle, 2005). Studies were largely structured around testing pre-existing theories or validating hypotheses. Glaser and Strauss recognized a gap in this approach: while existing theories were being refined, where were the new theories emerging from?

Grounding Theory in Empirical Reality

Both Glaser and Strauss had backgrounds in empirical research prior to the development of Grounded Theory. Glaser had been trained at Columbia University in quantitative methodology but was also exposed to qualitative research. Strauss, on the other hand, had his foundations in symbolic interactionism, a perspective that emphasized human interaction and its interpretative nature (Charmaz, 2014).

Their collaboration, which began in the late 1950s, was centered on studying the experiences of dying hospital patients. Throughout this study, they realized the need for a methodological approach that would allow theories to emerge directly from the data, rather than forcing data to fit into pre-existing theoretical frameworks (Glaser & Strauss, 1965).

At the crux of Grounded Theory (GT) lies its unwavering commitment to grounding theoretical constructs within the raw fabric of empirical data. This orientation towards empirical reality sets GT apart from many other research methodologies, emphasizing the organic evolution of theory from data, rather than a top-down imposition of pre-existing theoretical frameworks onto collected data.

Emergence from Empirical Engagements

Both Barney Glaser and Anselm Strauss, the architects of Grounded Theory, had their scholarly roots deeply embedded in empirical studies. Prior to conceptualizing GT, they were actively engaged in empirical research endeavors. Their collaboration on the study of dying hospital patients was seminal not only in its findings but in crystallizing their shared belief that meaningful theory could—and should—emerge directly from data (Glaser & Strauss, 1965).

A Paradigm Shift in Research Philosophy

Historically, many research approaches in the social sciences were structured around testing or refining existing theories. Data was often a secondary entity, used mainly as a means to confirm or refute established theoretical postulates (Bryant & Charmaz, 2007). Grounded Theory marked a departure from this. It proposed a bottom-up approach, where data wasn't just a tool for theory verification but the very bedrock on which new theories were to be constructed (Glaser & Strauss, 1967).

Data as the Starting Line, Not the Finish

In GT, the relationship between data and theory is dynamic. The research process is iterative, with continuous shuttling between data collection and analysis. As data is sifted and sorted, initial concepts emerge, which then guide subsequent data collection. This symbiotic relationship ensures that the evolving theory remains intimately connected to, and is a reflection of, the empirical world (Charmaz, 2014).

Beyond Mere Description: The Theoretical Sensitivity

While GT emphasizes empirical grounding, it's not synonymous with mere data description. Researchers are encouraged to develop what Glaser termed "theoretical sensitivity". This refers to a researcher's ability to see beyond the obvious in data, to intuitively grasp and articulate latent patterns, connections, and theoretical avenues (Glaser, 1978).

Addressing the Critique of Empiricism

Critics of GT have occasionally labeled it as being overly empirical, potentially sidelining the role of existing literature or the researcher's interpretative lens. However, proponents argue that GT's strength lies in its empirical roots. While it doesn't disregard existing theories, it ensures they don't overshadow the voices and patterns emerging from the data (Charmaz, 2006).

The grounding of theory in empirical reality is the defining hallmark of Grounded Theory. It underscores the belief that robust and relevant theories are those that emanate from the lived experiences, narratives, and nuances captured within data, ensuring a genuine resonance with the empirical world from which they are birthed

The Birth of a New Methodological Paradigm

In 1967, their groundbreaking book, *The Discovery of Grounded Theory*, was published. This work proposed an inductive methodology that emphasized the "grounding" of theory in empirical data. It was met with both acclaim and critique. For many, it opened up new vistas in research, providing a systematic approach to qualitative inquiry that could yield robust theoretical outcomes. For others, it deviated too far from traditional scientific rigor (Glaser & Strauss, 1967).

The Legacy and Evolution of Grounded Theory

As with many pioneering methodologies, Grounded Theory underwent significant evolution and branching after its inception. While Glaser and Strauss jointly introduced Grounded Theory, they later diverged in their interpretations and applications of the methodology. Glaser advocated for a more purist approach, emphasizing the emergence of theory without forcing data. Strauss, collaborating with Juliet Corbin, introduced more structured coding procedures and placed a greater emphasis on the interpretive capabilities of the researcher (Strauss & Corbin, 1990).

The historical context of Grounded Theory is emblematic of a broader shift in the 20th-century social sciences, marking a move from rigid, positivist paradigms to more flexible, interpretative methodologies. The legacy of Grounded Theory is evident in its continued prominence in qualitative research and its ongoing evolution to suit contemporary research needs.

Key Principles of Grounded Theory

1. **Data Collection and Analysis are Interconnected**: GT is characterized by its iterative process. Data collection, analysis, and theory development happen simultaneously, allowing the researcher to refine data collection strategies based on emerging insights (Charmaz, 2014).

One of the most distinctive attributes of Grounded Theory (GT) is its intertwined nature of data collection and analysis. This methodological principle, unlike more linear research processes, is iterative, ensuring that the theory remains deeply anchored in empirical data.

Historical Roots of the Principle

The tenet that data collection and analysis are interconnected can be traced back to the foundational work of Glaser and Strauss. In their landmark book, *The Discovery of Grounded Theory*, they emphasized the simultaneous interplay of collecting and analyzing data, positing this as a core strength of GT (Glaser & Strauss, 1967).

Breaking Down the Interconnected Process

a) **Cyclical, Not Linear**: Traditional research often follows a sequential path: first data collection, then data analysis, followed by theory development. GT, conversely, sees these stages as dynamically interconnected, often cycling back and forth multiple times within a single study (Charmaz, 2014).
b) **Emerging Directions**: Initial rounds of data collection in GT guide preliminary analysis, which then, in turn, directs subsequent data collection. This ensures that data gathering is always purposeful and aligned with emerging theoretical insights (Corbin & Strauss, 2015).
c) **Constant Comparative Method**: Central to this interconnected process is the "constant comparative method." As new data is collected, it is immediately compared with existing data, aiding in refining categories, identifying patterns, and guiding subsequent data collection (Glaser, 1965).
d) **Development of Theoretical Sensitivity**: The iterative process enhances the researcher's "theoretical sensitivity" or the ability to discern theoretical directions in the data. With each cycle, the researcher becomes more attuned to the subtleties and nuances of the data, enriching the analysis and emerging theory (Glaser, 1978).

Advantages of the Interconnected Approach

i. **Grounding of Theory**: By continuously weaving between data collection and analysis, the resultant theory remains closely tied to empirical evidence, reinforcing GT's commitment to groundedness (Charmaz, 2006).

ii. **Flexibility**: The dynamic nature of the approach allows researchers to adapt their inquiries based on emergent findings. This flexibility often leads to more nuanced and comprehensive theoretical outcomes (Bryant & Charmaz, 2007).
iii. **Enhanced Validity**: The iterative comparison of data strengthens the validity of the findings. As patterns are continuously tested against new data, the emergent theory becomes more robust and reliable (Strauss & Corbin, 1998).

The intertwined relationship between data collection and analysis in Grounded Theory isn't just a methodological step but is emblematic of the philosophy of GT itself. It encapsulates the belief that theory should be inextricably linked to empirical reality, dynamically shaped and reshaped by the rhythm of the data.

2. **Theoretical Sampling**: Unlike random or convenience sampling, in GT, sampling decisions are made to refine emergently developed theoretical concepts. This means that researchers decide what data sources to explore based on evolving theoretical insights (Glaser & Strauss, 1967).

Theoretical sampling stands as one of the cardinal pillars underpinning the methodological rigor of Grounded Theory (GT). Distinct from conventional sampling methods, theoretical sampling in GT revolves around the purposeful collection of data to refine emergently developed theoretical constructs. Its inclusion as a central principle underscores GT's unwavering commitment to ensuring that theory remains, at all times, deeply rooted in empirical evidence.

Historical Emergence of Theoretical Sampling

The conceptualization of theoretical sampling can be traced back to the seminal work of Glaser and Strauss in their foundational text, *The Discovery of Grounded Theory*. They delineated how data collection in GT is not premised on a pre-established plan, but is constantly modulated based on the evolving analytical needs of the research (Glaser & Strauss, 1967).

Defining Theoretical Sampling

a) **Purposeful Evolution**: Unlike traditional research designs where sampling criteria are set in advance, theoretical sampling in GT is emergent. Initial data informs the researcher's understanding, which then dictates the subsequent data to be collected (Charmaz, 2014).
b) **Iterative Process**: The process is cyclical. As analysis advances, researchers identify gaps, ambiguities, or new dimensions of the theory, which then guide further data collection, ensuring that theory remains comprehensive and robust (Corbin & Strauss, 2015).
c) **Saturation as the Endpoint**: Theoretical sampling continues until 'theoretical saturation' is achieved. This is the point where additional data no longer offers fresh insights or refines existing categories, signaling that the theory is well-developed and exhaustive (Glaser & Strauss, 1967).

Significance of Theoretical Sampling

I. **Ensuring Groundedness**: By continually aligning data collection with emerging theoretical needs, theoretical sampling guarantees that the resultant theory is intimately tethered to empirical reality (Charmaz, 2006).
II. **Enhancing Depth and Nuance**: This sampling method enables researchers to explore categories in depth, unearthing nuances, contradictions, and intricacies, thereby enhancing the richness and multifaceted nature of the developed theory (Bryant & Charmaz, 2007).
III. **Optimizing Resource Use**: As theoretical sampling is directed and purposeful, it ensures that researchers' efforts are optimized, focusing on data that genuinely enriches the theory, thereby avoiding superfluous data collection (Dey, 1999).

Critiques and Considerations

i. **Potential for Subjectivity**: Some critics highlight that the emergent nature of theoretical sampling could introduce subjectivity, as researchers may inadvertently steer data collection based on personal biases (Hammersley, 1992).
ii. **Logistical Challenges**: The unpredictability inherent in theoretical sampling can pose logistical challenges, especially in terms of research timelines and resources, which might be difficult in constrained research environments (Morse, 1991).

Theoretical sampling is emblematic of Grounded Theory's adaptive and responsive nature. It encapsulates the belief that for a theory to truly resonate with empirical realities, the process of its construction must be deeply dialogic, allowing data and theory to continually shape and inform one another.

3. **Constant Comparative Method**: Central to GT is the continuous comparison of data, codes, and categories. This iterative process allows the researcher to identify patterns and variations within the data and helps in refining conceptual categories (Charmaz, 2014).

At the heart of Grounded Theory (GT) resides the Constant Comparative Method, a dynamic analytical tool that distinguishes GT from many other qualitative research methodologies. By continually juxtaposing pieces of data against each other, and against evolving conceptual categories, this method acts as a crucible where raw data is transformed into refined theoretical constructs.

Origins of the Constant Comparative Method

Introduced by Glaser and Strauss, the architects of Grounded Theory, the Constant Comparative Method emerged as a robust and systematic approach to data analysis. Their seminal work, *The Discovery of Grounded Theory*, propounded the method as a cornerstone of GT, highlighting its value in ensuring that theory remained inextricably tied to empirical data (Glaser & Strauss, 1967).

Key Aspects of the Constant Comparative Method

a) **Continuous Comparison**: As the name suggests, data is continuously compared at various levels - initial data with initial data, initial data with emerging categories, and categories with other categories (Charmaz, 2014).

b) **Iterative Nature**: The process is not static. As researchers delve deeper into the data and as categories evolve, they constantly loop back, revisiting and reanalyzing previous data in light of new insights (Corbin & Strauss, 2015).

c) **Identification of Patterns**: Through persistent comparison, patterns, and regularities in the data start to emerge. These patterns lay the foundation for developing broader categories and, eventually, theoretical constructs (Glaser, 1965).

d) **Enhancement of Theoretical Sensitivity**: Engaging deeply with the data, constantly comparing and contrasting, sharpens the researcher's "theoretical sensitivity", enhancing their ability to discern subtle, yet significant theoretical directions (Glaser, 1978).

Significance and Impact

I. **Depth and Nuance**: The Constant Comparative Method allows for a granular analysis of data. It unearths nuances and complexities, offering a rich, multi-dimensional view of the phenomenon under study (Charmaz, 2006).

II. **Groundedness of Theory**: By continuously anchoring analysis in the data, the method ensures that emergent theories are genuinely reflective of, and grounded in, empirical realities (Bryant & Charmaz, 2007).

III. **Refinement of Categories**: Constant comparison refines emerging categories, helping to delineate their properties, outline their boundaries, and identify their relationships to other categories (Dey, 1999).

Challenges and Critiques

i. **Potential for Overwhelm**: Given its iterative and intensive nature, the method can be overwhelming, especially for novice researchers, and may extend the duration of the research process (Hammersley, 1992).

ii. **Risk of Fragmented Analysis**: Detractors suggest that there's a risk of becoming lost in the details, potentially leading to a fragmented or overly complex analysis (Morse, 1991).

The Constant Comparative Method is not merely a technique but encapsulates the spirit of Grounded Theory. It underscores the belief in a continuous dialogue between data and theory, ensuring that emergent theories are both empirically grounded and analytically sharp.

4. **Memo Writing**: As data is collected and analyzed, researchers write memos to capture their thoughts, reflections, and emerging insights. Memos play a pivotal role in refining analysis and guiding subsequent data collection (Corbin & Strauss, 2014).

Memo writing stands as an integral component of Grounded Theory (GT), serving as an intermediary space where raw data begins its transformative journey towards becoming theory. Beyond its mere instrumental value, memo writing encapsulates the reflective spirit of GT, where researchers continually engage with, and are informed by, their data.

Historical Context of Memo Writing in GT

Glaser and Strauss, in their foundational work, *The Discovery of Grounded Theory*, underscored the value of memo writing as an essential cog in the GT machinery. They posited memos as a bridge between the empirical world of data and the conceptual realm of theory, highlighting their role in grounding emergent theories (Glaser & Strauss, 1967).

Key Facets of Memo Writing in GT

a) **Reflective Space**: Memos offer researchers a space to reflect, ponder, and engage with their data. It is here that initial hunches, questions, and connections are explored, allowing for a deeper immersion into the data (Charmaz, 2014).
b) **Documenting Analytical Progress**: As research progresses, memos trace the researcher's analytical journey. They capture the evolution of ideas, offering a chronological map of how categories and theories have developed, changed, or merged over time (Corbin & Strauss, 2015).
c) **Enhancing Theoretical Sensitivity**: The act of memo writing sharpens the researcher's theoretical sensitivity. Engaging deeply with data through memos cultivates an intuitive grasp of its theoretical potentials and implications (Glaser, 1978).
d) **Facilitating Integration**: Memos assist in integrating diverse pieces of data, revealing patterns, contrasts, and underlying structures. They play a pivotal role when moving from fragmented ideas to a cohesive, integrated theory (Charmaz, 2006).

The Process and Utility of Memo Writing

I. **Dynamism**: Memos are not static. They evolve as the research progresses, mirroring the dynamic nature of GT itself. Earlier memos can be revisited and revised in light of new data or insights (Birks & Mills, 2015).
II. **Flexibility in Format**: There's no prescriptive format for memos. They can range from short annotations to extensive reflective pieces, from diagrams to narrative texts, accommodating the researcher's preferences and the specific needs of the data (Lempert, 2007).
III. **A Catalyst for Theoretical Sampling**: Memos often highlight areas of ambiguity or gaps in understanding, guiding the next steps in data collection, and enhancing the process of theoretical sampling (Dey, 1999).

Challenges and Considerations

i. **Potential for Over-Analysis**: Given the reflective nature of memos, there's a risk of becoming overly introspective, potentially leading to over-analysis or deviation from the data (Hammersley, 1992).
ii. **Maintaining Balance**: Striking a balance between staying grounded in the data and indulging in abstract theorization can be challenging. Memos should ideally serve as anchors, ensuring that theoretical reflections remain rooted in empirical evidence (Morse, 1991).

Memo writing in Grounded Theory is more than a mere methodological step; it is a manifestation of the GT ethos. It emphasizes the continual dialogue between data and researcher, ensuring that emergent theories are both deeply grounded and analytically rigorous.

5. **Seeking Theoretical Saturation**: Data collection continues until "theoretical saturation" is reached – a point where new data no longer offers fresh theoretical insights. It's a signal that the developed theory is robust and comprehensive (Glaser & Strauss, 1967).

Within the Grounded Theory (GT) methodology, theoretical saturation occupies a central position, demarcating the point where the researcher's immersion in data transitions from exploration to consolidation. It signifies the stage at which the iterative dance between data collection and analysis reaches a crescendo, resulting in a rich, robust, and exhaustive theory.

Historical Context of Theoretical Saturation

The concept of theoretical saturation was introduced by Glaser and Strauss in their groundbreaking work, *The Discovery of Grounded Theory*. It was posited as a criterion to ensure that the emergent theory was not only grounded in data but also thoroughly comprehensive, leaving no aspect unexplored (Glaser & Strauss, 1967).

Core Aspects of Theoretical Saturation:

a) **Exhaustive Exploration**: Theoretical saturation implies that all facets of a category within the data have been thoroughly investigated. It ensures that categories are well-developed, with their properties and dimensions fully delineated (Charmaz, 2014).
b) **No New Insights**: A key marker of saturation is the point at which additional data collection or further analysis does not yield fresh insights or new dimensions related to the category under investigation (Corbin & Strauss, 2015).
c) **Integration and Density**: As categories are saturated, they become more integrated and dense, with clear inter-category connections and relationships emerging, leading towards a cohesive theoretical framework (Glaser, 1978).

Implications and Impact of Seeking Saturation

I. **Robustness of Theory**: Achieving theoretical saturation ensures that the resultant theory is not superficial but deeply rooted in data, offering a nuanced and comprehensive understanding of the phenomenon (Charmaz, 2006).

II. **Guidance for Data Collection**: The quest for saturation provides a clear directive for the researcher's data collection efforts. It enables them to identify areas requiring deeper exploration and guides the process of theoretical sampling (Bryant & Charmaz, 2007).

III. **Enhancing Validity**: Saturation strengthens the validity and transferability of the findings. Since all facets of the phenomenon are exhaustively explored, the theory's applicability to similar contexts or populations is enhanced (Dey, 1999).

Challenges and Critiques

i. **Ambiguity of Saturation Point**: One critique of theoretical saturation is its inherent ambiguity. Determining the exact point of saturation can be subjective, with potential variations across researchers (Hammersley, 1992).

ii. **Practical Constraints**: Achieving complete saturation can be resource-intensive, especially in terms of time and funding. In constrained research environments, striking a balance between depth and feasibility becomes crucial (Morse, 1991).

The pursuit of theoretical saturation within Grounded Theory is emblematic of the methodology's commitment to depth, rigor, and empirical grounding. It serves as a touchstone, ensuring that emergent theories are not just reflections of empirical reality but also comprehensive encapsulations of the studied phenomenon.

Applications and Variations

Over the years, Grounded Theory has seen various interpretations and adaptations:

1. **Constructivist Grounded Theory**: Kathy Charmaz's take on GT emphasizes the interpretative nature of the research process. It acknowledges the co-construction of meaning between researcher and participant, placing a premium on reflexivity (Charmaz, 2014).

Grounded Theory (GT) has seen several iterations and interpretations since its inception by Glaser and Strauss. Among these iterations, Constructivist Grounded Theory, championed predominantly by Kathy Charmaz, has carved a significant niche, emphasizing the interpretive nature of knowledge construction. This approach recognizes the subjective interplay between the researcher and the researched, underscoring the co-construction of meaning in qualitative inquiry.

Historical Overview and Emergence

Constructivist Grounded Theory can be seen as a reaction to the more positivist leanings of classical GT. Where traditional GT often postulated an objective reality to be discovered, constructivist GT embraces the idea that realities are multiple and co-constructed (Charmaz, 2000).

Core Tenets of Constructivist Grounded Theory

a) **Relational Co-construction**: Knowledge, in the constructivist paradigm, is not merely "discovered" but co-constructed between the researcher and the participants. The research process is, therefore, a dynamic interplay of shared and interpreted meanings (Charmaz, 2006).
b) **Subjective Interpretation**: Constructivist GT acknowledges the researcher's subjectivity. Rather than being a detached observer, the researcher is embedded within the research, bringing their own perspectives, values, and biases to the analytical process (Charmaz, 2014).
c) **Flexible Guidelines**: While classical GT is known for its methodological prescriptions, constructivist GT offers more flexible guidelines, emphasizing the importance of reflexivity and the iterative nature of research (Charmaz & Bryant, 2008).

Applications and Implications

I. **Diverse Research Fields**: Given its flexible and interpretative nature, constructivist GT has been applied across a plethora of disciplines, from health and social care to education and business studies, capturing the intricacies and nuances of human experience (Mills, Bonner, & Francis, 2006).
II. **Enhanced Reflexivity**: The emphasis on co-construction necessitates heightened reflexivity, prompting researchers to continually interrogate their own influence on data collection, analysis, and theory development (Breckenridge, Jones, Elliott, & Nicol, 2012).
III. **Richer, Layered Insights**: By acknowledging multiple realities and the role of interpretation, constructivist GT often yields richer, more layered insights, capturing the depth and breadth of human experiences (Charmaz, 2014).

Critiques and Challenges

i. **Claims to Objectivity**: Critics argue that by emphasizing co-construction and interpretation, constructivist GT may compromise the objectivity and rigor associated with traditional GT (Glaser, 2002).
ii. **Potential for Over-Relativism**: There's a concern that constructivist GT might drift into over-relativism, where any interpretation is deemed acceptable, potentially compromising the credibility of the research (Bryant, 2002).

Constructivist Grounded Theory offers a fresh epistemological lens, emphasizing the co-construction of meaning in research. While it has its critics, its focus on interpretation, reflexivity, and the relational dynamics of research provides researchers with tools to delve deeply into the complexities of human experience, yielding profound and nuanced insights.

2. **Situational Analysis**: Adele Clarke expanded GT by introducing situational analysis, which emphasizes the broader socio-material contexts within which phenomena occur. This approach widens the analytical lens, considering non-human actors and socio-political conditions (Clarke, 2005).

While Grounded Theory (GT) has always been anchored in exploring social processes and phenomena, the introduction of Situational Analysis (SA) marks a distinct shift, expanding the analytical lens to more comprehensively explore the situational contexts and complexities in which these phenomena occur. SA, championed by Adele E. Clarke, bridges the classical GT with postmodern sensibilities, emphasizing the complexities, multiplicities, and contingencies in human affairs.

Historical Emergence of Situational Analysis

The genesis of Situational Analysis (SA) can be traced to Clarke's desire to advance and expand the scope of Grounded Theory, integrating it with postmodernist and constructivist elements. Clarke's seminal text, *Situational Analysis: Grounded Theory after the Postmodern Turn* (2005), laid down the foundational tenets of this approach.

Core Tenets of Situational Analysis

a) **Mapping Complexities**: At the heart of SA is the use of extensive mapping exercises, which include situational, social worlds/arenas, and positional maps. These maps visually represent the intricate relationships, actors, and discourses at play in a particular situation (Clarke, 2005).
b) **Beyond Individuals**: While classical GT often centers on individuals' actions and interactions, SA emphasizes the broader situational context, considering non-human elements, discourses, and historical contingencies as equally significant (Clarke, Friese, & Washburn, 2018).
c) **Acknowledging Multiple Realities**: SA operates from a standpoint that there are multiple, often competing realities and truths in any given situation. It encourages the researcher to embrace and represent this complexity, rather than distilling it into a singular narrative (Clarke & Charmaz, 2014).

Applications and Implications

I. **Diverse Research Areas**: Given its comprehensive and expansive lens, SA has been applied across a myriad of disciplines, including health care, public policy, environmental studies, and technology. It is particularly adept at unraveling complex, multifaceted issues where many actors and factors intersect (Clarke & Star, 2008).
II. **Visual Representation**: The mapping exercises intrinsic to SA provide visual clarity to intricate scenarios. They help in elucidating connections, power dynamics, and areas of contestation, offering both researchers and readers a clearer understanding of the situation (Clarke, 2005).

III. **Expanding the Analytical Horizon**: By emphasizing situational complexities, SA pushes researchers to think beyond individual actors or actions, considering broader socio-political, historical, and even material contexts, thereby offering a richer, more nuanced analysis (Clarke & Charmaz, 2014).

Critiques and Considerations

i. **Analytical Complexity**: One of the critiques directed towards SA is its level of complexity, particularly for novice researchers. The extensive mapping exercises, while valuable, can be daunting and require a steep learning curve (Ramos & Ferreira, 2017).
ii. **Potential Over-Extension**: There's a risk of over-extending the analysis, trying to map every conceivable element of the situation. Researchers must strike a balance between comprehensive analysis and focused relevance (Reed, 2018).

Situational Analysis offers a nuanced extension to Grounded Theory, urging researchers to move beyond individual narratives and delve into the broader, intricate webs of situational contexts. Its emphasis on visual mapping and recognizing multiple realities provides a robust tool for analyzing complex phenomena in our increasingly interconnected world.

Critiques of Grounded Theory

Like any methodology, GT has its share of critiques:

1. **Ambiguity of Theoretical Saturation**: Critics argue that the concept of theoretical saturation can be subjective, with different researchers potentially arriving at saturation at different points (Dey, 1999).

Grounded Theory (GT) stands as one of the most influential qualitative research methodologies. Yet, like all methodologies, it is not exempt from critique. Among the various aspects of GT, the concept of theoretical saturation remains particularly contested. It symbolizes the point at which new data no longer brings forth fresh insights, signaling the readiness for theory development. However, determining the moment of this saturation is debated for its ambiguity and subjectivity.

Historical Context of Theoretical Saturation

Theoretical saturation, as introduced by Glaser and Strauss in *The Discovery of Grounded Theory* (1967), was posited as a measure of completeness in data collection and analysis. The proposition seemed straightforward: researchers should continue collecting data until no new properties or dimensions of a category emerge.

Central Critiques Surrounding Ambiguity of Theoretical Saturation

a) **Subjective Determination**: One of the primary critiques is the inherent subjectivity in discerning the point of saturation. What might appear as saturation to one researcher might be seen as an area requiring further exploration to another, raising questions about consistency and replicability (Hammersley, 2010).

b) **Scope and Depth Dilemma**: While the broad scope of inquiry is celebrated in qualitative research, it does pose challenges. Determining saturation might vary depending on whether one aims for breadth or depth in their analysis. A topic might seem saturated from a broader perspective, but new nuances can emerge when looked at in depth (O'Reilly & Parker, 2013).

c) **Fluidity of Social Realities**: Given that social phenomena are dynamic and evolving, the notion of ever achieving a full saturation is debated. Can a point of saturation be definitive when the very essence of the subject under study is its fluidity? (Charmaz, 2006).

d) **Practical Constraints**: Often, external factors, such as time, funding, or access, can influence the declaration of saturation. Researchers might announce saturation due to these practical limitations rather than genuine analytical completeness (Bowen, 2008).

Implications of These Critiques

I. **Validity Concerns**: If saturation is prematurely or inaccurately claimed, it could lead to a lack of depth and richness in the resultant theory, thus affecting the validity and robustness of the research findings (Mason, 2010).

II. **Inconsistencies in Application**: Different researchers, guided by their subjective interpretations, might apply the concept of theoretical saturation inconsistently. This can pose challenges for those attempting to compare or build upon prior studies (Saunders et al., 2018).

The ambiguity surrounding theoretical saturation in Grounded Theory underscores the complexities inherent in qualitative research. While the concept serves as a valuable guide, prompting rigorous and comprehensive data engagement, its subjectivity poses challenges. Recognizing and navigating this ambiguity, researchers can better harness the strengths of Grounded Theory while ensuring the depth, rigor, and credibility of their analyses.

2. **Issues of Representation**: Some scholars raise concerns about the extent to which GT findings, being grounded in specific datasets, can be generalized or applied to broader populations (Hammersley, 1992).

Grounded Theory (GT) has etched its significance in qualitative research, aiding scholars in the grounded development of theory from data. Yet, amidst its accolades, GT faces critiques regarding representation. This involves concerns over how participants, contexts, and cultures are portrayed, interpreted, and framed, potentially influencing the authenticity and integrity of research findings.

Historical Context of Representation in Qualitative Research

Representation in qualitative methodologies, including GT, traces back to concerns about the power dynamics in research. This dialogue gained traction especially with the postcolonial and postmodern critiques of anthropology and sociology in the late 20th century, questioning the positioning of researchers vis-à-vis their subjects and the constructed narratives emanating from their work (Said, 1978; Clifford & Marcus, 1986).

Key Critiques on Representation in Grounded Theory

a) **Researcher's Lens Over Participant's Voice**: GT, in its coding and categorizing processes, risks imposing the researcher's interpretations and biases onto the data, potentially overshadowing the authentic voices and lived experiences of participants. The final emergent theory may lean more towards the researcher's worldview than that of the participants (Denzin & Lincoln, 2000).
b) **Potential for Oversimplification**: In GT's quest for patterns and universal themes, there's a risk of oversimplifying or homogenizing diverse experiences. This could result in the erasure of outliers, nuances, or minority perspectives that don't fit into the primary emergent categories (Geertz, 1973).
c) **Cultural Misrepresentation**: Especially in cross-cultural research, GT could inadvertently perpetuate stereotypes or misinterpret cultural practices, given the researcher's outsider perspective. Without a deep, immersive understanding, there's a risk of producing a distorted or superficial representation of the culture under study (Said, 1978).
d) **The Static Nature of Constructed Theory**: Sociocultural realities are dynamic. A theory grounded in a specific temporal and spatial context may not accurately represent the evolving nature of the phenomena, leading to potential misrepresentations when applied to newer contexts (Charmaz, 2014).

Implications of These Critiques

I. **Ethical Implications**: Misrepresentation or oversimplification in GT can have ethical implications. It can inadvertently perpetuate harm, particularly if misrepresented communities are already marginalized or vulnerable (Mauthner & Doucet, 2003).
II. **Impact on Policy and Practice**: Grounded theories, when adopted in policy or practice, can have real-world impacts. Misrepresentations can lead to misguided interventions, policies, or strategies, potentially exacerbating issues rather than alleviating them (Harding, 2004).

While Grounded Theory offers a robust framework for qualitative inquiry, the challenges surrounding representation remind researchers of the ethical and methodological considerations intrinsic to the research process. To navigate these complexities, GT researchers must engage deeply with reflexivity, continually questioning and adjusting their positionality, interpretations, and representations to produce authentic, ethically sound research.

3. **Rigidity in Coding**: Overemphasis on the systematic coding process can sometimes overshadow the narrative or holistic understanding of the data (Bryant, 2002).

Grounded Theory (GT), a qualitative methodology acclaimed for its potential in theory development from empirical data, is often identified by its characteristic coding processes. However, its structured approach to coding—ranging from open coding to axial and selective coding—has been both a hallmark of its rigor and a subject of critique. Concerns arise surrounding the potentially rigid structure of this coding process, which some argue may constrain the explorative nature intrinsic to qualitative research.

Understanding Coding in Grounded Theory

Central to GT is a systematic approach to coding, which typically involves:

a) **Open Coding**: Fragmenting data into discrete parts, closely examining, and comparing for similarities and differences.
b) **Axial Coding**: Organizing data in new ways after open coding, often around a "core" category.
c) **Selective Coding**: Developing the main theory by relating categories to the core category (Strauss & Corbin, 1998).

Key Critiques Concerning Rigidity in Coding

I. **Over-Standardization**: The structured coding procedure, while ensuring consistency, may lead to an over-standardized approach to data analysis. Critics argue that this could curtail the organic, fluid exploration of data, pushing researchers to fit data into pre-defined stages rather than allowing patterns to emerge naturally (Charmaz, 2000).
II. **Limiting Creativity and Intuition**: The systematic coding structure might sideline researchers' intuition, creativity, and imaginative insights, potentially overlooking the depth and nuance of participant experiences in favor of a more structured analytical path (Bryant & Charmaz, 2007).
III. **Potential for Superficial Analysis**: Adhering rigidly to the coding procedures without flexibility might lead to a more surface-level analysis. Some critics opine that the structured approach might prevent researchers from delving deeper into intricacies, complexities, and interconnectedness within the data (Dey, 1999).
IV. **The Danger of Forced Fit**: The multi-tiered coding system, if followed too rigidly, might press researchers to force data into categories, even when such categorizations don't naturally fit, leading to potential misrepresentations or oversimplifications (Glaser, 1992).

Implications of These Critiques

i. **Threat to Authenticity**: Over-reliance on structured coding, at the expense of a more open, intuitive exploration, might compromise the authenticity of the research findings, potentially overlooking subtleties and variances in participant narratives (Charmaz, 2006).

ii. **Scope for Methodological Evolution**: The critiques serve as a call for the evolution of GT methodology, incorporating more flexibility and adaptability in the coding process, potentially leading to richer, more nuanced outcomes (Clarke, 2005).

While the systematic coding process within Grounded Theory offers a structured pathway to data analysis, it's essential to strike a balance between methodological rigor and the flexibility required for genuine qualitative exploration. Recognizing and addressing the potential limitations of rigidity in coding can ensure that GT continues to yield deep, authentic, and valuable insights from qualitative data.

CONCLUSION

Grounded Theory, with its commitment to theory generation from empirical data, has cemented its place as a cornerstone in qualitative research. Its iterative, flexible nature ensures that the resultant theories are deeply embedded in, and reflective of, the data from which they emerge.

Ethnography

Ethnography, derived from the Greek words 'ethnos' (people) and 'graphein' (to write), is a qualitative research methodology centered around understanding and describing human cultures and societies. Originating in the discipline of anthropology, ethnography has transformed over the years, transcending its traditional boundaries to become an interdisciplinary research tool (Geertz, 1973).

Historical Roots of Ethnography

Ethnography's roots can be traced back to early anthropological explorations, where researchers undertook long-term immersion in 'foreign' cultures to understand their ways of life. Renowned figures like Malinowski and Evans-Pritchard played pivotal roles in shaping the ethnographic methodologies, emphasizing deep immersion and participant observation (Malinowski, 1922; Evans-Pritchard, 1940).

Ethnography, as an intrinsic methodology in anthropological research, has witnessed profound transformations since its inception. Its roots, woven into the fabric of early anthropological endeavors, involve explorations into the nuances of diverse cultures and their practices. Understanding the historical trajectory of ethnography is essential to appreciate its contemporary applications and significance.

Early Beginnings

The infancy of ethnography can be traced back to the travel accounts of early explorers, missionaries, and colonial administrators. These individuals, while not anthropologists by today's definition, provided some of the earliest detailed descriptions of cultures and societies outside the European realm (Malinowski, 1922).

As we delve into the early roots of ethnography, it becomes evident that the methodology's conceptual foundation lies in broader historical contexts of exploration, colonization, and early encounters with the 'other.' The early stages of ethnographic writing were entwined with the first-person narratives of those who ventured beyond their familiar territories, inadvertently shaping the field of anthropology and the practice of ethnography itself.

Travelogues and Encounters

Long before the formalization of anthropology as an academic discipline, European explorers, traders, and missionaries ventured into uncharted territories, recording their observations and experiences of unfamiliar cultures. These accounts, often replete with descriptions of exotic customs, rituals, and social structures, can be seen as precursors to modern ethnography (Pratt, 1992).

Colonial Documentation

With the onset of European colonization, administrators and colonial officers took an interest in understanding the societies they sought to govern. This was not a purely academic endeavor; knowledge was power, and understanding local customs and structures was pivotal for effective governance and control. These colonial accounts, often detailed and extensive, laid groundwork for subsequent ethnographic studies, albeit from a position of domination and control (Asad, 1973).

Missionaries and Conversion

Christian missionaries, in their zeal to convert indigenous populations, also contributed to early ethnographic literature. In trying to understand the belief systems they sought to replace, they documented religious rituals, myths, and societal norms. Their accounts provide a complex blend of observation, interpretation, and judgment, making them valuable, albeit biased, sources of ethnographic data (Comaroff & Comaroff, 1991).

The 'Salvage Paradigm'

By the late 19th century, as industrialization and Western influence spread, there was a growing belief among scholars and researchers that indigenous cultures were on the brink of extinction. This gave rise to the 'salvage paradigm,' where anthropologists felt an urgency to document 'disappearing' cultures before they were forever lost to modernization. This mindset further spurred efforts to record, in detail, the customs, languages, rituals, and folklores of various societies (Kuper, 1973).

The early beginnings of ethnography were not situated in academic rigor or methodological frameworks as we understand them today. Instead, they were shaped by encounters, often asymmetrical in power dynamics, between the West and the rest. These initial interactions, observations, and documentations, regardless of their motivations, undeniably paved the way for the establishment of ethnography as a structured, academic practice.

Institutionalization of Anthropology

By the late 19th and early 20th centuries, anthropology began to institutionalize as an academic discipline. Scholars sought to document and understand 'primitive' societies, perceived as living relics of earlier evolutionary stages of human civilization. This perspective, while now considered problematic and ethnocentric, heavily influenced early ethnographic work (Stocking, 1987).

The historical trajectory of ethnography cannot be understood fully without addressing the institutionalization of anthropology. This process not only shaped the development and formalization of ethnographic methodology but also established anthropology as a legitimate and distinct academic discipline.

Background: A Science of Humanity

The late 19th and early 20th centuries witnessed increasing academic interest in understanding the diversity of human societies and cultures. As the Western world underwent significant scientific, political, and philosophical changes, there emerged a desire to study humanity systematically, classifying and analyzing cultures and societies much like species in the natural world (Kuklick, 1997).

University Programs and Specializations

The late 1800s saw the establishment of anthropology programs in prestigious universities. Institutions like the University of Oxford, Cambridge, and later, American universities such as Harvard and the University of Chicago began offering courses, eventually leading to specialized degrees in anthropology. This formal education inculcated a systematic approach to the study of societies, making fieldwork a crucial component of anthropological training (Stocking, 1987).

Foundational Figures and Their Contributions

Prominent scholars played pivotal roles in shaping the contours of academic anthropology:

- **Franz Boas**: Often deemed the 'father of modern anthropology,' Boas was critical in moving away from the evolutionary frameworks prevalent in the 19th century. He emphasized the importance of fieldwork and cultural relativism, arguing that cultures should be understood on their terms without hierarchical judgments. His students, including Margaret Mead and Ruth Benedict, continued his legacy, deepening the ethnographic tradition (Boas, 1940).
- **Émile Durkheim**: A foundational figure in sociology and anthropology, Durkheim introduced structural functionalism, emphasizing the interconnectedness of societal parts. He argued that cultural phenomena, including rituals and myths, played roles in maintaining societal cohesion (Durkheim, 1915).

Influence of Professional Societies

The establishment of anthropological societies, such as the Royal Anthropological Institute in the UK and the American Anthropological Association in the US, further solidified anthropology's academic stature. These societies facilitated dialogue, research dissemination, and set standards for anthropological research, further influencing the direction and methodology of ethnography (Barnard, 2011).

Ethical Standards and Reflection

With institutionalization came a growing recognition of the ethical complexities inherent in studying human societies. As anthropology matured, there emerged discussions about the responsibilities of anthropologists to the communities they studied. The implications of 'extracting' knowledge from cultures, often without reciprocation, became a point of contention and reflection (Fluehr-Lobban, 2002).

The institutionalization of anthropology marked a decisive turn in the evolution of ethnography. As anthropology gained academic rigor and structure, so too did ethnography, transitioning from loosely-structured observations to a methodologically robust tool for understanding human cultures.

The Fieldwork Revolution

Bronislaw Malinowski's pioneering work among the Trobriand Islanders marked a significant shift in ethnographic research. In his seminal work, "Argonauts of the Western Pacific" (1922), Malinowski championed the method of participant observation, urging researchers to immerse themselves deeply into the lives and cultures of the people they studied. This approach starkly contrasted with the armchair anthropologists of the previous era who relied on second-hand accounts (Malinowski, 1922).

Ethnography's turn towards immersive fieldwork, often deemed the 'Fieldwork Revolution,' marked a significant departure from earlier methodologies and forever changed the landscape of anthropological research. This revolution underscored the importance of firsthand experiences and insights, rather than reliance on second-hand accounts or mere speculation.

Malinowski and the Trobriand Islands

Central to the fieldwork revolution is Bronislaw Malinowski, a Polish-born British anthropologist. His extensive work in the Trobriand Islands during World War I set a new benchmark for anthropological research. Unlike his predecessors who primarily relied on informants or collected data through intermediaries, Malinowski lived among the Trobrianders, learning their language, participating in their daily activities, and immersing himself in their culture (Malinowski, 1922).

In his influential work, "Argonauts of the Western Pacific" (1922), Malinowski emphasized the method of participant observation. He advocated for a holistic approach, arguing that to grasp the full picture of a society, one must understand its customs, trade, kinship structures, folklore, and more, all interconnected and mutually constitutive.

Radical Implications for Ethnography

The implications of this immersive approach were manifold:

a) **Depth and Nuance**: Immersion allowed for deeper insights, capturing the nuances and intricacies of daily life, beliefs, rituals, and practices (Geertz, 1983).
b) **From 'Them' to 'Us'**: The earlier detached perspective that often othered studied societies was now replaced by a more intimate understanding, diminishing the distance between the observer and the observed (Rabinow, 1977).

c) **Ethics and Representation**: This approach required researchers to grapple with the ethics of representation, as they were now participants in the societies they were documenting. The dynamics of power, privilege, and responsibility became central to anthropological discourse (Fabian, 1983).

Beyond Malinowski: Expanding the Revolution

While Malinowski is often credited as a key figure in the fieldwork revolution, he was not alone in this transformative endeavor. Margaret Mead's work in Samoa, where she explored adolescent behavior, showcased the range of human cultural and social possibilities. E.E. Evans-Pritchard's research with the Nuer of Sudan emphasized the importance of understanding indigenous systems of thought and logic, setting a precedent for cultural relativism in ethnographic studies (Evans-Pritchard, 1940; Mead, 1928).

Challenges and Critiques

While the shift towards immersive fieldwork was groundbreaking, it was not without challenges:

i. **Personal and Physical Strain**: Extended fieldwork in remote or challenging environments could take a toll on the researcher's health and well-being.
ii. **Interpretative Challenges**: Being deeply embedded in a culture posed the risk of 'going native,' potentially losing objectivity and critical distance (Sanjek, 1990).
iii. **Ethical Concerns**: With closer relationships came the challenge of maintaining confidentiality and respecting the agency and autonomy of participants (Clifford, 1990).

The fieldwork revolution, with its emphasis on immersion and participant observation, reshaped ethnography, moving it from a somewhat detached study of the 'other' to a deeply engaged understanding of human societies. This transformation underscored the belief that to truly understand a culture, one must live it.

Functionalism and Structural Functionalism

The early 20th century witnessed the rise of functionalist perspectives in anthropology. Prominent figures like Radcliffe-Brown and Evans-Pritchard emphasized understanding societies as holistic systems, where each component (e.g., rituals, kinship structures) had specific functions that contributed to societal stability and continuity (Radcliffe-Brown, 1952; Evans-Pritchard, 1940).

The evolution of anthropology, and by extension ethnography, has been significantly influenced by various theoretical perspectives. Among these, functionalism and its later adaptation, structural functionalism, provided foundational lenses through which societies and cultures were analyzed during the early to mid-20th century.

Functionalism: Origins and Key Tenets

Functionalism emerged as a dominant theoretical perspective in British anthropology during the early 20th century. At its core, functionalism posits that every aspect of a society or culture, be it a ritual, institution, or custom, has a specific function or role that contributes to the overall stability and continuation of that society.

1. **Organic Analogy**: Drawing parallels with biological organisms, functionalists saw societies as complex systems where each part (akin to organs in a body) plays a vital role in ensuring the survival and well-being of the whole (Radcliffe-Brown, 1952).
2. **Holistic Perspective**: Functionalists believed in studying societies holistically, understanding that the various components of a society are interconnected and interdependent (Malinowski, 1944).

Bronislaw Malinowski: Needs and Functions

One of the most influential proponents of functionalism was Bronislaw Malinowski. He argued that cultural practices and institutions exist to satisfy the biological and psychological needs of individuals in a society. For Malinowski, the goal of ethnography was to discern these needs and the cultural mechanisms addressing them. His work in the Trobriand Islands is often cited as a prime example of functionalist ethnography, wherein he meticulously detailed the Kula ring's exchange system and its role in Trobriand society (Malinowski, 1922).

A.R. Radcliffe-Brown: Social Structures and Relations

A.R. Radcliffe-Brown, another key figure in functionalism, shifted the focus slightly from individual needs to the relationships and structures within society. He emphasized the recurring patterns of social relations and believed that these relations and structures were paramount in maintaining societal equilibrium. This systematic and relational approach would later give rise to structural functionalism (Radcliffe-Brown, 1952).

Structural Functionalism: Evolution and Emphasis

Structural functionalism, while rooted in functionalist thought, gave greater emphasis to the importance of social structures over individual practices or beliefs.

a) **Societal Equilibrium**: Structural functionalists posited that societies strive for equilibrium. When disruptions occur, societal mechanisms or structures work to restore balance (Nadel, 1951).
b) **Role of Institutions**: This perspective underscored the role of institutions (like family, religion, or law) in maintaining social order. Each institution plays a crucial part in the coherent functioning of society, addressing specific needs and potential conflicts (Evans-Pritchard, 1940).

Critiques and Beyond

Functionalism and structural functionalism, while foundational, were not without their critics:

i. **Static and Conservative**: Some critics argued that these frameworks presented societies as overly static and resistant to change, failing to account for societal transformations and revolutions (Leach, 1954).

ii. **Overemphasis on Cohesion**: The theories were critiqued for overly emphasizing societal cohesion, often overlooking conflicts, inequalities, and power dynamics inherent in societies (Turner, 1957).

Functionalism and structural functionalism significantly influenced ethnographic work throughout the 20th century, offering tools to decipher the intricate workings of societies and cultures. While they faced criticisms and gave way to newer theoretical paradigms, their impact on the discipline of anthropology and the practice of ethnography remains undeniable.

Reflexivity and Postmodern Turn

By the mid-20th century, ethnography and anthropology at large began to witness a profound paradigmatic shift. This transformation was characterized by increasing reflexivity and an embrace of postmodern sensibilities. Both changes bore significant implications for how ethnographic research was conducted, interpreted, and disseminated.

Reflexivity: Introspective Ethnography

Reflexivity involves a heightened self-awareness of the ethnographer's positionality, biases, and influence on the research process. The reflexive turn urged anthropologists to scrutinize their roles, recognizing that their presence, interpretations, and interactions invariably shape the ethnographic narrative.

a) **Acknowledging the Observer's Impact**: Reflexivity challenged the previously held notion of the 'objective' anthropologist. Ethnographers began to understand and articulate how their own backgrounds, beliefs, and relationships influenced the research process (Behar, 1996).
b) **Dialogic Approach**: Reflexive ethnographers often embraced a more collaborative approach, emphasizing dialogue with their participants and, at times, incorporating their feedback and perspectives into the final narrative (Rosaldo, 1989).

The Postmodern Turn

Postmodernism, with its skepticism towards grand narratives and its emphasis on the relativity of truth, deeply influenced ethnographic practice.

i. **Critique of Representation**: Postmodernism prompted ethnographers to question the authenticity and authority of their representations. This led to introspections about who gets to speak, whose voices are privileged, and whose narratives are marginalized (Clifford & Marcus, 1986).
ii. **Fragmented Narratives**: As a response to the postmodern critique, some ethnographies began presenting non-linear, fragmented narratives that reflected the complexities and multiplicities of human experiences, rather than a singular 'truth' (Marcus & Fischer, 1986).
iii. **Decolonizing Ethnography**: The postcolonial critique, an offshoot of postmodernism, emphasized the colonial roots and legacies inherent in ethnographic practices. Scholars like Said highlighted the often problematic representations of the 'Orient' by Western scholars, urging for more nuanced, egalitarian, and decolonized approaches to research (Said, 1978).

Notable Figures and Contributions

- **Clifford Geertz**: Advocated for "thick description" in ethnographic research, emphasizing the depth and nuance of cultural interpretations. Geertz's approach underscored the interpretive nature of anthropological understanding (Geertz, 1973).
- **James Clifford**: Emphasized the constructed nature of ethnographic accounts, viewing them as "fictions" that are co-created by the ethnographer and the studied community (Clifford, 1988).
- **Ruth Behar**: Her introspective and personal approach to ethnography blurred the lines between the ethnographer and the 'other,' emphasizing mutual vulnerability and humanity (Behar, 1996).

The reflexive and postmodern turns in ethnography were transformative, prompting deep methodological, epistemological, and ethical reflections within the discipline. These shifts underscored the complexities of representation, the intricacies of cross-cultural understanding, and the profound responsibilities shouldered by ethnographers.

The historical journey of ethnography reflects the broader intellectual currents and shifts within anthropology and the social sciences. From its early descriptive beginnings to its contemporary reflexive and critical stances, ethnography remains a powerful lens through which human societies are understood, represented, and interpreted.

Fundamental Tenets of Ethnography

1. **Holistic Perspective**: Ethnography seeks a comprehensive understanding of the cultural phenomena under study. It does not merely focus on isolated behaviors but explores them within the broader socio-cultural context (Spradley, 1979).

Central to ethnographic research is the holistic perspective, a foundational principle which posits that cultural phenomena must be understood in their entirety and in relation to their broader socio-cultural contexts. This holistic approach stands in contrast to reductionist methodologies that may extract and analyze elements of culture in isolation.

Origins of the Holistic Perspective

The holistic perspective's roots can be traced back to early anthropological endeavors. As anthropology sought to understand the entirety of human experience, it necessitated a comprehensive approach that considered all aspects of a given culture as interconnected and interdependent (Boas, 1940).

The Interwoven Fabric of Culture

a) **Interconnectedness**: A key tenet of the holistic perspective is the understanding that various elements of a culture – from rituals and kinship structures to economic systems and folklore – are interconnected, each influencing and being influenced by the others (Geertz, 1973).
b) **Against Isolation**: The holistic approach cautions against studying cultural elements in isolation. For instance, a religious ritual might be deeply tied to socio-economic structures, local ecologies,

historical narratives, and communal identities. To fully understand the ritual, one must consider these intersecting factors (Turner, 1967).

Benefits of the Holistic Approach

i. **Comprehensive Understanding**: The holistic perspective provides a richer, more nuanced understanding of cultural phenomena. It encourages the researcher to delve deeper, exploring the myriad influences and implications of a particular practice or belief (Malinowski, 1922).
ii. **Avoiding Misinterpretation**: By viewing cultural elements in context, the ethnographer is less likely to misinterpret or misrepresent practices, beliefs, or values. It allows for a more authentic representation of the studied culture (Evans-Pritchard, 1940).
iii. **Embracing Complexity**: The holistic perspective acknowledges the complexities and layers within cultures. It recognizes that cultures are not monolithic or static but are dynamic entities with myriad influences and factors at play (Bateson, 1936).

Critiques and Challenges

While the holistic perspective has been foundational to ethnography, it's not without its critiques:

I. **Overwhelming Complexity**: Some critics argue that the holistic approach, in its quest for comprehensiveness, can become overwhelming, making it challenging to discern central themes or patterns (Sanjek, 1990).
II. **Potential for Overgeneralization**: There's a risk that, in seeking holistic insights, ethnographers might overgeneralize or gloss over nuances and internal diversities within the studied culture (Appadurai, 1991).

The holistic perspective in ethnography underscores the intricate tapestry of human cultures. It encourages ethnographers to move beyond surface-level observations, seeking deeper, more comprehensive insights into the multifaceted worlds they study. By embracing this perspective, ethnographers can hope to capture the richness, depth, and complexity of human cultural experience.

2. **Participant Observation**: Central to ethnography is the method of participant observation, where the researcher immerses themselves in the daily lives of the study participants, sometimes for extended periods (DeWalt & DeWalt, 2011).

Participant observation stands as one of the cornerstone methodologies in ethnographic research. By actively immersing oneself within the studied community, an ethnographer attempts to understand the intricacies of daily life, social interactions, rituals, and behaviors, offering insights that might remain obscured in a more detached research approach.

Origins of Participant Observation

The origins of participant observation can be located in the early days of anthropological fieldwork, where scholars realized the value of firsthand experience and immersion within communities as opposed to relying solely on second-hand reports or superficial observations (Malinowski, 1922).

Characteristics of Participant Observation

a) **Immersive Research**: At the heart of participant observation is immersion. Ethnographers spend extended periods within the community, actively participating in daily routines, rituals, and events (Geertz, 1973).
b) **Dual Role**: Ethnographers find themselves in the dual role of observer and participant. This balance is crucial, allowing for engagement without completely losing the ability to analyze and reflect on observed phenomena (Spradley, 1980).
c) **Building Rapport**: Effective participant observation necessitates building trust and rapport with community members. This trust can lead to more candid conversations and deeper insights (Bernard, 2011).

Benefits of Participant Observation

I. **Depth of Insight**: Immersion allows ethnographers to understand nuances, contexts, and subtleties that might otherwise be missed. It offers a richness of data unparalleled by more distant methodologies (Wolcott, 1995).
II. **Challenging Preconceptions**: By living within the community, ethnographers often confront and challenge their own preconceptions and biases, leading to a more authentic representation (Rosaldo, 1989).
III. **Dynamic Understanding**: Cultures are not static. Participant observation, by virtue of its extended nature, can capture the dynamism and fluidity of cultural practices and beliefs (Clifford, 1997).

Challenges and Critiques

i. **Subjectivity**: Critics argue that the deep immersion characteristic of participant observation might compromise objectivity, making findings more susceptible to the ethnographer's personal biases (Atkinson et al., 2001).
ii. **Ethical Dilemmas**: The close relationships built during participant observation can lead to ethical challenges, especially concerning informed consent, confidentiality, and potential power imbalances (Murphy & Dingwall, 2001).
iii. **Physical and Emotional Strain**: Extended immersion can be physically demanding and emotionally taxing, leading to challenges like culture shock or burnout (Kulick & Willson, 1995).

Participant observation, with its emphasis on immersion and engagement, remains a defining feature of ethnographic research. It offers a window into the lived realities of communities, capturing the intricacies of daily life and social dynamics. By navigating the challenges and embracing the depth of insight it offers, ethnographers can produce rich, nuanced, and authentic accounts of human cultural experience.

3. **Thick Description**: Coined by Geertz (1973), 'thick description' emphasizes capturing the layers of cultural meaning and the contextual nuances that shape human behaviors.

The concept of "thick description" has been pivotal in ethnographic research, emphasizing not just the documentation of behaviors and events but also the layers of context, meaning, and interpretation that surround them. This methodological principle underscores the depth and richness that ethnographic inquiries strive for in representing human cultures.

Origins of Thick Description

The term "thick description" is most famously associated with Clifford Geertz, an American anthropologist, who introduced and elaborated on the idea in his seminal work, "The Interpretation of Cultures" (Geertz, 1973). Drawing from the philosophy of Gilbert Ryle, Geertz positioned thick description against "thin description" – the mere recording of behaviors devoid of context or meaning.

Defining Features of Thick Description

a) **Beyond Surface Observations**: Thick description goes beyond mere observations to delve into the deeper layers of context, intent, and meaning. It seeks to capture not just what people do, but why they do it and how they understand their actions (Geertz, 1973).
b) **Symbolic Interpretation**: For Geertz, human behavior is a web of symbols. Thick description, therefore, involves interpreting these symbolic actions, unraveling the cultural scripts and narratives that give behaviors their significance (Geertz, 1973).
c) **Narrative Richness**: Thick descriptions are often characterized by their narrative depth, offering readers a vivid and detailed portrayal of the studied phenomena. This enables outsiders to gain a sense of the cultural and social realities being described (Geertz, 1983).

Implications for Ethnographic Research

I. **Holistic Understanding**: Thick description emphasizes a holistic approach to ethnography. By weaving together behaviors, emotions, contexts, and interpretations, it offers a comprehensive picture of cultural phenomena (Turner, 1986).
II. **Challenge to Positivism**: Geertz's advocacy for thick description challenged positivist approaches in social sciences that often sought objective, detached, and quantifiable data. Instead, it recognized the importance of subjectivity, interpretation, and the co-construction of knowledge (Marcus & Fischer, 1986).
III. **Ethical Considerations**: With its focus on depth and detail, thick description raises ethical considerations around representation, consent, and the potential for over-exposure of participants' lives and identities (Abu-Lughod, 1991).

Critiques and Debates

I. **Potential for Over-interpretation**: Critics have pointed out the risk of over-interpretation in the quest for thick description, suggesting that ethnographers might sometimes read more into a situation than what might actually exist (Sanjek, 1990).

II. **Generalizability**: The deep, context-specific nature of thick description raises questions about the generalizability of its findings. Can insights derived from such detailed descriptions of particular settings be applied to broader populations or contexts? (Ortner, 1995).

Thick description remains a foundational tenet in ethnographic research, emphasizing the depth, context, and richness of human cultural expressions. By focusing on the intricate layers of meaning, it offers a counter-narrative to reductionist approaches, inviting readers to immerse themselves in the complexities and nuances of human experience.

4. **Emic and Etic Perspectives**: Ethnographers aim to understand cultural phenomena both from an insider's (emic) perspective and an outsider's (etic) perspective, ensuring a well-rounded understanding (Pike, 1967).

Ethnographic research often grapples with the tension between the "inside" and "outside" perspectives when studying a cultural group. These orientations are encapsulated by the terms "emic" and "etic," concepts that have profound implications for how ethnographers understand, interpret, and represent cultures.

Origins of Emic and Etic

The terms "emic" and "etic" were popularized in the realm of anthropology by Kenneth Pike, a linguist who initially introduced them in the context of linguistic analysis. Pike (1967) borrowed the terms from the linguistic concepts of "phonemic" and "phonetic," but they were soon co-opted into cultural anthropology to denote differing approaches to understanding human behavior and thought.

Defining the Emic Perspective

a) **Insider's View**: The emic perspective seeks to understand a culture from the "inside." It prioritizes the viewpoints, beliefs, values, and understandings of the people within the cultural group itself (Pike, 1967).

b) **Deep Dive**: This approach typically necessitates prolonged engagement with the community, allowing the ethnographer to grasp the nuances, symbolic meanings, and indigenous interpretations of cultural phenomena (Headland, 1990).

c) **Subjective and Contextual**: The emic perspective is inherently subjective, rooted in the particularities of the studied culture. It may not always be generalizable but offers rich, contextual insights (Harris, 1976).

Characteristics of the Etic Perspective

I. **Outsider's Lens**: The etic perspective approaches a culture from the "outside." It employs categories and concepts that might be part of the ethnographer's own cultural framework or those of academia at large, rather than those indigenous to the studied group (Pike, 1967).
II. **Comparative Potential**: Since etic approaches often use standardized categories or criteria, they can facilitate cross-cultural comparisons. This allows for the identification of patterns, similarities, and differences across societies (Levinson, 1998).
III. **Objective Tendencies**: The etic perspective tends to strive for objectivity, sometimes at the risk of missing out on the deeper, subjective, and symbolic layers of a culture (Headland, 1990).

Interplay and Balance

Many ethnographers recognize the value of both emic and etic perspectives and attempt to integrate both in their research:

i. **Iterative Process**: Ethnographers might start with an etic approach to get a broad understanding, then delve into the emic perspective for depth, and revert to the etic for analysis and representation (Bernard, 2011).
ii. **Enhancing Validity**: Combining both perspectives can enhance the validity and richness of the ethnographic account, offering a balance between depth and breadth, subjectivity and objectivity (Morris, Leung, & Iyengar, 2004).

The tension and interplay between emic and etic perspectives underscore the complexities inherent in ethnographic research. They remind ethnographers of the challenges in representing cultures—balancing the insider's rich, subjective experience with the outsider's broader, analytical gaze. Integrating both perspectives can lead to ethnographic accounts that are both deeply insightful and broadly illuminating.

Key Methodological Steps in Ethnography

1. **Selecting a Field Site**: This involves identifying a suitable community, organization, or setting that aligns with the research question (Fetterman, 2019).

Choosing a field site is a pivotal step in the ethnographic research process. This decision not only determines the context of the study but can also influence the researcher's access, relationships, and insights. A meticulously selected field site provides a foundation for rich, nuanced, and meaningful ethnographic exploration.

Historical Overview of Field Site Selection

In the early days of anthropology, field sites were often "exotic" locales, driven by Western academia's interest in documenting societies perceived as untouched by Western influence. This approach, however, evolved over time, with contemporary ethnographers also focusing on urban settings, institutional contexts, and online environments (Marcus, 1995).

Criteria for Selecting a Field Site

a) **Relevance to Research Question**: The foremost criterion is the alignment of the site with the overarching research questions and objectives. The field site should offer opportunities to explore the phenomenon of interest in depth (Bernard, 2011).
b) **Accessibility**: Logistical considerations, such as geographical accessibility, permissions required, and potential risks, play a significant role. Ethnographers must ensure that they can gain access to, and spend prolonged periods within, the chosen site (Clifford, 1997).
c) **Richness of Context**: Sites that offer multi-dimensional insights, characterized by diverse interactions, practices, and narratives, are often preferred. Such richness enables a more holistic and nuanced understanding of the cultural context (Geertz, 1973).
d) **Ethical Considerations**: Potential field sites should be evaluated for ethical concerns, including the implications of the researcher's presence, issues of representation, and potential impacts on the community (Fluehr-Lobban, 1991).

Challenges in Field Site Selection

I. **Overfamiliarity**: Sometimes, being too familiar with a site can lead to assumptions or overlook nuances. Conversely, being an outsider might make access and rapport-building more challenging (Agar, 1980).
II. **Dynamic Nature of Sites**: Field sites are not static; they evolve over time due to various socio-cultural, political, and economic factors. These changes can influence the course of the research (Amit, 2000).
III. **Multiplicity of Sites**: Contemporary ethnographic research often necessitates multi-sited ethnography, complicating the process of field site selection (Marcus, 1995).

Benefits of Thoughtful Field Site Selection

i. **Stronger Rapport**: A well-chosen site facilitates the establishment of trust and rapport with participants, vital for deeper ethnographic insights (Schensul et al., 1999).
ii. **Feasibility and Sustainability**: Ensuring logistical and practical feasibility ensures that researchers can sustain their immersion, leading to richer data collection (Bernard, 2011).
iii. **Ethical Research**: Thoughtful site selection, cognizant of ethical dimensions, ensures the well-being of participants and the integrity of the research process (Ortner, 2016).

The selection of a field site, while a fundamental step in ethnographic research, is fraught with considerations, challenges, and responsibilities. A judiciously chosen field site, aligned with research objectives, and approached with ethical considerations, forms the bedrock of successful ethnographic inquiry.

2. **Building Rapport**: Establishing trust and mutual respect with participants is fundamental for gaining deep insights (Bernard, 2011).

Central to the ethnographic endeavor is the process of building rapport. Establishing trustful, meaningful relationships with participants not only paves the way for deeper insights but also ethically grounds the research in mutual respect and understanding. Building rapport remains pivotal for any ethnographer hoping to capture the essence of lived experiences.

Understanding the Essence of Rapport

Rapport, derived from the French word meaning 'relationship' or 'connection,' encapsulates the quality of the relationship between the ethnographer and the participants (Bernard, 2011). It's not merely about friendship but about establishing a trustful, respectful, and ethical bond that facilitates open dialogue and sharing.

Significance of Building Rapport

a) **Access to Rich Data**: Genuine rapport can lead participants to share deeply personal, sensitive, or nuanced aspects of their lives, resulting in richer and more comprehensive data (Kawulich, 2005).
b) **Overcoming Barriers**: Trust can help in overcoming cultural, linguistic, or social barriers, which are often inherent in ethnographic fieldwork (Spradley, 1980).
c) **Ethical Conduct**: Rapport ensures that the research is anchored in mutual respect, making the research process more ethically sound (Ellis, 2007).

Strategies for Building Rapport

I. **Prolonged Engagement**: Spending extended time in the field is crucial. This allows relationships to develop organically over time and demonstrates the researcher's genuine commitment (Lincoln & Guba, 1985).
II. **Active Listening**: Ethnographers should prioritize listening over speaking. Actively attending to participants' narratives fosters trust and validates their experiences (Rubin & Rubin, 2011).
III. **Participation**: Engaging in local customs, rituals, and everyday practices can endear the researcher to the community and show respect for their way of life (DeWalt & DeWalt, 2010).
IV. **Transparency**: Being open about the research goals, methods, and intentions helps in demystifying the research process for participants (Murphy & Dingwall, 2007).
V. **Feedback and Reciprocity**: Sharing findings with the community and giving back in meaningful ways can further strengthen rapport (Brettell, 2017).

Challenges in Building Rapport

i. **Power Dynamics**: Ethnographers need to be aware of and navigate power imbalances that may exist, ensuring participants do not feel coerced or unduly influenced (Ortner, 2016).
ii. **Over-familiarity**: Being too close can blur the lines between the personal and professional, potentially clouding objectivity (Hammersley & Atkinson, 2007).

iii. **Cultural Misunderstandings**: Differences in cultural norms and values can sometimes lead to misinterpretations or breaches in trust (Agar, 1996).

Building rapport is more than a methodological step; it's the ethical heart of ethnography. By fostering genuine, trustful relationships, ethnographers can hope to delve deeper into the complexities of human experience, capturing narratives with sensitivity, respect, and depth.

3. **Gathering Data**: Through observations, informal interviews, and participation in daily activities, ethnographers collect rich and diverse data (LeCompte & Schensul, 2010).

Within the ethnographic research framework, data gathering is a multifaceted process, demanding both rigorous techniques and a nuanced, culturally sensitive approach. Ethnographers immerse themselves within the lifeworlds of their participants, adopting methods that allow them to capture the richness and complexity of lived experiences.

Understanding Ethnographic Data Collection

Ethnography prioritizes thick, descriptive data that captures the intricacies of social and cultural phenomena. This type of data sheds light on people's behaviors, beliefs, rituals, social norms, and the meanings they ascribe to their experiences (Geertz, 1973).

Primary Techniques for Data Gathering

a) **Participant Observation**: A cornerstone of ethnography, participant observation requires the ethnographer to engage directly in the activities of the community being studied. This deep immersion offers insights into tacit practices, norms, and dynamics that might not emerge in formal interviews (Malinowski, 1922).
b) **In-depth Interviews**: Ethnographers conduct open-ended, often unstructured or semi-structured interviews to glean insights into participants' perspectives, narratives, and interpretations. This method provides depth, allowing exploration of individual experiences and viewpoints (Spradley, 1979).
c) **Focus Group Discussions**: Bringing together a group of participants, focus group discussions can highlight communal narratives, shared beliefs, and social dynamics. They're particularly useful for exploring consensus and divergences within the community (Morgan, 1996).
d) **Document Analysis**: Ethnographers might examine texts, artifacts, or other cultural products to understand historical context, communal narratives, or shared symbols. This includes examining local literature, media, historical records, or even everyday artifacts like posters and advertisements (Hodder, 2000).
e) **Visual Ethnography**: With the advent of technology, methods like photography, video recordings, or even digital ethnography have gained traction. These tools can capture non-verbal cues, spatial dynamics, and rituals in rich detail (Pink, 2007).

Challenges in Data Gathering

I. **Positionality and Reflexivity**: Ethnographers must remain constantly aware of their own positionality—how their background, beliefs, and biases might influence their observations and interpretations. Reflexivity, or the practice of reflecting upon one's influence on the research, is crucial (Bourdieu & Wacquant, 1992).

II. **Cultural Sensitivity and Ethics**: Respect for participants' privacy, beliefs, and boundaries is paramount. Ethnographers must navigate these ethical terrains, ensuring they don't inadvertently harm or misrepresent their subjects (Fluehr-Lobban, 1991).

III. **Overwhelm and Selectivity**: The sheer volume and depth of data in ethnography can be overwhelming. Ethnographers must be selective, discerning which data is most pertinent to their research questions, without losing sight of the holistic context (Emerson, Fretz, & Shaw, 1995).

Gathering data in ethnography is an intricate dance between immersion and detachment, subjectivity and objectivity. With the goal of capturing the essence of human experience in its full cultural and social context, ethnographers employ a diverse toolkit, ensuring their findings resonate with authenticity, depth, and respect.

4. **Data Analysis**: Ethnographers analyze data iteratively, looking for recurring patterns, themes, and cultural meanings (Emerson, Fretz, & Shaw, 2011).

Data analysis is a pivotal phase in ethnographic research. It involves the transformation of raw, often voluminous, field notes and observations into coherent, insightful interpretations that elucidate cultural patterns, beliefs, and practices. Ethnographers strive to unravel the underlying meanings of observed behaviors and articulated narratives to provide a holistic understanding of a culture or social group.

Understanding the Nature of Ethnographic Data

Ethnographic data is inherently rich, multifaceted, and context-dependent. It comprises field notes, interview transcripts, visual artifacts, and sometimes, even personal reflections. Such data, rooted in prolonged engagement, captures the nuances, contradictions, and depths of human experiences (Emerson, Fretz, & Shaw, 1995).

Key Steps in Ethnographic Data Analysis

a) **Organizing and Familiarization**: Before diving into deeper analysis, ethnographers need to organize their amassed data meticulously. This often involves reading and re-reading field notes, transcripts, and other data sources to familiarize themselves with the content (Agar, 1980).

b) **Coding**: Ethnographers categorize segments of data into codes, which are labels that identify key themes, patterns, concepts, or phenomena. This process aids in chunking data and laying the foundation for deeper interpretation (Saldana, 2015).

c) **Thematic Analysis**: Post coding, themes emerge. Themes are broader than codes and represent patterns or topics recurrently found in the data. Ethnographers identify these themes and examine their relevance, interconnections, and implications (Braun & Clarke, 2006).

d) **Interpretation and Theoretical Linkage**: With themes in hand, ethnographers interpret their findings, relating them to broader anthropological or sociological theories. This step positions their insights within larger academic discourses and provides a theoretical grounding for their observations (Geertz, 1973).
e) **Memo Writing**: Throughout the analysis, ethnographers often write memos—reflective notes that capture their evolving thoughts, insights, and interpretations. These memos are integral to the iterative nature of ethnographic analysis and aid in the development of nuanced interpretations (Charmaz, 2006).
f) **Synthesis and Representation**: The culmination of analysis involves synthesizing the insights and crafting a coherent narrative or representation of the studied culture. This could be a descriptive account, a theoretical exposition, or even a visual representation (Van Maanen, 1988).

Challenges in Ethnographic Data Analysis

I. **Subjectivity and Reflexivity**: Ethnographic analysis is inherently subjective. Ethnographers must continually reflect on their biases, preconceptions, and influence on both data collection and analysis (Davies, 2008).
II. **Volume and Depth**: The sheer depth and volume of ethnographic data can be overwhelming. Ethnographers must strike a balance between comprehensive analysis and focused, relevant insights (Lecompte & Schensul, 1999).
III. **Cultural Sensitivity**: Interpreting data demands a culturally sensitive lens, ensuring that insights don't inadvertently misrepresent or stereotype the studied group (Madison, 2005).

Data analysis in ethnography is a complex, iterative journey, moving from raw observations to profound cultural insights. Ethnographers, through rigorous and reflective analysis, weave together the threads of human experience, crafting narratives that resonate with depth, authenticity, and cultural sensitivity.

5. **Representation and Writing**: Crafting a compelling ethnographic account that accurately portrays the culture under study is vital. This narrative must be both descriptive and interpretative (Van Maanen, 2011).

Ethnographic representation is the culmination of the ethnographer's journey. It translates the painstakingly gathered and analyzed data into a cohesive narrative that brings to life the intricacies, nuances, and complexities of the culture or group under study. The act of representation, however, is not without its challenges, debates, and methodological considerations.

The Significance of Representation in Ethnography

At its core, ethnography aims to provide a "thick description" (Geertz, 1973) of a specific cultural or social group. This requires an articulation that captures both the overt behaviors and the underlying cultural codes, meanings, and interpretations. Representation is the medium through which these observations and interpretations are conveyed to a broader audience.

Facets of Ethnographic Representation

a) **Narrative Construction**: Ethnographers weave together participant accounts, observations, and contextual information to form a compelling narrative. The goal is to transport the reader into the studied world, facilitating an empathetic understanding (Van Maanen, 1988).
b) **Theoretical Linkages**: Beyond mere description, ethnographers link their findings to broader theoretical constructs, ensuring that their observations are anchored in established anthropological or sociological discourses (Clifford & Marcus, 1986).
c) **Voice and Reflexivity**: Ethnographers grapple with the question of voice – whose story is being told and from whose perspective? They often employ reflexivity, consciously reflecting on their positionality and influence on both the data and its representation (Behar, 1996).
d) **Multimodal Representations**: With technological advancements, representation has evolved beyond textual accounts. Today, visual ethnography, including photography and film, and digital ethnography, which utilizes online platforms, have gained prominence (Pink, 2007).

Challenges and Debates in Ethnographic Representation

I. **Authenticity vs. Interpretation**: Striking a balance between representing the 'authentic' voices of participants and providing interpretations is challenging. Ethnographers walk a tightrope, ensuring their interpretations don't overshadow or distort participant voices (Denzin, 1997).
II. **Ethics of Representation**: Ethnographers must navigate the ethical terrain of representation. Misrepresentation or over-interpretation can lead to stereotyping or inadvertently harming the community under study (Fluehr-Lobban, 1991).
III. **The Crisis of Representation**: The postmodern turn in anthropology brought about intense debates about the very possibility of accurate cultural representation, given the inherent subjectivity and power dynamics in writing and interpreting culture (Marcus & Fischer, 1986).

Representation in ethnography is both an art and a methodological endeavor. Ethnographers are storytellers, tasked with the responsibility of crafting narratives that are at once compelling, respectful, insightful, and grounded in rigorous research. As they pen down their accounts, they not only bring to life a culture but also engage in ongoing methodological and ethical reflections that shape the discipline's evolving landscape.

Ethnography in Contemporary Research

Beyond anthropology, ethnography has found relevance in various disciplines such as sociology, education, health studies, and business. In these contexts, it has been employed to understand organizational cultures, healthcare practices, classroom dynamics, and consumer behaviors, to name a few (Murphy & Dingwall, 2007).

In the rapidly shifting terrain of contemporary research, ethnography has continued to evolve, adapting to new challenges and integrating innovative methodologies. From its roots in traditional anthropological explorations of 'distant' cultures, ethnography now navigates an array of settings, from urban landscapes to virtual realms. This evolution reflects the discipline's resilience and its commitment to capturing the complex textures of human experience.

Contemporary Domains of Ethnographic Inquiry

A. **Urban Ethnography**: As the global populace becomes increasingly urbanized, ethnographers have turned their gaze to cities. Urban ethnography seeks to understand the nuances of urban life, exploring issues of migration, identity, inequality, and sociocultural dynamics in metropolitan contexts (Low, 1999).

B. **Virtual Ethnography**: The rise of digital technologies and online communities has birthed virtual or digital ethnography. Researchers explore online interactions, cyber cultures, and digital identities, navigating spaces from social media platforms to online gaming worlds (Hine, 2000).

C. **Business and Organizational Ethnography**: Contemporary ethnographers often penetrate the world of businesses and organizations. They explore corporate cultures, organizational dynamics, and consumer behaviors, providing insights beneficial to both academia and industry (Czarniawska, 2007).

D. **Medical Ethnography**: Situated at the intersection of anthropology and healthcare, medical ethnography delves into the cultural contexts of health, illness, and medical practices, offering a nuanced understanding of health beliefs, patient-provider dynamics, and medical systems (Good, 1994).

Emerging Methodological Approaches

I. **Collaborative Ethnography**: Breaking from the traditional model of the lone ethnographer, collaborative ethnography emphasizes partnership with participants, fostering co-authorship and joint interpretation of findings (Lassiter, 2005).

II. **Autoethnography**: Here, ethnographers turn the analytical lens on themselves, exploring their own experiences as primary data. This approach challenges traditional objectivity and emphasizes personal narrative and reflection (Ellis, Adams, & Bochner, 2011).

III. **Sensory Ethnography**: Moving beyond the textual, sensory ethnography prioritizes other senses – sound, sight, touch, taste, and smell – in data collection and representation, resulting in multisensory accounts of cultural experiences (Pink, 2009).

Challenges and Debates in Contemporary Ethnography

i. **Ethical Dilemmas**: In an interconnected world, issues of consent, privacy, and representation gain complexity. Virtual ethnography, for instance, grapples with questions about online privacy and the public versus private nature of digital spaces (Boellstorff et al., 2012).

ii. **Postcolonial and Feminist Critiques**: Ethnography faces critiques from postcolonial and feminist scholars who point out power imbalances, Eurocentric biases, and gendered perspectives inherent in traditional practices (Mohanty, 1988; Abu-Lughod, 1991).

iii. **The Challenge of Rapid Change**: In our fast-paced world, cultures and contexts shift rapidly, posing challenges for ethnographers who need to capture these moving targets without losing depth or nuance (Marcus, 1998).

As ethnography continues to adapt to contemporary challenges, it showcases its enduring relevance and versatility. Ethnographers today, while grounded in traditional practices, are innovatively reimagining methods and domains of study to capture the multifaceted tapestry of contemporary human experience.

Critiques and Challenges of Ethnography

1. **Subjectivity and Bias**: The deeply immersive nature of ethnography makes it vulnerable to accusations of subjectivity, potentially skewing the representation (Atkinson, Coffey, & Delamont, 2003).

Ethnography's value lies in its deep immersion and intimate understanding of cultural phenomena. Yet, this very strength makes it susceptible to critiques related to subjectivity and bias. Unlike quantitative methods which often prioritize objectivity and distance, ethnography embraces the researcher's involvement, but this closeness comes with its own set of methodological and ethical challenges.

The Nature of Subjectivity in Ethnography

Subjectivity in ethnographic research refers to the influence of the researcher's personal feelings, experiences, and perspectives on their observations, interpretations, and representations. Since ethnographers are central instruments in data collection and analysis, their backgrounds, identities, and beliefs can shape the research outcomes (Clifford & Marcus, 1986).

Critiques of Ethnographic Subjectivity

A. **Lack of Neutrality**: Critics argue that the deep immersion of ethnographers makes neutrality almost impossible. Their personal biases, consciously or unconsciously, can influence their observations, leading to skewed interpretations (Hammersley & Atkinson, 2007).

B. **Reproduction of Stereotypes**: Personal biases can lead ethnographers to unwittingly reproduce existing stereotypes, misrepresenting and potentially harming the communities they study (Spivak, 1988).

C. **Reliability and Validity Concerns**: The influence of subjectivity raises questions about the reliability and validity of ethnographic findings. Two ethnographers studying the same community might produce vastly different accounts due to their personal perspectives (Fetterman, 2010).

Challenges Posed by Bias

I. **Confirmation Bias**: Ethnographers, like all humans, are susceptible to confirmation bias—the tendency to search for, interpret, and remember information that confirms one's pre-existing beliefs (Nickerson, 1998).

II. **Cultural Bias**: Researchers might unconsciously impose their own cultural norms and values onto the studied community, interpreting behaviors through their cultural lens rather than understanding them within the community's own context (Said, 1978).
III. **Power Dynamics**: Ethnographers often come from positions of relative power, whether due to their educational status, nationality, or other factors. This can influence interactions with participants and shape the narrative, potentially silencing or marginalizing certain voices (Abu-Lughod, 1991).

Addressing Subjectivity and Bias

i. **Reflexivity**: Ethnographers are encouraged to practice reflexivity—constantly reflecting on their feelings, reactions, and biases, and considering how these might influence the research (Davies, 2008).
ii. **Triangulation**: By employing various data collection methods or collaborating with other researchers, ethnographers can compare and cross-check findings, mitigating potential biases (Denzin, 1978).
iii. **Member Checks**: Sharing findings with community members and seeking their feedback can provide corrective perspectives and ensure the representation is accurate and respectful (Lincoln & Guba, 1985).

While subjectivity and bias present challenges in ethnography, they are also inherent in the method's richness and depth. Rather than discarding or sidelining these issues, modern ethnographers engage with them, considering them integral to the research process and striving for a balance between immersion and critical distance.

2. **Generalizability**: Given its in-depth focus on specific contexts, generalizing ethnographic findings to broader populations remains debated (Marcus, 1995).

Ethnography's strength lies in its profound depth and detailed exploration of cultural or social phenomena within specific contexts. However, this in-depth focus often sparks critiques concerning the generalizability of ethnographic findings. While the intent of ethnography is rarely to produce universally applicable results, the issue of generalizability remains a significant point of contention within qualitative research debates.

The Nature of Generalizability in Research

Generalizability, often associated with quantitative research, refers to the extent to which research findings can be applied to broader populations, settings, or situations beyond the studied sample (Bryman, 2016). In contrast, ethnography typically seeks to provide a nuanced, context-rich understanding of specific groups or settings.

Critiques Concerning Ethnographic Generalizability

a) **Limited Scope**: Critics argue that because ethnographic studies focus on specific groups, communities, or settings, the findings are context-bound and cannot be easily generalized to other populations or contexts (Gobo, 2008).

b) **Lack of Replicability**: The deep immersion and personal engagement central to ethnography mean that another researcher might not have the same experiences or derive the same insights in a repeated study. This unique relationship between researcher and context further complicates generalizability (Hammersley, 1992).

c) **Absence of Statistical Representation**: Unlike quantitative studies, which often use random sampling to ensure a representative sample, ethnographies do not typically aim for statistical generalization. This can lead to critiques about the broader applicability of their findings (Brewer, 2000).

Addressing the Challenge of Generalizability in Ethnography

I. **Theoretical or Analytic Generalization**: Instead of empirical generalization, ethnographers often aim for theoretical or analytic generalization. This involves generating broader theories or insights based on the specific data gathered, which can be applied to similar settings or populations (Yin, 2013).

II. **Rich, Thick Description**: By providing detailed, context-rich accounts, ethnographers allow readers to determine the extent to which findings might be transferable to other contexts. This "thick description" enables a depth of understanding that can inform interpretations in similar settings (Geertz, 1973).

III. **Multi-Sited Ethnography**: Some ethnographers conduct research across multiple sites to enhance generalizability. By studying a phenomenon across various settings, they can draw broader insights and identify common patterns (Marcus, 1995).

IV. **Case-to-Case Transfer**: Ethnographers can facilitate the transfer of findings from one case to another by providing sufficient detail and context, allowing readers or other researchers to make connections between different yet comparable situations (Stake, 1995).

While traditional notions of generalizability might not align with ethnographic aims, ethnographers do contribute to broader understandings through theoretical insights, rich descriptions, and methodological adaptations. Recognizing the value of depth over breadth, ethnography offers insights that, while rooted in specificity, can resonate across varied contexts and enrich our understanding of the human experience.

3. **Time and Resource Intensive**: Conducting comprehensive ethnographies requires a significant time commitment and may strain resources (Hammersley & Atkinson, 2007).

Ethnography's intricate methodology, which often involves prolonged immersion in the field, extensive data collection, and detailed data analysis, makes it one of the most time and resource-intensive research methods. While the depth and nuance achieved by ethnographic studies are unparalleled, this commitment often draws criticisms and poses practical challenges for researchers and stakeholders.

The Nature of Ethnography's Time and Resource Intensiveness

Ethnography often involves lengthy fieldwork periods, sometimes extending over months or even years. The ethnographer engages deeply with the studied community, observing, participating, and conducting interviews. Post-fieldwork, data transcription, coding, analysis, and writing can also be extensive (Atkinson & Hammersley, 1994).

Critiques Pertaining to the Time and Resource Intensiveness

a) **Prolonged Fieldwork Demands**: Extensive fieldwork might not always be feasible due to personal, financial, or institutional constraints. Academics, especially early-career researchers, might find it challenging to dedicate long durations to single projects (Wolcott, 2001).
b) **Financial Implications**: Funding for long-term ethnographic projects can be hard to secure, particularly when outcomes are unpredictable, and research questions might evolve in the field (Nader, 1972).
c) **Shifting Contexts**: In fast-changing settings, findings might be outdated by the time they are published, questioning the utility of such prolonged studies (Rabinow, 1977).
d) **Scope Limitations**: Given the time-intensive nature, ethnographers often focus on one specific group, community, or setting at a time, potentially limiting the breadth of their studies (Bernard, 2006).

Challenges Posed by Time and Resource Intensiveness

i. **Personal Strain**: The immersive nature of ethnography can have emotional and psychological implications for the researcher, as they often form deep bonds in the field and navigate various cultural, ethical, and personal challenges (Coffey, 1999).
ii. **Analytical Overload**: Ethnographers typically collect vast amounts of data, which can be overwhelming and challenging to sift through during analysis (Emerson, Fretz, & Shaw, 1995).
iii. **Logistical Challenges**: Extended fieldwork requires careful logistical planning, including securing appropriate accommodations, managing local permissions, and addressing health and safety concerns (Sanjek, 1990).

Navigating the Challenges

I. **Focused Ethnographies**: Some researchers adopt 'focused ethnographies', which are shorter-term and centered around specific issues, events, or questions while maintaining the core principles of ethnography (Knoblauch, 2005).
II. **Collaborative Approaches**: By collaborating with other researchers or local partners, ethnographers can share responsibilities, ensuring efficient data collection and analysis (Lassiter, 2005).
III. **Digital Tools**: The use of digital tools, like software for data coding and analysis, can streamline certain stages of the ethnographic process, saving time and resources (Fielding & Lee, 1991).

While the time and resource intensiveness of ethnography can be daunting, many argue that the unparalleled depth, nuance, and richness it offers justify the investment. As the field evolves, researchers continue to find innovative ways to navigate these challenges, ensuring ethnography's continued relevance and vitality.

CONCLUSION

Ethnography, with its emphasis on capturing the richness of human cultures, remains a potent tool in qualitative research. While it presents methodological challenges, its strengths in revealing deep insights and nuanced understandings of human behaviors and societies are unparalleled.

Narrative Research

Narrative research stands as a pillar within qualitative inquiry, emphasizing the exploration and understanding of human experience through stories. Narratives, whether oral, written, or visual, have been central to human culture, aiding our understanding of the world, and our place within it. This chapter delves deep into the theoretical and methodological facets of narrative research.

Narrative research, emerging as a pivotal approach in the qualitative domain, emphasizes the deep-rooted human proclivity for storytelling. Stories are not just a means to convey events but are tools to express experiences, make sense of our surroundings, and manifest our personal and cultural identities. Such is the power of narrative that its echoes can be traced back to ancient civilizations, where oral traditions and mythologies shaped societies, cultures, and even entire civilizations. This chapter seeks to illuminate the theoretical landscapes of narrative research, from its inception and evolution to its current-day manifestations and challenges, as scholars recognize the profound potential narratives hold in illuminating the complexities of the human experience (Bruner, 1991).

Evolution of Narrative Research

From Homer's epics to contemporary biographies, the narrative has been a vehicle for humans to encapsulate experiences, values, beliefs, and emotions. With the sociocultural turn in the humanities and social sciences, narratives evolved from mere accounts or chronicles to intricate webs of meaning and representation. As postmodernism and constructivist paradigms took root in academia, the emphasis shifted from objective realities to subjective experiences, thereby catapulting narratives to a preeminent research tool (Lyotard, 1984; Polkinghorne, 1988).

Narrative research has undergone significant transformation over millennia. Its progression traces not only the maturation of narrative as a tool of expression but also its emerging prominence as a method of inquiry in the academic realm. By understanding its evolutionary trajectory, we can appreciate the richness of the narrative methodology and its vast potential for multidisciplinary explorations.

Oral Traditions: The Genesis of Narratives

At the inception of human civilization, narratives were orally exchanged. Every community had its storytellers, who orally passed down legends, myths, and historical accounts from one generation to the next. These oral narratives served as the primary repository of collective wisdom, traditions, and cultural memories (Ong, 1982). Through these stories, societies could make sense of their environment, understand the cosmos, and pass down essential life lessons.

The Written Word: A Shift Towards Permanence

With the advent of writing, narratives underwent a transformational shift. What were once fluid and evolving oral tales now found a more permanent form in written texts. Ancient civilizations – from the cuneiform inscriptions of the Sumerians to the hieroglyphics of the Egyptians – began to record their narratives, conferring them with a sense of authority and longevity (Goody & Watt, 1963).

Literature and the Art of Storytelling

As societies advanced, so did their narratives. The literary realm blossomed with epics, poems, dramas, and prose that explored a myriad of human experiences. Works like Homer's *Iliad* and the Indian epic *Mahabharata* showcased the profundity of narratives, which could encapsulate everything from grand battles to intricate human emotions. These literary masterpieces, beyond their artistic merit, provided invaluable insights into the cultures and epochs they emerged from (Frye, 1957).

Narratives in the Social Sciences

The 19th and 20th centuries saw a growing interest in narratives within the domain of social sciences. Anthropologists began collecting folk tales, myths, and oral histories from communities they studied, recognizing the immense cultural and sociological value these narratives held (Malinowski, 1922). Similarly, sociologists tapped into personal life stories, understanding that individual narratives could shed light on broader societal patterns and shifts (Mills, 1959).

Postmodern Turn and the Rise of Narrative Inquiry

The postmodern era, with its emphasis on subjective realities and skepticism towards grand narratives, played a pivotal role in the ascent of narrative research as a distinct qualitative methodology. Researchers in fields ranging from psychology to education began to employ narratives as both data and method, acknowledging that personal stories could provide nuanced, in-depth insights unattainable through quantitative measures (Lyotard, 1984; Bruner, 1991).

Digital Era: New Frontiers in Narrative Research

The dawn of the digital age brought forth novel narrative forms. Blogs, social media posts, digital storytelling, and online memoirs expanded the narrative repertoire. For researchers, these digital narratives presented both opportunities and challenges. While they enabled access to diverse, global narratives, they also introduced questions about authenticity, ephemerality, and the digital divide (Jenkins, 2006).

From its origins in oral tales to its contemporary digital avatars, narrative research has continually evolved, mirroring the societal, technological, and academic changes it has traversed. As we embark on future explorations, it remains imperative to appreciate the rich tapestry of this evolution, as it informs our understanding and practice of narrative research.

Narratives: Windows to Experience

Narrative research isn't about sheer data collection but venturing into the heart of lived experiences. Every narrative encapsulates layers of culture, identity, emotion, and cognition. They are accounts steeped in context, providing deep insights into individuals' worlds and the sociocultural matrix they navigate (Riessman, 2008).

Within the vast academic panorama, narrative research stands as a poignant reminder of humanity's age-old affinity for storytelling. Through narratives, qualitative researchers seek not just to inform but to transform, understanding that behind every story lies an intricate tapestry of emotions, socio-cultural contexts, and introspections. Delving deeper into the idea of narratives as "windows to experience" elucidates why this methodological approach resonates deeply within the academic and broader societal landscape (Riessman, 2008).

The Epistemological Foundations of Narrative Research

Narratives serve as foundational epistemological tools. They represent the complex interplay between objective events and subjective interpretations. Every story, while rooted in real experiences, is mediated through personal cognition, emotions, and cultural filters. This interplay provides researchers with rich, nuanced insights that transcend mere factual recounting (Bruner, 1986).

Situating Narratives within Socio-Cultural Paradigms

Every narrative, while personal, is embedded within a larger socio-cultural fabric. For narrative researchers, stories offer gateways into understanding societal norms, collective memories, and shared values. They act as microcosms, encapsulating the broader cultural, historical, and socio-political contexts from which they emanate (Geertz, 1973).

Temporal Dimensions and the Fluidity of Narratives

Narratives are intrinsically temporal, capturing the flux of time. They trace personal and collective evolutions, chronicling transformations, continuities, and ruptures. For a narrative researcher, understanding this temporal dimension is crucial as it reveals the dynamics of change and stability within personal and collective histories (Freeman, 2010).

Emotions, Cognition, and the Human Psyche

At the heart of narrative research is an acknowledgment of the profound emotional and cognitive layers underpinning stories. Narratives are reservoirs of emotions - they evoke, express, and encapsulate a spectrum of feelings. Simultaneously, they are cognitive endeavors, shaped by memory, perception, and interpretation. This duality provides researchers with a comprehensive view of the human psyche in action (McAdams, 1993).

Constructing and Deconstructing Narratives

Narrative research is an act of co-construction. While narrators construct their stories, researchers engage in the act of deconstructing and subsequently reconstructing these narratives to extract meanings, patterns, and themes. This iterative process emphasizes the active role both narrators and researchers play in shaping the narrative discourse (Mishler, 1995).

Narrative research, with its focus on stories as "windows to experience," offers a profound methodological approach that privileges human experiences in all their complexity. By honoring, studying, and interpreting narratives, researchers contribute to a deeper, more holistic understanding of the human condition, reaffirming the age-old adage: "In stories, we find ourselves."

Multidisciplinary Foundations

Though often associated with anthropology or sociology, narrative research's tentacles spread wide, drawing from diverse disciplines. Literature offers the richness of narratives in textual form, while psychology delves into how narratives shape cognition and identity. History uses narratives to construct timelines, and even fields like medicine employ narratives for patient histories, realizing that illnesses are not mere physiological events but experiences imbued with emotions, beliefs, and cultural nuances (Bamberg, 2012; Charon, 2006).

Narrative research, while rooted in the qualitative paradigm, stands not as an isolated island but rather as a converging point for numerous disciplines. The beauty of narrative inquiry lies in its ubiquity; stories are universal, traversing the boundaries of academic silos. By probing into the multidisciplinary foundations of narrative research, we can truly appreciate its breadth and depth, acknowledging the multifaceted perspectives that shape its core (Riessman, 2008).

Anthropology: Narratives as Cultural Portraits

Anthropology, with its focus on understanding human societies and cultures, has long recognized the potency of narratives. Ethnographers collect stories, myths, and oral histories to paint vivid portraits of communities. These narratives offer insights into rituals, societal structures, belief systems, and collective memories, capturing the essence of cultural identity (Geertz, 1973).

Psychology: Narratives and the Self

In the realm of psychology, narratives are seen as crucial constructs in shaping identity. Life stories, personal myths, and autobiographical memories serve as scaffolding upon which the self is built. By exploring these narratives, psychologists delve into the intricacies of personality, memory, and cognitive development, illuminating the intertwining threads of experience and identity (McAdams, 1993).

Sociology: Narratives and Collective Consciousness

Sociologists employ narratives to understand societal dynamics, collective memories, and group identities. Personal stories, while individualistic, often mirror broader societal narratives, reflecting common themes, shared experiences, and collective aspirations. Analyzing these stories allows sociologists to comprehend societal transformations, power structures, and group dynamics (Somers, 1994).

Literary Studies: The Art and Craft of Narratives

Literary studies provide a unique lens to narrative research. Here, narratives are both form and content. Literary scholars dissect narrative structures, styles, and motifs, uncovering layers of meaning and artistic craftsmanship. This discipline brings forth the nuances of narrative techniques, offering rich insights into the art of storytelling (Bal, 2009).

History: Narratives as Chronicles of Time

Historians, while often dealing with 'grand narratives' of civilizations and epochs, also recognize the value of personal narratives. Diaries, memoirs, and oral histories offer glimpses into individual lives, detailing lived experiences within broader historical landscapes. These personal chronicles, when pieced together, form a mosaic of historical epochs, providing depth to historical inquiry (White, 1981).

Education: Narratives in Pedagogy and Learning

Within educational settings, narratives play a dual role. They are tools for pedagogy, enabling educators to weave stories into lessons, making learning relatable and engaging. Simultaneously, educators employ narrative inquiry to understand students' experiences, learning journeys, and challenges, thus tailoring educational practices to cater to diverse needs (Connelly & Clandinin, 1990).

Narrative research, with its multidisciplinary foundations, stands as a testament to the universal appeal of stories. From anthropology to education, stories permeate every discipline, acting as windows to human experiences. By acknowledging and integrating these multidisciplinary perspectives, narrative researchers enrich their inquiries, ensuring a holistic, nuanced, and comprehensive exploration of the narrative realm.

Constructing and Deconstructing Narratives

As researchers engage with narratives, the task isn't just to listen or read but to decipher, interpret, and contextualize. Every story, verbalized or written, undergoes a process of construction by the narrator. In parallel, the researcher deconstructs and reconstructs these stories, finding patterns, themes, and underlying meanings. This dual construction-deconstruction continuum forms the crux of narrative analysis (Labov, 1972).

The art of narrative research is as much about the act of listening and gathering as it is about the intricate process of constructing and deconstructing narratives. It's a dance between the tangible, voiced stories and the interpretative, analytical efforts of the researcher. Deeply entwined in this process is the understanding that narratives are not fixed entities but fluid constructs, shaped by myriad forces, and ripe for exploration and reinterpretation (Riessman, 2008).

Constructing Narratives: The Art of Eliciting Stories

The narrative construction begins long before a story is penned or voiced. It starts with the lived experiences of individuals, their interpretations of these experiences, and the contexts within which they unfold. Researchers play an active role in this construction, carefully crafting their questions, fostering trust, and creating an environment conducive to honest, reflective storytelling (Clandinin & Connelly, 2000).

Roles, Relationships, and Reflexivity

The relationship between the researcher and the narrator is pivotal in narrative construction. This relationship, influenced by power dynamics, trust, and shared understanding, determines the depth and authenticity of the stories shared. Researchers must remain reflexive, constantly interrogating their roles, biases, and influences on the narrative being constructed (Etherington, 2004).

Narrative Structures and Modalities

Once voiced or written, narratives take on specific structures and modalities. These could range from linear chronicles to fragmented recollections, from prose to poetry. The choice of structure often reflects the narrator's intent, cultural background, and the nature of the experience being relayed (Labov & Waletzky, 1967).

Deconstructing Narratives: Analysis and Interpretation

The process of deconstruction begins when researchers immerse themselves in the narratives, seeking patterns, anomalies, and underlying themes. This phase is not merely analytical but interpretative, demanding a delicate balance between staying true to the narrator's voice and introducing scholarly interpretations (Mishler, 1995).

Tools and Techniques of Deconstruction

Several methodologies guide the deconstruction of narratives. Thematic analysis might be employed to discern recurring themes, structural analysis could shed light on the narrative's form, and dialogical analysis might probe the multiple voices within a single narrative. The choice of technique often aligns with the research question and the nature of the narratives (Riessman, 1993).

Narratives Within Larger Narratives

Deconstruction also involves situating individual stories within broader social, cultural, or historical narratives. This macro-micro interplay offers insights into how personal stories are influenced by—and in turn, influence—collective narratives (Andrews, Squire, & Tamboukou, 2013).

The dual processes of constructing and deconstructing narratives in research are neither linear nor formulaic. They represent a dynamic interplay of methodological rigor, interpersonal relationships, and interpretative finesse. By appreciating the complexity of these processes, researchers can harness the true potential of narrative inquiry, unveiling the richness of human experiences encapsulated in stories.

Methodological Challenges and Innovations

Narrative research, for all its profundity, isn't without challenges. The very subjectivity that adds richness also raises concerns about validity, generalizability, and bias. Moreover, narratives can be both illuminative and deceptive, requiring researchers to tread with caution and reflexivity. Yet, innovations abound. Digital narratives, multimedia storytelling, and collaborative narrative methods are reshaping the narrative research landscape, promising richer, more diverse, and interactive avenues of exploration (De Fina & Georgakopoulou, 2012).

Narrative research, while providing a rich tapestry of human experiences and profound insights, is not devoid of challenges. The methodology invites complex dialogues, weaving together the voices of narrators, researchers, and larger socio-cultural discourses. However, with these challenges have come a myriad of innovative solutions, adding depth and dynamism to the field. Let's delve into the multifaceted challenges and the consequent innovations in narrative research (Czarniawska, 2004).

Methodological Challenges in Narrative Research

a) **Authenticity and Representation**: One of the primary challenges is ensuring the authenticity of narratives. How does one ensure that the stories told are genuine reflections and not shaped by external influences or researcher bias? How can a researcher avoid over-interpretation or misrepresentation? (Mishler, 1986)
b) **Positionality and Reflexivity**: The researcher's position—shaped by their background, beliefs, and biases—can influence the collection, interpretation, and representation of narratives. Maintaining reflexivity, or constant self-awareness, becomes imperative (Etherington, 2004).
c) **Temporal and Contextual Fluidity**: Narratives, being deeply personal and temporal, can change over time or differ based on context. Capturing a narrative at one point might yield a different story than at another time (Rosenwald & Ochberg, 1992).
d) **Ethical Concerns**: Issues of confidentiality, consent, and potential harm arise, especially when dealing with sensitive or traumatic narratives. Protecting participants while staying true to their stories poses a challenge (Josselson, 2007).

Innovations Addressing These Challenges

I. **Collaborative Storytelling**: Instead of merely collecting stories, some researchers actively collaborate with participants. This co-construction approach ensures that narratives are jointly reviewed, interpreted, and represented, increasing authenticity (Gergen & Gergen, 2002).
II. **Digital Storytelling**: Leveraging multimedia—videos, audio recordings, images—allows for a richer, more holistic representation of narratives. Digital tools offer participants alternative, often more expressive, ways to share their stories (Lambert, 2013).
III. **Iterative Interviews**: To capture the fluidity of narratives, researchers may opt for iterative interviews, speaking to participants multiple times across different contexts and time frames. This method helps in capturing the evolving nature of stories (Clandinin & Connelly, 2000).

IV. **Ethical Frameworks**: Specialized ethical frameworks for narrative research have been developed, offering guidelines on obtaining informed consent, ensuring confidentiality, and navigating sensitive topics with care and empathy (Ellis, 2007).
V. **Meta-narrative Analysis**: Recognizing that narratives are embedded within larger socio-cultural and historical discourses, some researchers have employed meta-narrative analysis. This method scrutinizes the overarching narratives that shape and are shaped by individual stories (Greenhalgh, Robert, Macfarlane, Bate, & Kyriakidou, 2004).

The challenges posed by narrative research, while intricate, have spurred methodological innovations that enhance the robustness and richness of the approach. By acknowledging, navigating, and innovating around these challenges, researchers ensure that narrative research remains a powerful, authentic, and ethically grounded methodology, capturing the essence of human experiences.

Narrative research, at its core, is about honoring the human experience. As we navigate an increasingly complex world, stories remain our compass, guiding us through challenges, triumphs, ambiguities, and transformations. This chapter sets the stage, beckoning readers to delve deeper into the theoretical tapestries, methodological intricacies, and transformative potentials of narrative research.

Origins of Narrative Research

Narrative inquiry traces its roots to multiple disciplines. Historians, literary scholars, and anthropologists have long recognized the power of stories in encoding and transmitting cultural knowledge, values, and beliefs (Bruner, 1986). With the rise of postmodernism and critiques of positivism in the 20th century, social scientists increasingly turned to narratives as vehicles to understand individual and collective experiences (Lyotard, 1984).

Since time immemorial, humans have been storytellers. From the fireside tales of ancient tribes to sophisticated narratives in postmodern literature, stories have acted as vessels that carry the essence of humanity's existence, experiences, and epistemologies. Delving into the origins of narrative research is akin to tracing the evolution of mankind's engagement with narrative as a tool for understanding, representation, and communication. This exploration not only unravels the historical tapestry of narrative traditions but also elucidates the theoretical foundations that underpin contemporary narrative research.

Ancient Beginnings: The Primacy of Oral Tradition

In preliterate societies, narratives were primarily oral. Stories, myths, and legends passed down from generation to generation served multiple purposes: preserving history, imparting moral lessons, explaining natural phenomena, and cementing community bonds (Ong, 1982). These narratives were fluid, evolving with each retelling, reflecting both the core values of a culture and the individual nuances of the storyteller.

Written Narratives: The Shift to Permanence

The invention of writing transformed narratives from ephemeral oral tales to enduring written records. Ancient civilizations, like the Mesopotamians, Egyptians, and Greeks, began to document stories, religious texts, and historical events, moving narratives from the mutable domain of memory to the static realm of stone, clay, and parchment (Goody, 1987).

Religious Texts and Moral Narratives

Many of the world's foundational religious texts, be it the *Bhagavad Gita*, the *Bible*, or the *Quran*, are imbued with narratives. These stories, rich in metaphor and allegory, were not only religious doctrines but also intricate tales that explored human nature, ethics, and existential queries (Armstrong, 2005).

Narrative as Art: The Rise of Literature

Narratives began to be recognized not just as records or moral tools but as art forms. Epics like Homer's *Iliad* and *Odyssey*, the Mahabharata, and Beowulf epitomize this transition, where storytelling was elevated to an art, and the narrative became a mirror reflecting societal values, conflicts, and transformations (Frye, 1957).

The Philosophical Turn: Narratives and Identity

The Enlightenment era heralded a new relationship between individuals and narratives. Philosophers like Rousseau and Voltaire employed narrative both as a means of self-exploration and as a tool to critique society. This era underscored the belief that our understanding of the self is inherently tied to narratives, a sentiment later echoed by existentialists like Sartre and Camus (Ricoeur, 1991).

Narrative in Social Sciences: An Evolving Tool

By the 19th and early 20th century, narratives were being employed as tools in anthropology and sociology. Early anthropologists like Malinowski and Boas gathered narratives from indigenous communities, recognizing their potential in offering cultural insights (Geertz, 1988). Meanwhile, sociologists highlighted the role of narratives in shaping collective identities and memories (Halbwachs, 1950).

Postmodernism and the Rise of Narrative Inquiry

The postmodern era, with its emphasis on relativism and skepticism of grand narratives, elevated narrative research to a distinct methodological approach. Researchers began to understand that stories were not just reflective of realities but were instrumental in constructing them. Narratives were now seen as lenses through which individuals and communities interpret, negotiate, and give meaning to their experiences (Lyotard, 1979; Bruner, 1991).

The Contemporary Canvas: Digital Narratives

The digital revolution has once again transformed the landscape of narrative. The internet birthed new narrative forms - blogs, vlogs, podcasts, and social media stories. These digital narratives, democratized and decentralized, are reshaping how we engage with stories and, in turn, how narrative research is conducted in the contemporary world (Jenkins, 2006).

Tracing the origins of narrative research is a journey across epochs, cultures, disciplines, and mediums. It offers a testament to the enduring and evolving relationship between humans and stories. As we stand on the cusp of a new era, with artificial intelligence and virtual realities, one wonders how narratives will morph and what new avenues they will open for researchers. Yet, one truth remains – our intrinsic connection with stories, as tellers, listeners, and scholars.

Narrative as Constructed Reality

Central to narrative research is the idea that reality is constructed and represented through stories (Riessman, 2008). Narratives don't just reflect experiences; they shape and organize them. In telling stories, individuals give meaning to their experiences, make sense of events, and navigate their identities (Polkinghorne, 1991).

Narrative research, as a qualitative endeavor, moves beyond the surface level of mere data collection. It delves into understanding narratives as not just representations but as active constructions of reality. This constructivist viewpoint acknowledges that narratives, while based on lived experiences, are shaped, molded, and constructed through a myriad of internal and external factors. Analyzing narratives as constructed realities provides an enriched, multi-dimensional understanding of human experiences and the ways they are articulated and comprehended (Bruner, 1991).

Narrative Construction: Between Fact and Meaning

While narratives are rooted in real events and experiences, they do not serve as mere factual recountings. Instead, they weave together objective occurrences with subjective interpretations, molding them into coherent, meaningful wholes. This synthesis, influenced by individual cognition, emotions, cultural contexts, and historical backgrounds, results in narratives that are more than the sum of their parts (Polkinghorne, 1988).

Narrative research has always stood at the crossroads where objective events meet subjective interpretations. This juncture is where the raw material of lived experience transforms into the crafted edifice of story. The act of narrating is, therefore, not merely recounting; it involves constructing, interpreting, and imbuing events with meaning. The transformation of facts into narratives becomes an arena where personal, cultural, and historical contexts converge and intertwine (Bruner, 1991).

The Fluidity of Fact in Narratives

At a fundamental level, every narrative has its roots in actual events or facts. However, the mere occurrence of an event does not a narrative make. Narratives involve selecting from a multitude of events, arranging them, giving them context, and embedding them in a flow of meaning. In this process, the starkness of fact undergoes modification, interpretation, and sometimes even omission, tailoring it to fit the emerging narrative (Polkinghorne, 1988).

The Interpretative Essence of Narration

To narrate is inherently to interpret. Each event, when recounted, carries the weight of the narrator's perceptions, emotions, biases, and desires. These internal factors play pivotal roles in shaping how facts are presented, prioritized, and perceived. The same event, when narrated by different individuals or even by the same individual at different times, can yield diverse narratives, each colored by its unique interpretative hue (Ricoeur, 1984).

Cultural and Historical Lenses

Narrative construction does not occur in a vacuum. It is deeply influenced by the cultural and historical milieu in which the narrator is embedded. Cultural norms, societal values, and historical events provide frameworks that guide the selection, emphasis, and interpretation of facts. These external factors often act as lenses, magnifying certain events, blurring others, and sometimes even introducing distortions (Bamberg, 2004).

Meaning-Making and the Quest for Coherence

Central to narrative construction is the human need for coherence and meaning. Faced with the chaos of myriad events, individuals seek patterns, causal relationships, and overarching themes. This drive for coherence often leads to the crafting of narratives that not only recount but also make sense, provide explanations, and offer resolutions. In this quest, facts might be molded, reinterpreted, or even reshaped to fit the emerging coherent whole (McAdams, 2001).

In narrative research, appreciating the intricate dance between fact and meaning is crucial. Recognizing that narratives, while rooted in real events, are shaped by a complex interplay of internal interpretations and external influences allows for a more nuanced, empathetic, and holistic engagement with the stories being told. This recognition also underscores the profound power of narratives: to transform the raw material of life into crafted tales that resonate, reflect, and reveal the depths of human experience.

Socio-Cultural Influences on Narrative Construction

Narratives are deeply embedded within socio-cultural frameworks. The stories individuals tell and the ways they tell them are shaped by societal norms, values, and collective histories. Cultural idioms, societal expectations, and shared belief systems can guide narrative structures, themes, and motifs, influencing how personal experiences are constructed into narratives (Gergen & Gergen, 1986).

The narratives we craft and share are not self-contained entities but emerge from, and are deeply embedded within, broader socio-cultural landscapes. These narratives, while personal in nature, often reflect, resist, or negotiate the larger cultural narratives and social norms that envelop them. Engaging with narrative research necessitates an appreciation of these socio-cultural influences that act as scaffolding, shaping the ways stories are constructed, told, and received (Bourdieu, 1990).

Cultural Narratives as Backdrops

Every culture carries with it dominant narratives, stories that encapsulate shared values, beliefs, and histories. These cultural narratives, often passed down through generations, provide templates against which individual stories are crafted. Personal narratives, therefore, are frequently influenced by these broader tales, either aligning with them or offering counter-narratives that challenge prevailing norms (Geertz, 1973).

Social Norms and Taboos

Societal norms and taboos exert significant influence on narrative construction. The stories individuals choose to tell, the ways they frame them, and the aspects they emphasize or downplay often reflect these norms. For instance, certain topics might be silenced, while others are amplified, based on what is deemed acceptable or taboo within a given society (Foucault, 1977).

The Role of Language and Discourse

Language, an essential tool for narrative construction, is culturally specific and carries with it inherent meanings and structures that influence storytelling. Furthermore, dominant discourses, or ways of talking about and understanding phenomena, can guide the shaping of narratives. These discourses, deeply embedded in culture, act as molds, guiding or even constraining the ways stories are formed and understood (Fairclough, 1992).

Socio-Economic and Power Dynamics

Narratives are also influenced by the socio-economic contexts and power dynamics within which they emerge. Stories from marginalized communities might reflect experiences of oppression, resistance, or resilience, while narratives from dominant groups might reinforce prevailing power structures or offer insights into the privileges they enjoy. The socio-economic landscape, thus, deeply influences the fabric of personal stories (Lorde, 1984).

Historical Context and Collective Memory

The historical backdrop of a culture or society shapes its collective memory and, in turn, influences personal narratives. Events of historical significance, shared traumas, or collective achievements can find echoes in individual stories, making narratives not just personal chronicles but also reflections of larger historical narratives (Halbwachs, 1992).

In narrative research, the interplay between personal stories and socio-cultural influences is profound. Recognizing the powerful role of culture, social norms, language, socio-economic structures, and history in shaping narratives allows researchers to appreciate the multilayered tapestry of stories. This deep understanding facilitates a more nuanced engagement with narratives, celebrating them as intersections of personal experiences and societal forces.

The Temporality of Narrative Construction

Narratives, by their nature, are bound up with time. However, the temporality of narrative construction is far from linear; it is a multidimensional phenomenon that encompasses past experiences, present contexts, and future aspirations. In narrative research, understanding the intricate ways in which time interacts with stories provides insights into the complexities of human memory, the fluidity of experience, and the evolving nature of personal identities (Ricoeur, 1984).

Narrative Time vs. Chronological Time

While chronological time is linear, moving from the past to the present and into the future, narrative time is malleable. Narrators can shift between timeframes, reorganize events, and blend memories with aspirations. This temporal flexibility allows for the creation of stories that are not bound by the strictures of chronological sequence but by the needs for meaning and coherence (Bruner, 1991).

Memory and Its Influence on Narrative Temporality

Memory plays a pivotal role in narrative construction. However, memories are not static imprints; they are dynamic entities, susceptible to changes, reinterpretations, and even erasures over time. As narrators recollect and recount their pasts, their present contexts, emotions, and knowledge can shape and color these memories, leading to narratives that evolve and change with time (Halbwachs, 1992).

Projection and Anticipation: The Future in Narratives

Narratives are not confined to recounting the past; they also encompass future aspirations, hopes, and fears. This forward-looking aspect of narratives, rooted in projection and anticipation, can significantly shape the construction of stories. The future can provide a lens through which the past is viewed, interpreted, or even reimagined (McAdams, 2001).

Temporal Coherence: Making Sense Across Time

The need for coherence drives the temporal construction of narratives. Narrators seek to create stories that, while spanning different timeframes, remain consistent, meaningful, and unified. This desire for temporal coherence can lead to the selective inclusion of events, the reinterpretation of memories, or the crafting of narratives that align past experiences with present identities and future aspirations (Polkinghorne, 1988).

Cultural and Societal Influences on Narrative Temporality

Cultural norms, societal expectations, and collective histories can shape the temporality of narratives. In some cultures, cyclical understandings of time, ancestral stories, or communal memories might deeply influence personal narratives. These external temporal frameworks can guide, constrain, or enrich the ways in which individual stories are constructed and understood (Zerubavel, 1997).

In narrative research, engaging with the temporality of narrative construction is crucial. Recognizing the nuanced interplay between chronological time, memory, anticipation, and cultural temporalities allows for a more profound understanding of narratives. This appreciation underscores the rich tapestry of stories, revealing them as intricate constructions that span, yet transcend, time.

Narrative Agency and Selfhood

Individuals exercise agency in constructing their narratives. They choose which events to highlight, which to downplay, and which to omit. This narrative agency is intertwined with notions of selfhood and identity. The act of narrative construction becomes an act of self-construction, wherein individuals craft versions of themselves, portraying identities they align with or aspire to (McAdams, 2001).

Narrative research recognizes that individuals do not passively recount their lives but actively construct narratives that hold significant implications for their sense of self. The term 'narrative agency' underscores this proactive role of individuals in shaping their stories, emphasizing the intertwined nature of narrative construction and self-construction. Engaging deeply with this aspect provides insights into how identities are formed, maintained, negotiated, and transformed through storytelling (McAdams, 2001).

Narrative Choices: Crafting a Coherent Self

Every act of narration involves a multitude of choices. Individuals decide which events to highlight, which details to emphasize, and which episodes to omit or downplay. These choices are not arbitrary. They reflect an individual's quest for coherence, the need to weave disparate events into a cohesive, meaningful narrative that aligns with their sense of self (Bruner, 1991).

Stories as Mirrors and Molds of Identity

Narratives act as mirrors, reflecting how individuals perceive themselves, their roles, and their places in the world. Concurrently, these stories also serve as molds, actively shaping and solidifying identities. The stories one tells, the heroes and villains one identifies, the challenges faced, and the resolutions achieved—all contribute to an evolving sense of selfhood (Ricoeur, 1992).

The Socio-Cultural Dimensions of Narrative Agency

Narrative agency is not exercised in a vacuum. The stories individuals craft are deeply influenced by socio-cultural contexts. Cultural norms, societal expectations, and communal narratives can either enable or constrain narrative choices, guiding the ways in which identities are articulated and understood. This interplay between personal agency and societal forces adds layers of complexity to the narrative construction of selfhood (Bourdieu, 1990).

Narrative Resistance and Reclamation

In many instances, individuals might find themselves confronting dominant narratives that marginalize, stereotype, or negate their experiences. In such contexts, the act of storytelling becomes one of resistance and reclamation. Crafting counter-narratives, challenging prevailing stories, or reinterpreting events allows individuals to reclaim their narrative agency and assert their identities (Bamberg & Andrews, 2004).

Narrative Fluidity and the Evolving Self

Just as narratives are malleable, so too are identities. As individuals navigate different life stages, face new challenges, or encounter diverse experiences, their narratives evolve. This fluidity of storytelling, driven by changing understandings, emotions, and aspirations, reflects the dynamic nature of selfhood, emphasizing identity as a work in progress (Gergen, 2001).

Within narrative research, delving into the intricate relationship between narrative agency and selfhood reveals the profound power of stories. Recognizing narratives as both mirrors and molds of identity illuminates the multifaceted processes through which individuals understand themselves, negotiate their places in the world, and chart their paths forward.

Audience and Narrative Co-Construction

The presence of an audience, be it a researcher, a community, or society at large, plays a role in narrative construction. Stories are often told with an audience in mind, and their reactions, anticipations, or expectations can influence the narrative's shape. This makes narrative construction a collaborative endeavor, a dialogue between the storyteller and the listener (Bamberg, 2004).

The act of narrating is inherently relational. While narratives are personal constructions, they are often crafted with an audience in mind, whether real or imagined. This audience, be it a single listener, a community, or a broader societal collective, plays a significant role in shaping the narrative. Recognizing this dyadic relationship underscores the co-constructive nature of storytelling, emphasizing narratives as dynamic interactions between the teller and the receiver (Bakhtin, 1981).

The Audience as Reflective Mirror

For narrators, the audience often acts as a reflective surface, mirroring back reactions, interpretations, and emotions. This reflective feedback can shape the storytelling process, influencing the choices of details, pacing, and emphasis. The narrator, anticipating or gauging the audience's reactions, adjusts the narrative in a bid to resonate, inform, persuade, or elicit empathy (Ricoeur, 1984).

Dialogical Nature of Narration

Drawing from Bakhtin's notion of dialogism, narration can be understood as a dialogue, not just a monologue. The audience's presence, even if silent, introduces a dialogical dimension. Their queries, affirmations, challenges, or mere attentive listening become part of the narrative fabric, co-constructing the story in tandem with the narrator (Bakhtin, 1981).

Cultural and Societal Narratives as Collective Audiences

Beyond immediate listeners, narrators are also influenced by broader societal and cultural narratives. These collective narratives, encapsulating shared beliefs, norms, and values, act as implicit audiences, guiding or constraining the ways individual stories are crafted. Conforming to or challenging these overarching narratives becomes an integral part of the storytelling endeavor (Anderson, 2006).

Narrative Co-construction in Research Settings

In narrative research, the researcher is a key audience, actively participating in the co-construction of narratives. The questions posed, the reactions given, and the interpretations made by the researcher can guide, shape, or even redirect the narrator's storytelling process. This dynamic interplay emphasizes the importance of reflexivity in narrative research, acknowledging the researcher's role in the co-construction of narratives (Mishler, 1986).

The Power Dynamics of Audience Influence

It's essential to recognize that audiences, especially in collective forms, can exert power dynamics, either enabling or constraining narrative agency. Dominant groups, institutions, or societal norms can act as powerful audiences, whose anticipated reactions might lead narrators to silence, modify, or amplify certain parts of their narratives (Foucault, 1977).

Understanding the intertwined roles of the audience and the narrator in narrative research deepens our appreciation of storytelling's complexities. Recognizing narratives as co-constructed endeavors, shaped through dynamic interactions between tellers and listeners, illuminates the profound influences of reception, dialogue, and relationship in the act of narrating.

Viewing narratives as constructed realities is to recognize the intricate dance between fact and interpretation, between lived experiences and the act of storytelling. Narrative research, with this lens, can appreciate the richness, depth, and dynamism of stories, understanding them as active constructions that provide insights into both individual lives and broader human conditions.

Types of Narrative Research

1. **Biographical and Auto-Biographical Studies**: Focusing on life stories, these studies either prioritize the subject's account of their own life or are rendered by a researcher (Denzin, 1989).

Narrative research, in its diverse applications, offers rich opportunities to delve into the complexities of human experiences. Two particularly poignant approaches within this paradigm are biographical and auto-biographical studies. Both aim to chronicle life stories, but they differ in their focal points, methodologies, and interpretative nuances. These studies offer invaluable insights into individual lives, social contexts, historical moments, and the intricate art of life-storytelling (Atkinson, 1998).

Biographical Studies: Narrating Others' Lives

Biographical studies focus on understanding an individual's life through narratives, typically gathered from sources other than the individual in question.

a) Sources and Data Collection: Biographical research often relies on diverse sources such as interviews with close associates, archival materials, letters, and even media accounts. This multi-source approach enriches the understanding of the subject's life (Plummer, 2001).
b) Contextual Interpretation: Biographies aim to situate individual life stories within broader social, cultural, and historical contexts. This alignment offers insights into the societal structures, cultural norms, and historical events that influence, or are influenced by, individual lives (Roberts, 2002).
c) Challenges: Biographical research faces challenges of representation and interpretation. Ensuring that the narrative remains faithful to the subject's life, while also being interpretatively insightful, requires a delicate balance (Smith & Watson, 2010).

Auto-Biographical Studies: Self-Narratives

Auto-biographical studies involve individuals narrating their own life stories, offering deeply personal insights into their lived experiences.

i. First-Person Perspective: The strength of auto-biographical research lies in its first-person perspective. The direct voice of the individual brings authenticity, depth, and emotional resonance to the narrative (Stanley, 1992).
ii. Reflexivity: Auto-biographies demand a high degree of reflexivity. Narrators engage in introspection, examining their memories, emotions, and interpretations. For researchers, understanding this introspective process is as crucial as the narrative itself (Ellis & Bochner, 2000).
iii. Challenges: Auto-biographical narratives can sometimes blur the lines between fact and interpretation, memory and imagination. Navigating this ambiguity and understanding the fluidity of self-narratives is an inherent challenge (Bruner, 1997).

Intersections of the Biographical and Auto-Biographical

While distinct in their methodologies, biographical and auto-biographical studies often intersect. Biographical accounts might include auto-biographical elements, where subjects reflect on their own lives. Conversely, auto-biographies might be supplemented with biographical inputs from other sources to provide a more rounded portrait (Smith & Watson, 2010).

Biographical and auto-biographical studies, as types of narrative research, offer profound insights into the intricacies of human lives. Through them, we not only access individual stories but also understand broader socio-cultural landscapes, historical epochs, and the intricate processes of memory, identity, and self-construction.

2. **Oral Histories**: Grounded in spoken testimonies, oral histories capture personal experiences and reflect broader cultural and societal contexts (Portelli, 1997).

Oral histories, a specialized form of narrative research, provide rich opportunities to capture the voices of individuals and communities, weaving them into broader historical and socio-cultural tapestries. Rooted in the tradition of verbal storytelling, oral histories prioritize personal accounts, memories, and interpretations of events, shedding light on aspects of history that might otherwise remain undocumented or marginalized (Portelli, 1991).

Defining Oral Histories

Oral histories involve collecting testimonies, stories, and memories from individuals about their past, their experiences, and the communities or societies they lived in. These narratives are often recorded and preserved for future generations, offering firsthand accounts of historical periods, events, and everyday life (Ritchie, 2003).

Key Characteristics of Oral Histories

a) *Personal and Collective:* While oral histories are deeply personal, they also encompass collective experiences, bridging individual memories with communal or societal events (Thompson, 1988).
b) *Voice and Agency:* Oral histories prioritize the voices of the narrators, allowing them to shape, present, and interpret their own experiences. This fosters a sense of agency, especially for communities that have been historically marginalized or silenced (Grele, 1985).
c) *Contextual Interpretation:* Oral histories recognize the importance of context. Narrators' accounts are situated within their socio-cultural, economic, and political milieus, offering layered and textured interpretations of events (Portelli, 1991).

Methodological Considerations

I. *Interview Techniques:* Successful oral history projects rely on skillful interview techniques. Open-ended questions, active listening, and a genuine interest in the narrator's story are paramount (Ritchie, 2003).
II. *Ethical Considerations:* Given the personal nature of oral histories, ethical considerations are paramount. Consent, confidentiality, and respect for the narrator's rights and sensitivities are foundational principles (Yow, 2005).
III. *Preservation and Access:* Oral histories often aim to create archives for future generations. Hence, considerations regarding recording quality, transcription, storage, and public access become significant (Frisch, 1990).

Applications and Significance

i. *Documenting Marginalized Voices:* Oral histories have been instrumental in capturing the experiences of marginalized communities, offering insights into perspectives that might be absent from official historical records (Bornat, 1989).
ii. *Complementing Written Records:* Oral histories complement existing written or documentary evidence, providing richer, more nuanced accounts of historical periods and events (Perks & Thomson, 2006).
iii. *Education and Public History:* Oral histories have been utilized in educational settings, museums, and public history projects, helping to bridge the gap between academic history and popular understandings (Rosenberg, 2011).

Oral histories, as a form of narrative research, are powerful tools in understanding both the granular and broad strokes of history. By giving voice to individual memories and placing them within larger socio-cultural frameworks, they enrich our collective historical consciousness, reminding us that history is not just about grand events but also about everyday lives and stories.

3. **Diaries and Journals**: These offer personal, contemporaneous accounts, providing insights into individuals' reflections, perceptions, and emotions over time (Zimmerman & Wieder, 1977).

Diaries and journals, often considered as deeply private and reflective tools, have carved a significant space within narrative research. These introspective accounts, penned with authenticity and immediacy, offer invaluable insights into personal experiences, emotions, and the evolving self over time. Beyond individual introspection, diaries and journals have the potential to mirror broader societal, cultural, and historical landscapes, providing researchers with textured and layered narratives (Bishop, 2004).

Defining Diaries and Journals

Diaries and journals, while similar, have subtle distinctions. Diaries often contain daily accounts and may be more factual, logging events, encounters, or observations. Journals, on the other hand, might be written intermittently and often emphasize reflection, thoughts, and feelings, delving deeper into personal interpretations and responses (Lejeune, 2009).

Key Characteristics of Diaries and Journals

a) *Immediacy and Authenticity:* As they are generally written close to the occurrence of events, diaries and journals capture raw emotions, immediate reactions, and unfiltered perceptions, ensuring a level of authenticity (Zinsser, 1989).
b) *Evolution Over Time:* Regular entries chronicle the evolution of thoughts, emotions, and interpretations over time, showcasing personal growth, shifts in perspective, or changing circumstances (Plummer, 2001).
c) *Private Narratives:* Given their personal nature, these accounts are usually not influenced by external audiences, lending a sense of purity and genuine self-reflection to the narratives (Bunkers & Huff, 1996).

Methodological Considerations

I. *Access and Ethics:* Gaining access to personal diaries and journals requires trust and often comes with ethical considerations. Respecting privacy, ensuring anonymity where needed, and navigating sensitive content are essential (Blair, 2004).
II. *Interpretation:* Interpreting diaries and journals demands a nuanced approach, recognizing the personal contexts, underlying emotions, and the possibility of fluctuating self-representations (Stanley, 2007).
III. *Authenticity and Factuality:* While diaries and journals provide authentic emotional insights, researchers should approach them as subjective interpretations rather than objective factual accounts (Motz, 1983).

Applications and Significance

i. *Personal Histories:* Diaries and journals offer rich insights into individual lives, illuminating personal challenges, achievements, relationships, and milestones (Kuhn, 2002).

ii. *Socio-Cultural Chronicles:* Beyond personal narratives, these tools often reflect broader societal events, cultural norms, and historical epochs, making them invaluable for sociologists, historians, and cultural researchers (Procter, 2009).
iii. *Psychological Insights:* For psychologists, diaries and journals can be windows into individual psyches, revealing patterns of thought, coping mechanisms, emotional landscapes, and personal growth trajectories (Pennebaker, 1997).

Diaries and journals, as forms of narrative research, serve as treasured chronicles of human experience. Their deeply personal and authentic narratives offer researchers unique avenues to delve into intimate corners of human life, mirroring both individual journeys and broader societal landscapes.

Key Principles of Narrative Research

1. **Chronology**: Narratives are structured around sequences of events, though not always linearly. The emphasis is on understanding sequences and their meanings (Labov & Waletzky, 1997).

Narrative research, with its emphasis on stories and lived experiences, inherently interacts with the concept of time. One of the fundamental principles underlying narrative inquiry is chronology, the sequence in which events are remembered, recounted, and interpreted. Chronology is not just about temporality; it shapes the narrative structure, the coherence of life stories, and provides meaning to personal and collective histories (Bruner, 1991).

The Significance of Chronology

a) *Temporal Ordering:* At its most basic level, chronology represents the temporal ordering of events in a narrative. This linear progression lends a story its flow and coherence, guiding listeners or readers through the progression of events or experiences (Labov & Waletzky, 1967).
b) *Causality and Connection:* Chronology often goes beyond mere sequencing; it provides a structure that elucidates causal connections. Understanding the chronological order can illuminate how one event leads to another, offering insights into cause-effect relationships within the narrative (Riessman, 1993).
c) *Framework for Meaning:* Chronology creates a scaffolding upon which the meanings of events are hung. The temporal placement of events can influence their significance, with certain life moments taking on greater importance due to their timing or sequence (Freeman, 2010).

Challenges in Chronological Narration

I. *Memory and Accuracy:* Relying on human memory can introduce inconsistencies or gaps in the chronological recounting of events. As researchers, understanding these inconsistencies is as crucial as the events themselves, offering insights into the processes of memory and selective recall (Schacter, 1999).

II. *Non-linear Narratives:* Not all narratives are told linearly. Flashbacks, foreshadowing, or circular narrative structures can disrupt strict chronology, reflecting the complexities of human cognition and the non-linear nature of memory recall (Eakin, 2006).

Chronology in Research Contexts

i. *Life Histories:* In life history narratives, chronology provides a backbone, mapping the trajectory from early life to later years. Here, temporal sequencing is crucial in understanding life transitions, milestones, and turning points (Bertaux & Kohli, 1984).
ii. *Historical Narratives:* In accounts that tie personal narratives to broader historical events, chronology plays a dual role. It not only sequences personal events but aligns them with larger socio-cultural or political timelines, offering a dual lens of micro and macro chronologies (Abrams, 2010).
iii. *Disruptions in Chronology:* In some narratives, especially those recounting trauma or transformative experiences, chronological disruptions can be significant. These disruptions, or temporal disjunctions, can be indicative of the impact of events on cognitive structures and memory recall (Caruth, 1996).

Chronology, while ostensibly a simple sequencing of events, plays a profound role in narrative research. It provides narratives with structure, coherence, and meaning. As researchers engage with chronology, they encounter not just the linearity of events but the intricate web of memory, cognition, and interpretation that shapes human storytelling.

2. **Emplotment**: This refers to how events are organized within a plot, bringing coherence to diverse experiences (Ricoeur, 1981).

Narrative research revolves around the intricate art of storytelling, and a central pillar in this realm is the principle of emplotment. Building on Paul Ricoeur's seminal work on narrative identity, emplotment refers to the process by which diverse and sometimes disjointed events are woven into a coherent, unified storyline (Ricoeur, 1984). It's the narrative thread that gives meaning to individual events, contextualizing them within broader life stories.

The Essence of Emplotment

a) *Structuring Chaos:* Life events often unfold in non-linear, unpredictable ways. Emplotment serves as the narrative mechanism that organizes these events into a logical, structured, and comprehensible order, transforming mere occurrences into meaningful episodes (White, 1981).
b) *Bridging Gaps:* Emplotment seamlessly integrates disparate events, bridging temporal, spatial, or causal gaps. This integration creates a continuity, rendering the narrative both fluid and cohesive (Bruner, 1991).
c) *Generating Meaning:* Beyond sequencing, emplotment assigns significance to events. Through this narrative process, moments gain context, relationships are highlighted, and broader themes or life lessons may emerge (Polkinghorne, 1988).

Emplotment in Action

I. *Character and Agency:* Characters, often central to any narrative, gain depth and dimension through emplotment. Their desires, actions, and transformations become integral to the unfolding plot, emphasizing the interplay between individual agency and external events (Bal, 1997).

II. *Conflict and Resolution:* Emplotment often introduces a central conflict or challenge that drives the narrative forward. The trajectory from conflict introduction to eventual resolution creates a dynamic tension, engaging readers or listeners and enhancing the story's impact (Abbott, 2008).

III. *Temporal Dimensions:* Time, as manipulated through emplotment, is more than linear progression. Flashbacks, foreshadowing, and pacing play pivotal roles in shaping the narrative's rhythm, tone, and emphasis (Ricoeur, 1984).

Research Implications

i. *Interpretive Depth:* For researchers, understanding emplotment offers deeper insights into participants' worldviews. How individuals choose to plot their stories can shed light on their values, priorities, and interpretations of experiences (Clandinin & Connelly, 2000).

ii. *Addressing Discontinuities:* Emplotment helps researchers navigate and interpret narrative inconsistencies or discontinuities, understanding them not as errors but as intrinsic elements of personal storytelling (Mattingly, 1998).

iii. *Contextual Relevance:* Recognizing emplotment patterns can anchor individual stories within broader socio-cultural, historical, or communal narratives, providing a layered and contextualized understanding of personal accounts (Somers, 1994).

Emplotment, in the realm of narrative research, transcends mere event sequencing. It's a transformative process, one that crafts raw life events into resonant, meaningful tales. By appreciating the nuances of emplotment, researchers can delve deeper into the heart of personal stories, unraveling the intricacies of human experience.

3. **Voice and Perspective**: The narrative voice, be it first-person or third-person, shapes how a story is told and received. Recognizing whose voice is heard (and whose is silenced) is critical (Bamberg, 2004).

Central to the nature of narratives is the articulation of voice and the presentation of perspective. Voice, in narrative research, refers to the distinct articulation of an individual's or group's lived experiences, beliefs, and emotions. Perspective delves deeper into the angle or viewpoint from which these narratives are told, revealing underlying biases, socio-cultural contexts, and cognitive frameworks. Together, voice and perspective shape how stories are formed, heard, and interpreted (Bakhtin, 1981).

The Intricacies of Voice

a) *Authenticity and Authority:* Voice lends authenticity to a narrative, making it resonant and genuine. Moreover, it confers a sense of authority to the narrator, asserting the legitimacy of their lived experiences and their right to recount them (Cavarero, 2000).
b) *Diversity of Voices:* Narratives can encompass a spectrum of voices – from individual to collective, from mainstream to marginalized. Recognizing this diversity is essential for an inclusive understanding of experiences (Couldry, 2010).
c) *Silences and Subtexts:* Listening to a voice also involves recognizing its silences, hesitations, and what remains unsaid. These subtexts can offer as much meaning as overt narratives, revealing underlying tensions, conflicts, or societal pressures (Spivak, 1988).

Nuances of Perspective

a) *Emic and Etic Views:* An emic perspective offers an insider's viewpoint, detailing experiences from within a cultural, social, or individual context. The etic view, on the other hand, approaches the narrative from an outsider's standpoint, often analytical and comparative in nature (Pike, 1967).
b) *Shifting Perspectives:* A single narrative can encompass multiple perspectives, shifting between past and present, self and other, or different socio-cultural viewpoints. Recognizing these shifts is crucial for a holistic understanding of the story (Bamberg, 2004).
c) *Socio-Cultural Influences:* Perspectives are often shaped by broader socio-cultural contexts. Historical epochs, societal norms, or cultural beliefs can influence how an event is viewed, interpreted, and recounted (Bruner, 1993).

Implications for Narrative Research

a) *Ethical Considerations:* Respecting the voice of participants is an ethical imperative in narrative research. Researchers must ensure that voices are not unduly influenced, altered, or silenced during the research process (Etherington, 2007).
b) *Reflexivity:* Researchers must be aware of their own voice and perspective, recognizing how it can influence the collection, interpretation, and representation of narratives (Finlay, 2002).
c) *Multivocality:* Embracing multivocality, the presence of multiple voices within a narrative, can offer richer, layered insights into complex socio-cultural phenomena or multifaceted individual experiences (Chase, 2005).

Voice and perspective serve as twin pillars in the architecture of narrative research. They offer both the raw material and the lens through which stories are crafted, understood, and shared. Engaging deeply with these elements allows researchers to unearth the complexities of human experiences, told in all their textured, nuanced glory.

Methodological Steps in Narrative Research

1. **Choosing Narratives**: Whether seeking personal stories, folklore, or other forms of narrative, researchers must decide which narratives to explore, considering their relevance and significance (Clandinin & Rosiek, 2007).

Narrative research, anchored in the exploration of stories and lived experiences, embarks on its journey with the vital step of choosing which narratives to focus on. This choice isn't arbitrary but is framed by the research question, theoretical underpinnings, and methodological considerations. The decision on which narratives to engage with significantly influences the depth, direction, and outcomes of the research (Riessman, 2008).

Determinants in Choosing Narratives

a) *Research Objectives:* Central to the choice of narratives is the research's overarching objective. For studies aiming to understand specific phenomena, the narratives chosen would ideally provide rich insights into that particular domain (Mishler, 1995).
b) *Sociocultural Context:* The broader sociocultural milieu can influence narrative selection. Researchers might prioritize narratives that shed light on societal issues, cultural dynamics, or historically marginalized voices (Chase, 2005).
c) *Ethical Considerations:* The decision might also be governed by ethical considerations. Narratives involving vulnerable populations, for instance, might necessitate more delicate handling and robust informed consent processes (Etherington, 2007).

The Process of Choosing Narratives

a) *Preliminary Exploration:* Before settling on specific narratives, researchers might engage in a preliminary exploration, reviewing potential stories to discern which would be most illustrative or insightful (Clandinin & Connelly, 2000).
b) *Criterion-Based Selection:* Some researchers employ specific criteria to guide narrative selection. Such criteria could range from demographic factors to specific experiences or perspectives that are pertinent to the research question (Polkinghorne, 1995).
c) *Snowball Sampling:* Especially when exploring lesser-known or marginalized narratives, researchers might leverage snowball sampling. Here, initial participants introduce researchers to other potential narrators, broadening the research horizon (Goodson & Sikes, 2001).
d) *Purposeful Random Sampling:* To ensure diverse perspectives, a researcher might select narratives randomly from a pool of stories that meet certain essential criteria (Patton, 2002).

Implications of Choosing Narratives

a) *Depth vs. Breadth:* The choice of narratives often hinges on the desired depth of exploration. Focusing on a few detailed narratives might offer depth, while a broader selection could provide a more comprehensive overview (Smith & Sparkes, 2008).
b) *Representation and Bias:* The narratives chosen play a significant role in the research's representativeness. Researchers must be wary of potential biases, ensuring that the narratives selected don't unintentionally skew findings or interpretations (Squire, 2008).
c) *Ethical and Practical Challenges:* Certain narratives, while rich in insights, might also present ethical or practical challenges. These could range from concerns about confidentiality to the complexities of navigating sensitive or traumatic tales (Jones, 2007).

In narrative research, the stories chosen for exploration lay the groundwork for the ensuing investigative journey. Their selection, far from being a mere preliminary step, is pivotal in shaping the research's direction, depth, and ethical stance. As researchers weave through the intricate tapestry of human tales, the narratives they choose become the threads that color, enrich, and define their scholarly endeavors.

2. **Gathering Narratives**: This can be done through interviews, observations, archival research, or soliciting written accounts.

Gathering narratives, a pivotal step in narrative research, goes beyond mere data collection. It entails entering the intimate spaces of human experiences, bearing witness to their stories, and ensuring their authentic representation. Navigating this process requires methodological rigor, ethical sensitivity, and a deep appreciation for the sanctity of personal stories (Clandinin & Connelly, 2000).

Techniques in Gathering Narratives

i. *In-Depth Interviews:* Often the primary method in narrative research, in-depth interviews enable researchers to delve into participants' experiences, feelings, and reflections. It allows for an interactive engagement, where clarifications, probes, and elaborations can be sought (Riessman, 2008).
ii. *Life Histories:* Drawing on longer engagements with participants, life histories provide comprehensive accounts of individuals' lives, exploring significant events, turning points, and long-term evolutions (Atkinson, 1998).
iii. *Journals and Diaries:* Offering a more personal and introspective lens, journals and diaries capture individuals' daily experiences, thoughts, and feelings over extended periods (Jones, 2000).
iv. *Digital Narratives:* With technological advancements, digital platforms, such as blogs, podcasts, and video diaries, have emerged as potent tools to capture and share narratives in more dynamic and multimedia formats (Gubrium & Holstein, 2009).

Considerations in Gathering Narratives

i. *Establishing Trust:* The depth and authenticity of narratives often hinge on the rapport and trust between the researcher and the participant. Building a safe, empathetic, and non-judgmental space is crucial (Denzin, 2001).
ii. *Cultural Sensitivity:* Understanding and respecting cultural nuances, traditions, and values can shape the narrative-gathering process, ensuring that stories are contextualized and culturally anchored (Geertz, 1973).
iii. *Ethical Deliberations:* Beyond informed consent, narrative researchers must be attuned to issues of confidentiality, potential harm, or distress to participants, and the responsibilities of representing someone else's story (Etherington, 2007).
iv. *Reflexivity:* Researchers must continually reflect on their role, biases, and influence in the narrative-gathering process. Being aware of one's own beliefs, feelings, and reactions can enhance the integrity and depth of the collected narratives (Finlay, 2002).

Challenges in Gathering Narratives

i. *Representational Dilemmas:* The act of gathering often involves translation – from lived experiences to articulated stories. This translation can introduce discrepancies between the experienced and the narrated (Spivak, 1988).
ii. *Power Dynamics:* The narrative process can inadvertently reproduce power hierarchies, with researchers holding undue influence over which stories are told and how (Fine, 1994).
iii. *Emotional Labor:* Delving into deeply personal, and sometimes traumatic, stories can exact an emotional toll, both on participants and researchers. Recognizing and managing this emotional labor is vital (Dickson-Swift et al., 2009).

Gathering narratives is a journey of co-construction, where researchers and participants collaboratively delve into the depths of human experience. Navigating this terrain with sensitivity, respect, and methodological rigor ensures that the stories gathered are not just data points but resonate as authentic, profound testimonies of lived realities.

3. **Analyzing Narratives**: This involves decoding the structure of narratives, understanding themes, and interpreting meanings (Gee, 1986). It's a dynamic process, balancing between respecting the narrator's voice and offering scholarly insight.

Narrative analysis stands at the intersection of raw stories and scholarly interpretations. It involves delving into the rich tapestries of narratives, examining their structures, contents, and contexts, and distilling deeper meanings and insights. While narratives encapsulate personal experiences, analyzing them unfolds broader socio-cultural, psychological, and historical dimensions (Riessman, 2008).

Approaches to Narrative Analysis

i. *Structural Analysis:* Rooted in the works of scholars like Labov (1972), this approach examines the structural components of narratives. By breaking down stories into abstracts, orientations, complicating actions, evaluations, resolutions, and codas, researchers can discern the underlying patterns and emphases within narratives.
ii. *Thematic Analysis:* Focusing more on the content than structure, thematic analysis identifies recurring themes, motifs, or concepts within narratives. Braun and Clarke's (2006) six-phase guide offers a structured method to extract, code, and categorize these themes.
iii. *Dialogic/Performative Analysis:* Building on Bakhtin's (1981) notion of dialogism, this approach views narratives as dynamic, situated, and interactive. It emphasizes how narratives are performed and how they engage with multiple voices, audiences, and socio-cultural contexts.
iv. *Visual and Multimodal Analysis:* With the proliferation of digital narratives, researchers like Jewitt and Oyama (2001) have emphasized analyzing visual, auditory, and other multimodal elements of stories, recognizing their capacity to convey meanings beyond words.

Key Considerations in Narrative Analysis

i. *Contextual Sensitivity:* Narratives don't exist in isolation. Understanding the socio-cultural, historical, and personal contexts within which they are embedded is crucial for a nuanced analysis (Bruner, 1991).
ii. *Reflexivity:* Researchers need to be cognizant of their own biases, interpretations, and influences throughout the analysis process, continually questioning and reflecting on their interpretations (Finlay, 2002).
iii. *Ethical Responsibility:* Analyzing narratives entails re-presenting someone else's story. This brings an ethical duty to represent narratives authentically, avoiding misinterpretations or distortions (Etherington, 2007).

Challenges in Narrative Analysis

i. *Subjectivity:* Every narrative analysis is, to some extent, subjective. Balancing personal interpretations with fidelity to the narrative is an ongoing challenge (Smith & Osborn, 2003).
ii. *Complexity of Narratives:* Stories can be multi-layered, contradictory, and evolving. Capturing their complexity without oversimplifying or overcomplicating is a nuanced task (Phoenix, 2008).
iii. *Navigating Silences:* What remains unsaid in narratives often holds as much meaning as what is vocalized. Deciphering these silences and their implications requires discernment (Spivak, 1988).

Analyzing narratives is akin to embarking on a scholarly expedition. Each narrative is a rich landscape of meaning, waiting to be explored, mapped, and understood. Through careful, thoughtful, and ethical analysis, researchers can unveil the profound insights and universals that individual stories often encapsulate, contributing to broader academic discourses and human understanding.

4. **Representing Narratives**: Deciding how to present narratives – as direct quotations, summarized accounts, or thematic extracts – is crucial. It impacts how readers engage with and understand the researched experiences (Richardson, 2000).

Representing narratives, a critical juncture in narrative research, involves the delicate task of presenting the gathered stories in a manner that is both meaningful and respectful. While capturing the essence of individual experiences, this step also interweaves broader academic, cultural, and interpretive contexts, offering a platform for the voices of participants to resonate within wider audiences (Riessman, 2008).

Techniques in Representing Narratives

i. *Verbatim Transcriptions:* One of the most authentic ways to represent narratives is through verbatim transcriptions, capturing the exact words, pauses, inflections, and emotions of participants, thus preserving the raw texture of the narrative (Polkinghorne, 1995).
ii. *Reconstructed Narratives:* For some studies, the raw narratives are restructured or reconstructed to enhance clarity, flow, or thematic coherence, ensuring that the core essence remains intact (Josselson, 2006).
iii. *Composite Narratives:* Particularly useful when protecting identities or when representing common themes across multiple participants, composite narratives weave together elements from different stories to construct a unified narrative (Sandelowski, 1995).
iv. *Visual Representations:* In an era of multimedia, narratives are increasingly represented through visual means such as documentaries, photo essays, or digital storytelling, allowing for a richer, multi-sensory engagement (Gubrium & Harper, 2013).

Ethical and Practical Considerations in Representation

i. *Participant Collaboration:* In the spirit of co-creation, researchers often collaborate with participants during the representation phase, seeking feedback, validation, or co-authorship (Lambert, 2010).
ii. *Anonymity and Confidentiality:* Even while seeking authenticity, ensuring the privacy and anonymity of participants is paramount. Names, locations, and identifiable details might be altered without compromising the story's integrity (Ellis, 2007).
iii. *Emphasizing Context:* Represented narratives should provide readers or viewers with sufficient context, enabling a deeper understanding of the socio-cultural, temporal, and personal backdrops against which the narratives unfold (Bruner, 1991).
iv. *Reflexivity in Representation:* Researchers must reflect upon and disclose their own interpretive roles, biases, and influences in the act of representing narratives, underscoring the co-constructed nature of the outcome (Finlay, 2002).

Challenges in Representing Narratives

i. *Balancing Authenticity and Coherence:* Striving for a coherent and engaging representation can sometimes risk oversimplifying or altering the raw authenticity of a narrative (Riessman, 1993).

ii. *Navigating Power Dynamics:* Representation is inherently a process of selection, emphasis, and interpretation, and researchers must remain vigilant about not imposing their own voices or overshadowing those of the participants (Fine, 1994).
iii. *Meeting Academic Rigor:* Representing narratives in a way that marries personal stories with academic rigor, theoretical underpinnings, and methodological robustness is a nuanced endeavor (Chase, 2005).

Representation is not merely a final step in narrative research but is the bridge connecting personal stories to broader academic, cultural, and social landscapes. By navigating the act of representation with sensitivity, respect, and reflexivity, researchers ensure that each narrative, in its re-presented form, continues to echo with authenticity, depth, and profound human resonance.

Contemporary Applications of Narrative Research

Today's digital age offers new avenues and challenges. Online platforms birth digital stories, blending text, images, and sounds, and are rich sites for narrative research. However, issues like authenticity, ephemerality, and digital divides require nuanced approaches (Jones, Chik, & Hafner, 2015).

Narrative research, with its emphasis on understanding human experiences through the lens of personal stories, has witnessed a surge in its applications across various domains in recent times. Rooted in the belief that individual narratives can illuminate larger societal, cultural, and historical patterns (Riessman, 2008), contemporary applications are diverse, dynamic, and increasingly interdisciplinary.

1. **Healthcare and Medicine:**

Narrative medicine, as introduced by Charon (2006), underscores the significance of patients' stories in diagnostic processes, therapeutic interventions, and fostering physician-patient relationships. Understanding patients' narratives helps healthcare professionals to see the person behind the disease, facilitating holistic care.

2. **Education Teachers and educators:**

utilize narrative research to reflect on their teaching practices, understand students' learning experiences, and navigate the dynamics of classroom environments (Clandinin & Connelly, 2000). Student narratives also offer insights into the challenges and triumphs of learning, shedding light on diverse educational needs.

3. **Social Justice and Activism:**

Narratives of marginalized and oppressed groups provide powerful testimonies that challenge dominant discourses and bring attention to issues of justice, equity, and rights (Andrews, Squire, & Tamboukou, 2013). They act as tools for advocacy, consciousness-raising, and instigating social change.

4. **Organizational Studies:**

Organizational narratives shed light on corporate cultures, leadership dynamics, and employee experiences, offering a more humanistic perspective on organizational life (Boje, 1991). These narratives help in understanding change processes, decision-making dynamics, and the intricacies of workplace relationships.

5. **Digital Media and Virtual Realities:**

With the advent of the digital age, narrative research has ventured into the realms of online blogs, virtual worlds, and social media platforms, exploring how identities are constructed, communities formed, and experiences articulated in virtual spaces (Page, 2012).

6. **Migration and Diaspora Studies:**

Narratives of migrants, refugees, and diaspora communities chronicle journeys of displacement, settlement, identity negotiations, and cultural amalgamations, providing nuanced insights into global migration patterns and their human implications (Eastmond, 2007).

Challenges and Innovations

Contemporary applications of narrative research are not without challenges. The proliferation of narratives, especially in digital formats, raises questions about authenticity, representation, and interpretation (De Fina & Georgakopoulou, 2015). Ethical dilemmas, especially when dealing with vulnerable populations or sensitive topics, remain paramount (Etherington, 2007).

However, innovations in narrative methodologies, including the integration of visual, auditory, and digital elements, are expanding the horizons of what can be explored and how (Gubrium & Harper, 2013). The versatility of narrative research lies in its ability to bridge the personal and the universal. As contemporary applications continue to evolve, the essence remains consistent: to understand the complexities of human experience and to value the power and sanctity of personal stories.

Critiques and Challenges

Narrative research isn't without its critics. Concerns are raised about its subjectivity, its perceived lack of generalizability, and the ethical implications of interpreting and representing others' stories (Atkinson & Delamont, 2006). Yet, the potency of narratives in illuminating human experiences remains indisputable.

Narrative research, despite its profound contributions to understanding human experience, is not without its critiques and challenges. By delving into the intricacies of personal stories, scholars confront a multitude of issues related to methodology, representation, interpretation, and the very nature of narrative itself (Riessman, 2008).

1. **Subjectivity and Interpretation:**

At the heart of narrative research lies the process of interpretation. However, the subjectivity inherent in interpreting personal narratives is a persistent critique. How can researchers ensure that their interpretations are faithful to participants' intended meanings? Whose interpretation – the narrator's, the researcher's, or the reader's – is paramount? (Mishler, 1995).

2. **Authenticity and Truth:**

Narratives, being subjective accounts, raise questions about authenticity and truth. Can narratives be treated as factual accounts of events or merely as representations of a perceived reality? How can researchers discern embellishments, selective memories, or even intentional distortions in narratives? (Spence, 1982).

3. **Representation and Ethical Concerns:**

The act of representing narratives, especially of marginalized or vulnerable groups, brings forth ethical concerns. How are these stories presented without exoticizing, essentializing, or misrepresenting? How are participants' identities protected, and how is their agency respected in the representation? (Ellis, 2007).

4. **Generalizability and Transferability:**

Given that narrative research delves deeply into individual stories, a common critique pertains to its lack of generalizability. How can insights from one individual's narrative be applicable to broader populations or contexts? (Polkinghorne, 2007).

5. **Temporality and Fluidity:**

Narratives, being temporal in nature, might change over time. The story someone tells today might differ from the one they tell tomorrow, raising questions about the stability and reliability of narrative data (Bruner, 1991).

6. **Power Dynamics:**

Narrative research is not immune to power dynamics. Researchers hold power in the process of selection, emphasis, interpretation, and representation. How is this power negotiated, and how are participants' voices truly prioritized? (Fine, 1994).

7. **Methodological Rigor:**

Given its qualitative nature, narrative research often faces challenges in ensuring methodological rigor. How are narratives selected, analyzed, and represented in a manner that meets academic standards while preserving their inherent richness and complexity? (Chase, 2005).

8. **Digital Narratives and Authenticity:**

In the age of social media and digital communication, narrative research confronts new challenges. How do researchers navigate the authenticity and veracity of digital narratives, which might be curated or influenced by online personas and dynamics? (Page, 2012).

While narrative research offers profound insights into human experiences, its journey is riddled with complexities and challenges. By acknowledging, reflecting upon, and navigating these critiques, researchers can harness the true potential of narratives – providing a platform for voices to be heard, stories to resonate, and human experiences to be deeply understood.

CONCLUSION

Narrative research offers profound insights into human experiences, societies, and cultures. By respecting, understanding, and analyzing stories, researchers bridge personal experiences and broader sociocultural contexts, advancing our collective understanding.

Case Study Research

The exploration of complex phenomena often requires methodologies that can capture intricacies, contextual variations, and deep insights. The case study methodology stands out as a versatile and rigorous approach, allowing researchers to delve into the nuances of a particular "case" within its real-life context. As defined by Yin (2003), a case study is an empirical inquiry that "investigates a contemporary phenomenon in depth and within its real-world context." It is particularly beneficial when the boundaries between the phenomenon and context are not clearly evident.

Delving into the depths of human experience, organizational intricacies, or multifaceted societal phenomena necessitates an approach that is both rigorous and flexible. The case study research design, distinguished for its capacity to offer rich insights into the intricacies of real-world contexts, stands as a paramount tool in the arsenal of qualitative researchers (Yin, 2003). It represents a bridge, marrying empirical investigation with theoretical exploration, capturing the depth and nuance often missed by broader quantitative designs (Stake, 1995).

Origins and Evolution

The genesis of the case study approach is as eclectic as its application. Historically rooted in clinical medicine, where physicians documented anomalies in patient conditions (Smith, 1978), it expanded its horizons to social sciences, business, education, and beyond. Anthropologists and sociologists documented individual and community life, seeking to understand cultural, social, and behavioral intricacies (Geertz, 1973).

The case study approach, today a hallmark of qualitative research, traces its roots across multiple disciplines and epochs, reflecting the persistent human inclination to delve deeply into specific phenomena to derive broader understanding.

1. The Clinical Origins in Medicine:

Historically, the foundations of the case study approach are firmly anchored in clinical medicine. Physicians meticulously documented individual patient cases, especially anomalies or unexplained medical conditions, to inform future practice and enhance understanding (Smith, 1978). These case studies were essentially detailed observations of patients over time, looking into symptoms, responses to treatments, and eventual outcomes. The primary aim was twofold: understanding the particularities of individual cases and, when possible, extrapolating lessons for broader medical knowledge.

2. Expanding Horizons: Anthropology and Sociology:

From the tightly confined spaces of clinical examination rooms, the case study approach found resonance in the open terrains of human societies and cultures. Anthropologists, often immersed in foreign cultures for extended periods, used the approach to document and interpret social phenomena, rituals, and traditions (Malinowski, 1922). In sociology, the case study became instrumental in understanding group behaviors, social roles, and the structures of societies, offering profound insights into human interactions and hierarchies (Park, Burgess, & McKenzie, 1925).

3. The Advent in Education and Psychology:

In the realms of education and psychology, case studies emerged as powerful tools for understanding individual learning behaviors, cognitive developments, and emotional intricacies (Piaget, 1952). They were essential in studying unique phenomena, such as exceptional talents or specific learning disabilities, often revealing patterns and strategies applicable in broader educational or psychological contexts.

4. Business, Public Administration, and Beyond:

As the modern world evolved, businesses and public administrations faced complex challenges requiring nuanced understanding. Whether it was an organization grappling with a unique challenge or a government initiative with unforeseen consequences, the case study approach became a vehicle to dissect, understand, and learn from specific instances, contributing significantly to management and organizational theories (Yin, 2003).

5. The Shift Towards Rigor and Methodology:

The transition from merely descriptive narratives to a rigorous research method is a significant evolution in the case study approach. This shift, significantly influenced by the works of researchers like Robert K. Yin and Robert E. Stake, sought to establish case studies as systematic inquiries with robust methodological frameworks, ensuring both depth and rigor (Stake, 1995; Yin, 2014).

The journey of the case study approach is a testament to its adaptability, depth, and enduring relevance. From bedside observations in medical clinics to boardroom analyses in global corporations, this methodology remains an indispensable tool for in-depth, contextual understanding across disciplines.

Defining the Case Study

While definitions vary, most agree that a case study is an in-depth exploration of a bounded system (a case) or multiple bounded systems (cases) over time, using detailed, in-depth data collection involving multiple sources of information (Creswell, 2013). This 'case' could be an individual, a group, an organization, or an event.

The case study, as a qualitative research design, has carved a distinct niche within the academic and professional landscape, offering scholars and practitioners a potent instrument for in-depth exploration. But what precisely constitutes a case study, and how do its characteristics set it apart?

➢ The Essence of a Case Study

A case study, at its core, represents an intensive analysis of a single unit with the aim to generalize across a larger set of units (Gerring, 2004). This "unit" can be multifaceted—a person, a group, an event, an organization, a community, or even an entire culture. The objective is to derive an in-depth understanding of that unit within its real-world context (Yin, 2003). The case study, as an investigative method, stands out not merely for its procedural uniqueness but for the essence it seeks to capture. This essence, intrinsically intertwined with the nature of qualitative inquiry, warrants a deeper examination.

1. A Microcosm of the Macrocosm

At its very heart, a case study is a lens focusing on a singular entity or phenomenon, but it endeavors to reveal insights that resonate far beyond the specific boundaries of that case (Stake, 1995). It captures a microcosm with the ambition of shedding light on the broader macrocosm, making it a reflection of the larger reality (Geertz, 1973).

2. The Intensive Exploration

A case study dives deep. Unlike broad spectrum surveys or wide-ranging experiments, it narrows its gaze but intensifies its scrutiny. This in-depth exploration provides rich, detailed insights into the intricacies of the subject matter, often illuminating aspects that broader studies might overlook (Flyvbjerg, 2006).

3. The Interplay of Particular and General

While a case study hones in on the particular, it is not confined by it. It seeks to understand the individual case in all its unique specificity, yet it continuously grapples with the play between the specific instance and the broader generalizations it might suggest (Eisenhardt & Graebner, 2007). It treads the line between the uniqueness of the 'instance' and the universality of the 'principle'.

4. A Holistic Understanding

A case study is not just about isolating variables or understanding fragments; it is about grasping the whole. This holistic endeavor aims to understand the case in its entirety, considering all its facets, nuances, and complexities (Yin, 2014).

5. Dynamic, Not Static

While some research methodologies capture a snapshot in time, the case study often captures a dynamic process. It not only considers the 'what' and 'where' but also delves deeply into the 'how' and 'why', recognizing that entities and phenomena are not static but evolve, change, and transform (Merriam, 1998).

The essence of a case study is, in many ways, a reflection of the essence of qualitative inquiry itself – it seeks depth, nuance, holistic understanding, and resonance. It is a bridge that connects the profound intricacies of individual instances with the broader strokes of universal principles, making it an indispensable tool in the qualitative researcher's toolkit.

➢ Bounded System

A quintessential characteristic of a case study is its focus on a 'bounded system', emphasizing clear boundaries around what the case is and what it isn't (Stake, 1995). These boundaries help concentrate the study, ensuring depth and clarity.

The concept of the "bounded system" is pivotal in the realm of case study research. It fundamentally addresses the issue of demarcation, outlining the clear confines within which the investigation unfolds. Let's delve deeper into this concept, appreciating its nuances and implications.

1. The Boundary Defined:

A bounded system delineates the "case" under study, demarcating it from its broader context or environment (Stake, 1995). This boundary is essential for ensuring clarity of focus. Without it, the case's unique attributes might dissolve into the vast sea of broader phenomena, making it indistinguishable and defying the very purpose of a case study.

2. Fluid Yet Firm Boundaries:

While the need for a boundary is paramount, its nature is not always rigid. Depending on the research question and the evolving nature of the case, these boundaries might adjust as the study progresses (Merriam, 2009). They are, paradoxically, both fluid and firm – adaptable to the demands of the inquiry yet steadfast in maintaining the case's distinct identity.

3. Ensuring Depth Over Breadth:

A case study's strength lies in its depth. By establishing a bounded system, researchers can channel their efforts, time, and resources into delving deeply into the chosen case rather than spreading too thin across a broader spectrum (Creswell, 2013).

4. Variety in Boundedness:

The nature and scope of the bounded system can vary widely. It can be as narrow as a single individual in a psychological case study or as broad as an entire community in an anthropological inquiry. The boundary's scale and scope are determined by the research objectives and the nature of the phenomenon under investigation (Yin, 2014).

5. Contextual Relevance:

While the bounded system isolates the case for in-depth study, it doesn't detach it from its context. In fact, the boundary serves to highlight the interplay between the case and its environment, offering insights into how external factors influence, shape, or are shaped by the case (Flyvbjerg, 2006).

The bounded system is more than a mere methodological requisite in case study research; it's a philosophical stance underscoring the essence of the case study approach. It advocates for the deep, holistic understanding of specific phenomena within their unique contexts, making every case study a voyage into the intricacies of particularity.

➢ The Significance of Context

Unlike some research methodologies that might abstract a phenomenon from its environment, case studies immerse deeply into the real-life context. The phenomena under scrutiny are often inextricably tied to their context, making the latter not just a backdrop, but a vital component of the research (Flyvbjerg, 2006).

Central to the essence of case study research is its intimate relationship with context. Rather than perceiving phenomena in isolation, case studies view them as embedded in a nexus of spatial, temporal, cultural, and social dimensions. Delving into context provides a deeper understanding, helping the researcher see nuances otherwise overlooked.

1. The Relational Aspect of Context:

Every case resides within a contextual sphere, interwoven in a network of relationships (Flyvbjerg, 2006). Context not only situates the case but also contributes to its very definition. Understanding this interconnectedness becomes pivotal, as it elucidates how variables interact, influence, and are influenced within a system.

2. Rich Descriptions and Contextual Anchoring:

By anchoring findings in context, case study research provides "thick descriptions" (Geertz, 1973). These descriptions offer a nuanced portrayal of the case, embedding it in its socio-cultural, historical, and environmental milieu, thereby allowing readers to derive meaningful interpretations.

3. Complexity and Holism:

Case studies recognize that phenomena don't operate in a vacuum. By emphasizing context, they cater to the complexity of real-world scenarios, often shedding light on unexpected interactions, emergent patterns, and multifaceted dynamics (Stake, 1995).

4. Transferability and Contextual Insights:

While generalizability is a debated aspect in qualitative research, context-rich insights from case studies enhance transferability (Lincoln & Guba, 1985). When researchers delineate the intricate details of the context, it aids others in drawing parallels and discerning the applicability of findings to different settings.

5. Evolutionary Nature of Context:

Context isn't static; it evolves, and with it, the case itself might transform (Yin, 2014). Capturing these evolutionary trajectories allows researchers to understand changes, adaptations, and developments over time.

6. Contextual Layers: From Micro to Macro:

Context isn't monolithic. Case studies often unravel layers, from the immediate settings (micro-context) to broader societal, historical, or even global dimensions (macro-context). This multi-layered approach facilitates a comprehensive understanding of the case from different vantage points (Merriam, 2009).

In the realm of case study research, context is not a peripheral backdrop; it's central to the narrative. It furnishes depth, adds layers of meaning, and accentuates the intricate dance of variables in their natural habitat. Without context, the story of the case remains incomplete, lacking depth and resonance.

➢ Multiple Sources of Evidence

Case studies aren't reliant on a single source of data. They flourish on multiple sources – interviews, documents, archival records, observations, artifacts – each offering a piece of the holistic picture (Creswell, 2013). This variety not only ensures comprehensiveness but also reinforces the study's validity through triangulation (Denzin, 1978).

A defining hallmark of case study research is its reliance on multiple sources of evidence. Rather than being restricted by a single viewpoint or data source, case studies, by their very nature, are methodologically eclectic, weaving together various data strands to create a comprehensive, nuanced portrayal of the case in question.

1. Triangulation for Enhanced Validity:

Drawing from diverse data sources is a form of triangulation, a technique aiming to cross-verify findings and thereby enhance the validity and robustness of the research (Denzin, 1978). With each additional data point or perspective, the picture becomes clearer, revealing facets that might have been overlooked or misrepresented by a singular source.

2. Depth and Breadth of Understanding:

Multiple sources not only corroborate findings but also add depth and breadth to the understanding of the phenomenon under study (Yin, 2014). While one source might highlight overarching themes or patterns, another might delve into intricate details or present alternative viewpoints, enriching the overall narrative.

3. Flexibility in Data Collection:

Case study research is inherently flexible, allowing researchers to harness a range of evidence types, from interviews, observations, and documents, to artifacts, audio-visual materials, and more (Stake, 1995). This flexibility ensures that the research is adaptive, moulding itself to the contours of the case rather than imposing rigid methodological constraints.

4. Complementarity and Synergy:

Different sources often complement each other. For instance, while observational data might offer real-time insights into behaviors, interviews can provide subjective perceptions, feelings, and motivations. Together, these sources create a synergistic effect, where the combined insights surpass what each source might offer in isolation (Merriam, 2009).

5. Addressing Potential Biases:

Relying on multiple sources can mitigate potential biases or distortions inherent in any single data type (Creswell, 2013). By juxtaposing and reconciling different viewpoints, researchers can arrive at a more balanced, objective, and holistic understanding of the case.

6. Enhancing Transferability and Generalizability:

Rich, multi-sourced data can augment the transferability of case study findings. Detailed, varied evidence enables readers or subsequent researchers to better gauge the applicability of insights to other contexts or settings (Lincoln & Guba, 1985).

In case study research, the adage 'the whole is greater than the sum of its parts' rings especially true. By weaving together multiple sources of evidence, researchers construct a rich tapestry that captures the multifaceted essence of the case, making it a deeply insightful and robust form of qualitative inquiry.

➢ Versatility in Purpose

Case studies can be exploratory, descriptive, or explanatory (Yin, 2014). An exploratory case study might aim to define the questions and hypotheses of a subsequent study. Descriptive case studies aim to describe a phenomenon and its context in detail. In contrast, explanatory case studies examine causal links in complex interventions, shedding light on the "how" and "why" of a particular phenomenon.

At the intersection of empirical inquiry and real-world phenomena, case study research stands out for its remarkable versatility in purpose. Spanning descriptive, exploratory, and explanatory endeavors, case studies are adaptable tools that cater to a broad spectrum of research questions and scholarly objectives.

1. Descriptive Case Studies:

Descriptive case studies primarily focus on presenting a detailed account of a phenomenon within its context. Their goal is to offer a comprehensive portrayal, often delving into the "how" and "why" of the observed events, processes, or behaviors. Such studies, in essence, seek to paint a vivid picture, bringing the reader into the thick of the situation (Yin, 2014).

2. Exploratory Case Studies:

When the landscape of research is marked by uncertainty or lacks clarity, exploratory case studies come to the fore. Their intent is to probe, discover, and gain initial insights into relatively uncharted territories. Often, they serve as a precursor to more extensive research endeavors, setting the stage for subsequent inquiries by identifying patterns, hypotheses, or pertinent questions (Stake, 1995).

3. Explanatory Case Studies:

Positioned to tackle intricate causal relationships, explanatory case studies dive deep into the underlying mechanisms that give rise to observed outcomes. They aim to unpack the multilayered causative factors, tracing the pathways from cause to effect. In complex systems where linear investigations fall short, these case studies shed light on the intricate web of interactions (Merriam, 2009).

4. Intrinsic Versus Instrumental:

Stake (1995) delineates between intrinsic case studies, where the case itself is of primary interest, and instrumental case studies, where the case serves as a conduit to understand broader issues. This distinction underscores the adaptability of case study research in catering to both specific and generalized inquiries.

5. Constructing Theory from Cases:

Especially in the realm of social sciences, case studies often serve as a foundation for theory construction. By delving deep into the nuances of specific instances, researchers can extrapolate broader theoretical insights, crafting generalizable models from detailed empirical observations (Eisenhardt & Graebner, 2007).

The inherent versatility in the purpose of case study research underscores its expansive utility in the academic domain. Whether aiming to describe, explore, explain, or theorize, case studies offer a methodological lens that is both deep and broad, making them invaluable tools in the researcher's repertoire.

➢ Rich and Thick Descriptions

Grounded in the belief of offering a comprehensive view, case studies provide "thick" descriptions, a term coined by Geertz (1973), which means they describe a phenomenon in sufficient detail that the reader can derive meaning and understanding from it.

One of the defining strengths of case study research is its capacity to provide "rich" and "thick" descriptions. These detailed, contextually grounded portrayals of phenomena offer readers an immersive understanding, fostering insights that are deeply embedded in the nuances and intricacies of the studied cases.

1. Delving Beyond the Surface:

Unlike superficial summaries or cursory overviews, rich descriptions dive deep into the heart of the matter, capturing the intricate details, subtleties, and dynamics that define the case (Geertz, 1973). By doing so, they convey a sense of lived experience, allowing the reader to virtually "walk in the shoes" of those within the study.

2. Contextual Immersion:

Thick descriptions, on the other hand, go beyond detailing the phenomena itself, providing a comprehensive understanding of its context. This encompasses the socio-cultural, historical, environmental, and even emotional landscapes that surround the case. Such depth of contextual embedding ensures that the phenomena are not detached from the very elements that shape and define them (Lincoln & Guba, 1985).

3. Multiplicity of Voices:

Rich and thick descriptions often encapsulate a multitude of perspectives. By weaving together diverse narratives from various stakeholders or participants, case study research fosters a polyphonic, multi-voiced account, capturing the complexity of real-world phenomena (Stake, 1995).

4. Bridging the Empirical with the Interpretative:

While the rich descriptions offer an empirical, data-driven account, the thick descriptions bridge this empirical data with interpretative insights. This fusion helps in constructing meaning, making sense of the data within the broader tapestry of its context (Merriam, 2009).

5. Facilitating Transferability:

Detailed, context-rich portrayals enhance the transferability of case study findings. When readers or subsequent researchers encounter the thick descriptions, they can discern the similarities and differences between the studied case and other contexts, thus gauging the potential applicability of insights across settings (Creswell, 2013).

At its core, the strength of case study research lies in its commitment to deep, holistic understanding. By offering rich and thick descriptions, it presents a layered, multifaceted view of phenomena, allowing for insights that resonate with the complexity and richness of the real world. The case study, by its very definition, seeks depth, nuance, and contextual richness. It thrives in the gray areas, the intricacies, and the multifaceted realms of real-life situations. As researchers continuously aim for a blend of breadth and depth, the case study remains a beacon, elucidating the shadows and contours of the complex realities they explore.

Why Opt for a Case Study Design?

I. **Depth over Breadth:** Case studies allow for profound exploration. They dig deep, providing a multi-faceted understanding of the phenomenon within its real-life context (Flyvbjerg, 2006).

In the realm of research methodologies, the decision to choose one approach over another is often shaped by the goals and nuances of the research question at hand. One of the distinctive features of case study design is its profound emphasis on depth over breadth. But what does this entail, and why is this focus on depth so crucial for certain types of inquiries?

1. Unraveling Complexity

Case studies, by their nature, are designed to investigate complex phenomena that cannot be adequately understood with a superficial glance (Yin, 2014). By opting for depth, researchers can unearth the multifaceted dynamics, intricate relationships, and underlying mechanisms that characterize the phenomenon in question.

2. A Deep Dive into the 'Why' and 'How'

While some research designs might answer the 'what', case studies strive to uncover the 'why' and 'how'. This involves a nuanced exploration into the causative factors, processes, and contextual variables that give rise to observed outcomes (Stake, 1995).

3. Embracing Holism

By choosing depth, case study research often adopts a holistic approach. Rather than fragmenting a phenomenon into isolated variables, the design aims to understand it in its entirety, valuing the interplay of elements within the system (Merriam, 2009).

4. Richness of Data

Depth invariably leads to richer data. This richness, marked by detailed descriptions, multiple data sources, and varied perspectives, allows for a comprehensive and nuanced understanding of the case, facilitating both empirical robustness and interpretative insights (Geertz, 1973).

5. Contextual Sensitivity

Prioritizing depth means delving into the thick of the context in which the phenomenon resides. This emphasis ensures that findings are not decontextualized, but are deeply embedded within the socio-cultural, temporal, and spatial dimensions that shape them (Lincoln & Guba, 1985).

In research, breadth has its value, offering overarching views and generalized patterns. However, when the objective is to penetrate the layers of complexity, to weave together the threads of meaning, and to capture the essence of real-world phenomena in their intricate glory, then depth takes precedence. It is in these contexts that a case study design, with its profound emphasis on depth over breadth, truly shines.

II. **Real-World Context:** Case studies thrive in real-world settings. They do not extract the phenomenon from its environment; they study it in situ, embracing its complexity (Merriam, 1998).

In the vast landscape of research methodologies, case study design uniquely positions itself as a bridge between empirical rigor and the unfolding realities of the world we inhabit. Central to this is its intrinsic alignment with real-world context. But why is this contextual orientation so pivotal, and how does it enrich the inquiry process?

1. Grounded Insights:

Case study research is not conducted in a vacuum; it is deeply rooted in real-world scenarios. This grounding ensures that the insights derived are not abstract or decontextualized but are relevant and reflective of the actualities of life (Yin, 2014).

2. Context as a Lens:

In case study design, context isn't a mere backdrop; it's an active lens through which phenomena are viewed and understood. The socio-cultural, economic, historical, and spatial dimensions of the context shape the case and influence its dynamics, offering researchers a richer, more nuanced perspective (Stake, 1995).

3. Understanding Complexity:

Real-world scenarios are marked by complexity, with multifaceted interactions, varied stakeholders, and evolving dynamics. A case study's contextual focus allows researchers to embrace this complexity, delving into the intricacies and the interplay of variables that other designs might overlook (Flyvbjerg, 2006).

4. Enhancing Transferability:

While generalizability is often a point of contention in case study research, its deep contextual grounding enhances transferability. Detailed contextual descriptions allow readers or subsequent researchers to gauge the relevance and applicability of insights to other settings or similar contexts (Lincoln & Guba, 1985).

5. Bridging Theory and Practice:

Case study research often serves as a conduit between theoretical constructs and real-world applications. By immersing in the context, researchers can test, refine, or expand theoretical models, ensuring they resonate with, and are validated by, real-world instances (Merriam, 2009).

In the dance of research, where abstraction meets reality, case study design stands out for its graceful embrace of the real-world context. This orientation not only enriches the depth and relevance of the inquiry but ensures that the generated insights resonate with the pulsating rhythms of lived experiences.

III. **Theoretical and Practical Implications:** While rich in empirical detail, case studies also contribute to theory building, challenging or supporting existing paradigms (Eisenhardt, 1989).

The decision to employ a case study design in research often stems from a desire to delve deeper into a particular phenomenon, aiming not just for empirical understanding but also for the discernment of both theoretical and practical implications. But how does this design uniquely cater to these dual objectives, and why is this synthesis so crucial?

1. Unearthing Theoretical Insights:

At the heart of every case study lies a quest to understand. This understanding often leads to the formulation or refinement of theoretical constructs. By closely examining a specific case within its real-world context, researchers can discern patterns, relationships, and mechanisms that might be obscured in broader analyses (Eisenhardt & Graebner, 2007).

2. Testing Theoretical Propositions:

While case studies can give rise to theory, they also serve as a platform to test and validate existing theoretical propositions. The depth and contextual richness of case studies provide an ideal setting for such rigorous theoretical scrutiny, allowing for nuances and complexities to emerge (Yin, 2014).

3. Informing Practice:

Beyond theoretical insights, case study research often bears significant practical implications. By studying phenomena in their natural settings, researchers gain insights that are immediately relevant and applicable, aiding practitioners in their decision-making processes, interventions, or policy formulations (Stake, 1995).

4. Bridging Theory and Practice:

Perhaps one of the most notable strengths of case study design is its inherent capability to bridge the chasm between theory and practice. The insights gleaned from the context-rich, in-depth exploration of real-world cases can simultaneously enrich academic discourse and guide practical action, ensuring that neither realm remains isolated (Flyvbjerg, 2006).

5. Enhancing Reflexivity and Responsiveness:

The immersion into real-world contexts, which case study research necessitates, often demands a high degree of reflexivity from researchers. This reflexivity, in turn, ensures that the theoretical and practical implications drawn are both rigorous and responsive to the evolving dynamics of the context (Lincoln & Guba, 1985).

In the intricate dance between the abstract world of theory and the tangible realm of practice, case study design offers a harmonizing melody. Its capacity to discern deep theoretical insights while simultaneously shedding light on practical implications makes it an invaluable tool for researchers striving for impactful, relevant scholarship.

Key Characteristics

- **Particularistic:** Focuses on a particular phenomenon, situation, or entity (Stake, 1995).
- **Descriptive:** Offers a rich and thick description of the phenomenon being studied (Geertz, 1973).
- **Heuristic:** Can illuminate the understanding of the phenomenon, revealing new meanings and insights (Merriam, 1998).
- **Inductive:** Often operationalizes without a fixed hypothesis, letting patterns emerge from data (Eisenhardt, 1989).

Designing a Robust Case Study: Steps and Considerations

I. **Defining and Bounding the Case:** Determine what the case is and what it isn't. The boundaries help maintain the study's manageability (Yin, 2014).

Crafting a robust case study requires meticulous planning, with a critical step being the clear definition and bounding of the case. This seemingly straightforward task is layered with intricacy, demanding careful consideration and precision. Understanding what constitutes a 'case' and how to appropriately bound it can determine the depth, clarity, and relevance of the study.

1. The Essence of a 'Case':

A 'case' is not merely a topic or a general subject area. Instead, it is a specific, complex unit of analysis situated within real-life contexts (Stake, 1995). It could be an individual, a group, an institution, an event, or a phenomenon. Before delving deeper, researchers must be clear about what their case is and, equally important, what it isn't.

2. Contextualizing the Case:

Each case exists within a particular context, and understanding this context is pivotal (Yin, 2014). Whether it's the cultural backdrop of an ethnographic study or the historical timeline of a specific event, the surroundings in which the case resides can shape, influence, and even define it.

3. Bounding the Case: Time, Place, and Activity:

Bounding refers to defining the limits or edges of the case (Merriam, 2009). This often involves determining:

- Temporal bounds: When does the case start and end?
- Spatial bounds: Where is the case located?
- Activity or process bounds: What activities or processes are central to the case?

By explicitly setting these boundaries, researchers can focus their study and avoid becoming overwhelmed by extraneous information.

4. Considering the Multifaceted Nature of Cases:

Cases are rarely one-dimensional. A single case can encompass multiple sub-units or embedded units of analysis (Creswell & Poth, 2018). For instance, a case study of a school might include sub-units like teachers, students, and administrative staff. Recognizing and defining these can enrich the research, allowing for layered insights.

5. Clarity and Rationale:

It is imperative for researchers to clearly articulate the rationale behind their choice of case and its boundaries. This ensures that the study is guided by a deliberate, purposeful logic rather than arbitrary decisions (Flyvbjerg, 2006).

Defining and bounding a case are foundational steps in case study research, laying the groundwork for all subsequent stages. A well-defined and appropriately bounded case not only strengthens the methodological rigor but also ensures that the resulting insights are sharp, focused, and deeply resonant with the complexities of real-world phenomena.

II. **Determining the Type:** Decide between intrinsic (focus on the case itself), instrumental (focus on an external theoretical question or issue), or collective (multiple cases) (Stake, 1995).

Within the broad umbrella of case study research lies a spectrum of different types, each with its distinct features, purposes, and methodologies. Deciding on the most appropriate type of case study is not just a procedural step, but a strategic choice that can influence the direction, depth, and outcome of the research.

1. Exploratory Case Studies:

Exploratory case studies are typically preliminary in nature, conducted to define the parameters of a main study or to clarify research questions (Yin, 2014). They serve as a prelude, offering insights that can shape subsequent, more detailed inquiries.

2. Explanatory Case Studies:

These delve into causal questions, seeking to explain the underlying mechanisms and reasons behind a particular phenomenon or event. Explanatory case studies are particularly suited for complex situations where multiple factors may interact in non-trivial ways (Yin, 2014).

3. Descriptive Case Studies:

As the name suggests, descriptive case studies aim to provide a detailed, holistic portrayal of a phenomenon within its real-life context (Stake, 1995). These studies can serve as comprehensive accounts, narrating the 'story' of a case.

4. Intrinsic Versus Instrumental Case Studies:

Stake (1995) differentiates between intrinsic and instrumental case studies. In intrinsic case studies, the focus is on the case itself, often because it possesses some unique or unusual attributes. Instrumental case studies, on the other hand, utilize the case primarily as a means to provide insights into a broader issue.

5. Collective or Multiple Case Studies:

Sometimes, researchers may choose to study multiple cases in one research project to understand a phenomenon, population, or general condition (Creswell & Poth, 2018). This approach, akin to multiple experiments, provides a more robust basis for generalizing results.

6. Critical Instance Case Studies:

These examine a single situation of unique interest or test a well-formulated theory. The purpose is to challenge universal or taken-for-granted assertions that might stand in "critical" contradiction to the case (Flyvbjerg, 2006).

The determination of case study type is a pivotal step in the research design process, influencing not only the methodological approach but also the depth, breadth, and contours of the inquiry. As researchers navigate this decision, clarity of purpose, understanding of the research question's nuances, and awareness of the potential implications of each type are paramount.

III. **Data Collection:** Multiple sources, including interviews, observations, documents, and physical artifacts, enhance the study's validity through triangulation (Denzin, 1978).

A cornerstone of case study research, data collection is not a mere procedural step, but a series of deliberate and strategic choices that can significantly influence the study's validity, depth, and richness. Given the inherent complexity and context-specific nature of case studies, data collection is multi-faceted and requires a nuanced approach.

1. Diversity of Data Sources:

One of the hallmarks of case study research is the use of multiple data sources, which can provide a comprehensive view of the case. These might include interviews, documents, archival records, direct observations, artifacts, and more (Yin, 2014).

2. Interviews:

Interviews, particularly semi-structured or open-ended ones, are commonly used in case study research. They allow for in-depth exploration and provide access to participants' experiences, perceptions, and interpretations (Stake, 1995).

3. Observations:

Direct observations offer researchers a chance to gather data in situ, witnessing events, behaviors, and interactions firsthand. Such observations can provide context and lend texture to the study (Merriam, 2009).

4. Document Analysis:

Analyzing documents—ranging from formal reports to informal correspondence—can offer insights into the case's historical context, ongoing developments, or participants' perspectives (Bowen, 2009).

5. Triangulation:

Triangulation involves using multiple sources or methods to gather data. The goal is not necessarily to validate a singular truth but to deepen understanding and enhance the study's richness by examining the case from various angles (Denzin, 1978).

6. Access and Rapport:

Gaining access to the desired data sources, especially in sensitive or closed contexts, is a crucial consideration. Building rapport with participants can facilitate more candid, insightful data sharing (Yin, 2014).

7. Use of Technology:

Modern case study research can harness various technologies, from recording devices for interviews to software for data management and analysis. Such tools can enhance accuracy and efficiency (Baxter & Jack, 2008).

8. Ethical Considerations:

Researchers must navigate issues of consent, confidentiality, and potential harm, ensuring that participants' rights and well-being are safeguarded throughout the data collection process (Creswell & Poth, 2018).

Data collection in case study research is both an art and a science, demanding meticulous planning, adaptability, and ethical rigor. The choices made during this phase lay the foundation for subsequent analysis and interpretation, playing a pivotal role in determining the study's depth, breadth, and overall quality.

IV. **Data Analysis:** Continuous, iterative, and can be done using various strategies such as pattern matching, explanation building, or time-series analysis (Yin, 2014).

Data analysis within the context of a case study is a nuanced and intricate process. Unlike some other research methodologies, which might have clearly delineated analytical steps, case study data analysis requires iterative, in-depth, and holistic approaches. The way researchers engage with the data can significantly influence the insights derived from the study.

1. Immersion in the Data:

To begin, researchers often immerse themselves in the data, revisiting transcripts, notes, and records multiple times. This deep engagement fosters familiarity and allows for initial patterns or themes to emerge (Stake, 1995).

2. Pattern Recognition:

Once sufficiently immersed, researchers look for recurring themes, concepts, or patterns within the data. These patterns can serve as the foundational constructs around which the case is analyzed and understood (Yin, 2014).

3. Creating Descriptive Frameworks:

For some case studies, especially those that are more descriptive in nature, researchers may create detailed narratives or frameworks that chronologically present the case's details (Merriam, 2009).

4. Use of Analytical Tools:

While manual coding and analysis are common, many researchers leverage software tools like NVivo or Atlas.ti to manage, code, and categorize their data. Such tools can streamline the analytical process, particularly for large data sets (Bazeley & Jackson, 2013).

5. Cross-Case Synthesis:

In studies involving multiple cases, a cross-case synthesis can be undertaken. This involves comparing and contrasting data across different cases to identify shared patterns or divergent themes (Eisenhardt, 1989).

6. Theoretical Propositions and Case Study Data:

If the study was initiated with theoretical propositions, these propositions would guide the analysis. The data can be examined in light of these propositions, thereby testing, refining, or potentially discarding them (Yin, 2014).

7. Seeking Alternative Explanations:

To bolster the robustness of the analysis, researchers should actively seek out alternative explanations for the patterns or phenomena observed. By challenging initial interpretations, a richer and more nuanced understanding can be developed (Eisenhardt & Graebner, 2007).

8. Triangulation for Validation:

Drawing upon multiple data sources, or triangulation, can enhance the validity of the analysis. By corroborating evidence from different sources, researchers can ensure that their interpretations are grounded in the data (Denzin, 1978).

9. Maintaining a Chain of Evidence:

Throughout the analysis, maintaining a clear chain of evidence is crucial. This ensures that the research process is transparent and that findings can be traced back to specific data sources, enhancing credibility (Yin, 2014).

Data analysis in case study research is as much an art as it is a science. It requires not only rigorous methodological approaches but also creativity, intuition, and reflexivity. Through meticulous and thoughtful analysis, the complexities and nuances of the case can be unveiled, contributing rich insights to both theory and practice.

V. **Reporting:** Crucial for communicating findings. The format might be illustrative (descriptive), exploratory (investigative), cumulative (comparative), or critical (challenging mainstream thought) (Merriam, 1998).

The reporting phase of a case study encompasses the final presentation of findings, conclusions, and potential recommendations. It's the culmination of the entire research process, where insights derived from data are communicated to the intended audience. This phase demands clarity, coherence, and a judicious use of evidence to make the narrative persuasive and grounded.

1. Structuring the Report:

A well-structured report aids in presenting information logically and coherently. Typically, case study reports might include an introduction, background information, the presentation of findings, discussion, and conclusion (Yin, 2014).

2. Crafting a Narrative:

Given the in-depth nature of case studies, crafting a compelling narrative is pivotal. This involves weaving together data, analysis, and interpretation into a coherent story that brings the case to life (Stake, 1995).

3. Using Evidence Judiciously:

Throughout the report, assertions and interpretations should be backed by evidence. Direct quotations, observational notes, or visual data can be used to illustrate points and ground the narrative in empirical reality (Merriam, 2009).

4. Triangulation in Reporting:

Drawing upon multiple data sources (triangulation) not only enhances analysis but also reporting. By presenting evidence from different sources, researchers can enhance the robustness and credibility of their findings (Denzin, 1978).

5. Addressing Potential Limitations:

It's essential to be transparent about the study's limitations, whether these pertain to data collection, potential biases, or constraints of the methodology. Addressing these upfront enhances the report's trustworthiness (Flyvbjerg, 2006).

6. Reflecting on the Research Process:

Including a reflexive section where researchers contemplate their role, biases, and influence on the research can add depth and honesty to the report. This acknowledges the inherently interpretive nature of case study research (Creswell & Poth, 2018).

7. Visual Aids:

Graphs, charts, tables, or photographs can enhance the comprehensibility and richness of the report. Such visual aids can offer readers alternative ways to engage with and understand the data (Baxter & Jack, 2008).

8. Conclusions and Implications:

The report should culminate in clear conclusions drawn from the findings. Where appropriate, implications for theory, practice, or further research should be outlined, providing readers with a sense of the study's broader relevance (Yin, 2014).

9. Feedback and Peer Review:

Before finalizing the report, seeking feedback from peers or experts can provide valuable insights. Peer review can help identify gaps, ambiguities, or areas of improvement, enhancing the report's quality (Eisenhardt & Graebner, 2007).

Reporting in case study research is not just about presenting findings but crafting a persuasive, evidence-backed narrative that does justice to the richness and complexity of the case. A judicious blend of structured presentation, narrative depth, and empirical grounding makes for a compelling case study report.

Criticisms and Rebuttals

Despite its strengths, case study research has faced criticism, most notably for its lack of generalizability (Guba & Lincoln, 1994). However, proponents argue that the strength of case studies lies in their depth and nuance, providing transferability rather than broad generalizability (Stake, 1995). Additionally, concerns about subjectivity are addressed through rigorous data collection and validation methods, including member checks and peer debriefing, ensuring the reliability and validity of findings (Creswell, 2013).

Case study research, despite its rich contributions to multiple fields, has not been immune to criticism. Over the years, several concerns have been raised about its validity, generalizability, and rigor. However, scholars of case study research have also offered robust rebuttals to these critiques, emphasizing the unique strengths and contributions of the method.

1. Criticism: Lack of Scientific Rigor:

Many posit that case studies lack the systematic approach found in other research methods, potentially leading to biased results (Campbell & Stanley, 1966).

Rebuttal

Yin (2014) argues that the rigor in case studies stems from a rigorous design, the use of multiple data sources (triangulation), and the maintenance of a clear chain of evidence. Thus, when designed properly, case studies can maintain a high level of scientific rigor.

2. Criticism: Limited Generalizability:

Case studies, particularly single-case designs, have faced critiques regarding their generalizability given their deep focus on one or a few instances (Stoecker, 1991).

Rebuttal

Stake (1995) and Flyvbjerg (2006) note that the primary goal of case studies is often to achieve "analytical" or "theoretical" generalization rather than "statistical" generalization. That is, insights from case studies can be used to inform and refine theory, offering deep, context-specific insights.

3. Criticism: Subjectivity and Bias:

Some critics argue that the interpretive nature of case study research can introduce researcher bias, thereby compromising objectivity (Becker, 1998).

Rebuttal

Guba and Lincoln (1989) argue that complete objectivity is an illusion in any qualitative research. Instead, they advocate for reflexivity, where researchers actively reflect on and disclose their biases and positions, enhancing the transparency and credibility of the research.

4. Criticism: Not Reproducible:

Given the deep contextual nature of case studies, replicating them can be challenging, leading to questions about their reliability (Kirk & Miller, 1986).

Rebuttal

Yin (2014) contends that the goal isn't necessarily to reproduce the results in another setting but to replicate the methodological approach in a different case to see if similar patterns emerge.

5. Criticism: Overly Descriptive

There's a belief that case studies can sometimes be overly descriptive, lacking in analytical depth (Lijphart, 1971).

Rebuttal

Eisenhardt (1989) posits that while some case studies might be exploratory and descriptive in nature, many employ rigorous analytical frameworks, leading to theory development and refinement.

While criticisms of case study research persist, it's essential to understand the unique strengths and potentialities of this approach. When used appropriately, with careful consideration of its strengths and weaknesses, the case study method can offer unparalleled depth, nuance, and contextual richness. The case study research design, in its versatility and depth, offers a distinctive approach to understanding the complexities of real-world phenomena. It captures the stories, challenges, victories, and lessons, illuminating them not just as isolated incidents but as resonant narratives intertwined with context, offering both scholars and practitioners nuanced insights and profound learnings.

Distinctiveness of Case Study Research

I. **Holistic Understanding:** Case studies provide a comprehensive view of the phenomenon under study, allowing for multiple data sources and methods to weave a holistic picture (Creswell, 2013).

Case study research, by its nature, offers an approach distinct from many other research methodologies. One of its most notable attributes is the capacity to provide a holistic understanding of a phenomenon within its real-life context. This holistic perspective is a defining strength of the case study method and sets it apart from more segmented or decontextualized research approaches.

1. Depth Over Breadth:

Rather than aiming for broad generalizations, case studies delve deeply into specific contexts, providing rich, detailed insights into the complexities of a particular case. This depth enables a comprehensive understanding of the multi-faceted nature of the phenomenon under study (Stake, 1995).

2. Contextual Sensitivity:

Case studies are intrinsically sensitive to context. They recognize that phenomena do not exist in isolation but are influenced by, and interact with, their environments. This contextual focus brings out the intricate interplay between a phenomenon and its surrounding environment, highlighting factors that might be overlooked in more controlled research settings (Yin, 2014).

3. Integrative Approach:

Unlike methodologies that might separate or isolate specific variables or factors for study, case studies integrate various elements to present a full picture. This integrative approach allows for an understanding of how different factors intersect, influence, and shape each other within the case (Merriam, 2009).

4. Real-world Relevance:

By studying phenomena in their natural settings, case studies ensure high ecological validity. The holistic insights derived are grounded in real-world contexts, making the findings particularly relevant and applicable to practitioners and policymakers (Flyvbjerg, 2006).

5. Dynamic Perspective:

Case studies often adopt a dynamic perspective, examining how phenomena evolve over time. This temporal dimension, combined with the holistic approach, allows researchers to understand processes, transitions, and developments in depth (Eisenhardt & Graebner, 2007).

The holistic understanding provided by case study research is its distinct hallmark. By embracing the complexity and contextuality of real-world phenomena, case studies offer insights that are rich, nuanced, and deeply embedded in actual practice. While this approach may not lend itself to broad statistical generalization, it provides invaluable depth and integrative insights that many other research methodologies cannot.

II. **Contextual Sensitivity:** Unlike many research methodologies that strive for context-neutrality, case studies deeply embed the context, believing it to be intrinsically linked with the phenomenon (Merriam, 1998).

One of the most defining characteristics of case study research is its profound sensitivity to context. Unlike certain research methods that seek to strip away context to focus on specific variables in controlled environments, case studies are deeply entrenched in real-world settings, valuing the richness and intricacies of contextual nuances.

1. Embracing the Complexity of Real-world Settings:

Case study research does not shy away from the messy, intertwined realities of the real world. It acknowledges that phenomena are influenced by multiple, often overlapping, contextual factors (Geertz, 1973). This makes case studies especially valuable for understanding complex systems, organizations, or societal phenomena.

2. High Ecological Validity:

Given the deep-rooted commitment to real-world exploration, the findings from case studies often boast high ecological validity. This means that the results are more likely to reflect true, real-world behaviors, attitudes, and dynamics, making them especially relevant for practitioners (Bronfenbrenner, 1979).

3. Context as an Active Contributor:

In case study research, context is not just a passive backdrop but an active contributor to the phenomenon under investigation. The interplay between the phenomenon and its context can reveal patterns, influences, and dynamics that might be obscured in more decontextualized research designs (Yin, 2014).

4. Temporal and Spatial Dimensions:

The sensitivity to context in case study research encompasses both temporal and spatial dimensions. Researchers consider historical, cultural, socio-economic, and other relevant factors, recognizing that events or phenomena do not occur in isolation but are embedded in a web of contextual elements spanning time and space (Stake, 1995).

5. Flexible and Adaptive Research Design:

Given the emphasis on context, case study researchers often approach their work with flexibility, ready to adapt their methods, questions, or focus based on emerging contextual insights. This adaptability ensures that the research remains relevant and aligned with the complexities of the real-world setting (Flyvbjerg, 2006).

The contextual sensitivity of case study research sets it apart from many other methodologies. By deeply valuing and incorporating context, case studies offer multifaceted, nuanced insights that resonate with the complexities of real-world scenarios, making them invaluable tools for both scholars and practitioners.

III. **Exploratory, Explanatory, and Descriptive:** Case studies can serve various purposes. They can explore unknown territories, explain the intricacies of known phenomena, or describe a phenomenon in depth (Yin, 2014).

Case study research is prized for its versatility. Depending on the research question and objectives, it can serve exploratory, explanatory, or descriptive purposes. Each of these orientations offers unique advantages, making case studies a highly adaptable tool for investigators across disciplines.

1. Exploratory Case Studies:

Purpose

Exploratory case studies are generally preliminary, aiming to identify the exact nature of a phenomenon or to generate hypotheses for future research (Yin, 2014).

Strengths

- They're useful for investigating new or under-researched areas where established theories or explanations may be lacking.
- Exploratory designs can adapt as research progresses, accommodating new directions or areas of inquiry (Eisenhardt, 1989).

Applications: For example, an exploratory case study might examine a novel organizational practice in a single company before more extensive research is conducted.

2. Explanatory Case Studies:

Purpose

These delve into the underlying causes or mechanisms of a phenomenon, aiming to provide answers to "how" or "why" questions (Yin, 2014).

Strengths

- They allow for the deep investigation of complex phenomena within real-life contexts, integrating multiple data sources to offer robust explanations.
- Explanatory studies can challenge or refine existing theories, making them instrumental in theory development (George & Bennett, 2005).

Applications: An explanatory case study might investigate how specific management strategies influence corporate innovation, aiming to uncover underlying mechanisms.

3. Descriptive Case Studies

Purpose

Descriptive case studies aim to present a comprehensive portrayal of a phenomenon in its natural context, often answering "what" questions (Yin, 2014).

Strengths

- They offer rich, detailed accounts, providing a deep understanding of the case.
- Descriptive designs can capture the nuances and complexities of real-world phenomena, yielding insights that might be obscured in broader surveys or experiments (Stake, 1995).

Applications: For instance, a descriptive case study might detail the operations of a successful educational program, shedding light on its components, stakeholders, challenges, and successes.

The versatility of case study research in catering to exploratory, explanatory, and descriptive needs underscores its distinctiveness and appeal. By adapting to the demands of the research question and the intricacies of the case itself, case study research offers researchers a dynamic and holistic tool for understanding complex real-world phenomena.

Types of Case Studies

- ***Single-case designs***: These focus on an individual unit. A single case might be selected because it is critically important, unique, or representative (Flyvbjerg, 2006).

Case study research provides a rich, in-depth avenue to examine complex phenomena within their contexts. Within the broad umbrella of case study research, the single-case design stands out as a focused and detailed methodology, honed in on one specific instance or unit of study. This singular focus allows for comprehensive examination and deep understanding.

1. **Purpose and Rationale**

The single-case design aims to delve deeply into a specific case to understand its complexities and nuances. The choice to study a single case might stem from its uniqueness, its representational quality, or the opportunity it provides to explore or test a particular theory (Yin, 2014).

2. **Types of Single-case Designs**

• **Critical Case**: Used to test a well-formulated theory. The case is chosen because it is predicted to either confirm or refute the theory, making it a crucial test (Flyvbjerg, 2006).

- **Extreme or Unique Case**: Selected because it is unusual or rare. This design is particularly useful when the phenomena or circumstances being studied are not common and can provide unique insights (Stake, 1995).
- **Revelatory Case**: Here, the researcher has the opportunity to study a phenomenon previously inaccessible to scientific investigation. The case becomes a window to new understandings (Yin, 2014).
- **Representative or Typical Case**: Examines a case that embodies characteristics common to many cases, aiming to capture the conditions and features of a typical situation (George & Bennett, 2005).

3. **Strengths of Single-case Designs**

- **Depth**: The singular focus allows for exhaustive and detailed exploration, often resulting in rich, nuanced insights.
- **Flexibility**: Researchers can adjust their methods, tools, or focus based on emerging data, ensuring alignment with the evolving nature of the case (Stake, 1995).
- **Triangulation**: Given the depth of exploration, multiple sources of evidence can be used, enhancing the validity and reliability of the findings (Yin, 2014).

4. **Limitations and Considerations**

- **Generalizability**: Findings from a single-case design may not be generalizable to other settings or contexts. However, they can offer analytic generalization where the results provide a conceptual framework that can be tested in other situations (George & Bennett, 2005).
- **Overemphasis**: There's potential risk of placing undue emphasis on a singular instance, which might not be reflective of broader phenomena (Flyvbjerg, 2006).

Single-case designs, while focusing on one particular instance, offer a profound depth of understanding and insight into complex phenomena within specific contexts. Whether highlighting the typical, the unique, the critical, or the revelatory, these designs provide valuable contributions to academic discourse and practical implications.

- ***Multiple-case designs:*** This approach is used when the researcher decides to compare two or more cases to understand similarities, differences, or patterns (Eisenhardt & Graebner, 2007).

While the single-case design offers a detailed exploration of one particular instance, the multiple-case (or multicase) design broadens the scope to include several instances in a single study. By juxtaposing multiple cases, researchers can engage in a more robust examination, deriving insights from patterns, comparisons, and contrasts.

1. Purpose and Rationale

Multiple-case designs aim to investigate a particular phenomenon across several cases to capture a wider array of perspectives, settings, or conditions. This design offers replication logic; findings from one case can be compared and contrasted with others, enhancing the robustness of the conclusions (Yin, 2014).

2. Types of Multiple-case Designs
    - Comparative Case Studies: These involve comparing two or more cases to identify similarities and differences, often to uncover patterns or to test theories across varied settings (Bartlett & Vavrus, 2017).
    - **Replication Logic**: Cases are selected based on the expectation that they will either produce similar (literal replication) or contrasting (theoretical replication) results, given the underlying theory (Yin, 2014).
    - **Holistic Multiple-case Studies**: This approach considers each case as a whole, examining it in its entirety before moving on to the next (George & Bennett, 2005).
3. Strengths of Multiple-case Designs
    - **Robustness**: By examining multiple instances, researchers can generate findings that are more robust and compelling than those from single-case studies.
    - **Variability**: Multiple cases provide varied contexts, settings, or conditions, allowing for richer data and a deeper understanding of the phenomenon's scope and nuances (Stake, 2006).
    - **Analytic Generalizability**: While single-case designs might face challenges in generalizing findings, multiple-case designs can facilitate analytic generalizability, where the developed theory or framework can be applied to different situations (Yin, 2014).
4. Limitations and Considerations
    - **Resource Intensive**: Managing multiple cases can be more time-consuming, demanding more resources, and complicating logistics (Baxter & Jack, 2008).
    - **Complex Analysis**: Juggling data from multiple cases can make analysis more intricate, requiring meticulous data management and synthesis (Eisenhardt, 1989).

Multiple-case designs in case study research offer a comprehensive and multifaceted lens, allowing researchers to capture a richer breadth of insights. Whether the aim is to compare, contrast, or replicate findings across cases, this approach provides a solid foundation for building robust and generalizable theories or frameworks.

- ***Intrinsic, Instrumental, and Collective*:** Stake (1995) proposed this typology where intrinsic refers to studying a case for its inherent value, instrumental for understanding something broader through a specific case, and collective as studying multiple cases.

Stake (1995) provides a thoughtful categorization of case study research into three main types: intrinsic, instrumental, and collective. These classifications offer distinct approaches to the subject of investigation, each with its own purpose and emphasis. Distinguishing among these types can guide researchers in their methodological considerations and choices.

1. Intrinsic Case Study

Purpose: The intrinsic case study is characterized by an intrinsic interest in the case itself. The researcher's main concern isn't to generalize or to build a theory but to gain a deep understanding of that particular case.

Characteristics

- The case is unique, and there's something intriguing about its particularity.
- The focus is on exploring the case in all its complexity rather than trying to generalize findings.

Applications: For example, a researcher might conduct an intrinsic case study on a unique educational program that has gained significant attention.

Reference: Stake, R. E. (2000). *Case Studies.* In N. K. Denzin & Y. S. Lincoln (Eds.), *Handbook of Qualitative Research* (pp. 435-454). Sage.

2. Instrumental Case Study

Purpose: An instrumental case study provides insights into an issue or redraws a generalization. The case is of secondary interest; it plays a supportive role, facilitating our understanding of something else.

Characteristics

- The focus is not on the case itself but on a broader issue or phenomenon.
- The case becomes a means to answer a research question.

Applications: Investigating a particular school's approach to inclusive education not because the school is unique, but to understand the broader challenges and strategies of inclusive education.

Reference: Stake, R. E. (1995). *The art of case study research.* Sage.

3. Collective Case Study

Purpose: A collective case study (or multiple case study) investigates several cases to inquire into a particular phenomenon, population, or general condition.

Characteristics

- It is an extension of an instrumental case study, with multiple cases being chosen.
- The cases are studied jointly rather than sequentially.

Applications: Studying how various companies have adapted their operations in response to a global pandemic, aiming to identify common challenges and strategies.

Reference: Creswell, J. W. (2013). *Qualitative inquiry and research design: Choosing among five approaches.* Sage publications.

Stake's classification of case studies into intrinsic, instrumental, and collective offers researchers a nuanced understanding of the range of approaches within case study research. By discerning the primary interest—whether it's the case itself, an issue illuminated by the case, or a broader phenomenon across multiple cases—researchers can tailor their methodologies and analyses to best serve their research objectives.

Methodological Rigor in Case Study Research

While case studies are sometimes critiqued for being less "generalizable", it's essential to understand that they prioritize "transferability" (Lincoln & Guba, 1985). Ensuring rigor involves:

I. **Triangulation:** Employing multiple sources of data and methods enhances the validity of findings (Denzin, 1978).

Triangulation is a foundational element in ensuring methodological rigor in case study research. By employing multiple sources, methods, investigators, or theories, triangulation strengthens the validity and reliability of findings, allowing researchers to paint a comprehensive picture of the phenomenon under study. The concept of triangulation in qualitative research has its roots in navigation and surveying, where multiple reference points ensure accuracy. Similarly, in research, triangulation ensures that the study's findings are more than a reflection of a singular method or source.

1. Types of Triangulation in Case Study Research
    - **Data Triangulation**: This involves gathering information from various sources at different times, settings, and people, ensuring that the phenomenon under study is not narrowly observed (Denzin, 1978).
    - **Methodological Triangulation**: Here, researchers use more than one method to gather data, such as interviews, observations, and document analysis. By cross-checking across multiple methods, it ensures a well-rounded understanding (Carter, Bryant-Lukosius, DiCenso, Blythe, & Neville, 2014).
    - **Investigator Triangulation**: This involves multiple researchers analyzing the data, providing diverse perspectives, and ensuring a broader view that mitigates individual biases (Patton, 1999).

- **Theoretical Triangulation**: This requires examining the data from multiple theoretical perspectives, broadening the interpretative frame, and enhancing the depth of the analysis (Knafl & Breitmayer, 1989).

2. Advantages of Triangulation
    - **Enhanced Validity**: Triangulation lends credibility to the findings by demonstrating that they are consistent across various data sources or methods (Lincoln & Guba, 1985).
    - **Comprehensive Understanding**: It offers a multidimensional view of the phenomenon, ensuring no facet is overlooked (Cohen & Manion, 2000).
    - **Mitigation of Bias**: Using multiple sources, methods, or investigators reduces the chance that the findings will reflect singular biases or limitations (Mathison, 1988).
3. Challenges in Triangulation
    - **Resource Intensive**: Managing and analyzing data from multiple sources or methods can be time-consuming and demanding (Denzin, 1978).
    - **Potential Conflicts**: Varied methods or sources might produce differing or even conflicting results, necessitating complex interpretations and reconciliation (Fielding & Fielding, 1986).

Triangulation stands as a pillar of methodological rigor in case study research. By integrating diverse data sources, methods, perspectives, or theories, triangulation ensures that the derived insights are rich, reliable, and comprehensive, further establishing the case study's value in illuminating complex real-world phenomena.

II. **Member Checking:** Sharing findings with participants to ascertain accuracy (Creswell & Miller, 2000).

Methodological rigor in qualitative research is paramount, and in the realm of case study research, member checking stands as a critical method for ensuring the accuracy, credibility, and validity of findings. Member checking, also known as informant feedback or respondent validation, involves sharing research findings with participants to validate the accuracy and resonance of the researcher's interpretations.

1. Process of Member Checking

Member checking typically involves returning to participants with summaries, interpretations, or even entire drafts of findings, allowing them the opportunity to confirm, refute, or modify the interpretations (Lincoln & Guba, 1985). This iterative feedback loop serves to validate or refine the researcher's account.

2. Advantages of Member Checking:
    - **Enhanced Credibility**: Validation from participants lends credibility to the findings, ensuring they resonate with participants' experiences (Creswell & Miller, 2000).
    - **Mitigation of Researcher Bias**: By seeking participants' views on the accuracy of interpretations, researchers can identify and correct potential biases or misunderstandings (Birt, Scott, Cavers, Campbell, & Walter, 2016).
    - **Deepened Insights**: Participant feedback can provide deeper insights, nuanced understandings, or alternative perspectives that the researcher may have overlooked (Harper & Cole, 2012).

3. Challenges and Considerations in Member Checking
    - **Time-Intensive**: The iterative process of sharing findings and incorporating feedback can be lengthy (Morse, 2015).
    - **Potential for Conflict**: Discrepancies between researcher interpretations and participant views can arise, necessitating careful negotiation (Thomas, 2017).
    - **Representation Issues**: The risk exists that researchers might prioritize some voices over others, leading to an imbalanced representation (Maxwell, 2012).
4. Implementing Member Checking Effectively

To maximize the benefits and address potential challenges of member checking:

- Researchers should clearly explain the purpose of member checking to participants (Creswell & Poth, 2018).
- Feedback should be sought at various stages of the research, not just at the end (Smith & McGannon, 2018).
- It's essential to approach the feedback with an open mind, being receptive to critiques and willing to adapt interpretations accordingly (Cho & Trent, 2006).

Member checking stands as a cornerstone of methodological rigor in case study research. By actively engaging participants in the validation process, researchers not only enhance the credibility and accuracy of their findings but also uphold the ethical commitment to represent participants' experiences faithfully.

III. **Rich, Thick Descriptions:** Providing detailed accounts enhances the depth and credibility of the study (Geertz, 1973).

In qualitative research, and notably within the case study methodology, the presentation of data in the form of rich, thick descriptions is an essential component to ensure depth, breadth, and authenticity. These detailed accounts offer readers a vivid picture of the context, settings, interactions, and the nuances of participants' experiences.

1. The Concept of Rich, Thick Descriptions

Coined by the anthropologist Clifford Geertz (1973), the term "thick description" refers to a detailed account of field experiences in which the researcher makes explicit both their own interpretations and the deeper meaning of human actions they've observed. Whereas "thin description" merely records facts, "thick description" explains not just behaviors but their contexts as well, thereby interpreting the layers of meaning in those behaviors.

2. Importance in Case Study Research
    - **Contextual Relevance**: Rich descriptions provide the context necessary to understand the phenomena under study, ensuring the research findings' transferability (Lincoln & Guba, 1985).

- **Credibility**: Detailed descriptions enable readers to judge the authenticity and credibility of the research findings, allowing them to determine the relevance of the research to their own contexts (Creswell, 1998).
- **Transparency**: By offering in-depth insights, the researcher transparently showcases the data, enabling a clear path of interpretation from data collection to findings (Merriam, 1995).

3. Challenges in Crafting Thick Descriptions
    - **Balancing Detail with Relevance**: While depth is necessary, the researcher must discern which details are pertinent to the research questions and omit superfluous information (Stake, 1995).
    - **Maintaining Participant Anonymity**: Given the depth of description, ensuring that participants' identities are not inadvertently disclosed can be challenging (Holloway & Todres, 2003).
4. Employing Rich Descriptions Effectively
    - **Deep Engagement with Data**: Researchers should immerse themselves deeply in their data, ensuring they capture the nuances and intricacies of participants' experiences (Flyvbjerg, 2006).
    - **Iterative Writing Process**: Crafting thick descriptions requires multiple drafts and revisions, with researchers continually reflecting on the best ways to convey the depth of their observations (Geertz, 1973).

Rich, thick descriptions stand as a hallmark of methodological rigor in case study research. By offering in-depth and layered accounts of phenomena, researchers provide readers with the tools to understand, interpret, and evaluate the study's context, relevance, and implications.

IV. **Peer Debriefing:** Engaging peers in the review of the data and findings ensures objectivity (Lincoln & Guba, 1985).

Peer debriefing, also known as peer review or peer examination, serves as a critical tool for enhancing methodological rigor in case study research. By subjecting one's interpretations and insights to the scrutiny of peers, researchers can increase the credibility, validity, and reliability of their findings.

1. Understanding Peer Debriefing

Peer debriefing involves researchers engaging colleagues who can critically assess the research process and outcomes. These peers, having a distance from the direct research process, offer fresh perspectives, posing challenging questions, and suggesting alternative interpretations (Lincoln & Guba, 1985).

2. The Significance in Case Study Research
    - **Enhancing Credibility**: Peer debriefing aids in confirming the accuracy and resonance of the findings, ensuring that the interpretations align with the data (Creswell & Miller, 2000).
    - **Reducing Researcher Bias**: By engaging peers to challenge interpretations and methods, researchers can identify and rectify potential biases, blind spots, or overemphasis in the study (Erlandson et al., 1993).

- **Refining Analytical Processes**: As peers challenge assumptions and methodologies, they assist the primary researcher in refining their analytical process and enhancing the study's rigor (Morse et al., 2002).

3. Effective Implementation of Peer Debriefing
   - **Engage Diverse Perspectives**: Researchers should engage peers from varied backgrounds to get comprehensive feedback (Rodgers, 2008).
   - **Regular and Iterative Process**: Rather than a one-off consultation, debriefing should be regular and iterative, accompanying various research phases (Spall, 1998).
   - **Openness to Critique**: Researchers must approach the debriefing process with an openness to critique and be willing to reconsider aspects of their work based on feedback (Harper & Cole, 2012).
4. Potential Challenges in Peer Debriefing
   - **Time Constraints**: Both for the researcher and the peer, this process can be time-consuming, especially if conducted iteratively (Creswell, 2013).
   - **Varied Interpretations**: Multiple peers may provide diverse or even conflicting feedback, which can be challenging to navigate (Elliott & Luker, 2005).

Peer debriefing stands as a cornerstone of ensuring methodological rigor in case study research. By inviting external critique and reconsideration, researchers can ensure that their interpretations are grounded, unbiased, and resonate accurately with the data they have collected.

Case Study in Contemporary Research

With the advent of dynamic, complex, and interconnected global challenges, the case study methodology is witnessing renewed interest. Be it in understanding intricate business strategies in a volatile market, exploring classroom dynamics in multicultural settings, or analyzing the nuances of urban planning in fast-growing cities, case studies offer deep insights grounded in real-world contexts (Thomas, 2011).

In the rapidly changing landscape of the 21st century, case study research has undergone substantial transformations and adaptations, solidifying its importance in various academic and applied disciplines. From business to education, and from social sciences to health care, the case study methodology has been employed to illuminate complex phenomena within real-life contexts.

1. Interdisciplinary Expansion

Historically rooted in anthropology and sociology, the case study approach has now permeated disciplines like business management, psychology, education, and health sciences. In business, for instance, case studies offer in-depth analyses of corporate strategies, consumer behavior, and market dynamics (Eisenhardt & Graebner, 2007).

2. Technology and the Digital Age

Digital transformation has expanded the horizons of case study research. With advancements in data collection tools, researchers can now gather rich multimedia data, enabling more vivid and dynamic presentations of cases. Virtual and augmented reality platforms also offer immersive case experiences, revolutionizing traditional pedagogical case study applications (Denzin & Giardina, 2016).

3. Globalization and Comparative Case Studies

Global interconnectedness has spurred the growth of comparative case studies that span across different cultural, social, and geopolitical contexts. These studies provide nuanced insights into cross-cultural dynamics and global-local interactions (Bartlett & Ghoshal, 1989).

4. Emphasis on Pragmatic Applications

Contemporary case studies increasingly emphasize solving real-world problems. In fields like public health or urban planning, case studies are pivotal in informing policy decisions, planning interventions, and evaluating their outcomes (Yin, 2018).

5. Integration with Mixed Methods Research

Modern research often embraces methodological pluralism. Case studies are frequently integrated with other qualitative or quantitative methods, providing a comprehensive research strategy that captures both depth and breadth of inquiry (Teddlie & Tashakkori, 2009).

6. Postmodern and Decolonial Perspectives

Postmodernism, with its emphasis on multiple realities and subjective interpretations, has influenced the narrative and interpretative aspects of case studies. Additionally, decolonial approaches have called for case studies that respect indigenous knowledge systems and challenge Eurocentric epistemologies (Smith, 2012).

Contemporary case study research, while retaining its foundational principles, has evolved in response to global, technological, and epistemological shifts. As researchers continue to navigate the complexities of the modern world, the versatility and depth of the case study methodology will undoubtedly remain invaluable.

CONCLUSION

The case study, as a methodology, provides a lens to view the intricate tapestry of real-world phenomena, not as isolated variables, but as interwoven threads of contexts, challenges, and nuances. As researchers continue to confront complex and multifaceted questions, the case study stands as a beacon, guiding them through the labyrinth of real-life intricacies to the heart of understanding.

REFERENCES

Abbott, H. P. (2008). *The Cambridge introduction to narrative*. Cambridge University Press. doi:10.1017/CBO9780511816932

Abrams, L. (2010). *Oral history theory*. Routledge. doi:10.4324/9780203849033

Agar, M. (1980). *The Professional Stranger: An Informal Introduction to Ethnography*. Academic Press.

Agar, M. (1996). *The Professional Stranger: An Informal Introduction to Ethnography*. Academic Press.

Amit, V. (Ed.). (2000). *Constructing the field: Ethnographic fieldwork in the contemporary world*. Routledge.

Appadurai, A. (1991). Global ethnoscapes: Notes and queries for a transnational anthropology. In R. G. Fox (Ed.), *Recapturing Anthropology* (pp. 191–210). School of American Research Press.

Arendt, H. (1958). *The Human Condition*. University of Chicago Press.

Arendt, H. (1970). *On Violence*. Harcourt, Brace & World.

Asad, T. (Ed.). (1973). *Anthropology & the colonial encounter*. Ithaca Press.

Atkinson, P., Coffey, A., & Delamont, S. (2003). *Key themes in qualitative research: Continuities and changes*. Altamira Press.

Atkinson, P., Coffey, A., Delamont, S., Lofland, J., & Lofland, L. (2001). *Handbook of Ethnography*. SAGE Publications. doi:10.4135/9781848608337

Atkinson, P., & Delamont, S. (2006). Rescuing narrative from qualitative research. *Narrative Inquiry*, *16*(1), 164–172. doi:10.1075/ni.16.1.21atk

Atkinson, R. (1998). The life story interview. *Sage (Atlanta, Ga.)*.

Bakhtin, M. M. (1981). *The dialogic imagination: Four essays*. University of Texas Press.

Bal, M. (1997). *Narratology: Introduction to the theory of narrative*. University of Toronto Press.

Bamberg, M. (2004). Narrative discourse and identities. In J. C. Meister, T. Kindt, & W. Schernus (Eds.), Narratology beyond literary criticism (pp. 213-237). Walter de Gruyter.

Bartlett, C. A., & Ghoshal, S. (1989). *Managing Across Borders: The Transnational Solution*. Harvard Business Press.

Bartlett, L., & Vavrus, F. (2017). *Comparative case studies*. Routledge. doi:10.1590/2175-623668636

Bateson, G. (1936). *Naven: A survey of the problems suggested by a composite picture of the culture of a New Guinea tribe drawn from three points of view*. Stanford University Press.

Baxter, P., & Jack, S. (2008). Qualitative case study methodology: Study design and implementation for novice researchers. *The Qualitative Report*, *13*(4), 544–559.

Becker, H. S. (1998). *Tricks of the trade: How to think about your research while you're doing it*. University of Chicago Press. doi:10.7208/chicago/9780226040998.001.0001

Behar, R. (1996). *The Vulnerable Observer: Anthropology That Breaks Your Heart*. Beacon Press.

Bernard, H. R. (2006). *Research methods in anthropology: Qualitative and quantitative approaches*. AltaMira Press.

Bernard, H. R. (2011). *Research methods in anthropology: Qualitative and quantitative approaches*. AltaMira Press.

Bertaux, D., & Kohli, M. (1984). The life story approach: A continental view. *Annual Review of Sociology*, *10*(1), 215–237. doi:10.1146/annurev.so.10.080184.001243

Birks, M., & Mills, J. (2015). Grounded theory: A practical guide. *Sage (Atlanta, Ga.)*.

Birt, L., Scott, S., Cavers, D., Campbell, C., & Walter, F. (2016). Member Checking. *Qualitative Health Research*, *26*(13), 1802–1811. doi:10.1177/1049732316654870 PMID:27340178

Bishop, E. (2004). The art of the diary. In D. Finkelstein & A. McCleery (Eds.), *The book history reader* (pp. 407–415). Routledge.

Blair, C. (2004). The historian's craft, popular memory, and Wikipedia. In D. Boyd (Ed.), *Memory and popular film* (pp. 215–235). Manchester University Press.

Blattner, W. (2006). Heidegger's 'Being and Time': A Reader's Guide. *Continuum*.

Boas, F. (1940). *Race, language, and culture*. University of Chicago Press.

Boje, D. M. (1991). The storytelling organization: A study of story performance in an office-supply firm. *Administrative Science Quarterly*, *36*(1), 106–126. doi:10.2307/2393432

Bornat, J. (1989). Oral history as a social movement: Reminiscence and older people. *Oral History (Colchester)*, *17*(2), 16–24.

Braun, V., & Clarke, V. (2006). Using thematic analysis in psychology. *Qualitative Research in Psychology*, *3*(2), 77–101. doi:10.1191/1478088706qp063oa

Brentano, F. (1874). *Psychology from an Empirical Standpoint*. Routledge.

Brettell, C. (2017). *When they read what we write: The politics of ethnography*. Taylor & Francis.

Bruner, J. (1991). The narrative construction of reality. *Critical Inquiry*, *18*(1), 1–21. doi:10.1086/448619

Bruner, J. (1993). The autobiographical process. In R. Folkenflik (Ed.), *The culture of autobiography: Constructions of self-representation* (pp. 38–56). Stanford University Press. doi:10.1515/9781503622043-006

Bruner, J. (1997). A narrative model of self-construction. In J. G. Snodgrass & R. L. Thompson (Eds.), *The self across psychology* (pp. 145–161). NYU Press.

Bryant, A., & Charmaz, K. (Eds.). (2007). *The SAGE Handbook of Grounded Theory*. Sage. doi:10.4135/9781848607941

Bunkers, S. L., & Huff, C. A. (1996). Issues in studying women's diaries: A theoretical and critical introduction. In S. L. Bunkers & C. A. Huff (Eds.), *Inscriptions: Diaries and journals as sources in literary research* (pp. 1–26). Popular Press.

Butler, J. (1990). *Gender Trouble: Feminism and the Subversion of Identity*. Routledge.

Campbell, D. T., & Stanley, J. C. (1966). *Experimental and quasi-experimental designs for research*. Rand McNally.

Carman, T. (2008). *Merleau-Ponty*. Routledge. doi:10.4324/9780203461853

Carter, N., Bryant-Lukosius, D., DiCenso, A., Blythe, J., & Neville, A. J. (2014). The use of triangulation in qualitative research. *Oncology Nursing Forum*, *41*(5), 545–547. doi:10.1188/14.ONF.545-547 PMID:25158659

Caruth, C. (Ed.). (1996). *Trauma: Explorations in memory*. Johns Hopkins University Press.

Cavarero, A. (2000). *Relating narratives: Storytelling and selfhood*. Routledge.

Chalmers, D. J. (1996). The Conscious Mind. In *Search of a Fundamental Theory*. Oxford University Press.

Charmaz, K. (2006). Constructing grounded theory: A practical guide through qualitative analysis. *Sage (Atlanta, Ga.)*.

Charmaz, K. (2014). *Constructing grounded theory* (2nd ed.). Sage.

Charon, R. (2006). *Narrative medicine: Honoring the stories of illness*. Oxford University Press. doi:10.1093/oso/9780195166750.001.0001

Chase, S. E. (2005). Narrative inquiry: Multiple lenses, voices, and co-constructed narratives. In N. K. Denzin & Y. S. Lincoln (Eds.), *The Sage handbook of qualitative research* (pp. 651–679). Sage.

Cho, J., & Trent, A. (2006). Validity in qualitative research revisited. *Qualitative Research*, *6*(3), 319–340. doi:10.1177/1468794106065006

Clandinin, D. J., & Connelly, F. M. (2000). *Narrative inquiry: Experience and story in qualitative research*. Jossey-Bass.

Clandinin, D. J., & Rosiek, J. (2007). Mapping a landscape of narrative inquiry. In D. J. Clandinin (Ed.), *Handbook of narrative inquiry: Mapping a methodology* (pp. 35–75). Sage. doi:10.4135/9781452226552.n2

Clifford, J., & Marcus, G. E. (Eds.). (1986). *Writing culture: The poetics and politics of ethnography*. University of California Press. doi:10.1525/9780520946286

Coffey, A. (1999). The ethnographic self: Fieldwork and the representation of identity. *Sage (Atlanta, Ga.)*.

Cohen, L., & Manion, L. (2000). *Research methods in education* (5th ed.). RoutledgeFalmer.

Comaroff, J., & Comaroff, J. L. (1991). *Of revelation and revolution: Christianity, colonialism, and consciousness in South Africa* (Vol. 1). University of Chicago Press. doi:10.7208/chicago/9780226114477.001.0001

Corbin, J., & Strauss, A. (2015). *Basics of qualitative research: Techniques and procedures for developing grounded theory*. Sage publications.

Couldry, N. (2010). Why voice matters: Culture and politics after neoliberalism. *Sage (Atlanta, Ga.)*. Advance online publication. doi:10.4135/9781446269114

Crane, T. (2001). *Elements of Mind: An Introduction to the Philosophy of Mind*. Oxford University Press.

Creswell, J. W. (2013). *Research design: Qualitative, quantitative, and mixed methods approaches*. Sage publications.

Creswell, J. W., & Poth, C. N. (2018). *Qualitative inquiry and research design: Choosing among five approaches* (4th ed.). Sage.

Davies, C. A. (2008). *Reflexive ethnography: A guide to researching selves and others*. Routledge.

De Fina, A., & Georgakopoulou, A. (2015). *The handbook of narrative analysis*. John Wiley & Sons. doi:10.1002/9781118458204

Dennett, D. C. (1987). *The Intentional Stance*. MIT Press.

Denzin, N. K. (1978). *The research act: A theoretical introduction to sociological methods*. McGraw-Hill.

Denzin, N. K. (1989). *The research act: A theoretical introduction to sociological methods*. Prentice Hall.

Denzin, N. K. (1997). Interpretive ethnography: Ethnographic practices for the 21st century. *Sage (Atlanta, Ga.)*. Advance online publication. doi:10.4135/9781452243672

Denzin, N. K., & Giardina, M. D. (2016). Qualitative inquiry through a critical lens. *International Journal of Qualitative Studies in Education : QSE*, *29*(6), 731–753.

DeWalt, K. M., & DeWalt, B. R. (2011). *Participant observation: A guide for fieldworkers*. Rowman & Littlefield Publishers.

Dey, I. (1999). *Grounding grounded theory: Guidelines for qualitative inquiry*. Academic Press.

Dreyfus, H. L. (1991). *Being-in-the-world: A Commentary on Heidegger's Being and Time, Division I*. MIT Press.

Durkheim, É. (1915). *The elementary forms of the religious life*. George Allen & Unwin.

Eakin, P. J. (2006). *The ethics of life writing*. Cornell University Press.

Eastmond, M. (2007). Stories as lived experience: Narratives in forced migration research. *Journal of Refugee Studies*, *20*(2), 248–264. doi:10.1093/jrs/fem007

Eisenhardt, K. M., & Graebner, M. E. (2007). Theory building from cases: Opportunities and challenges. *Academy of Management Journal*, *50*(1), 25–32. doi:10.5465/amj.2007.24160888

Elliott, N., & Luker, K. (2005). The role of the researcher in the qualitative research process. A potential barrier to archiving qualitative data. *Forum Qualitative Sozialforschung / Forum: Qualitative. Social Research*, *6*(3).

Ellis, C. (2007). Telling secrets, revealing lives: Relational ethics in research with intimate others. *Qualitative Inquiry*, *13*(1), 3–29. doi:10.1177/1077800406294947

Ellis, C., & Bochner, A. P. (2000). Autoethnography, personal narrative, reflexivity. In N. K. Denzin & Y. S. Lincoln (Eds.), *Handbook of qualitative research* (2nd ed., pp. 733–768). Sage.

Emerson, R. M., Fretz, R. I., & Shaw, L. L. (1995). *Writing ethnographic fieldnotes*. University of Chicago Press. doi:10.7208/chicago/9780226206851.001.0001

Etherington, K. (2007). Ethical research in reflexive relationships. *Qualitative Inquiry*, *13*(5), 599–616. doi:10.1177/1077800407301175

Evans-Pritchard, E. E. (1940). *The Nuer*. Clarendon Press.

Fabian, J. (1983). *Time and the Other: How Anthropology Makes Its Object*. Columbia University Press.

Fielding, N. G., & Fielding, J. L. (1986). Linking data. *Sage (Atlanta, Ga.)*.

Fielding, N. G., & Lee, R. M. (1991). Using computers in qualitative research. *Sage (Atlanta, Ga.)*.

Fine, M. (1994). Working the hyphens: Reinventing self and other in qualitative research. In N. K. Denzin & Y. S. Lincoln (Eds.), *Handbook of qualitative research* (pp. 70–82). Sage.

Finlay, L. (2002). "Outing" the Researcher: The Provenance, Process, and Practice of Reflexivity. *Qualitative Health Research*, *12*(4), 531–545. doi:10.1177/104973202129120052 PMID:11939252

Fluehr-Lobban, C. (1991). Ethical Problems of Fieldwork in the Newly Recognized States. *Anthropology Today*, *7*(4), 5–7.

Fluehr-Lobban, C. (2002). *Ethics and the profession of anthropology: Dialogue for a new era*. Altamira Press.

Flyvbjerg, B. (2006). Five misunderstandings about case-study research. *Qualitative Inquiry*, *12*(2), 219–245. doi:10.1177/1077800405284363

Freeman, M. (2010). *Hindsight: The promise and peril of looking backward*. Oxford University Press.

Frisch, M. (1990). *A shared authority: Essays on the craft and meaning of oral and public history*. SUNY Press.

Gee, J. P. (1986). *Narrative, literacy, and face in interethnic communication*. Ablex Publishing.

Geertz, C. (1973). *The interpretation of cultures: Selected essays*. Basic Books.

George, A. L., & Bennett, A. (2005). *Case studies and theory development in the social sciences*. MIT Press.

Gerring, J. (2004). What is a case study and what is it good for? *The American Political Science Review*, *98*(2), 341–354. doi:10.1017/S0003055404001182

Giorgi, A. (2009). *The Descriptive Phenomenological Method in Psychology: A Modified Husserlian Approach*. Duquesne University Press.

Glaser, B. G. (1965). The constant comparative method of qualitative analysis. *Social Problems*, *12*(4), 436–445. doi:10.2307/798843

Glaser, B. G. (1978). *Theoretical sensitivity: Advances in the methodology of grounded theory*. Sociology Press.

Glaser, B. G., & Strauss, A. L. (1965). *Awareness of dying*. Aldine Publishing Company.

Grant, C., & Osanloo, A. (2014). Understanding, selecting, and integrating a theoretical framework in dissertation research. *Administrative Issues Journal: Connecting Education, Practice, and Research*, *4*(2), 12-26.

Grele, R. J. (1985). Movement without aim: Methodological and theoretical problems in oral history. In R. J. Grele (Ed.), *Envelopes of sound: The art of oral history* (pp. 38–59). Praeger.

Gubrium, A., & Harper, K. (2013). *Participatory visual and digital research in action*. Left Coast Press.

Hammersley, M. (1992). *What's Wrong with Ethnography? Methodological Explorations*. Routledge.

Hammersley, M. (2007). The issue of quality in qualitative research. *International Journal of Research & Method in Education*, *30*(3), 287–305. doi:10.1080/17437270701614782

Harper, M., & Cole, P. (2012). Member checking: Can benefits be gained similar to group therapy? *The Qualitative Report*, *17*(2), 510–517.

Heidegger, M. (1927/1962). *Being and Time*. Harper & Row.

Holloway, I., & Todres, L. (2003). The status of method: Flexibility, consistency and coherence. *Qualitative Research*, *3*(3), 345–357. doi:10.1177/1468794103033004

Horkheimer, M. (1982). *Critical theory*. Seabury Press.

Husserl, E. (2012). *Ideas: General Introduction to Pure Phenomenology*. Routledge. (Original work published 1913) doi:10.4324/9780203120330

Jewitt, C., & Oyama, R. (2001). Visual meaning: A social semiotic approach. In T. van Leeuwen & C. Jewitt (Eds.), *Handbook of visual analysis* (pp. 134–156). Sage.

Jones, K. (2007). *Narratives of identity and place*. Routledge.

Jones, R. H., Chik, A., & Hafner, C. A. (2015). *Discourse and digital practices: Doing discourse analysis in the digital age*. Routledge. doi:10.4324/9781315726465

Kawulich, B. B. (2005). Participant observation as a data collection method. *Forum Qualitative Social Research*, *6*(2).

Kelle, U. (2005). "Emergence" vs. "Forcing" of empirical data? A crucial problem of "Grounded Theory" reconsidered. *Forum Qualitative Sozialforschung/Forum: Qualitative. Social Research*, *6*(2).

Kirk, J., & Miller, M. L. (1986). Reliability and validity in qualitative research. *Sage (Atlanta, Ga.)*.

Knafl, K. A., & Breitmayer, B. J. (1989). *Triangulation in qualitative research: Issues of conceptual clarity and purpose*. Qualitative nursing research: A contemporary dialogue, 226-239.

Knoblauch, H. (2005). Focused ethnography. *Forum Qualitative Social Research*, *6*(3).

Kuhn, A. (2002). *Family secrets: Acts of memory and imagination*. Verso.

Kuklick, H. (1997). *The savage within: The social history of British anthropology, 1885-1945*. Cambridge University Press.

Kuper, A. (1973). *Anthropologists and anthropology: The British school 1922-1972*. Penguin.

Labov, W. (1972). *Language in the inner city: Studies in the Black English vernacular*. University of Pennsylvania Press.

Labov, W., & Waletzky, J. (1967). Narrative analysis: Oral versions of personal experience. In J. Helm (Ed.), *Essays on the verbal and visual arts* (pp. 12–44). University of Washington Press.

Labov, W., & Waletzky, J. (1997). Narrative analysis: Oral versions of personal experience. *Journal of Narrative and Life History*, *7*(1-4), 3–38. doi:10.1075/jnlh.7.02nar

Lambert, J. (2010). *Digital storytelling: Capturing lives, creating community*. Routledge.

Lassiter, L. E. (2005). *The Chicago guide to collaborative ethnography*. University of Chicago Press. doi:10.7208/chicago/9780226467016.001.0001

Leach, E. R. (1954). *Political Systems of Highland Burma*. Harvard University Press.

Lecompte, M. D., & Schensul, J. J. (1999). *Designing & conducting ethnographic research*. Rowman & Littlefield.

LeCompte, M. D., & Schensul, J. J. (2010). *Designing and conducting ethnographic research: An introduction*. AltaMira Press.

Lejeune, P. (2009). *On diary*. University of Hawai'i Press.

Lempert, L. B. (2007). Asking questions of the data: Memo writing in the grounded theory tradition. The SAGE Handbook of Grounded Theory, 245-264.

Lijphart, A. (1971). Comparative politics and the comparative method. *The American Political Science Review*, *65*(3), 682–693. doi:10.2307/1955513

Lincoln, Y. S., & Guba, E. A. (1985). *Naturalistic inquiry*. Sage. doi:10.1016/0147-1767(85)90062-8

Madison, D. S. (2005). Critical ethnography: Method, ethics, and performance. *Sage (Atlanta, Ga.)*.

Malinowski, B. (1922). *Argonauts of the Western Pacific*. Routledge & Kegan Paul.

Malinowski, B. (1944). *A Scientific Theory of Culture and Other Essays*. University of North Carolina Press.

Marcus, G. E. (1995). Ethnography in/of the world system: The emergence of multi-sited ethnography. *Annual Review of Anthropology*, *24*(1), 95–117. doi:10.1146/annurev.an.24.100195.000523

Marcus, G. E., & Fischer, M. M. (1986). *Anthropology as Cultural Critique: An Experimental Moment in the Human Sciences*. University of Chicago Press.

Mathison, S. (1988). Why triangulate? *Educational Researcher, 17*(2), 13–17. doi:10.2307/1174583

Mattingly, C. (1998). *Healing dramas and clinical plots: The narrative structure of experience*. Cambridge University Press. doi:10.1017/CBO9781139167017

Maxwell, J. A. (2012). The importance of qualitative research for causal explanation in education. *Qualitative Inquiry, 18*(8), 655–661. doi:10.1177/1077800412452856

Maxwell, J. A. (2013). *Qualitative research design: An interactive approach*. Sage Publications.

Mead, M. (1928). *Coming of Age in Samoa*. William Morrow & Company.

Merleau-Ponty, M. (2012). *Phenomenology of Perception*. Routledge. (Original work published 1945)

Merriam, S. B. (1998). *Qualitative research and case study applications in education*. Jossey-Bass.

Mishler, E. G. (1995). Models of narrative analysis: A typology. *Journal of Narrative and Life History, 5*(2), 87–123. doi:10.1075/jnlh.5.2.01mod

Moran, D. (2000). *Introduction to Phenomenology*. Routledge.

Morgan, D. L. (1996). *Focus groups as qualitative research* (2nd ed.). SAGE.

Morse, J. M. (1991). Strategies for sampling. In J. M. Morse (Ed.), *Qualitative nursing research: A contemporary dialogue* (pp. 127–145). Sage. doi:10.4135/9781483349015.n16

Morse, J. M. (2015). Critical analysis of strategies for determining rigor in qualitative inquiry. *Qualitative Health Research, 25*(9), 1212–1222. doi:10.1177/1049732315588501 PMID:26184336

Morse, J. M., Barrett, M., Mayan, M., Olson, K., & Spiers, J. (2002). Verification strategies for establishing reliability and validity in qualitative research. *International Journal of Qualitative Methods, 1*(2), 13–22. doi:10.1177/160940690200100202

Motz, A. (1983). The diary as a transitional object in female adolescent development. *The Psychoanalytic Study of the Child, 38*(1), 283–297.

Moustakas, C. (1994). Phenomenological research methods. *Sage (Atlanta, Ga.)*.

Mulhall, S. (2005). *Heidegger and Being and Time*. Routledge.

Murphy, E., & Dingwall, R. (2001). The Ethics of Ethnography. In P. Atkinson, A. Coffey, S. Delamont, J. Lofland, & L. Lofland (Eds.), *Handbook of Ethnography* (pp. 339–351). SAGE Publications. doi:10.4135/9781848608337.n23

Nadel, S. F. (1951). *The Foundations of Social Anthropology*. Cohen & West.

Nader, L. (1972). Up the anthropologist: Perspectives gained from studying up. In D. Hymes (Ed.), *Reinventing anthropology* (pp. 284–311). Pantheon Books.

Ortner, S. B. (2016). Dark anthropology and its others: Theory since the eighties. *HAU, 6*(1), 47–73. doi:10.14318/hau6.1.004

Page, R. (2012). *Stories and social media: Identities and interaction*. Routledge.

Park, R. E., Burgess, E. W., & McKenzie, R. D. (1925). *The city*. University of Chicago Press.

Patton, M. Q. (1999). Enhancing the quality and credibility of qualitative analysis. *Health Services Research*, *34*(5 Pt 2), 1189. PMID:10591279

Patton, M. Q. (2002). *Qualitative research and evaluation methods* (3rd ed.). Sage Publications.

Pennebaker, J. W. (1997). Writing about emotional experiences as a therapeutic process. *Psychological Science*, *8*(3), 162–166. doi:10.1111/j.1467-9280.1997.tb00403.x

Perks, R., & Thomson, A. (Eds.). (2006). *The oral history reader*. Routledge.

Phoenix, C. (2008). Analyzing narrative data. In P. Liamputtong & J. Rumbold (Eds.), *Knowing differently: Arts-based and collaborative research methods* (pp. 77–89). Nova Science Publishers.

Piaget, J. (1952). *The origins of intelligence in children*. International Universities Press. doi:10.1037/11494-000

Pike, K. L. (1967). Emic and etic standpoints for the description of behavior. In J. L. Helm (Ed.), *Essays in the verbal and visual arts* (pp. 32–38). University of Washington Press. doi:10.1037/14786-002

Pink, S. (2007). Doing visual ethnography. *Sage (Atlanta, Ga.)*.

Plummer, K. (2001a). *Documents of life 2: An invitation to a critical humanism*. Sage. doi:10.4135/9781849208888

Plummer, K. (2001b). The call of life stories in ethnographic research. In P. Atkinson, A. Coffey, S. Delamont, J. Lofland, & L. Lofland (Eds.), *Handbook of ethnography* (pp. 395–406). Sage. doi:10.4135/9781848608337.n27

Polkinghorne, D. E. (1988). *Narrative knowing and the human sciences*. State University of New York Press.

Polkinghorne, D. E. (1991). Narrative and self-concept. *Journal of Narrative and Life History*, *1*(2&3), 135–153. doi:10.1075/jnlh.1.2-3.04nar

Polkinghorne, D. E. (2007). Validity issues in narrative research. *Qualitative Inquiry*, *13*(4), 471–486. doi:10.1177/1077800406297670

Portelli, A. (1991). *The death of Luigi Trastulli and other stories: Form and meaning in oral history*. SUNY Press.

Portelli, A. (1997). What makes oral history different. In R. Perks & A. Thomson (Eds.), *The oral history reader* (pp. 63–74). Routledge.

Procter, L. (2009). Fashioning the diary as a literary genre, 1660–1800. *Literature Compass*, *6*(2), 431–441.

Rabinow, P. (1977). *Reflections on Fieldwork in Morocco*. University of California Press.

Radcliffe-Brown, A. R. (1952). *Structure and function in primitive society*. Free Press.

Ravitch, S. M., & Riggan, M. (2017). *Reason & rigor: How conceptual frameworks guide research*. Sage Publications.

Richardson, L. (2000). Writing: A method of inquiry. In N. K. Denzin & Y. S. Lincoln (Eds.), *Handbook of qualitative research* (pp. 923–948). Sage.

Ricoeur, P. (1981). Narrative time. In W. J. T. Mitchell (Ed.), *On narrative* (pp. 165–186). University of Chicago Press.

Ricoeur, P. (1984). *Time and narrative* (Vol. 1). University of Chicago Press.

Riessman, C. K. (1993). *Narrative analysis.* Sage Publications.

Riessman, C. K. (2008). *Narrative methods for the human sciences.* Sage Publications.

Ritchie, D. A. (2003). *Doing oral history: A practical guide.* Oxford University Press.

Roberts, B. (2002). *Biographical research.* Open University Press.

Rodgers, B. L. (2008). Audit trail. In L. M. Given (Ed.), *The Sage encyclopedia of qualitative research methods* (Vol. 1, pp. 44–45). Sage Publications.

Romdenh-Romluc, K. (2010). *Routledge Philosophy GuideBook to Merleau-Ponty and Phenomenology of Perception.* Routledge. doi:10.4324/9780203482896

Rosaldo, R. (1989). *Culture & Truth: The Remaking of Social Analysis.* Beacon.

Rosenberg, B. A. (2011). The voice of the past: Oral history. In *D. R. M. Oral history and digital humanities* (pp. 39–58). Palgrave Macmillan.

Rubin, H. J., & Rubin, I. S. (2011). *Qualitative interviewing: The art of hearing data* (3rd ed.). SAGE.

Saldaña, J. (2015). *The coding manual for qualitative researchers* (3rd ed.). Sage.

Sandelowski, M. (1995). Sample size in qualitative research. *Research in Nursing & Health, 18*(2), 179–183. doi:10.1002/nur.4770180211 PMID:7899572

Sanjek, R. (1990). On Ethnographic Validity. In R. Sanjek (Ed.), *Fieldnotes: The Makings of Anthropology* (pp. 385–418). Cornell University Press. doi:10.7591/9781501711954

Sartre, J.-P. (1943/1956). *Being and Nothingness: An Essay on Phenomenological Ontology.* Philosophical Library.

Sartre, J.-P. (1946). *Existentialism is a Humanism.* Yale University Press.

Schensul, J. J., Schensul, S. L., & LeCompte, M. D. (1999). *Essential ethnographic methods: Observations, interviews, and questionnaires.* Rowman Altamira.

Smith, B., & McGannon, K. R. (2018). Developing rigor in qualitative research: Problems and opportunities within sport and exercise psychology. *International Review of Sport and Exercise Psychology, 11*(1), 101–121. doi:10.1080/1750984X.2017.1317357

Smith, B., & Sparkes, A. C. (2008). Contrasting perspectives on narrating selves and identities: An invitation to dialogue. *Qualitative Research, 8*(1), 5–35. doi:10.1177/1468794107085221

Smith, J. A. (1978). The idea of health: A philosophical inquiry. *Advances in Nursing Science*. PMID:6782945

Smith, J. A. (2003). Qualitative psychology: A practical guide to research methods. *Sage (Atlanta, Ga.)*.

Smith, J. A., & Osborn, M. (2003). Interpretative phenomenological analysis. In J. A. Smith (Ed.), *Qualitative psychology: A practical guide to research methods* (pp. 51–80). Sage.

Smith, L. T. (2012). *Decolonizing methodologies: Research and indigenous peoples*. Zed Books Ltd.

Smith, S., & Watson, J. (2010). *Reading autobiography: A guide for interpreting life narratives* (2nd ed.). University of Minnesota Press.

Somers, M. R. (1994). The narrative constitution of identity: A relational and network approach. *Theory and Society*, *23*(5), 605–649. doi:10.1007/BF00992905

Spall, S. (1998). Peer debriefing in qualitative research: Emerging operational models. *Qualitative Inquiry*, *4*(2), 280–292. doi:10.1177/107780049800400208

Spence, D. P. (1982). *Narrative truth and historical truth: Meaning and interpretation in psychoanalysis*. WW Norton & Company.

Spivak, G. C. (1988). Can the subaltern speak? In C. Nelson & L. Grossberg (Eds.), Marxism and the interpretation of culture (pp. 271-313). Macmillan.

Spradley, J. P. (1979). *The ethnographic interview*. Holt, Rinehart, and Winston.

Squire, C. (2008). Experience-centred and culturally-oriented approaches to narrative. In M. Andrews, C. Squire, & M. Tamboukou (Eds.), *Doing narrative research* (pp. 41–63). Sage. doi:10.4135/9780857024992.d4

Stake, R. E. (1995). *The art of case study research*. Sage Publications.

Stanley, L. (Ed.). (1992). *The auto/biographical I: The theory and practice of feminist auto/biography*. Manchester University Press.

Stanley, L. (2007). The epistolarium: On theorizing letters and correspondences. *Auto/Biography*, *12*(2), 201–235.

Stoecker, R. (1991). Evaluating and rethinking the case study. *The Sociological Review*, *39*(1), 88–112. doi:10.1111/j.1467-954X.1991.tb02970.x

Strauss, A., & Corbin, J. (1990). *Basics of qualitative research: Grounded theory procedures and techniques*. Sage Publications, Inc.

Teddlie, C., & Tashakkori, A. (2009). Foundations of mixed methods research: Integrating quantitative and qualitative approaches in the social and behavioral sciences. *Sage (Atlanta, Ga.)*.

Thomas, D. R. (2017). Feedback from research participants: Are member checks useful in qualitative research? *Qualitative Research in Psychology*, *14*(1), 23–41. doi:10.1080/14780887.2016.1219435

Thomas, E., & Magilvy, J. K. (2011). Qualitative rigor or research validity in qualitative research. *Journal for Specialists in Pediatric Nursing*, *16*(2), 151–155. doi:10.1111/j.1744-6155.2011.00283.x PMID:21439005

Turner, V. (1957). *Schism and Continuity in an African Society*. Manchester University Press.

Turner, V. (1967). *The forest of symbols: Aspects of Ndembu ritual*. Cornell University Press.

Van Maanen, J. (1988). *Tales of the field: On writing ethnography*. University of Chicago Press.

Van Maanen, J. (2011). *Tales of the field: On writing ethnography*. University of Chicago Press. doi:10.7208/chicago/9780226849638.001.0001

Varela, F. J., Thompson, E., & Rosch, E. (1991). *The Embodied Mind: Cognitive Science and Human Experience*. MIT Press. doi:10.7551/mitpress/6730.001.0001

Vygotsky, L. S. (1978). *Mind in society: The development of higher psychological processes*. Harvard University Press.

White, H. (1981). The value of narrativity in the representation of reality. *Critical Inquiry*, *7*(1), 5–27. doi:10.1086/448086

Wolcott, H. F. (1995). *The art of fieldwork*. AltaMira Press.

Wolcott, H. F. (2001). Writing up qualitative research. *Sage (Atlanta, Ga.)*.

Yin, R. K. (2003). *Case study research: Design and methods* (3rd ed.). Sage Publications.

Yin, R. K. (2018). *Case study research and applications: Design and methods*. Sage publications.

Yow, V. R. (2005). *Recording oral history: A guide for the humanities and social sciences*. AltaMira Press.

Zahavi, D. (2005). *Husserl's Phenomenology*. Stanford University Press.

Zimmerman, D. H., & Wieder, D. L. (1977). The diary: Diary-interview method. *Urban Life*, *5*(4), 479–498. doi:10.1177/089124167700500406

Zinsser, W. (1989). *Inventing the truth: The art and craft of memoir*. Houghton Mifflin Harcourt.

Chapter 3
Research Design and Planning

ABSTRACT

In this chapter, the foundational blueprint of qualitative research is articulated, emphasizing the systematic strategies that guide researchers through the investigative journey. The chapter begins by elucidating the concept of research design, which encompasses the selection of appropriate methods of data collection and analysis techniques, and the structuring of the research timeframe and process. It delineates various types of research designs, including descriptive, experimental, observational, comparative, and longitudinal designs, highlighting their distinct objectives and applications.

INTRODUCTION

Research design and planning are pivotal stages in any research journey, serving as the blueprint that guides the entire investigative process. This chapter delineates key considerations and steps essential for constructing robust research designs and strategic planning.

1. Understanding Research Design

Research design is a systematic plan detailing how the research will be conducted. It encompasses choosing the research methods, deciding the data collection and analysis techniques, and determining the structure and timeframe of the study (Creswell, 2013).

Research design plays a pivotal role in shaping the trajectory of a research project, ensuring its credibility and guiding researchers to answer their questions systematically. The design is the blueprint for the collection, measurement, and analysis of data (De Vaus, 2001).

A. What is Research Design?

Research design refers to the overall strategy used to integrate the various components of a study in a coherent and logical manner, thereby ensuring the research problem is effectively addressed (Creswell, 2013).

DOI: 10.4018/979-8-3693-2414-1.ch003

A.1 Structure and Function: It provides the framework for the collection and analysis of data. The choice of design influences the types of questions asked, the data collection method, and how the results are interpreted (Maxwell, 2005).

B. Types of Research Design

B.1 Descriptive Designs: Aim to describe phenomena and their characteristics. They don't necessarily look for cause-and-effect relationships but are about capturing the "what" of a subject (Yin, 2017).

B.2 Experimental Designs: Often used in quantitative research, these designs explore cause-and-effect relationships by manipulating one or more variables and observing the effect on another (Shadish, Cook, & Campbell, 2002).

B.3 Observational Designs: Rather than manipulating variables, these designs observe and record behaviors in a natural setting, often used in ethnographic and some case study research (Angrosino, 2007).

B.4 Comparative and Cross-sectional Designs: Look at differences and similarities across groups or phenomena at a specific point in time (Bryman, 2015).

B.5 Longitudinal Designs: Examine changes over an extended period. They are crucial for understanding developments and evolutions (Farrington, 1991).

C. Components of Research Design

C.1 Research Questions: The crux of your study. Your design should be able to address these questions comprehensively (Ritchie, Lewis, Nicholls, & Ormston, 2013).

C.2 Theoretical Framework: Grounds your research in a particular theoretical perspective, guiding how you approach and interpret your study (Anfara & Mertz, 2015).

C.3 Methods and Data Collection: The tools and techniques used to gather data. This can be interviews, observations, surveys, etc. (Silverman, 2016).

C.4 Sampling: Determines who or what will be studied and how they will be accessed (Palys, 2008).

C.5 Analysis: How you will interpret and make sense of your collected data (Gibbs, 2007).

D. Importance of a Robust Research Design

D.1 Credibility: A well-thought-out design enhances the trustworthiness and validity of findings (Lincoln & Guba, 1985).

D.2 Feasibility: Ensures that the study can be realistically executed, considering time, cost, and other resources (Punch, 2013).

D.3 Ethical Assurance: Ensures that participants are treated with respect, and their integrity is maintained (Israel & Hay, 2006).

Research design is the backbone of any research project, providing the necessary structure and approach needed to address research questions accurately and ethically. It is a multi-faceted component that demands careful thought and consideration.

2. Components of Research Design

  2.1 Research Questions: Central to the research endeavor, they guide the methodology, data collection, and analysis (Yin, 2014).

  2.2 Research Approach: This can be either inductive, where theories are developed post data collection, or deductive, where hypotheses are tested through data (Saunders et al., 2009).

  2.3 Research Strategy: This refers to the overarching plan – whether the research will be a case study, survey, experiment, etc. (Denzin & Lincoln, 2011).

2.4 Time Horizons: Determines whether the study is cross-sectional (studied at one point in time) or longitudinal (studied over a prolonged period) (Bryman, 2016).

3. Paradigms in Research Design

Research paradigms shape how researchers view the world and approach their studies. Paradigms like positivism, interpretivism, and constructivism have distinct ontological, epistemological, and methodological underpinnings, influencing the chosen research methods (Mackenzie & Knipe, 2006).

Paradigms in research design are foundational belief systems or worldviews that researchers hold about the world and the nature of knowledge. These paradigms influence how researchers approach their studies, from the questions they ask to the methods they use to gather and interpret data. Understanding one's research paradigm is crucial as it can implicitly shape the research process (Kuhn, 1962).

A. Understanding Paradigms

A.1 Definition: A paradigm is a basic set of beliefs that guide action, whether in everyday life or in research (Guba & Lincoln, 1994).

A.2 Role in Research: Paradigms shape every aspect of the research process, including how questions are framed, how data is collected, and how findings are interpreted (Denzin & Lincoln, 2011).

B. Major Research Paradigms

B.1 Positivism:

- **Core Belief**: Reality is objective and singular, separating the researcher from the researched.
- **Methods**: Quantitative, experiments, and surveys.
- **Criticism**: Can be critiqued for being overly deterministic and not accounting for human agency (Bryman, 2015).

B.2 Constructivism:

- **Core Belief**: Multiple realities are constructed through human experience and interpretation.
- **Methods**: Qualitative, in-depth interviews, and participant observation.
- **Application**: Especially relevant in studies exploring perceptions, experiences, and interpretations (Charmaz, 2014).

B.3 Interpretivism:

- **Core Belief**: The understanding of social phenomena requires grasping the subjective meanings participants attach to them.
- **Methods**: Qualitative, ethnography, and hermeneutic techniques.
- **Significance**: Emphasizes understanding over explanation (Schwandt, 2000).

B.4 Critical Theory:

- **Core Belief**: Research should be transformative, addressing societal issues and imbalances.
- **Methods**: Both qualitative and quantitative, with an emphasis on revealing power dynamics.
- **Focus**: Often linked to social justice and emancipatory research (Kincheloe & McLaren, 2005).

B.5 Pragmatism:

- **Core Belief**: Research approaches should be driven by the research question, rather than a priori adherence to one paradigm.
- **Methods**: Can be mixed-method, employing both quantitative and qualitative techniques.
- **Relevance**: Focuses on problem-solving and what works in practice (Morgan, 2014).

C. Choosing a Paradigm

The choice of paradigm is often influenced by several factors:

- **Nature of the Research Problem**: Complex, multifaceted problems might require an interpretive or constructivist approach (Creswell & Poth, 2017).
- **Researcher's Philosophical Beliefs**: A researcher's worldview or personal beliefs can significantly influence paradigm choice (Guba & Lincoln, 1994).
- **Audience and Purpose**: If a researcher is addressing a specific audience or aiming for policy change, a more positivist or pragmatic approach might be appropriate.

Paradigms play an integral role in shaping research. Recognizing and understanding the paradigm underpinning a research project is crucial for its coherence, consistency, and overall success.

4. Choosing the Right Design

The choice of design is contingent on the research questions, objectives, available resources, and the researcher's philosophical stance. It's vital to ensure alignment between the research question and the chosen design to maintain validity (Punch, 2013).

Research design is the blueprint of a research project, defining the process by which researchers go about their investigation. Choosing the right design is a pivotal step in ensuring the validity and reliability of the findings. The right research design can illuminate understanding, while a flawed or inappropriate design can obscure the phenomena under study (Creswell, 2014).

A. Factors Influencing the Choice of Research Design

A.1 Nature of the Research Problem:

- Research questions that aim to describe a phenomenon might best be answered through a descriptive design, while those seeking to determine causality might necessitate an experimental or quasi-experimental design (Bryman, 2015).

A.2 Research Goals and Objectives:

- Research aiming to test an existing theory may lean towards quantitative methods, whereas exploring a new or under-researched topic may necessitate a qualitative design (Marshall & Rossman, 2014).

A.3 Available Resources:

- Time, funding, and accessibility to participants can greatly influence the feasibility of certain research designs over others (Miles, Huberman, & Saldaña, 2013).

A.4 Researcher's Expertise and Preferences:

- A researcher's training and comfort level with specific methodologies can influence design choices (Denzin & Lincoln, 2011).

A.5 Ethical Considerations:

- Ensuring participant well-being might preclude certain research designs, especially experimental ones that could potentially harm participants (Orb, Eisenhauer, & Wynaden, 2000).

B. Evaluating Suitability

B.1 Aligning with Research Questions:

- The chosen design should allow for the research questions to be answered in a comprehensive manner (Yin, 2013).

B.2 Validity and Reliability:

- The design should permit valid inferences from the data and offer consistent findings if replicated (Lincoln & Guba, 1985).

B.3 Feasibility:

- The practicality of conducting the research within the constraints of time, budget, and other resources should be assessed (Silverman, 2016).

C. Common Research Designs in Qualitative Research

C.1 Case Study Design:

- Suitable for in-depth exploration of a single case or multiple cases in real-life settings (Stake, 1995).

C.2 Ethnographic Design:

- Appropriate for studying cultural groups in naturalistic settings over a prolonged period (Geertz, 1973).

C.3 Grounded Theory Design:

- Chosen to develop a theory grounded in data from the field, typically in situations where existing theories do not adequately explain the phenomenon (Glaser & Strauss, 1967).

C.4 Phenomenological Design:

- Suitable for understanding lived experiences around a particular issue or phenomenon (Moustakas, 1994).

C.5 Narrative Research Design:

- Best for collecting detailed stories or life experiences and then analyzing them for key elements and themes (Clandinin & Connelly, 2000).

Choosing the right research design is instrumental in ensuring the study's robustness and the validity of its outcomes. Careful consideration of the research objectives, the nature of the research problem, available resources, and ethical concerns is vital in making an informed choice.

5. Ethical Considerations

Researchers must ensure the ethical soundness of their designs, safeguarding participants' rights, maintaining confidentiality, and ensuring informed consent (Orb et al., 2001).

The process of research planning and design, while primarily methodological, must be deeply rooted in ethical considerations. Adherence to ethical standards ensures not only the integrity and quality of the research but also the protection of participants' rights and dignity (Resnik, 2015). This section explores the multifaceted ethical dimensions that intersect with qualitative research design and planning.

A. Informed Consent
  A.1 Purpose:
  - Participants should be adequately and comprehensively informed about the purpose, methods, risks, benefits, and potential outcomes of the research (Beauchamp & Childress, 2013).

  A.2 Process:
  - Researchers must ensure that the consent is voluntary, free from coercion, and is obtained after potential participants have had ample opportunity to ask questions (Miller, Mauthner, & Birch, 2012).

  A.3 Documentation:
  - Depending on institutional or publication requirements, written or verbal consent should be documented (Faden, Beauchamp, & Kass, 2014).

B. Confidentiality and Anonymity
  B.1 Data Handling:
  - Measures should be put in place to ensure that collected data remains confidential and is stored securely (Guillemin & Gillam, 2004).

  B.2 Anonymizing Data:
  - Participant identities should be kept anonymous unless otherwise agreed upon, and pseudonyms or code numbers should be used where appropriate (Saunders, Kitzinger, & Kitzinger, 2015).

C. Avoiding Harm
  C.1 Emotional and Psychological Risks:
  - Researchers must be aware of potential emotional and psychological risks, especially in sensitive areas of research, and take steps to minimize harm (Dickson-Swift, James, Kippen, & Liamputtong, 2007).

  C.2 Physical Risks:
  - While less common in social science research, physical risks should also be considered and minimized (Israel & Hay, 2006).

D. Researcher's Positionality
  D.1 Reflexivity:
  - Researchers must engage in continual self-reflection to understand how their backgrounds, biases, and relationships might affect the research process and outcomes (Pillow, 2003).

  D.2 Power Dynamics:
  - The researcher must be aware of power dynamics and avoid exploiting participants, ensuring mutual respect throughout the research process (Bourke, 2014).

E. Cultural Sensitivity
  E.1 Respect for Local Norms and Practices:
  - Particularly in cross-cultural research, there's a need to respect and adhere to local norms, beliefs, and customs (Smith, 2012).

  E.2 Collaborative Approaches:
  - Engaging local communities in the research process, from design to dissemination, can enhance the ethical standing of the research (Chilisa, 2012).

Every step in research design and planning is intertwined with ethical considerations. Maintaining a strong ethical orientation ensures the credibility of the research and safeguards the rights and well-being of participants.

6. Practical Considerations in Research Planning
   6.1 Feasibility: Assessing available resources, including time, funding, and equipment.
   6.2 Access and Permissions: Securing appropriate permissions to access participants, data, or research sites.
   6.3 Timeline Construction: Outlining the various research phases and setting deadlines for each (Gantt, 1917).
   6.4 Contingency Planning: Preparing for unforeseen challenges and having backup plans.
7. Pilot Studies

Conducting pilot studies can be invaluable. They allow researchers to test their instruments, refine their methods, and ensure the viability of the main study (Van Teijlingen & Hundley, 2001).

A pilot study, sometimes referred to as a "feasibility study," is a small-scale version of a larger study, implemented to test and refine the various elements of the research design (van Teijlingen & Hundley, 2001). Such preliminary investigations provide invaluable insights that help to ensure the full-scale study is viable, efficient, and effective.

A. Rationale for Pilot Studies
   A.1 Validation of Instruments:
   - A pilot study can be utilized to test and refine data collection tools, such as questionnaires or interview schedules, ensuring their clarity, relevance, and reliability (Sampson, 2004).

   A.2 Feasibility Assessment:
   - It aids in gauging the feasibility of the main study, from recruitment procedures to data collection and analysis methods (Leon, Davis, & Kraemer, 2011).

   A.3 Estimating Resources:
   - Resource requirements such as time, budget, and staffing can be more accurately determined after conducting a pilot (Thabane et al., 2010).

B. Conducting a Pilot Study
   B.1 Scale:
   - Typically, a pilot study involves a smaller subset of participants from the target population. The exact number can vary but should be adequate to test the study's primary functions (Sim & Wright, 2000).

   B.2 Setting:
   - Ideally, the pilot should be conducted under conditions that closely mimic those of the primary study, ensuring that the results are relevant and actionable (Eldridge et al., 2016).

   B.3 Feedback Collection:
   - It's vital to gather feedback from participants and researchers involved in the pilot to understand challenges, discomforts, or suggestions for the main study (Hassan, Schattner, & Mazza, 2006).

C. Adjustments Post-Pilot

C.1 Refinement:

- Based on the outcomes and feedback from the pilot study, necessary adjustments can be made to the design, tools, and approach of the main study (Kim, 2011).

C.2 Decision Making:

- Sometimes, the pilot can reveal foundational issues with the research design, leading to more significant changes or even rethinking the feasibility of the entire study (Tickle-Degnen, 2013).

D. Limitations of Pilot Studies

D.1 Over- or Underestimation:

- Because of their small scale, pilot studies may not always accurately predict the challenges or outcomes of the larger study (Browne, 1995).

D.2 Misinterpretation:

- There's a risk of misinterpreting pilot results as conclusive findings, which can be misleading due to the limited sample size (Arain, Campbell, Cooper, & Lancaster, 2010).

While pilot studies add an additional layer of complexity and time to the research planning process, their contributions to refining research design and enhancing the credibility and feasibility of the main study are invaluable.

8. Review and Feedback

Before embarking on the research, it's prudent to seek feedback on the design and planning, either through peers, mentors, or formal review boards, to enhance the study's rigor and relevance (Baker, 1994).

Review and feedback are pivotal aspects of refining and ensuring the robustness of a research plan. Seeking external validation, critical perspectives, and diverse viewpoints enhances the overall credibility and rigor of the research proposal (Smith & Davis, 2017).

A. Importance of Review and Feedback

A.1 Enhancing Validity and Reliability:

- Feedback can help identify potential sources of bias or error in the research design, thereby enhancing its validity and reliability (Creswell & Miller, 2000).

A.2 Broadening Perspectives:

- Reviews, especially from diverse experts, ensure that the research is approached from multiple angles, enriching its depth and breadth (Lincoln & Guba, 1985).

A.3 Ethical Considerations:

- External reviews can highlight potential ethical issues that may not have been apparent to the primary researcher, aiding in preemptively addressing them (Guillemin & Gillam, 2004).

B. Sources of Review and Feedback

B.1 Peer Review:

- Having colleagues or experts in the field assess the research design can provide specific insights, given their understanding of the subject matter (Hammersley, 2007).

B.2 Research Committees:

- These committees, often present in academic institutions, evaluate the feasibility, ethical considerations, and significance of the study (Mertens & Ginsberg, 2009).

B.3 Stakeholder Feedback:

- Especially in applied research, obtaining feedback from stakeholders ensures the study's relevance and practical applicability (Greenwood & Levin, 2007).

C. Incorporating Feedback into the Research Design

C.1 Iterative Process:

- Feedback should be seen as part of an iterative process wherein the design is continually refined and improved based on inputs (Johnson & Onwuegbuzie, 2004).

C.2 Balancing Feedback with Original Intent:

- While feedback is invaluable, it's essential to discern which feedback aligns with the study's core objectives and which might steer it off course (Marshall & Rossman, 2014).

D. Challenges and Considerations

D.1 Potential for Conflicting Feedback:

- Multiple reviews can sometimes result in conflicting feedback, requiring the researcher to judiciously decide on the appropriate course of action (Hess, 2008).

D.2 Temporal and Resource Implications:

- Seeking extensive reviews can be time-consuming and may demand additional resources, which need to be factored into the research planning (Baxter & Jack, 2008).

Incorporating review and feedback processes into research design and planning is an indispensable step that enriches the study's depth, ethical grounding, and overall rigor. While challenges exist, their benefits in terms of enhancing the study's credibility and relevance are paramount.

Research design and planning form the nucleus of the entire research process. A coherent, thoughtful, and robust design ensures that the research remains purposeful, ethical, and valuable, offering meaningful insights into the studied phenomena.

The following are the steps needed to conduct qualitative research:

Formulating Research Questions

The formulation of research questions stands as the cornerstone of any research endeavor. These questions provide direction, focus the study, and determine the nature and scope of data collection, analysis, and interpretation. Properly articulated questions guide researchers in identifying what they seek to understand, ensuring their investigation remains methodologically and conceptually robust.

1. Importance of Research Questions

Research questions serve as the compass guiding the research voyage. They set the scope and boundaries of the investigation, directly influencing the research design, methodology, and the relevance of the findings (Creswell, 2013).

Research questions provide direction, focus, and depth to any research endeavor. They anchor the investigative process, guiding the methodology, data collection, analysis, and eventually the conclusions and implications drawn from the study. In qualitative research, they are especially vital, as the inquiry is often exploratory, seeking to understand phenomena in depth (Marshall & Rossman, 2016).

A. Providing Direction and Focus

A.1 Grounding the Research:

- Research questions lay the foundation for the study. They provide a roadmap for the researcher, outlining what needs to be explored and in what direction the inquiry should proceed (Maxwell, 2012).

A.2 Limiting Scope:

- Precisely formulated questions can prevent the research from becoming overwhelmingly vast, helping to limit the scope to manageable and meaningful boundaries (Creswell, 2013).

B. Facilitating Methodological Choices

B1 Determining Research Design:

- The nature and structure of the research questions often influence the design of the study, whether it be a case study, ethnography, phenomenology, etc. (Yin, 2014).

B.2 Influencing Data Collection and Analysis:

- Research questions, especially in qualitative studies, guide the choice of data collection tools (e.g., interviews, observations) and analysis techniques (e.g., thematic analysis) (Braun & Clarke, 2006).

C. Enhancing Relevance and Significance

C.1 Addressing Gaps in Literature:

- Well-crafted research questions are often derived from existing literature gaps, ensuring that the study's findings contribute valuable insights to the academic community (Boote & Beile, 2005).

C.2 Guiding Theoretical Frameworks:

- The questions can help in selecting or developing the theoretical framework that underpins the study, linking the research to broader concepts and theories (Ravitch & Riggan, 2017).

D. Providing Clarity to Stakeholders

D.1 Enhancing Understandability:

- Clearly articulated questions help stakeholders, including participants, peer researchers, and readers, understand the research's purpose and significance (Thomas, Silverman, & Nelson, 2015).

D.2 Aiding in Evaluation:

- For those evaluating the quality or relevance of the research, such as in peer-review processes, the research questions offer a clear lens to assess the study's merit (Eisenhart, 2009).

The formulation of relevant, clear, and concise research questions is crucial for ensuring the success and significance of a research endeavor. They not only provide direction but also enhance the quality and relevance of the study.

2. Characteristics of Good Research Questions

2.1 Clarity: Questions should be clear, devoid of ambiguity, ensuring the researcher and the audience understand what is being asked (Punch, 2013).

2.2 Focus: They should be specific, not overly broad, ensuring manageability and depth of investigation (Bryman, 2016).
2.3 Relevance: Questions should bear significance to the field of study and contribute to existing literature or practical applications (Denscombe, 2014).
2.4 Feasibility: Researchers should be able to address the questions within the confines of resources, time, and access (Maxwell, 2012).

3. Types of Research Questions
3.1 Descriptive: Aimed at portraying phenomena or situations. For instance, "How do high school students perceive online learning?"
3.2 Relational: Concerned with relationships between variables. E.g., "Is there a relationship between study habits and academic performance?"
3.3 Causal: Investigate the cause-effect dynamics between variables. E.g., "Does prolonged exposure to screens decrease sleep quality among adolescents?" (Yin, 2014).
4. From General to Specific: The Funnel Approach

A research endeavor often begins with broad questions or interests which are then narrowed down using a funnel approach. The general research interest undergoes progressive refinement, leading to more specific, focused, and researchable questions (Neuman, 2011).

Formulating research questions is a fundamental step in the research process, guiding the subsequent stages of data collection and analysis. The funnel approach offers a structured methodology for transitioning from a broad research topic to narrow, focused research questions, ensuring that the inquiry is systematic, coherent, and grounded (Creswell, 2013).

A. Starting with a Broad Topic
A.1 Identifying a Field of Interest:
- This step involves pinpointing a general area of study that intrigues the researcher. It is often driven by personal interest, observed gaps in literature, or pressing societal issues (Maxwell, 2012).

A.2 Scoping the Literature:
- An initial literature review helps understand the current state of knowledge, major debates, and existing gaps (Boote & Beile, 2005).

B. Narrowing the Focus
B.1 Pinpointing Subtopics:
- Based on preliminary reading and contemplation, researchers can identify specific subtopics or niches that warrant further investigation (Machi & McEvoy, 2016).

B.2 Formulating Preliminary Questions:
- Initial, somewhat broad questions can be framed that capture the essence of the subtopics identified. These aren't the final research questions but guide further refinement (Thomas, 2017).

C. Specifying the Research Questions
C.1 Operationalizing Concepts:
- It's crucial to define and operationalize central concepts, ensuring clarity and reducing ambiguity (Babbie, 2015).

C.2 Ensuring Feasibility:

- The questions should be formulated such that they are researchable within the constraints of time, resources, and accessibility (Yin, 2014).

C.3 Aligning with Research Objectives:

- The final questions should directly align with the objectives or aims of the research, offering a clear roadmap for the study (Silverman, 2016).

D. Reflecting and Refining

D.1 Peer Review and Feedback:

- Sharing the drafted questions with peers, mentors, or experts in the field can provide valuable feedback, helping refine and enhance their precision (Eisenhart, 2009).

D.2 Iterative Process:

- Often, as researchers delve deeper into data collection and early analysis, they find the need to slightly modify or fine-tune the research questions. Recognizing this as an iterative process is vital for qualitative studies (Tracy, 2013).

The funnel approach to formulating research questions ensures that the research process begins with a broad, holistic view of a topic and gradually narrows down to precise, researchable questions. This structured approach ensures depth, relevance, and feasibility in the research endeavor.

5. The Role of Literature Review

Engaging with existing literature is pivotal. It aids researchers in identifying gaps, refining their questions, and ensuring their inquiry adds value to the discipline. Previous studies can also guide the phrasing and focus of research questions (Hart, 1998).

The literature review plays a pivotal role in formulating research questions by offering researchers a comprehensive understanding of what is already known, revealing gaps, and suggesting potential avenues for exploration. Understanding the interplay between literature and research questions is crucial for creating a robust research design (Hart, 1998).

A. Contextualizing the Study

A.1 Grounding the Research:

- A thorough literature review enables researchers to situate their study within the broader academic conversation, ensuring that the research is building upon, rather than replicating, existing knowledge (Boote & Beile, 2005).

A.2 Mapping the Field:

- The review helps researchers delineate major themes, debates, theories, and methodologies in the field, serving as a roadmap for the uncharted territory they wish to explore (Randolph, 2009).

B. Identifying Gaps and Opportunities

B.1 Unanswered Questions:

- A meticulous review will reveal questions that prior research has raised but not adequately addressed, offering clear avenues for further inquiry (Creswell, 2013).

B.2 Novel Contexts:

- Even well-researched topics can be studied in new contexts, demographics, or time periods, breathing fresh life into established lines of inquiry (Maxwell, 2012).

C. Enhancing Relevance and Rigor

C.1 Aligning with Prior Findings:

- By being aware of existing research findings, researchers can frame their questions in ways that either support, contradict, or add nuance to established knowledge (Tranfield, Denyer, & Smart, 2003).

C.2 Improving Methodological Choices:

- Understanding the methodologies and tools used in prior research can guide researchers in selecting appropriate methods for their own inquiries (Ridder, 2017).

D. Crafting Focused and Feasible Questions

D.1 Refining Scope:

- By discerning what has been extensively studied and what has been overlooked, researchers can craft questions that are both focused and feasible (Onwuegbuzie, Leech, & Collins, 2012).

D.2 Ensuring Significance:

- Literature reviews ensure that the research questions posed are not just of interest to the researcher, but also of significance to the broader academic community (Littell, Corcoran, & Pillai, 2008).

A literature review is an indispensable ally in the formulation of research questions. It acts as a compass, guiding researchers through the vast landscape of existing knowledge towards their own unique contribution.

6. Iterative Process of Formulation

The development of research questions is rarely linear. As researchers delve deeper, questions might undergo revisions, reflecting a deeper understanding of the phenomena or logistic constraints (Miles, Silverman & Huberman, 2013).

The formulation of research questions is seldom a linear process. Instead, it can be visualized as a recursive, iterative cycle, wherein initial questions are continually refined based on emerging insights, feedback, and ongoing engagement with the literature (Creswell & Creswell, 2017).

A. The Spiral Nature of Question Formulation

A.1 Initial Curiosity:

- Research often begins with broad curiosities or general topics of interest, which are exploratory in nature and require further specification (Blaikie, 2009).

A.2 Refinement through Literature:

- As the researcher delves deeper into existing studies, initial questions can be honed to address more precise issues, factoring in what has already been explored and where gaps lie (Boote & Beile, 2005).

B. Feedback Loops in Formulation

B.1 Peer Feedback:

- Presenting initial research questions to colleagues or experts in the field can provide critical insights, highlighting overlooked areas or suggesting different angles of approach (Maxwell, 2012).

B.2 Pilot Studies:

- Preliminary investigations or smaller-scale studies can reveal the viability of a research question, potentially directing researchers to reframe or narrow down their inquiry (Yin, 2017).

C. Adapting to Emerging Data

C.1 Flexible Refinement:

- Especially in qualitative research, as data is collected, new patterns or themes might emerge, necessitating a revision or expansion of the original questions (Charmaz, 2006).

C.2 Staying Open to Serendipity:

- While it's essential to have a structured approach, remaining open to unexpected findings can lead to more profound insights and richer research outcomes (Stebbins, 2001).

D. Finalization and Validation

D.1 Convergence of Sources:

- As the iterative process unfolds, a point is reached where feedback, literature, and emerging data converge, solidifying the research questions (Miles, Huberman, & Saldaña, 2014).

D.2 Testing Validity:

- The final set of questions should be checked for validity, ensuring they align with the research objectives and have the potential to address the identified gaps or issues (Cohen, Manion, & Morrison, 2013).

Recognizing the iterative nature of formulating research questions underscores the dynamism inherent in the research process. It emphasizes the adaptability, openness, and critical thinking researchers must employ to ensure their inquiries are both rigorous and relevant.

CONCLUSION

Formulating research questions is an intricate dance of balancing ambition with feasibility, breadth with depth, and innovation with discipline. These questions serve as the foundation upon which the entire edifice of the research is built, determining its direction, rigor, and relevance.

Choosing a Qualitative Method

The selection of an appropriate qualitative method is paramount in ensuring the research objectives are met effectively. Each qualitative approach possesses its own set of principles, procedures, and philosophical underpinnings, all of which should align with the research questions and overall goals of the study.

1. Importance of Method Selection

Choosing the right qualitative method is a vital step in the research design, ensuring that the data collected is relevant, meaningful, and addresses the research questions with depth and richness (Creswell & Poth, 2017).

The selection of an appropriate qualitative method is a pivotal step in the research process. It provides the framework that guides the subsequent stages of data collection, analysis, and interpretation. Making an informed choice is crucial, as the method acts as a lens through which the research phenomena are examined, shaping both the insights gained and the resultant findings (Denzin & Lincoln, 2011).

A. Aligning with Research Objectives
   A.1 Addressing the Central Question:
   - The chosen method should inherently possess the capability to address the core research question. Different methods are adept at uncovering various aspects of human experience (Creswell & Poth, 2017).

   A.2 Meeting Study Goals:
   - Depending on whether the study seeks to describe, understand, or explore a phenomenon, the qualitative method selected should align with these overarching aims (Patton, 2015).

B. Consideration of Theoretical Framework
   B.1 Philosophical Underpinnings:
   - Every qualitative method is anchored in certain philosophical assumptions about the nature of reality, knowledge, and inquiry. Researchers must ensure that their personal and study's philosophical stance aligns with that of the method chosen (Guba & Lincoln, 1994).

   B.2 Paradigmatic Considerations:
   - Postpositivist, constructivist, transformative, or pragmatic paradigms might influence or even dictate method selection, depending on the dominant or preferred paradigm guiding the study (Mertens, 2014).

C. Practical Implications
   C.1 Resource Availability:
   - Some qualitative methods, like ethnography, may demand extended time commitments and intensive fieldwork. Such practical considerations should factor into method selection (Fetterman, 2019).

   C.2 Skillset of the Researcher:
   - Every qualitative method requires a specific set of skills. Researchers need to introspect whether they possess the necessary expertise or need additional training for the method they are considering (Seidman, 2013).

D. Authenticity and Ethical Considerations
   D.1 Ethical Alignment:
   - Certain methods might require more intimate and prolonged engagements with participants, leading to more profound ethical considerations. The chosen method should align with both the researcher's ethical beliefs and institutional ethical standards (Ellis & Bochner, 2000).

   D.2 Ensuring Authentic Representation:
   - The chosen method should facilitate an authentic and faithful representation of participants' experiences and views, minimizing the risks of misrepresentation or oversimplification (Tracy, 2010).

The importance of method selection in qualitative research cannot be overemphasized. It profoundly influences the study's trajectory, the depth of insights garnered, and the authenticity and validity of the findings.

2. Factors Influencing Method Selection:
   2.1 Nature of the Research Question: The research questions typically guide the method selection. For example, exploring lived experiences would direct a researcher towards phenomenology (Smith & Osborn, 2008).
   2.2 Theoretical and Philosophical Stance: The researcher's ontological and epistemological beliefs can influence the method chosen. Grounded theory, for instance, often aligns with a constructivist viewpoint (Charmaz, 2006).
   2.3 Access and Feasibility: Practical considerations, such as access to participants, time constraints, and resources, can impact method selection (Silverman, 2010).
   2.4 Depth and Breadth: Ethnography might be chosen for in-depth, extended study of a culture, while narrative research might be selected for depth into personal experiences (Geertz, 1973).
3. Overview of Common Qualitative Methods:
   3.1 Ethnography: Originating from anthropology, it seeks deep immersion in a culture or social group over extended periods (Atkinson & Hammersley, 1994).
   3.2 Grounded Theory: Aims to derive theory from data. Useful when existing theories do not adequately explain a phenomenon (Glaser & Strauss, 1967).
   3.3 Phenomenology: Seeks to understand the lived experiences of individuals concerning a phenomenon (Moustakas, 1994).
   3.4 Case Study: Investigates a specific instance or case in-depth within its real-world context (Yin, 2014).
   3.5 Narrative Research: Focuses on the stories of individuals, examining their sequence and consequences to derive meanings (Riessman, 2008).
4. Combining Qualitative Methods:

In some instances, employing multiple qualitative methods (method triangulation) can enrich data collection, providing a more holistic understanding of the phenomenon (Denzin, 1978).

The arena of qualitative research, replete with a multitude of methods, offers researchers a plethora of avenues to delve into the nuanced and multifaceted realm of human experiences. In recent years, there has been a growing trend towards adopting a combination of qualitative methods to enrich research insights (Creswell & Plano Clark, 2011). This approach, often termed "methodological triangulation," provides a more holistic, comprehensive, and robust understanding of research phenomena.

A. Rationale for Combining Qualitative Methods
   A.1. Comprehensiveness and Depth:
   - By juxtaposing multiple qualitative methods, researchers can capture various dimensions of a phenomenon, yielding richer, more comprehensive data (Denzin, 1978).

   A.2. Methodological Triangulation:
   - The synthesis of different methods can enhance the validity of research findings. Discrepancies, if any, between data sourced from different methods can lead to deeper introspection and richer interpretations (Patton, 1999).

B. Strategies for Combining Methods

B.1 Sequential Implementation:

- Researchers can opt for a phased approach, where one method follows another. This allows the results from the initial method to inform the subsequent one (Morse, 1991).

B.2 Concurrent Implementation:

- Here, multiple methods are deployed simultaneously. This concurrent approach is particularly valuable when studying complex phenomena requiring varied perspectives (Greene, Caracelli, & Graham, 1989).

C. Common Method Combinations

C.1 Ethnography and Interviews:

- While ethnographic observation provides a contextual understanding, interviews can offer in-depth personal narratives that elucidate observed behaviors (Tedlock, 2000).

C.2 Case Studies and Document Analysis:

- Case studies can be enriched through the analysis of relevant documents, adding layers of evidence to the study's assertions (Yin, 2009).

D. Considerations and Challenges

D.1 Ensuring Cohesiveness:

- Combining methods requires researchers to synergize different types of data, ensuring that the resultant narrative is coherent and logically structured (Bryman, 2006).

D.2 Resource Constraints:

- Multiple methods can be time-consuming and might demand more resources. Researchers must weigh the benefits against the potential increase in resource commitments (O'Cathain, Murphy, & Nicholl, 2010).

D.3 Analytical Complexity:

- Data from different methods might require varied analytical strategies. Integrating these insights demands a robust analytical framework and a deep understanding of each method's nuances (Thurmond, 2001).

Incorporating multiple qualitative methods in research design can substantially augment the depth, breadth, and robustness of the study. While this approach offers a myriad of advantages, it also brings forth challenges that researchers must anticipate and navigate adeptly.

5. Evolving Nature of Method Selection

The dynamic and iterative nature of qualitative research often means that method selection may be revisited and refined as the research progresses, ensuring alignment with evolving insights (Morrow, 2005).

The process of selecting a qualitative method has evolved over the years, reflecting shifts in epistemological stances, advancements in research techniques, and changing societal contexts. As researchers embrace diverse ways of knowing and interrogating the world, method selection has transitioned from being a static decision to a more dynamic and nuanced one.

A. Historical Perspectives

Historically, qualitative research was primarily rooted in disciplines like anthropology and sociology, with ethnography and participant observation being the cornerstone methods (Denzin & Lincoln, 1994). The choice was relatively linear, hinging mainly on the research question and the studied community.

B. The Pluralistic Turn

With the pluralistic turn in the social sciences, methodological choices expanded, introducing a range of methods, each with its own philosophical underpinning (Mason, 2002). This led to the acknowledgment that method selection should resonate with the study's ontological and epistemological stance.

C. Reflexivity in Method Selection

The 1990s and 2000s saw the rise of reflexivity, emphasizing the researcher's positionality and its influence on method selection (Finlay, 2002). Here, researchers started to consider their biases, backgrounds, and relationships with participants when opting for a method.

D. Pragmatic Approaches

Contemporary approaches to method selection lean towards pragmatism. Instead of strict adherence to one method or philosophy, researchers prioritize the research question and the problem's practical aspects (Morgan, 2007). This approach appreciates the flexibility in deploying multiple methods or adapting methods as the research progresses.

E. Tailored and Hybrid Methods

In recent times, the boundaries between established qualitative methods have become porous, leading to tailored or hybrid methodologies (Hesse-Biber & Leavy, 2011). These are often bespoke designs catering to unique research questions, emphasizing the evolving nature of method selection.

F. The Ethical Dimension

The evolving nature of method selection also includes a heightened focus on ethics. Researchers are now more cognizant of the power dynamics inherent in certain methods and the imperative to ensure participants' dignity, rights, and well-being (Guillemin & Gillam, 2004).

The evolving nature of qualitative method selection signifies the field's adaptive and progressive character. While foundational principles remain, the landscape of choice has expanded, embodying the complexities and diversities of human experiences and societal changes.

Conclusion

Choosing a qualitative method is a nuanced decision that requires a balance between the theoretical, the practical, and the empirical. It demands a clear understanding of the research's purpose, an appreciation for the strengths and limitations of each method, and a commitment to rigor and depth.

Sampling Strategies: Purposeful Sampling and Snowball Sampling

Sampling strategies in qualitative research differ from those in quantitative studies. The focus is not on the representativeness of the larger population but on the depth, richness, and relevance of the information obtained. Two common sampling strategies in qualitative research are purposeful sampling and snowball sampling.

1. Purposeful Sampling

Purposeful sampling, also known as purposive or judgmental sampling, is a strategy where researchers intentionally select specific individuals or sites that can provide an in-depth understanding of the phenomenon in question.

1.1 Characteristics and Rationale

Purposeful sampling allows for a detailed exploration of cases that are "information-rich" regarding the phenomenon of interest (Patton, 1990). The key lies in selecting participants who are particularly knowledgeable about or have experience with the research topic.

1.2 Types of Purposeful Sampling

- **Extreme or Deviant Case**: Learning from highly unusual manifestations of the phenomenon of interest (Flyvbjerg, 2006).
- **Typical Case**: Illustrating what's normal or average.
- **Homogeneous Sampling**: Focusing on a particular subgroup.
- **Criterion Sampling**: Picking cases that meet some predetermined criterion of importance (Patton, 2002).

1.3 Advantages and Limitations

The strength of purposeful sampling lies in its ability to yield deep insights from cases that can provide valuable information. However, the findings aren't generally considered generalizable in the statistical sense (Coyne, 1997).

2. Snowball Sampling

Snowball sampling is particularly useful when researching hard-to-reach or specialized populations. Starting with a few known participants, researchers ask these initial participants to refer others.

2.1 Process and Rationale

The process resembles a snowball effect: as more participants are involved, they recommend more, causing the sample size to grow exponentially. This method is especially relevant when the community is tight-knit or the population is hidden or hard to access (Biernacki & Waldorf, 1981).

2.2 Strengths and Limitations

The primary strength is its ability to access otherwise challenging-to-reach populations. However, since the sample is not random, there are potential biases: the sample can be limited to a specific network or community, potentially excluding diverse voices (Atkinson & Flint, 2001).

3. Considerations for Both Strategies

Regardless of the sampling strategy, it's crucial for researchers to maintain reflexivity, constantly reflecting on their choices and recognizing the potential implications of these decisions on the research outcomes (Finlay, 2002).

Purposeful and snowball sampling strategies offer qualitative researchers tools to delve deeply into specific phenomena, accessing rich, detailed data from participants. While these methods may not offer generalizability in the traditional sense, they provide depth, nuance, and insight that are hallmarks of qualitative inquiry.

Ethical Considerations in Qualitative Research

Ethics is a cornerstone of any research endeavor, particularly in qualitative research, where researchers often engage deeply with participants, entering their worlds, and interpreting their experiences. This closeness amplifies the need for rigor in ethical considerations, which range from consent to confidentiality.

1. Informed Consent

Arguably the foundational principle of research ethics, informed consent ensures participants are fully aware of the nature of the research, its purpose, the procedures involved, and any potential risks or benefits (Beauchamp & Childress, 2001).

1.1 Ongoing Consent: Given the iterative nature of qualitative research, consent is not a one-time event. It's a process requiring ongoing dialogue and reevaluation as the research unfolds (Guillemin & Gillam, 2004).

1.2 Reflexivity in Consent: Researchers must remain reflexive, ensuring power dynamics don't impede genuine, voluntary consent (Hammersley & Atkinson, 2007).

2. Confidentiality and Anonymity

Maintaining participant confidentiality is crucial. Researchers should rigorously protect personal data and, where necessary, anonymize identifiers in outputs (Saunders, Kitzinger, & Kitzinger, 2015).

2.1 Challenges in Small Settings: In tight-knit or unique settings, ensuring anonymity becomes challenging, demanding creative solutions (Eide & Kahn, 2008).

3. Emotional and Psychological Risks

Qualitative research can often involve sensitive topics, potentially eliciting strong emotional reactions.

3.1 Researcher's Role: The researcher has a responsibility to anticipate potential emotional risks and ensure participants have access to support if necessary (Dickson-Swift, James, Kippen, & Liamputtong, 2007).
3.2 Debriefing: Offering debrief sessions post-interview can help participants process their feelings and the researcher ascertain their well-being (Orb, Eisenhauer, & Wynaden, 2001).

4. Power Dynamics

The dynamics between researcher and participant can inadvertently replicate societal power imbalances. Recognizing and mitigating these dynamics is crucial (Bourke, 2014).

Ethical considerations in qualitative research have garnered substantial attention in recent decades, with power dynamics emerging as a prominent concern. Power dynamics, referring to the differential power relations that exist between the researcher and participants, can influence participants' willingness to share, the depth of their sharing, and even the authenticity of the narratives they present (Mauthner, Birch, Jessop, & Miller, 2002).

A. The Historical Context

Historically, qualitative research was conducted with a top-down approach, often by researchers from privileged backgrounds studying marginalized populations. This invariably created a power differential where participants had little to no agency (Denzin & Lincoln, 2011).

B. The Researcher-Participant Relationship

The relational dimension in qualitative inquiries makes power dynamics particularly salient. Often, researchers are in positions of authority or expertise, which may inadvertently intimidate or influence participants, potentially leading to biased or suppressed responses (Fine, Weis, Weseen, & Wong, 2000).

C. Gatekeepers and Power

Gatekeepers, or those who control access to participants, add another layer to the power dynamics. Their influence can limit who participates, what information is shared, and the narratives' nature, potentially biasing the research (Liamputtong, 2007).

D. Reflexivity and Acknowledgment

Modern qualitative methodologies advocate for reflexivity – a process where researchers continuously reflect upon and articulate their biases, positioning, and influence on the research process (Rose, 1997). This involves recognizing inherent power dynamics and striving to minimize or balance the disparities.

E. Co-constructed Knowledge

In rebalancing power dynamics, there's a push towards viewing research knowledge as co-constructed, valuing participants not merely as subjects but as active contributors (Berger, 2015).

F. Ensuring Voice and Representation

Ethical considerations also involve ensuring participants' voices are authentically represented, devoid of the researcher's undue influence or interpretation. This involves giving participants opportunities to review, amend, or corroborate findings (Harper & Thompson, 2012).

G. The Ethical Imperative

Given these power dynamics, the ethical imperative for researchers is not merely to avoid harm but to proactively foster an environment of trust, mutual respect, and shared authority, recognizing and addressing the asymmetries in the research relationship (Ellis & Bochner, 2000).

Power dynamics in qualitative research present complex ethical challenges. By foregrounding these dynamics, acknowledging them, and adopting strategies to redress imbalances, qualitative researchers can work towards more equitable, authentic, and ethically sound practices.

5. Cultural Sensitivity

When researching across cultures, extra care must be taken to ensure that the research does not impose dominant cultural norms or misinterpret cultural practices (Smith, 1999).

Cultural sensitivity in qualitative research refers to a conscious effort made by researchers to recognize, respect, and appropriately respond to the differences and nuances inherent in diverse cultural contexts (Tilley-Lubbs, 2009). Ethical research within various cultural settings requires not just a theoretical understanding but a genuine appreciation for the myriad worldviews, customs, and values that shape human experiences.

A. Historical Context

Historically, qualitative research, particularly in anthropology and sociology, often followed a pattern of 'outsider' researchers studying 'exotic' cultures. This sometimes led to ethnocentric biases, where researchers judged other cultures based on the values and norms of their own (Said, 1978).

B. Understanding Ethnocentrism

Ethnocentrism, the tendency to view one's own culture as superior, can inadvertently affect how research questions are formulated, how data are interpreted, and how findings are presented. Recognizing and counteracting this bias is pivotal in culturally sensitive research (Berry, 1989).

C. Emic and Etic Approaches

Researchers must balance between the emic (insider) and etic (outsider) perspectives. While the emic approach delves into the internal worldview of the participants, the etic approach provides a comparative or outsider's viewpoint. Striking a balance ensures cultural respect while maintaining analytical breadth (Pike, 1967).

D. Participatory Research Models

Participatory research models, where community members actively contribute to the research process, have emerged as a culturally sensitive method. This approach acknowledges local expertise and challenges the traditional power dynamics in research (Cammarota & Fine, 2008).

E. Informed Consent in Cultural Contexts

Obtaining informed consent, a cornerstone of ethical research, can present challenges in certain cultural contexts. What is deemed appropriate disclosure or comprehension may vary across cultures, necessitating adaptive strategies (Marshall & Koenig, 2004).

F. The Role of Local Intermediaries

Local intermediaries or cultural brokers can assist researchers in navigating cultural intricacies, ensuring respectful engagement, and minimizing potential misunderstandings (Breen, 2007).

G. Cultural Reflexivity

Cultural reflexivity involves researchers' active reflection on their own cultural backgrounds, biases, and the influence these might have on the research process. It demands a continuous interrogation of one's own positionality in relation to the cultural context being studied (Ratner, 2008).

Cultural sensitivity is not just an ethical imperative but also crucial for the validity and depth of qualitative research. By adopting a stance of humility, openness, and respect, researchers can ensure that their investigations genuinely honor the richness and diversity of human experiences across cultures.

6. Data Storage and Management

Given the personal nature of qualitative data, its secure storage, and handling is imperative to maintain participant privacy (Van den Hoonaard, 2002).

In the realm of qualitative research, ethical considerations go beyond participant consent and treatment, permeating into how researchers handle, store, and manage collected data. Ensuring the confidentiality and privacy of participant information is a fundamental responsibility of researchers, especially in an age where data breaches are increasingly common (Richards & Morse, 2013).

A. The Imperative for Confidentiality

Confidentiality in qualitative research encompasses the responsibility to protect participants' identities, responses, and any other identifiable information. This obligation becomes even more pronounced when dealing with sensitive subjects where a breach can have dire consequences for the participants (Esterberg, 2002).

B. Digital Security Measures

As more researchers digitize their data, utilizing encrypted storage devices and secure cloud-based platforms becomes crucial. Password protection and regular security software updates can act as the first line of defense against potential breaches (Markham & Buchanan, 2012).

C. Physical Storage Considerations

For physical data such as interview transcripts, locked file cabinets in restricted areas can provide a secure storage solution. It's essential that access to these materials is limited and regularly audited (Orb, Eisenhauer, & Wynaden, 2000).

D. Data Anonymization

An effective strategy to maintain participant confidentiality is to anonymize data at the earliest stage possible. This involves replacing identifiable details with pseudonyms or codes, ensuring that raw data remains unlinked to specific individuals (Saunders, Kitzinger, & Kitzinger, 2015).

E. Data Retention and Destruction

Ethical guidelines often dictate a period for which research data must be retained. After this, the secure and irreversible destruction of data is essential. Whether digitally using specialized software or physically via shredding, the destruction process must be thorough and verifiable (British Psychological Society, 2018).

F. Transparent Communication with Participants

Participants should be informed about how their data will be stored, managed, and eventually destroyed. Transparency in this process builds trust and ensures participants are aware of their rights regarding their provided information (Mason, 2017).

G. Addressing Challenges in Collaborative and Longitudinal Studies

In projects involving multiple researchers or spanning extended periods, consistent data management protocols become crucial. Regular team training and audits can ensure that everyone involved adheres to agreed-upon ethical standards (Fielding & Fielding, 2008).

Ethical considerations related to data storage and management are foundational to qualitative research. By ensuring that participant information remains confidential and secure, researchers uphold the integrity of the research process and protect the well-being and rights of their participants.

7. Reflexivity in Ethical Practice

Given the deep engagement of qualitative researchers with their participants, practicing reflexivity—where researchers continuously reflect on their role, impact, and influence in the research—is paramount for ethical integrity (Finlay, 2002).

Reflexivity is an integral component of qualitative research, informing not just methodology but also ethical considerations. By recognizing and critically assessing their own biases, influences, and interactions with the research context and participants, researchers can better navigate the ethical landscape of their studies (Pillow, 2003).

A. Understanding Reflexivity

At its core, reflexivity involves a continuous process of self-reflection where researchers examine their roles, responsibilities, and effects on the research (Finlay & Gough, 2003). This practice promotes transparency and ethical rigor, ensuring the study's integrity and the participants' well-being.

B. Reflexivity and Power Dynamics

Power imbalances between researchers and participants can impact the data collection process and the narratives derived from it. Reflexivity demands researchers to be cognizant of these imbalances and actively work towards minimizing them, ensuring participants' voices are truly heard (Bourke, 2014).

C. Positionality and Personal Biases

Every researcher brings their personal beliefs, values, and experiences into their work. Through reflexive practices, researchers can identify and address their biases, ensuring that these do not unduly influence the study's findings (Rose, 1997).

D. Ethical Decisions in the Field

Ethical guidelines provide a foundation, but real-world research often presents unexpected challenges. Reflexivity enables researchers to navigate these dilemmas by considering their actions and decisions in the broader context of the study and its ethical implications (Guillemin & Gillam, 2004).

E. Data Interpretation and Representation

Reflexive practices are crucial in data analysis, where researcher biases can influence findings. By continuously reflecting on their roles in data interpretation, researchers can strive for a more authentic representation of participants' experiences (Etherington, 2004).

F. Addressing Potential Harms

A reflexive approach to ethics involves ongoing assessments of potential risks and harms. By actively considering the implications of their actions and choices, researchers can better anticipate and mitigate potential adverse effects on participants (Råheim et al., 2016).

Conclusion: Ethical considerations in qualitative research are multi-faceted, demanding a commitment not just to procedural ethics but to an ongoing, reflexive engagement with the ethical dimensions and implications of the research.

Chapter 4: Data Collection Methods

Data Collection is a pivotal step in research, providing the empirical basis upon which scholarly insights are built. It is particularly crucial in qualitative research where the richness and depth of data drive understanding and interpretation. This chapter delves deeply into the diverse methods of data collection in qualitative research, exploring their nuances, applications, advantages, and challenges.

Foundational Importance of Data Collection

In qualitative research, data collection methods are foundational, as they allow for the nuanced, detailed, and rich exploration of phenomena (Merriam & Tisdell, 2015). The chosen method significantly influences the breadth and depth of insights gained, requiring thoughtful selection and application to appropriately address the research questions and align with the study's theoretical framework.

Diversity of Methods

Qualitative research is marked by a wide array of data collection methods, ranging from interviews and observations to document analysis and visual methods. Each method offers unique perspectives and insights, enabling researchers to explore different dimensions of their research questions (Creswell & Creswell, 2017).

Aligning Methods With Research Questions

The alignment of data collection methods with research questions is pivotal to the validity of qualitative research. Researchers must carefully select methods that allow them to delve deeply into the phenomena under study, ensuring congruence with the research objectives, questions, and theoretical framework (Silverman, 2016).

Ensuring Ethical Integrity

Ethical considerations permeate the process of data collection. Researchers are entrusted with the responsibility to respect participants' rights, dignity, and well-being, ensuring the integrity and ethical soundness of the research process (Orb, Eisenhauer & Wynaden, 2001).

Challenges and Solutions in Data Collection

While qualitative data collection methods offer rich, detailed insights, they also present unique challenges. Issues related to access, rapport, reflexivity, and data saturation require thoughtful consideration and strategic solutions to ensure the rigor and reliability of the collected data (Patton, 2014).

Structure of the Chapter

This chapter systematically explores various qualitative data collection methods, starting with a detailed overview of each method, its applications, benefits, and inherent challenges. Subsequent sections delve deeper into the practical aspects of applying these methods, including design considerations, implementation strategies, and ethical concerns. Case studies and practical examples are integrated throughout the chapter to illustrate the real-world application of different data collection methods, offering readers a holistic understanding of qualitative data collection processes.

Objective of the Chapter

The objective of this chapter is to equip readers with a comprehensive understanding of the diverse data collection methods in qualitative research. By exploring the theoretical underpinnings, practical applications, ethical considerations, and challenges of different methods, the chapter aims to provide readers with the knowledge and skills needed to design and implement effective data collection strategies in their research endeavors.

Finally, understanding the multitude of data collection methods and their appropriate application is crucial for conducting rigorous qualitative research. This chapter seeks to illuminate the intricacies of qualitative data collection methods, offering scholarly insights and practical guidance to both novice and experienced researchers in the field.

Observations

Observation as a method of data collection is a fundamental practice within qualitative research. Rooted in the practices of early anthropologists and sociologists who engaged in fieldwork to understand cultures and social groups, observation allows for an immersive, first-hand account of the phenomenon under study (DeWalt & DeWalt, 2002).

Nature of Observational Research

Observational research transcends mere "looking" to involve systematic noting, recording, and analysis of phenomena as they occur naturally. Rather than relying on participants' accounts or self-reports, observations provide researchers with direct data, capturing behaviors, interactions, events, and contexts as they unfold (Angrosino, 2007).

Types of Observations

1. **Participant Observation:** The researcher immerses themselves in the setting, often participating in the activities of the group under study. This deep involvement provides rich, inside perspectives but can raise questions about objectivity (Emerson, Fretz & Shaw, 1995).

Participant observation has long been foundational in qualitative research, acting as a bridge that allows researchers to immerse themselves in the lived experiences of the subjects under study. Originally rooted in anthropological traditions, participant observation involves the researcher taking an active role within the community or group they are studying, thereby gaining a first-hand, insider's view of the culture, practices, and interactions of that community (Malinowski, 1922).

Characteristics of Participant Observation

a) **Immersion and Involvement:** The researcher becomes a part of the community, sometimes living with them, participating in their daily activities, rituals, and ceremonies (Geertz, 1973).
b) **Dual Role:** The researcher has to balance the role of an observer and a participant, making sure they are active in community activities while also maintaining the detachment necessary for objective observation (Whyte, 1943).
c) **Holistic Understanding:** By being part of the community, the researcher gets an all-round view, understanding not just what is said but the unsaid – the implicit cultural norms, values, and beliefs (Spradley, 1980).
Strengths of Participant Observation:
    a) **Rich Data Collection:** As the researcher is directly involved, they can gather nuanced, in-depth data that might not be apparent from an external perspective (Wolcott, 1995).
    b) **Contextual Insights:** Participant observation allows the researcher to understand the context in which actions, interactions, and decisions take place (Emerson, Fretz & Shaw, 1995).
    c) **Building Trust:** Prolonged engagement can lead to deeper trust, making it more likely for community members to share authentic, sensitive information (Bernard, 2006).

Challenges of Participant Observation

a) **Maintaining Objectivity:** Being deeply involved might lead to 'going native', where the researcher becomes too attached and loses their objective viewpoint (Gold, 1958).
b) **Ethical Considerations:** Being an insider might mean that community members forget the researcher's official capacity, leading to potential ethical concerns about informed consent and confidentiality (Kawulich, 2005).
c) **Emotional and Physical Toll:** Living and engaging deeply in another community, especially if it is very different from the researcher's own, can be physically and emotionally taxing (Agar, 1996).

Participant observation, though demanding, offers a unique methodological approach that provides rich, contextually grounded insights. It requires a delicate balance of involvement and detachment and poses distinct ethical and practical challenges. However, when done judiciously, it can yield profound understanding and contribute significantly to qualitative research.

2. **Non-participant Observation:** The researcher remains a passive observer, without becoming involved in the activities of the group. This allows for a more detached perspective, but may limit the depth of understanding (Spradley, 1980).

Non-participant observation, distinct from participant observation, is a method wherein researchers immerse themselves in the study setting but do not actively participate in the activities of the community or group being studied. This method aims to provide an objective and undisturbed account of the observed events, interactions, and behaviors (Kawulich, 2005). Historically, this method has been invaluable in various fields of study, allowing for an undiluted perspective of events in their natural context.

Characteristics of Non-Participant Observation

a) **Detached Role:** The researcher, while present in the environment, does not intervene or engage in the ongoing activities, ensuring minimal disruption and influence on the observed events (Gold, 1958).
b) **Unobtrusive Stance:** Often, researchers will aim to be as inconspicuous as possible, preventing their presence from altering the natural flow of events (Patton, 2015).
c) **Objective Lens:** Given the detached nature of this method, researchers can capture data without the potential biases that may come from direct involvement (Lofland & Lofland, 1995).

Advantages of Non-Participant Observation

a) **Minimized Researcher Influence:** By staying detached, researchers can ensure that their presence doesn't significantly alter the behaviors or events they're observing (Spradley, 1980).
b) **Authentic Data:** Capturing events and interactions as they naturally occur ensures the authenticity of the data collected (Bryman, 2016).
c) **Enhanced Objectivity:** The distance maintained allows for a more neutral standpoint, which can be crucial in various research contexts (DeWalt & DeWalt, 2002).

Limitations of Non-Participant Observation

a) **Lack of Deep Understanding:** The detached position might result in missing out on the deeper, intrinsic cultural meanings, emotions, or perspectives experienced by insiders (Emerson, Fretz & Shaw, 1995).
b) **Potential Ethical Issues:** Being a silent observer might sometimes lead to observing without informed consent, especially in public spaces, leading to ethical dilemmas (Bryman, 2016).
c) **Missed Contextual Nuances:** Not engaging actively might lead to misinterpretations or overlooking certain subtle, yet significant, aspects of the observed events (Kawulich, 2005).

Non-participant observation serves as an essential tool in the qualitative research toolkit, especially when seeking to minimize influence on the setting or maintain a clear objective perspective. While it presents certain challenges, its merits in providing an authentic snapshot of events in their natural state are undeniable. A balanced approach, weighing its strengths and limitations, can lead to robust, credible insights.

The Process of Conducting Observations

1. **Gaining Access:** Before observing, researchers often need permission from gatekeepers, especially in private or sensitive settings (Berg, 2009).

Observational research, while insightful, often presents researchers with the initial challenge of accessing the field or site of study. Gaining access not only requires methodological considerations but also touches upon ethical concerns and interpersonal dynamics.

Why Is Gaining Access Critical?

Before immersing oneself in the observational field, it is crucial to secure the necessary permissions and foster relationships to ensure the research progresses smoothly (Dewalt & Dewalt, 2002). Gaining access can influence the richness of data, determine the quality of researcher-participant interactions, and influence the longevity of the research.

Strategies to Gain Access

a) **Leveraging Existing Networks:** Using personal or institutional contacts can be instrumental in navigating entry points into the research setting (Emerson, Fretz & Shaw, 1995).
b) **Building Trust:** Prioritizing the establishment of trust with gatekeepers or community leaders can open doors to richer observational opportunities (Adler & Adler, 1994).
c) **Formal Permissions:** Especially in institutional settings or private spaces, formal written permissions might be required to ensure the legality and ethicality of the observation (Bryman, 2016).
d) **Engaging in Pre-visit Dialogues:** Before formal observations, engaging in conversations or preliminary visits can familiarize the participants with the researcher's presence and objectives (Spradley, 1980).
e) **Transparency:** Clearly articulating the purpose, methods, and goals of the research can ease concerns and foster a more welcoming environment (Kawulich, 2005).

Challenges in Gaining Access

a) **Resistance or Skepticism:** Potential participants or gatekeepers might resist or view the research with skepticism, fearing misrepresentation or intrusion (Dewalt & Dewalt, 2002).
b) **Bureaucratic Hurdles:** Institutional settings might pose bureaucratic challenges that can delay or restrict observational opportunities (Bryman, 2016).
c) **Cultural or Social Barriers:** Differences in language, culture, or social norms can influence the ease of access and require researchers to demonstrate cultural sensitivity (Emerson, Fretz & Shaw, 1995).

Gaining access is a nuanced process, central to the success of observational research. It requires a blend of methodological astuteness, ethical considerations, and interpersonal skills. By effectively navigating this initial phase, researchers can set a positive tone for the entire observational journey, ensuring richer insights and meaningful engagements.

2. **Taking Field Notes:** Detailed notes during or immediately after observations capture data, insights, and initial interpretations. These notes may include sketches, maps, and diagrams (Sanjek, 1990).

Field notes are a cornerstone of observational research. Their purpose is to capture the lived experiences, events, interactions, and nuances that a researcher witnesses in the field. These notes serve as the primary data for analysis and interpretation later in the research process (Emerson, Fretz & Shaw, 1995).

The Significance of Field Notes

a) **Memory Augmentation:** Human memory is fallible. Field notes serve as a record, ensuring that subtle details and sequences of events aren't forgotten or misremembered (Sanjek, 1990).
b) **Contextual Richness:** They help capture the context, providing depth to the events and aiding in understanding the deeper cultural or social meanings (Clifford, 1990).
c) **Analytical Foundation:** Field notes are the foundational texts upon which coding, analysis, and interpretation are built in later stages (Ottenberg, 1990).

Techniques for Effective Field Note Taking

a) **Jotted Notes:** During observation, brief notes or keywords can be jotted down to be expanded upon later, ensuring minimal interruption to the observation process (Malinowski, 1967).
b) **Descriptive Detail:** Notes should capture sensory details, including sights, sounds, smells, etc., to convey a rich picture of the setting (Geertz, 1973).
c) **Direct Quotes:** Whenever possible, verbatim speech should be recorded, as it provides authentic insight into participants' thoughts and perspectives (Spradley, 1980).
d) **Reflexive Notes:** Researchers should also note their feelings, interpretations, and reactions to observed events, ensuring that personal reflexivity is part of the data (Davies, 2008).
e) **Use of Diagrams:** Spatial layouts, interaction diagrams, and other visual aids can supplement written notes to provide a clearer understanding of the physical and interactional context (Bernard, 2011).

Challenges in Taking Field Notes

a) **Selective Attention:** The impossibility of capturing everything means that researchers need to be selective, which introduces the possibility of bias (Jackson, 1983).
b) **Intrusiveness:** Taking notes can sometimes be viewed as intrusive or disruptive by participants (Gold, 1958).
c) **Memory Limitations:** Even with jotted notes, the challenge of recalling and expanding on them post-observation is ever-present (Sanjek, 1990).

Field notes, when taken with care, attention, and reflexivity, become a rich tapestry of data. They offer a window into the field, capturing the complexity, richness, and depth of the lived experiences and events researchers are privileged to witness.

3. **Reflecting and Debriefing:** Post-observation, it's beneficial for the researcher to reflect on what was observed, identifying patterns, anomalies, or emerging themes (Mulhall, 2003).

Reflection and debriefing are pivotal processes in observational research, acting as mechanisms that augment understanding, add depth to raw data, and facilitate the researcher's personal connection with the observed events.

Significance of Reflection and Debriefing

a) **Self-awareness and Reflexivity:** Reflection allows researchers to become aware of their biases, emotions, and preconceptions that may affect data collection and interpretation (Finlay, 2002).
b) **Data Enrichment:** Debriefing sessions, especially with peers or mentors, can offer additional insights or alternative interpretations of observed events (Borkan, 1999).
c) **Emotional Catharsis:** Observational research, particularly in sensitive settings, can be emotionally taxing. Reflecting and debriefing serve as outlets for emotional processing (Dickson-Swift et al., 2007).

Strategies for Effective Reflection and Debriefing

a) **Maintaining a Reflexive Journal:** Apart from field notes, a separate journal dedicated to personal reflections can capture the researcher's emotions, biases, and evolving insights (Ortlipp, 2008).
b) **Peer Debriefing Sessions:** Regular meetings with peers or mentors to discuss observations can enhance understanding and mitigate subjective biases (Lincoln & Guba, 1985).
c) **Scheduled Reflection Time:** Setting aside dedicated time after each observation for introspection ensures consistent and thorough reflection (Schön, 1983).
d) **Using Prompts:** Reflective prompts can guide the introspection process, encouraging researchers to explore various dimensions of their experience (Boud et al., 1985).

Challenges in Reflection and Debriefing

a) **Overwhelming Emotions:** Intense field experiences might lead to strong emotional reactions, which can cloud judgment and interpretation (Dickson-Swift et al., 2007).
b) **Time Intensity:** The process requires a significant time commitment, especially in long-term observational projects (Finlay, 2002).
c) **Potential for Bias:** While debriefing can offer multiple perspectives, it can also introduce new biases, necessitating careful consideration and discernment (Lincoln & Guba, 1985).

Reflection and debriefing, when practiced with diligence and sincerity, not only enhance the quality and depth of observational data but also contribute to the personal and professional growth of the researcher.

Strengths of Observational Research

Observations allow for:

1. **Capturing Context:** Observations encompass not only the actions but also the environment, allowing for a rich understanding of context (Gold, 1958).

Observational research offers a unique opportunity to deeply embed oneself within the natural settings of participants, capturing the intricate dynamics and layers of context that other methods might overlook or misrepresent.

A. Depth of Understanding:

Being present in the environment where behaviors and interactions occur, researchers gain a comprehensive understanding of socio-cultural, environmental, and personal factors that influence the phenomenon of interest (Spradley, 1980).

B. Nuances and Subtleties:

Direct observations enable researchers to pick up on non-verbal cues, silent pauses, and subtle interactions that are often missed in self-reports or interviews (Angrosino, 2007).

C. Dynamics of Interactions:

By observing how individuals interact with their environment and with others, a researcher can better comprehend social norms, roles, and relationships, and their effects on behaviors and decisions (Emerson et al., 2011).

D. Temporality and Sequence:

Observational methods allow the researcher to capture events in real-time, offering insights into sequences, patterns, and causality that are hard to deduce from retrospective accounts (DeWalt & DeWalt, 2011).

E. Naturalistic Setting:

Observations conducted in natural settings reduce the Hawthorne effect, where participants might change their behavior due to the awareness of being observed. Hence, the data is likely to be more authentic (Adler & Adler, 1994).

F. Bridging the Say-Do Gap:

What people say they do and what they actually do can be quite different. Observations help in bridging this gap by providing a real-world check on self-reported behaviors (Fetterman, 2019).

Capturing context is one of the hallmarks of observational research. Through direct, prolonged engagement, researchers can delve deeper into the intricate tapestry of events, behaviors, and interactions, offering a rich, nuanced understanding that is grounded in the reality of everyday life.

2. **Dynamic Data Collection:** Observing real-time events and interactions captures the fluidity and dynamics of situations (Kawulich, 2005).

Observational research is inherently dynamic. Unlike static methods that capture a snapshot of participant perspectives at a single point in time, observational methods delve into the ever-evolving flow of human behaviors and interactions, providing rich insights into the complexities of real-world phenomena.

A. Continuous Engagement:

Observational research allows for continuous and prolonged engagement in the field. This deep immersion enables researchers to observe changes, transitions, and developments as they occur, granting them a comprehensive understanding of processes over time (Gold, 1958).

B. Real-time Responses:

Observing events and behaviors as they unfold provides an opportunity to immediately recognize and probe unexpected occurrences, contradictions, or anomalies, thereby enhancing the depth and breadth of data collection (Patton, 2015).

C. Adaptability:

The dynamic nature of observational research allows researchers to adjust their focus based on emerging patterns, insights, or areas of interest. Such flexibility ensures that the research remains relevant and aligned with the evolving context (Sanjek, 1990).

D. Capturing the Unplanned:

While structured interviews or surveys might miss spontaneous events or unplanned incidents, observational methods are uniquely positioned to capture these, offering a fuller picture of the phenomenon under study (Lofland et al., 2006).

E. Embracing Complexity:

Human behaviors and interactions are intricate and multi-faceted. Observational research, with its dynamic data collection approach, embraces this complexity, allowing for a layered, multifaceted exploration of phenomena (Agar, 1980).

F. Contextual Interactions:

Observing how individuals dynamically interact with changing environments, circumstances, or other individuals provides rich insights into the fluidity of human behavior and the numerous factors that influence it (Bernard, 2011).

Dynamic data collection is an inherent strength of observational research. The method's adaptability, real-time engagement, and ability to capture the unplanned and unexpected provide researchers with a holistic, in-depth understanding of the intricate dance of human behaviors and interactions within their natural settings.

3. **Reduced Recall Bias:** Since data is collected as events unfold, observations are not susceptible to the recall biases that might affect interviews or surveys (Savage, 2006).

Recall bias is a common challenge in research methodologies that rely on participants' memory of past events, potentially leading to skewed or inaccurate results. Observational research, by design, curtails this bias as it captures behaviors and events in real-time, thus safeguarding the integrity and accuracy of the data collected.

A. Present-Moment Recording:

In observational research, the immediacy of data capture ensures that information is documented as events unfold, mitigating the need for participants to remember past occurrences. This real-time recording not only assures data accuracy but also reduces the distortions often introduced when relying on recollections (Kawulich, 2005).

B. Circumventing Memory Distortions

Memory is not a perfect recorder of events. Over time, individuals might forget, exaggerate, or minimize details, thereby introducing errors into their recollections. Observational research bypasses these memory distortions by focusing on real-time observations rather than retrospective accounts (Bernard, 2011).

C. Authenticity of Spontaneous Reactions

Observational research captures spontaneous reactions, behaviors, and interactions, which are less influenced by post-event reflections or reinterpretations. These authentic responses provide a more genuine representation of how individuals naturally behave in specific situations (Angrosino, 2007).

D. Reducing Social Desirability Bias

When participants are asked to recall their behaviors, they might inadvertently or intentionally depict themselves in a more favorable light, a phenomenon known as social desirability bias. Observations can reduce this bias by documenting actual behaviors, rather than relying on participants' potentially skewed self-reports (Silverman, 2010).

E. Verifying Self-Reports

Observations can act as a validating tool. When combined with self-report measures, researchers can compare observed behaviors with participants' accounts, offering a more holistic understanding and potentially identifying inconsistencies arising from recall bias (Jorgensen, 1989).

Reducing recall bias is a paramount strength of observational research. By focusing on real-time data capture, observational methods provide a more authentic, undistorted, and reliable account of events, behaviors, and interactions, enhancing the robustness and credibility of research findings.

Challenges in Observational Research

1. **Observer Effect:** The mere presence of an observer can influence participants' behavior, potentially skewing data (Webb et al., 1966).

One of the critical challenges in observational research is the potential for the "observer effect" or "Hawthorne effect." This phenomenon refers to the potential changes in behavior exhibited by research participants when they are aware they are being observed (Landsberger, 1958).

A. Conscious Alteration of Behaviors

When individuals know they are being observed, they might modify their actions, either consciously or subconsciously, to present themselves in a manner they deem favorable or acceptable. Such behavior alterations can skew research findings, as they might not represent the participants' typical or natural actions (Adair, 1984).

B. Observer Expectations

The observer's expectations, whether explicitly communicated or not, can influence participant behaviors. Participants might attempt to behave in ways they believe align with the observer's expectations, even if these behaviors diverge from their usual actions (Rosenthal, 1966).

C. Duration of Observation

While the observer effect is often most pronounced at the beginning of a study, the duration of observation can impact its persistence. In extended studies, participants might become desensitized to the observer's presence over time, leading to more genuine behaviors as the study progresses (McCambridge, Witton, & Elbourne, 2014).

D. Minimizing the Observer Effect

Several strategies can mitigate the observer effect. These include unobtrusive observation where participants may not be aware they're being observed, building trust and rapport to reduce the artificiality of the observation setting, and employing participant observers who are members of the group being studied, thereby reducing the external observer's influence (Gold, 1958).

E. The Inevitability of the Observer Effect

Despite attempts to minimize the observer effect, its presence to some degree is often inevitable in observational research. Recognizing and acknowledging this potential bias is crucial, and researchers should factor it into their analyses and interpretations of the observed data (Patton, 2015).

The observer effect presents a significant challenge in observational research, potentially altering the authenticity of observed behaviors. By understanding this phenomenon and employing strategies to reduce its influence, researchers can capture more genuine insights into participant behaviors and interactions.

2. **Subjectivity:** Researchers' backgrounds, beliefs, and experiences can influence what they notice and how they interpret it (Atkinson & Hammersley, 1994).

Observational research, by its very nature, is prone to the influence of the observer's personal beliefs, perceptions, and interpretations. This influence, often referred to as subjectivity, can present challenges in maintaining the objectivity and reliability of observational findings (Angrosino, 2005).

A. Observer Bias

Every observer brings with them a set of personal beliefs, experiences, and cultural perspectives that can influence how they interpret and record observed behaviors. This can lead to observer bias, where observations are skewed in a particular direction or manner based on these personal inclinations (Goldberg, 1965).

B. Influence of Preconceptions

Researchers entering a field with preconceived notions or hypotheses might unconsciously focus more on behaviors or events that align with these notions, potentially overlooking or downplaying other significant observations (DeWalt & DeWalt, 2002).

C. Variability Among Observers

When multiple observers are involved in a study, each may interpret and record the same events differently. This inter-observer variability can introduce inconsistencies in the collected data, complicating analysis and interpretation (Bogdan & Biklen, 2007).

D. Reflexivity in Observation

Reflexivity involves the researcher's awareness of and reflection upon their role, influence, and subjectivity in the research process. By acknowledging and deliberating on their biases and influences, researchers can aim for a more balanced and transparent observational approach (Guillemin & Gillam, 2004).

E. Strategies to Mitigate Subjectivity

While complete elimination of subjectivity in observational research is improbable, several strategies can reduce its impact. Triangulation, involving the use of multiple observers or methods, can provide a more holistic view. Training sessions for observers, aimed at standardizing observation and note-taking techniques, can reduce inter-observer variability. Regular debriefing sessions can also help in highlighting and addressing potential biases (Lincoln & Guba, 1985).

Subjectivity presents both a challenge and an inherent characteristic of observational research. While it can introduce biases and influence findings, when acknowledged and addressed, it can also contribute depth, nuance, and richness to qualitative research insights.

3. **Time-Intensive:** Comprehensive observations often require significant time in the field, making them resource-intensive (Tedlock, 2000).

Observational research, by virtue of its depth and detail, often demands considerable time from researchers. This time-intensive nature, though beneficial for rich data collection, poses challenges, especially when resources, including time, are constrained (Mulhall, 2003).

A. Extended Fieldwork

In-depth observations typically require prolonged periods of immersion in the field. Extended stays facilitate familiarity and rapport building, and allow researchers to witness a variety of events or behaviors (Fetterman, 2019). This duration, however, can strain resources and be physically and mentally exhausting.

B. Detailed Note-taking

Capturing the essence of observed events requires meticulous note-taking. This process can be laborious, especially when researchers strive to note down verbatim dialogues, spatial arrangements, or sequential actions (Emerson, Fretz, & Shaw, 1995).

C. Data Overload

Long-term observations can lead to vast amounts of data. Managing, coding, and analyzing such extensive data sets can be overwhelming, making the process of distilling meaningful patterns and insights challenging (Savage, 2006).

D. Time and Evolving Contexts

Over extended periods, contexts can change – participants may leave, environmental conditions may shift, or unforeseen events may occur. Such changes can impact the consistency of data and introduce additional variables (LeCompte & Goetz, 1982).

E. Strategies to Manage Time-Intensiveness

Researchers can employ several strategies to manage the time demands of observational research. Prioritizing observations, using technology to assist in data collection (e.g., audio/video recordings), and collaborating with co-researchers to share observational duties can optimize time. Furthermore, periodic data analysis, even during the data collection phase, can guide researchers, allowing them to focus on particularly significant areas (Baxter & Jack, 2008).

The time-intensive nature of observational research, while providing an avenue for rich and in-depth data collection, poses significant challenges. A thoughtful approach, combining rigorous planning with flexibility, can help researchers navigate these demands, ensuring quality and depth in their findings.

Ethical Considerations in Observational Research

Ensuring ethical integrity is crucial. This includes obtaining informed consent when required, ensuring anonymity, and being sensitive to intrusive observations, especially in private or sensitive contexts (Wiles et al., 2008).

Observational research poses a unique set of ethical challenges because of its immersive and often covert nature. Below, we will explore the primary ethical issues inherent in observational studies and propose strategies to address them (Kawulich, 2005).

A. Informed Consent

A hallmark of ethical research, informed consent can be particularly problematic in observational research. When observing in public spaces or community settings, obtaining explicit consent from every individual can be impractical.

Strategies:

- **Passive consent**: Notify community members about the research and provide opt-out opportunities (Erickson, 1989).
- **Deferred consent**: Gather data covertly but approach participants afterward for their consent to use the collected data (Bulmer, 1982).

B. Privacy and Anonymity

Ensuring the privacy and anonymity of observed participants is paramount, given the potential for unintended harm or misrepresentation.

Strategies

- **Anonymizing data**: Replace names and identifying details with pseudonyms or codes (Hammersley & Atkinson, 2007).
- **Secure storage**: Use encrypted storage solutions and restrict access to data (Wiles, Crow, & Pain, 2011).

C. Intrusiveness

The presence of an observer can influence participants' behavior, potentially causing discomfort or altering the dynamics of the observed setting.

Strategies

- **Minimize disruption**: Use unobtrusive methods and try to blend into the setting (Gold, 1958).
- **Rapport building**: Spend time getting to know participants to reduce the 'observer effect' (Jorgensen, 1989).

D. Misinterpretation and Bias

Subjectivity in observations can lead to misinterpretations, which can be exacerbated by researchers' biases.

Strategies

- **Triangulation**: Employ multiple data sources or methods to validate findings (Denzin, 1978).
- **Peer review**: Engage colleagues to review and critique observations to identify potential biases (Emerson, Fretz, & Shaw, 2011).
  E. The Issue of Covert Observation

While covert observation can offer genuine insights by minimizing the observer's influence, it raises significant ethical concerns about deception and non-consent.

Strategies

- **Weigh benefits**: Covert methods should only be used if the knowledge gained is of significant value and could not be obtained otherwise (Bulmer, 1982).
- **Post-study debrief**: Inform participants after the observation, explaining the reasons for secrecy and offering them the option to withdraw their data (Lee, 1993).

Ethical considerations in observational research are intricate due to its immersive and sometimes covert nature. Navigating these ethical waters requires a balance of methodological rigor, respect for participants, and a commitment to producing knowledge that justifies the potential ethical ambiguities.

Observations offer a dynamic, rich method of data collection, providing insights unattainable through other methods. Though they come with challenges, with careful planning and ethical considerations, observations can be an invaluable tool in qualitative research.

Interviews: Structured, Semi-Structured, and Unstructured

Interviews have been integral to qualitative research, providing in-depth insights into participants' experiences, perspectives, and the meanings they ascribe to them (Denzin & Lincoln, 2005). Based on the level of structure and standardization, interviews can be classified into three primary categories: structured, semi-structured, and unstructured.

1. Structured Interviews:

Structured interviews, rooted in positivist paradigms, are foundational in research methodologies aiming for replicability and generalizability. These interviews, heavily influenced by quantitative survey methods, have been tailored to fit qualitative research needs (Bryman, 2012). Below is a detailed examination of structured interviews:

Definition

Structured interviews involve a predetermined set of questions administered to every interviewee in precisely the same order and manner, ensuring consistency and standardization across data collection sessions (Fontana & Frey, 2000).

Key Characteristics

- **Standardization**: Every participant encounters the same questions in an identical sequence, minimizing the potential for interviewer bias (Berg, 2007).
- **Closed-ended Questions**: Predominantly, the questions posed are closed-ended, allowing interviewees to select from preset responses (Denzin & Lincoln, 2005).
- **Scalability**: Given their design, structured interviews are easily administered to large sample sizes (Ritchie & Lewis, 2003).

Advantages

a) **Consistency**: The standardization ensures uniform data, facilitating comparability across participants (Silverman, 2006).
b) **Efficiency**: Structured format often results in faster data collection and subsequent analysis (King & Horrocks, 2010).
c) **Reduced Bias**: The interviewer's discretion in phrasing or sequencing questions is minimized, potentially reducing variability in responses attributed to interviewer effects (Bryman, 2012).

Limitations

a) **Lack of Depth**: The rigidity can prevent interviewers from probing deeper into the participant's responses, potentially missing out on nuanced information (Patton, 2002).
b) **Flexibility**: Since questions are fixed, they might not account for all individual experiences or unforeseen nuances in a topic (Mason, 2002).
c) **Response Limitation**: Predetermined answers might not encapsulate participants' genuine feelings or experiences (Seidman, 2013).

Applications

Structured interviews are often favored in research settings where:

- The research question revolves around specific, predetermined aspects.
- There's a necessity for cross-comparisons between different study participants or even across different studies.
- The study aims for a broader sweep rather than an in-depth exploration (Creswell, 2013).

While structured interviews offer a range of benefits in terms of consistency and scalability, they come with the trade-off of depth and flexibility. Researchers must weigh these strengths and weaknesses against their research goals to determine if this method is the most appropriate for their study.

2. Semi-Structured Interviews

Semi-structured interviews occupy a pivotal space within qualitative research, bridging the gap between the rigidity of structured interviews and the free-flowing nature of unstructured formats. These interviews offer a balanced approach to data collection (Bryman, 2012). Let's delve deeper into this method:

Definition

Semi-structured interviews entail an interview guide comprised of open-ended questions, which provides direction to the interview. However, the interviewer retains the flexibility to probe further or redirect the conversation based on the participant's responses (Brinkmann & Kvale, 2015).

Key Characteristics

- **Interview Guide**: Predetermined questions form the backbone of the interview, ensuring coverage of key areas (DiCicco-Bloom & Crabtree, 2006).
- **Flexibility**: Interviewers can modify questions, change their order, or explore emergent themes during the interview (Rubin & Rubin, 2012).
- **Open-ended Questions**: Responses aren't restricted to predefined categories, allowing participants to express their views comprehensively (Fontana & Frey, 2000).

Advantages

a) **Depth and Breadth**: Semi-structured interviews provide comprehensive insights while ensuring that core topics are addressed (Mason, 2002).
b) **Responsiveness**: Interviewers can adapt to the participant's perspective, ensuring a participant-centered approach (Seidman, 2013).
c) **Standardization with Flexibility**: While there's a structure to follow, there's room for spontaneous exploration (Denzin & Lincoln, 2005).

Limitations

a) **Time-Consuming**: The open-ended nature of responses can prolong interview duration and subsequent data analysis (King & Horrocks, 2010).
b) **Requires Skill**: Achieving a balance between following the guide and exploring emergent topics demands skilled interviewing (Brinkmann & Kvale, 2015).

c) **Consistency Challenges**: Different interviews can vary significantly based on the direction taken during the session (Patton, 2002).

Applications

Semi-structured interviews are particularly useful when:

- The topic under study is multifaceted and may require adaptability during data collection.
- Researchers aim for depth and detail but also wish to ensure specific themes are consistently addressed.
- There's a necessity to allow participants the freedom to share experiences or viewpoints without stringent confines (Creswell, 2013).

Semi-structured interviews amalgamate structure with adaptability, making them a versatile tool in qualitative research. This method, while demanding in terms of time and skill, offers a unique blend of depth, breadth, and participant responsiveness.

3. Unstructured Interviews

Unstructured interviews, an essential instrument in the qualitative researcher's toolbox, prioritize the participant's voice and the fluid co-construction of knowledge. Unlike their structured or semi-structured counterparts, they lack a strict format, providing ample room for participants to direct the conversation (Fontana & Frey, 2005). Here's an in-depth look into this method:

Definition

Unstructured interviews are conversational and open-ended, often initiated with broad questions or topics. They allow the interviewee maximum freedom to express their experiences and perspectives without the limitations of pre-defined questions (Warren & Karner, 2010).

Key Characteristics

- **Open Initiation**: Interviews usually begin with open-ended questions, such as "Can you tell me about...?", letting the participant steer the direction (Bernard, 2006).
- **Flowing Conversation**: They resemble a casual conversation more than a formal interview, where topics emerge organically (Spradley, 1979).
- **Absence of a Set Framework**: There's no strict interview guide or predetermined set of questions (Lofland & Lofland, 1995).

Advantages

a) **Depth of Understanding**: The format allows for deep dives into experiences, feelings, and perceptions (Minichiello et al., 1990).

b) **Flexibility**: Provides room to explore unforeseen avenues and emergent themes (Roulston, 2010).
c) **Empowerment**: Gives participants a voice, often making them feel valued and heard (Fontana & Frey, 2000).

Limitations

a) **Requires Expertise**: The researcher needs strong interpersonal and active listening skills (Rubin & Rubin, 2011).
b) **Time-Consuming**: Both the interviews and subsequent analysis can be lengthy due to their open-ended nature (Kvale & Brinkmann, 2009).
c) **Lack of Consistency**: Comparing data across interviews can be challenging because of the varied content in each session (Seidman, 2013).

Applications

Unstructured interviews are especially suitable when:

- The research aims to understand deeply personal or sensitive topics.
- The research context is novel, and the researcher has limited prior understanding.
- The emphasis is on giving marginalized or silenced groups a voice (Ellis & Berger, 2003).

Embodying the essence of qualitative inquiry, unstructured interviews offer profound insights into human experiences, capturing nuances and subtleties. While they come with challenges, their potential for deep understanding is unparalleled, especially when the researcher's goal is to immerse fully in the participant's world.

Choosing the appropriate type of interview depends on the research question, the nature of the topic, the characteristics of the participants, and the desired data's depth and breadth. While structured interviews provide consistency, semi-structured and unstructured interviews offer depth and flexibility, respectively.

Focus Groups

Focus groups, a popular qualitative data collection method, involve interactive group discussions centered on a particular topic or set of topics. They've been extensively used in market research and are now a mainstay in various academic disciplines. In this section, we delve into the nuances of focus groups.

Definition: A focus group consists of a small, diverse group of participants who are gathered to discuss a predefined topic, guided by a skilled moderator (Krueger & Casey, 2014).

Key Characteristics

- **Interactive Dialogue**: The emphasis is on participant interaction, generating a rich understanding of people's experiences and beliefs (Morgan, 1996).
- **Guided Discussion**: A moderator uses a semi-structured guide to navigate the conversation, ensuring key topics are covered (Stewart, Shamdasani, & Rook, 2007).

- **Group Dynamics**: Participant reactions, disagreements, and consensus provide insights into group norms and values (Fern, 2001).

Advantages

a) **Rich Data**: Multiple perspectives offer a layered understanding of the topic (Krueger & Casey, 2014).
b) **Efficiency**: They provide the opportunity to gather data from multiple participants simultaneously (Merton, Fiske, & Kendall, 1990).
c) **Natural Environment**: Group settings can often mirror natural social interactions, making discussions more organic (Catterall & Maclaran, 1997).

Limitations

a) **Groupthink**: Participants might conform to dominant views, suppressing dissent (Janis, 1982).
b) **Moderator Influence**: The moderator's actions might influence participant responses (Smithson, 2000).
c) **Logistics**: Organizing groups, especially diverse ones, can be challenging (Morgan, 1996).

Applications

Focus groups are ideal for:

- Exploring new phenomena.
- Testing new products or ideas.
- Understanding group norms and cultural values (Kitzinger, 1994).

Ethical Considerations

- Ensuring confidentiality in group settings.
- Addressing power dynamics and potential conflicts.
- Obtaining informed consent for group discussions and any subsequent recordings (Farquhar & Das, 1999).

Focus groups, when designed and moderated effectively, offer an in-depth understanding of collective perspectives, social norms, and group dynamics. While they come with challenges, their interactive nature and the richness of data they provide make them invaluable in qualitative research.

Document and Content Analysis

Document and content analysis have long been employed as systematic methodologies to interpret and analyze textual information. These approaches enable researchers to discern patterns and themes within a broad range of documents, from historical manuscripts to modern digital content.

Definition: Document and content analysis refers to the systematic examination of textual, visual, or audio materials to identify patterns, themes, or biases (Krippendorff, 2012). This approach can be applied to diverse sources, including books, newspapers, digital media, and more.

Key Characteristics

- **Systematic Approach**: An organized method is applied to ensure replicability and validity (Neuendorf, 2016).
- **Objective Analysis**: The aim is to reduce subjectivity by adhering to a clear set of guidelines (White & Marsh, 2006).
- **Contextual Understanding**: Recognizing the context in which documents were created is crucial (Prior, 2008).

Advantages

1. **Unobtrusive**: This method does not require direct involvement with the study subjects, thereby eliminating the risk of influencing participant behavior (Bowen, 2009).
2. **Cost-effective**: Often, public documents are readily accessible and free, reducing research costs (Altheide, 2000).
3. **Temporal Analysis**: Enables tracking changes over time by analyzing documents from different periods (Hodder, 2000).

Limitations

1. **Data Limitations**: Restricted to existing content; researchers cannot probe or ask follow-up questions (Silverman, 2013).
2. **Validity Concerns**: The authenticity and credibility of the documents can sometimes be questionable (Scott, 1990).
3. **Potential Biases**: Published content may reflect the biases of authors, publishers, or prevailing societal norms (Fairclough, 1995).

Applications

Document and content analysis are invaluable for:

- Historical research.
- Media studies and bias exploration.

- Policy analysis and evaluation (Altheide & Schneider, 2013).

Ethical Considerations

- Ensuring privacy and confidentiality, especially when analyzing personal or sensitive documents.
- Recognizing and accounting for potential biases.
- Properly citing all sources and respecting copyright regulations (Mauthner et al., 1998).

Conclusion

Document and content analysis serve as powerful tools for interpreting and understanding vast amounts of textual data. By providing insights into historical shifts, societal norms, and media portrayals, these methods offer a rich and layered understanding of content across different contexts.

Visual Methods: Photographs, Drawings, and Videos

Visual methodologies, comprising photographs, drawings, and videos, have grown significantly in prominence as valuable tools within the research landscape. These methods enable rich, multi-layered insights by capturing the complexities of human experience in visual form, thus transcending linguistic barriers and offering unique perspectives.

Definition

Visual methods refer to the utilization of visual materials, either researcher-generated or participant-generated, as a means to gather, analyze, and present data in research (Rose, 2016). These visual tools can encapsulate emotions, experiences, and phenomena that might be challenging to articulate verbally.

Types of Visual Methods

1. **Photographs**: Still images capturing moments, scenes, or objects.
2. **Drawings**: Hand-drawn illustrations often used to understand individual perspectives, emotions, or experiences.
3. **Videos**: Moving images that capture sequences of events, interactions, or environments.

Key Characteristics

- **Multidimensional Data**: Visual tools can capture nuances, emotions, and subtleties that may be overlooked in textual descriptions (Prosser & Loxley, 2008).
- **Cross-cultural Utility**: Visual data can transcend language barriers, making it valuable in diverse cultural settings (Banks, 2007).
- **Empowerment**: Allowing participants to capture their worlds can give them agency in the research process (Mitchell, 2008).

Advantages

1. **Rich Insights**: Offers a deeper, multifaceted understanding of the subject matter (Pink, 2013).
2. **Engagement**: Engages participants actively, making the research process collaborative (Harper, 2002).
3. **Versatility**: Suitable for various research settings and participant groups, including children and marginalized populations (Thomson, 2008).

Limitations

1. **Interpretative Challenges**: Deciphering the meaning behind images can be subjective and complex (Rose, 2016).
2. **Ethical Concerns**: Issues of privacy, consent, and potential misuse of visual data (Wiles et al., 2008).
3. **Technical Barriers**: Requires familiarity with cameras, software, and other equipment (Gubrium & Harper, 2013).

Applications

Visual methods are used extensively in:

- Anthropological and ethnographic studies.
- Educational research, especially with young participants.
- Studies focusing on urban planning and environmental changes (Radley & Taylor, 2003).

Ethical Considerations

- Obtaining informed consent for capturing and using visual data.
- Respecting privacy rights and blurring identifiable features when necessary.
- Secure storage and handling of visual materials to prevent misuse or unauthorized access (Clark-Ibáñez, 2007).

Conclusion

Embracing visual methods in research provides an avenue to tap into the richness of human experiences that might remain unarticulated in solely verbal or written formats. As research becomes increasingly interdisciplinary, the inclusion of visual tools further enriches the methodological repertoire.

REFERENCES

Adair, J. G. (1984). The Hawthorne effect: A reconsideration of the methodological artifact. *The Journal of Applied Psychology*, *69*(2), 334–345. doi:10.1037/0021-9010.69.2.334

Anfara, V. A., & Mertz, N. T. (2015). Theoretical frameworks in qualitative research. *Sage (Atlanta, Ga.)*.

Angrosino, M. (2007). Doing ethnographic and observational research. *Sage (Atlanta, Ga.)*.

Angrosino, M. V. (2005). Recontextualizing observation: Ethnography, pedagogy, and the prospects for a progressive political agenda. In N. Denzin & Y. Lincoln (Eds.), The Sage handbook of qualitative research (pp. 729-745). Sage.

Arain, M., Campbell, M. J., Cooper, C. L., & Lancaster, G. A. (2010). What is a pilot or feasibility study? A review of current practice and editorial policy. *BMC Medical Research Methodology*, *10*(1), 67. doi:10.1186/1471-2288-10-67 PMID:20637084

Atkinson, P., & Hammersley, M. (1994). Ethnography and participant observation. In N. K. Denzin & Y. S. Lincoln (Eds.), *Handbook of qualitative research* (pp. 248–261). Sage.

Babbie, E. (2015). *The practice of social research*. Cengage Learning.

Baker, T. L. (1994). *Doing Social Research*. McGraw-Hill.

Banks, M. (2007). *Using visual data in qualitative research*. SAGE. doi:10.4135/9780857020260

Baxter, P., & Jack, S. (2008). Qualitative case study methodology: Study design and implementation for novice researchers. *The Qualitative Report*, *13*(4), 544–559.

Beauchamp, T. L., & Childress, J. F. (2013). *Principles of biomedical ethics*. Oxford University Press.

Berg, B. L. (2007). *Qualitative research methods for the social sciences* (6th ed.). Pearson.

Bernard, H. R. (2006). *Research methods in anthropology: Qualitative and quantitative approaches*. AltaMira Press.

Blaikie, N. (2009). Designing social research. *Polity*.

Bogdan, R. C., & Biklen, S. K. (2007). *Qualitative research for education: An introduction to theories and methods*. Allyn & Bacon.

Bourke, B. (2014). Positionality: Reflecting on the research process. *The Qualitative Report*, *19*(33), 1–9.

Bowen, G. A. (2009). Document analysis as a qualitative research method. *Qualitative Research Journal*, *9*(2), 27–40. doi:10.3316/QRJ0902027

Brinkmann, S., & Kvale, S. (2015). *InterViews: Learning the craft of qualitative research interviewing* (3rd ed.). SAGE.

Catterall, M., & Maclaran, P. (1997). Focus group data and qualitative analysis programs: Coding the moving picture as well as the snapshots. *Sociological Research Online*, *2*(1), 42–54. doi:10.5153/sro.67

Charmaz, K. (2006). Constructing grounded theory: A practical guide through qualitative analysis. *Sage (Atlanta, Ga.)*.

Charmaz, K. (2014). *Constructing grounded theory* (2nd ed.). Sage.

Chilisa, B. (2012). *Indigenous research methodologies*. Sage Publications.

Clandinin, D. J., & Connelly, F. M. (2000). *Narrative inquiry: Experience and story in qualitative research*. Jossey-Bass.

Clark-Ibáñez, M. (2007). Inner-city children in sharper focus: Sociology of childhood and photo elicitation interviews. *Visual Studies*, *22*(2), 70–89.

Creswell, J. W. (2013). *Research design: Qualitative, quantitative, and mixed methods approaches*. Sage publications.

Creswell, J. W. (2014). *Research design: Qualitative, quantitative, and mixed methods approaches*. Sage Publications.

Creswell, J. W., & Creswell, J. D. (2017). *Research design: Qualitative, quantitative, and mixed methods approaches*. Sage Publications.

Creswell, J. W., & Miller, D. L. (2000). Determining validity in qualitative inquiry. *Theory into Practice*, *39*(3), 124–130. doi:10.1207/s15430421tip3903_2

Creswell, J. W., & Plano Clark, V. L. (2011). *Designing and conducting mixed methods research* (2nd ed.). Sage.

Creswell, J. W., & Poth, C. N. (2017). *Qualitative inquiry and research design: Choosing among five approaches* (3rd ed.). Sage.

Denzin, N. K. (1978). *Sociological methods: A sourcebook*. McGraw-Hill.

Denzin, N. K., & Lincoln, Y. S. (1994). *Handbook of qualitative research*. Sage Publications.

DeWalt, K. M., & DeWalt, B. R. (2002). *Participant observation: A guide for fieldworkers*. AltaMira Press.

DiCicco-Bloom, B., & Crabtree, B. F. (2006). The qualitative research interview. *Medical Education*, *40*(4), 314–321. doi:10.1111/j.1365-2929.2006.02418.x PMID:16573666

Eisenhart, M. (2009). Generalization from qualitative inquiry. In *Qualitative research* (pp. 51–66). Routledge.

Eldridge, S. M., Lancaster, G. A., Campbell, M. J., Thabane, L., Hopewell, S., Coleman, C. L., & Bond, C. M. (2016). Defining feasibility and pilot studies in preparation for randomised controlled trials: Development of a conceptual framework. *PLoS One*, *11*(3), e0150205. doi:10.1371/journal.pone.0150205 PMID:26978655

Ellis, C., & Berger, L. (2003). Their story/my story/our story: Including the researcher's experience in interview research. In J. F. Gubrium & J. A. Holstein (Eds.), *Postmodern interviewing* (pp. 157–183). SAGE. doi:10.4135/9781412985437.n9

Erickson, F. (1989). Ethical considerations in qualitative research. In R. Jaeger (Ed.), *Complementary methods for research in education*. American Educational Research Association.

Faden, R. R., Beauchamp, T. L., & Kass, N. E. (2014). Informed consent, comparative effectiveness, and learning health care. *Journal of the American Medical Association, 311*(4), 403–404. PMID:24552325

Farquhar, C., & Das, R. (1999). Are focus groups suitable for 'sensitive' topics? In R. S. Barbour & J. Kitzinger (Eds.), *Developing Focus Group Research: Politics, Theory and Practice*. SAGE. doi:10.4135/9781849208857.n4

Farrington, D. P. (1991). Longitudinal research strategies: Advantages, problems, and prospects. *Journal of the American Academy of Child and Adolescent Psychiatry*, *30*(3), 369–374. doi:10.1097/00004583-199105000-00003 PMID:2055872

Fern, E. F. (2001). *Advanced focus group research*. SAGE. doi:10.4135/9781412990028

Finlay, L. (2002). "Outing" the Researcher: The Provenance, Process, and Practice of Reflexivity. *Qualitative Health Research*, *12*(4), 531–545. doi:10.1177/104973202129120052 PMID:11939252

Fontana, A., & Frey, J. H. (2000). The interview: From structured questions to negotiated text. In N. K. Denzin & Y. S. Lincoln (Eds.), *Handbook of qualitative research* (2nd ed., pp. 645–672). SAGE.

Gantt, H. L. (1917). *Work, Wages, and Profits*. Engineering Magazine.

Geertz, C. (1973). *The interpretation of cultures: Selected essays*. Basic Books.

Gibbs, G. R. (2007). Analyzing qualitative data. *Sage (Atlanta, Ga.)*.

Glaser, B. G., & Strauss, A. L. (1967). *The Discovery of Grounded Theory: Strategies for Qualitative Research*. Aldine Transaction.

Gold, R. L. (1958). Roles in sociological field observations. *Social Forces*, *36*(3), 217–223. doi:10.2307/2573808

Goldberg, L. R. (1965). Diagnosticians vs. diagnostic signs: The diagnosis of psychosis vs. neurosis from the MMPI. *Psychological Monographs*, *79*(9), 1–23. doi:10.1037/h0093885 PMID:14322679

Grant, C., & Osanloo, A. (2014). Understanding, selecting, and integrating a theoretical framework in dissertation research. *Administrative Issues Journal: Connecting Education, Practice, and Research*, *4*(2), 12-26.

Guillemin, M., & Gillam, L. (2004). Ethics, reflexivity, and "ethically important moments" in research. *Qualitative Inquiry*, *10*(2), 261–280. doi:10.1177/1077800403262360

Hammersley, M. (2007). The issue of quality in qualitative research. *International Journal of Research & Method in Education*, *30*(3), 287–305. doi:10.1080/17437270701614782

Harper, D. (2002). Talking about pictures: A case for photo elicitation. *Visual Studies*, *17*(1), 13–26. doi:10.1080/14725860220137345

Hart, C. (1998). *Doing a literature review: Releasing the research imagination*. Sage Publications.

Hassan, Z., Schattner, P., & Mazza, D. (2006). Doing a pilot study: Why is it essential? *Malaysian Family Physician : the Official Journal of the Academy of Family Physicians of Malaysia*, *1*(2-3), 70–73. PMID:27570591

Hess, D. R. (2008). What is evidence-based practice? *Respiratory Care*, *53*(10), 1317–1318. PMID:12962552

Hesse-Biber, S. (2011). Qualitative approaches to mixed methods practice. *Qualitative Inquiry, 17*(6), 455–467. doi:10.1177/1077800410364611

Horkheimer, M. (1982). *Critical theory*. Seabury Press.

Husserl, E. (2012). *Ideas: General Introduction to Pure Phenomenology*. Routledge. (Original work published 1913) doi:10.4324/9780203120330

Israel, M., & Hay, I. (2006). Research ethics for social scientists. *Sage (Atlanta, Ga.)*.

Janis, I. L. (1982). *Groupthink: Psychological studies of policy decisions and fiascoes*. Houghton Mifflin.

Johnson, R. B., & Onwuegbuzie, A. J. (2004). Mixed methods research: A research paradigm whose time has come. *Educational Researcher, 33*(7), 14–26. doi:10.3102/0013189X033007014

Kawulich, B. B. (2005). Participant observation as a data collection method. *Forum Qualitative Social Research, 6*(2).

Kim, Y. (2011). The pilot study in qualitative inquiry. *Qualitative Social Work: Research and Practice, 10*(2), 190–206. doi:10.1177/1473325010362001

Kincheloe, J. L., & McLaren, P. (2005). Rethinking critical theory and qualitative research. In N. K. Denzin & Y. S. Lincoln (Eds.), *The Sage handbook of qualitative research* (3rd ed., pp. 303–342). Sage.

King, N., & Horrocks, C. (2010). *Interviews in qualitative research*. SAGE.

Kitzinger, J. (1994). The methodology of focus groups: The importance of interaction between research participants. *Sociology of Health & Illness, 16*(1), 103–121. doi:10.1111/1467-9566.ep11347023

Kuhn, T. S. (1962). *The structure of scientific revolutions*. University of Chicago press.

Kvale, S., & Brinkmann, S. (2009). *InterViews: Learning the craft of qualitative research interviewing* (2nd ed.). SAGE.

Landsberger, H. A. (1958). *Hawthorne Revisited*. Cornell University.

LeCompte, M. D., & Goetz, J. P. (1982). Problems of reliability and validity in ethnographic research. *Review of Educational Research, 52*(1), 31–60. doi:10.3102/00346543052001031

Leon, A. C., Davis, L. L., & Kraemer, H. C. (2011). The role and interpretation of pilot studies in clinical research. *Journal of Psychiatric Research, 45*(5), 626–629. doi:10.1016/j.jpsychires.2010.10.008 PMID:21035130

Lincoln, Y. S., & Guba, E. A. (1985). *Naturalistic inquiry*. Sage. doi:10.1016/0147-1767(85)90062-8

Machi, L. A., & McEvoy, B. T. (2016). *The literature review: Six steps to success*. Corwin Press. doi:10.4135/9781071939031

Mackenzie, N., & Knipe, S. (2006). Research dilemmas: Paradigms, methods and methodology. *Issues in Educational Research, 16*(2), 193–205.

Marshall, C., & Rossman, G. B. (2014). *Designing qualitative research*. Sage Publications.

Mason, J. (2002). *Qualitative researching* (2nd ed.). SAGE.

Maxwell, J. A. (2005). Qualitative research design: An interactive approach. *Sage (Atlanta, Ga.)*.

Maxwell, J. A. (2013). *Qualitative research design: An interactive approach*. Sage Publications.

McCambridge, J., Witton, J., & Elbourne, D. R. (2014). Systematic review of the Hawthorne effect: New concepts are needed to study research participation effects. *Journal of Clinical Epidemiology*, *67*(3), 267–277. doi:10.1016/j.jclinepi.2013.08.015 PMID:24275499

Mertens, D. M., & Ginsberg, P. E. (2009). Deep in ethical waters: Transformative perspectives for qualitative social work research. *Qualitative Social Work: Research and Practice*, *8*(4). Advance online publication. doi:10.1177/1473325008097142

Merton, R. K., Fiske, M., & Kendall, P. L. (1990). *The focused interview: A manual of problems and procedures* (2nd ed.). The Free Press.

Miles, M. B., Silverman, D., & Huberman, A. M. (2013). Qualitative Data Analysis: A Sourcebook. *Sage (Atlanta, Ga.)*.

Miller, T., Mauthner, M., & Birch, M. (2012). Ethics in qualitative research. *Sage (Atlanta, Ga.)*.

Minichiello, V., Aroni, R., Timewell, E., & Alexander, L. (1990). *In-depth interviewing: Researching people*. Longman Cheshire.

Mitchell, C. (2008). Getting the picture and changing the picture: Visual methodologies and educational research in South Africa. *South African Journal of Education*, *28*(3), 365–383. doi:10.15700/saje.v28n3a180

Morgan, D. L. (2007). Paradigms lost and pragmatism regained: Methodological implications of combining qualitative and quantitative methods. *Journal of Mixed Methods Research*, *1*(1), 48–76. doi:10.1177/2345678906292462

Morrow, S. L. (2005). Quality and trustworthiness in qualitative research in counseling psychology. *Journal of Counseling Psychology*, *52*(2), 250–260. doi:10.1037/0022-0167.52.2.250

Morse, J. M. (1991). Approaches to qualitative-quantitative methodological triangulation. *Nursing Research*, *40*(2), 120–123. doi:10.1097/00006199-199103000-00014 PMID:2003072

Moustakas, C. (1994). Phenomenological research methods. *Sage (Atlanta, Ga.)*.

Mulhall, A. (2003). In the field: Notes on observation in qualitative research. *Journal of Advanced Nursing*, *41*(3), 306–313. doi:10.1046/j.1365-2648.2003.02514.x PMID:12581118

O'Cathain, A. (2010). Assessing the quality of mixed methods research: Toward a comprehensive framework. In A. Tashakkori & C. Teddlie (Eds.), *Handbook of mixed methods in social & behavioral research* (pp. 531–555). Sage. doi:10.4135/9781506335193.n21

Onwuegbuzie, A. J., Leech, N. L., & Collins, K. M. (2012). Qualitative analysis techniques for the review of the literature. *The Qualitative Report*, *17*(56), 1–28.

Orb, A., Eisenhauer, L., & Wynaden, D. (2000). Ethics in qualitative research. *Journal of Nursing Scholarship*, *33*(1), 93–96. doi:10.1111/j.1547-5069.2001.00093.x PMID:11253591

Orb, A., Eisenhauer, L., & Wynaden, D. (2001). Ethics in qualitative research. *Journal of Nursing Scholarship*, *33*(1), 93–96. doi:10.1111/j.1547-5069.2001.00093.x PMID:11253591

Palys, T. (2008). *Purposive sampling*. The Sage encyclopedia of qualitative research methods, 2, 697-8.

Patton, M. Q. (1999). Enhancing the quality and credibility of qualitative analysis. *Health Services Research*, *34*(5 Pt 2), 1189. PMID:10591279

Patton, M. Q. (2002). Two decades of developments in qualitative inquiry. *Qualitative Social Work: Research and Practice*, *1*(3), 261–283. doi:10.1177/1473325002001003636

Patton, M. Q. (2015). *Qualitative research & evaluation methods: Integrating theory and practice*. Sage publications.

Pillow, W. (2003). Confession, catharsis, or cure? Rethinking the uses of reflexivity as methodological power in qualitative research. *International Journal of Qualitative Studies in Education : QSE*, *16*(2), 175–196. doi:10.1080/0951839032000060635

Pink, S. (2013). *Doing visual ethnography*. SAGE.

Prosser, J., & Loxley, A. (2008). Introducing visual methods. *NCRM Research Methods Review Paper*, *10*, 1–11.

Punch, K. F. (2013). Introduction to social research: Quantitative and qualitative approaches. *Sage (Atlanta, Ga.)*.

Radley, A., & Taylor, D. (2003). Images of recovery: A photo-elicitation study on the hospital ward. *Qualitative Health Research*, *13*(1), 77–99. doi:10.1177/1049732302239412 PMID:12564264

Randolph, J. J. (2009). A guide to writing the dissertation literature review. *Practical Assessment, Research, and Evaluation*, 14(13), 1-13.

Ravitch, S. M., & Riggan, M. (2017). *Reason & rigor: How conceptual frameworks guide research*. Sage Publications.

Resnik, D. B. (2015). *What is ethics in research & why is it important?* National Institute of Environmental Health Sciences.

Ridder, H. G. (2017). The theory contribution of case study research designs. *Business Research*, *10*(2), 281–305. doi:10.1007/s40685-017-0045-z

Riessman, C. K. (2008). *Narrative methods for the human sciences*. Sage Publications.

Ritchie, J., Lewis, J., Nicholls, C. M., & Ormston, R. (Eds.). (2013). *Qualitative research practice: A guide for social science students and researchers*. Sage.

Rose, G. (2016). *Visual methodologies: An introduction to researching with visual materials* (4th ed.). SAGE.

Rosenthal, R. (1966). *Experimenter effects in behavioral research*. Appleton-Century-Crofts.

Roulston, K. (2010). *Reflective interviewing: A guide to theory and practice*. SAGE. doi:10.4135/9781446288009

Rubin, H. J., & Rubin, I. S. (2011). *Qualitative interviewing: The art of hearing data* (3rd ed.). SAGE.

Rubin, H. J., & Rubin, I. S. (2012). *Qualitative interviewing: The art of hearing data* (3rd ed.). SAGE.

Sampson, H. (2004). Navigating the waves: The usefulness of a pilot in qualitative research. *Qualitative Research, 4*(3), 383–402. doi:10.1177/1468794104047236

Sanjek, R. (1990). On Ethnographic Validity. In R. Sanjek (Ed.), *Fieldnotes: The Makings of Anthropology* (pp. 385–418). Cornell University Press. doi:10.7591/9781501711954

Saunders, B., Kitzinger, J., & Kitzinger, C. (2015). Anonymising interview data: Challenges and compromise in practice. *Qualitative Research, 15*(5), 616–632. doi:10.1177/1468794114550439 PMID:26457066

Seidman, I. (2013). *Interviewing as qualitative research: A guide for researchers in education and the social sciences*. Teachers College Press.

Shadish, W. R., Cook, T. D., & Campbell, D. T. (2002). *Experimental and quasi-experimental designs for generalized causal inference*. Houghton Mifflin.

Silverman, D. (2006). *Interpreting qualitative data: Methods for analyzing talk, text, and interaction* (3rd ed.). SAGE.

Silverman, D. (2010). Doing qualitative research: A practical handbook. *Sage (Atlanta, Ga.)*.

Silverman, D. (2016). *Qualitative research* (4th ed.). Sage.

Sim, J., & Wright, C. C. (2000). *Research in health care: Concepts, designs and methods*. Stanley Thornes.

Smith, J. A., & Davis, J. M. (2017). The role of external peer review in improving research design: An example from the qualitative literature. *Journal of Advanced Nursing, 73*(11), 2573–2579.

Smith, J. A., & Osborn, M. (2008). Interpretative phenomenological analysis. In J. A. Smith (Ed.), *Qualitative psychology: A practical guide to research methods* (2nd ed., pp. 53–80). Sage.

Smith, L. T. (2012). *Decolonizing methodologies: Research and indigenous peoples*. Zed Books Ltd.

Smithson, J. (2000). Using and analysing focus groups: Limitations and possibilities. *International Journal of Social Research Methodology, 3*(2), 103–119. doi:10.1080/136455700405172

Spradley, J. P. (1979). *The ethnographic interview*. Holt, Rinehart, and Winston.

Stake, R. E. (1995). *The art of case study research*. Sage Publications.

Stewart, D. W., Shamdasani, P. N., & Rook, D. W. (2007). *Focus groups: Theory and practice* (2nd ed.). SAGE. doi:10.4135/9781412991841

Tedlock, B. (2000). Ethnography and ethnographic representation. Handbook of Qualitative Research, 2, 455-486.

Thabane, L., Ma, J., Chu, R., Cheng, J., Ismaila, A., Rios, L. P., Robson, R., Thabane, M., Giangregorio, L., & Goldsmith, C. H. (2010). A tutorial on pilot studies: The what, why, and how. *BMC Medical Research Methodology*, *10*(1), 1. doi:10.1186/1471-2288-10-1 PMID:20053272

Thomas, D. R. (2017). Feedback from research participants: Are member checks useful in qualitative research? *Qualitative Research in Psychology*, *14*(1), 23–41. doi:10.1080/14780887.2016.1219435

Thomson, P. (2008). *Doing visual research with children and young people*. Routledge.

Thurmond, V. A. (2001). The point of triangulation. *Journal of Nursing Scholarship*, *33*(3), 253–258. doi:10.1111/j.1547-5069.2001.00253.x PMID:11552552

Tickle-Degnen, L. (2013). Nuts and bolts of conducting feasibility studies. *The American Journal of Occupational Therapy*, *67*(2), 171–176. doi:10.5014/ajot.2013.006270 PMID:23433271

Tracy, S. J. (2013). *Qualitative research methods: Collecting evidence, crafting analysis, communicating impact*. John Wiley & Sons.

Tranfield, D., Denyer, D., & Smart, P. (2003). Towards a methodology for developing evidence-informed management knowledge by means of systematic review. *British Journal of Management*, *14*(3), 207–222. doi:10.1111/1467-8551.00375

Van Teijlingen, E. R., & Hundley, V. (2001). The importance of pilot studies. *Social Research Update*, *35*(1), 1–4. PMID:11328433

Vygotsky, L. S. (1978). *Mind in society: The development of higher psychological processes*. Harvard University Press.

Warren, C. A. B., & Karner, T. X. (2010). *Discovering qualitative methods: Field research, interviews, and analysis* (2nd ed.). Oxford University Press.

Wiles, R., Charles, V., Crow, G., & Heath, S. (2008). Researching researchers: Lessons for research ethics. *Qualitative Research*, *8*(2), 283–299.

Yin, R. K. (2009). *Case study research: Design and methods*. Sage Publications.

Yin, R. K. (2013). *Case study research: Design and methods*. Sage publications.

Yin, R. K. (2014). *Case study research design and methods* (5th ed.). Sage Publications.

Yin, R. K. (2017). *Case study research and applications: Design and methods*. Sage publications.

Chapter 4
Data Collection Methods

ABSTRACT

This chapter delves into the intricate processes of data collection in qualitative research, highlighting the foundational importance and diverse methods used to capture the nuances and complexities of human behavior and societal trends. The core focus lies on how different data collection techniques—ranging from interviews and observations to document analysis and visual methods—can effectively address varied research questions within the qualitative paradigm. The alignment of specific methods with research questions is underscored as pivotal for ensuring the validity of the research outcomes.

Data Collection is a pivotal step in research, providing the empirical basis upon which scholarly insights are built. It is particularly crucial in qualitative research where the richness and depth of data drive understanding and interpretation. This chapter delves deeply into the diverse methods of data collection in qualitative research, exploring their nuances, applications, advantages, and challenges.

FOUNDATIONAL IMPORTANCE OF DATA COLLECTION

In qualitative research, data collection methods are foundational, as they allow for the nuanced, detailed, and rich exploration of phenomena (Merriam & Tisdell, 2015). The chosen method significantly influences the breadth and depth of insights gained, requiring thoughtful selection and application to appropriately address the research questions and align with the study's theoretical framework.

Diversity of Methods

Qualitative research is marked by a wide array of data collection methods, ranging from interviews and observations to document analysis and visual methods. Each method offers unique perspectives and insights, enabling researchers to explore different dimensions of their research questions (Creswell & Creswell, 2017).

DOI: 10.4018/979-8-3693-2414-1.ch004

Aligning Methods With Research Questions

The alignment of data collection methods with research questions is pivotal to the validity of qualitative research. Researchers must carefully select methods that allow them to delve deeply into the phenomena under study, ensuring congruence with the research objectives, questions, and theoretical framework (Silverman, 2016).

Ensuring Ethical Integrity

Ethical considerations permeate the process of data collection. Researchers are entrusted with the responsibility to respect participants' rights, dignity, and well-being, ensuring the integrity and ethical soundness of the research process (Orb, Eisenhauer & Wynaden, 2001).

Challenges and Solutions in Data Collection

While qualitative data collection methods offer rich, detailed insights, they also present unique challenges. Issues related to access, rapport, reflexivity, and data saturation require thoughtful consideration and strategic solutions to ensure the rigor and reliability of the collected data (Patton, 2014).

Structure of the Chapter

This chapter systematically explores various qualitative data collection methods, starting with a detailed overview of each method, its applications, benefits, and inherent challenges. Subsequent sections delve deeper into the practical aspects of applying these methods, including design considerations, implementation strategies, and ethical concerns. Case studies and practical examples are integrated throughout the chapter to illustrate the real-world application of different data collection methods, offering readers a holistic understanding of qualitative data collection processes.

Objective of the Chapter

The objective of this chapter is to equip readers with a comprehensive understanding of the diverse data collection methods in qualitative research. By exploring the theoretical underpinnings, practical applications, ethical considerations, and challenges of different methods, the chapter aims to provide readers with the knowledge and skills needed to design and implement effective data collection strategies in their research endeavors.

Finally, understanding the multitude of data collection methods and their appropriate application is crucial for conducting rigorous qualitative research. This chapter seeks to illuminate the intricacies of qualitative data collection methods, offering scholarly insights and practical guidance to both novice and experienced researchers in the field.

Observations

Observation as a method of data collection is a fundamental practice within qualitative research. Rooted in the practices of early anthropologists and sociologists who engaged in fieldwork to understand cultures and social groups, observation allows for an immersive, first-hand account of the phenomenon under study (DeWalt & DeWalt, 2002).

Nature of Observational Research

Observational research transcends mere "looking" to involve systematic noting, recording, and analysis of phenomena as they occur naturally. Rather than relying on participants' accounts or self-reports, observations provide researchers with direct data, capturing behaviors, interactions, events, and contexts as they unfold (Angrosino, 2007).

Types of Observations

1. **Participant Observation:** The researcher immerses themselves in the setting, often participating in the activities of the group under study. This deep involvement provides rich, inside perspectives but can raise questions about objectivity (Emerson, Fretz & Shaw, 1995).

Participant observation has long been foundational in qualitative research, acting as a bridge that allows researchers to immerse themselves in the lived experiences of the subjects under study. Originally rooted in anthropological traditions, participant observation involves the researcher taking an active role within the community or group they are studying, thereby gaining a first-hand, insider's view of the culture, practices, and interactions of that community (Malinowski, 1922).

Characteristics of Participant Observation

a) **Immersion and Involvement:** The researcher becomes a part of the community, sometimes living with them, participating in their daily activities, rituals, and ceremonies (Geertz, 1973).
b) **Dual Role:** The researcher has to balance the role of an observer and a participant, making sure they are active in community activities while also maintaining the detachment necessary for objective observation (Whyte, 1943).
c) **Holistic Understanding:** By being part of the community, the researcher gets an all-round view, understanding not just what is said but the unsaid – the implicit cultural norms, values, and beliefs (Spradley, 1980).

Strengths of Participant Observation

a) **Rich Data Collection:** As the researcher is directly involved, they can gather nuanced, in-depth data that might not be apparent from an external perspective (Wolcott, 1995).
b) **Contextual Insights:** Participant observation allows the researcher to understand the context in which actions, interactions, and decisions take place (Emerson, Fretz & Shaw, 1995).
c) **Building Trust:** Prolonged engagement can lead to deeper trust, making it more likely for community members to share authentic, sensitive information (Bernard, 2006).

Challenges of Participant Observation

a) **Maintaining Objectivity:** Being deeply involved might lead to 'going native', where the researcher becomes too attached and loses their objective viewpoint (Gold, 1958).
b) **Ethical Considerations:** Being an insider might mean that community members forget the researcher's official capacity, leading to potential ethical concerns about informed consent and confidentiality (Kawulich, 2005).
c) **Emotional and Physical Toll:** Living and engaging deeply in another community, especially if it is very different from the researcher's own, can be physically and emotionally taxing (Agar, 1996).

Participant observation, though demanding, offers a unique methodological approach that provides rich, contextually grounded insights. It requires a delicate balance of involvement and detachment and poses distinct ethical and practical challenges. However, when done judiciously, it can yield profound understanding and contribute significantly to qualitative research.

2. **Non-participant Observation:** The researcher remains a passive observer, without becoming involved in the activities of the group. This allows for a more detached perspective, but may limit the depth of understanding (Spradley, 1980).

Non-participant observation, distinct from participant observation, is a method wherein researchers immerse themselves in the study setting but do not actively participate in the activities of the community or group being studied. This method aims to provide an objective and undisturbed account of the observed events, interactions, and behaviors (Kawulich, 2005). Historically, this method has been invaluable in various fields of study, allowing for an undiluted perspective of events in their natural context.

Characteristics of Non-Participant Observation

a) **Detached Role:** The researcher, while present in the environment, does not intervene or engage in the ongoing activities, ensuring minimal disruption and influence on the observed events (Gold, 1958).
b) **Unobtrusive Stance:** Often, researchers will aim to be as inconspicuous as possible, preventing their presence from altering the natural flow of events (Patton, 2015).
c) **Objective Lens:** Given the detached nature of this method, researchers can capture data without the potential biases that may come from direct involvement (Lofland & Lofland, 1995).

Advantages of Non-Participant Observation

a) **Minimized Researcher Influence:** By staying detached, researchers can ensure that their presence doesn't significantly alter the behaviors or events they're observing (Spradley, 1980).
b) **Authentic Data:** Capturing events and interactions as they naturally occur ensures the authenticity of the data collected (Bryman, 2016).
c) **Enhanced Objectivity:** The distance maintained allows for a more neutral standpoint, which can be crucial in various research contexts (DeWalt & DeWalt, 2002).

Limitations of Non-Participant Observation

a) **Lack of Deep Understanding:** The detached position might result in missing out on the deeper, intrinsic cultural meanings, emotions, or perspectives experienced by insiders (Emerson, Fretz & Shaw, 1995).
b) **Potential Ethical Issues:** Being a silent observer might sometimes lead to observing without informed consent, especially in public spaces, leading to ethical dilemmas (Bryman, 2016).
c) **Missed Contextual Nuances:** Not engaging actively might lead to misinterpretations or overlooking certain subtle, yet significant, aspects of the observed events (Kawulich, 2005).

Non-participant observation serves as an essential tool in the qualitative research toolkit, especially when seeking to minimize influence on the setting or maintain a clear objective perspective. While it presents certain challenges, its merits in providing an authentic snapshot of events in their natural state are undeniable. A balanced approach, weighing its strengths and limitations, can lead to robust, credible insights.

The Process of Conducting Observations

1. **Gaining Access:** Before observing, researchers often need permission from gatekeepers, especially in private or sensitive settings (Berg, 2009).

Observational research, while insightful, often presents researchers with the initial challenge of accessing the field or site of study. Gaining access not only requires methodological considerations but also touches upon ethical concerns and interpersonal dynamics.

Why Is Gaining Access Critical?

Before immersing oneself in the observational field, it is crucial to secure the necessary permissions and foster relationships to ensure the research progresses smoothly (Dewalt & Dewalt, 2002). Gaining access can influence the richness of data, determine the quality of researcher-participant interactions, and influence the longevity of the research.

Strategies to Gain Access

a) **Leveraging Existing Networks:** Using personal or institutional contacts can be instrumental in navigating entry points into the research setting (Emerson, Fretz & Shaw, 1995).
b) **Building Trust:** Prioritizing the establishment of trust with gatekeepers or community leaders can open doors to richer observational opportunities (Adler & Adler, 1994).
c) **Formal Permissions:** Especially in institutional settings or private spaces, formal written permissions might be required to ensure the legality and ethicality of the observation (Bryman, 2016).
d) **Engaging in Pre-visit Dialogues:** Before formal observations, engaging in conversations or preliminary visits can familiarize the participants with the researcher's presence and objectives (Spradley, 1980).
e) **Transparency:** Clearly articulating the purpose, methods, and goals of the research can ease concerns and foster a more welcoming environment (Kawulich, 2005).

Challenges in Gaining Access

a) **Resistance or Skepticism:** Potential participants or gatekeepers might resist or view the research with skepticism, fearing misrepresentation or intrusion (Dewalt & Dewalt, 2002).
b) **Bureaucratic Hurdles:** Institutional settings might pose bureaucratic challenges that can delay or restrict observational opportunities (Bryman, 2016).
c) **Cultural or Social Barriers:** Differences in language, culture, or social norms can influence the ease of access and require researchers to demonstrate cultural sensitivity (Emerson, Fretz & Shaw, 1995).

Gaining access is a nuanced process, central to the success of observational research. It requires a blend of methodological astuteness, ethical considerations, and interpersonal skills. By effectively navigating this initial phase, researchers can set a positive tone for the entire observational journey, ensuring richer insights and meaningful engagements.

2. **Taking Field Notes:** Detailed notes during or immediately after observations capture data, insights, and initial interpretations. These notes may include sketches, maps, and diagrams (Sanjek, 1990).

Field notes are a cornerstone of observational research. Their purpose is to capture the lived experiences, events, interactions, and nuances that a researcher witnesses in the field. These notes serve as the primary data for analysis and interpretation later in the research process (Emerson, Fretz & Shaw, 1995).

The Significance of Field Notes

a) **Memory Augmentation:** Human memory is fallible. Field notes serve as a record, ensuring that subtle details and sequences of events aren't forgotten or misremembered (Sanjek, 1990).
b) **Contextual Richness:** They help capture the context, providing depth to the events and aiding in understanding the deeper cultural or social meanings (Clifford, 1990).
c) **Analytical Foundation:** Field notes are the foundational texts upon which coding, analysis, and interpretation are built in later stages (Ottenberg, 1990).

Techniques for Effective Field Note Taking

a) **Jotted Notes:** During observation, brief notes or keywords can be jotted down to be expanded upon later, ensuring minimal interruption to the observation process (Malinowski, 1967).
b) **Descriptive Detail:** Notes should capture sensory details, including sights, sounds, smells, etc., to convey a rich picture of the setting (Geertz, 1973).
c) **Direct Quotes:** Whenever possible, verbatim speech should be recorded, as it provides authentic insight into participants' thoughts and perspectives (Spradley, 1980).
d) **Reflexive Notes:** Researchers should also note their feelings, interpretations, and reactions to observed events, ensuring that personal reflexivity is part of the data (Davies, 2008).

e) **Use of Diagrams:** Spatial layouts, interaction diagrams, and other visual aids can supplement written notes to provide a clearer understanding of the physical and interactional context (Bernard, 2011).

Challenges in Taking Field Notes

a) **Selective Attention:** The impossibility of capturing everything means that researchers need to be selective, which introduces the possibility of bias (Jackson, 1983).
b) **Intrusiveness:** Taking notes can sometimes be viewed as intrusive or disruptive by participants (Gold, 1958).
c) **Memory Limitations:** Even with jotted notes, the challenge of recalling and expanding on them post-observation is ever-present (Sanjek, 1990).

Field notes, when taken with care, attention, and reflexivity, become a rich tapestry of data. They offer a window into the field, capturing the complexity, richness, and depth of the lived experiences and events researchers are privileged to witness.

3. **Reflecting and Debriefing:** Post-observation, it's beneficial for the researcher to reflect on what was observed, identifying patterns, anomalies, or emerging themes (Mulhall, 2003).

Reflection and debriefing are pivotal processes in observational research, acting as mechanisms that augment understanding, add depth to raw data, and facilitate the researcher's personal connection with the observed events.

Significance of Reflection and Debriefing

a) **Self-awareness and Reflexivity:** Reflection allows researchers to become aware of their biases, emotions, and preconceptions that may affect data collection and interpretation (Finlay, 2002).
b) **Data Enrichment:** Debriefing sessions, especially with peers or mentors, can offer additional insights or alternative interpretations of observed events (Borkan, 1999).
c) **Emotional Catharsis:** Observational research, particularly in sensitive settings, can be emotionally taxing. Reflecting and debriefing serve as outlets for emotional processing (Dickson-Swift et al., 2007).

Strategies for Effective Reflection and Debriefing

a) **Maintaining a Reflexive Journal:** Apart from field notes, a separate journal dedicated to personal reflections can capture the researcher's emotions, biases, and evolving insights (Ortlipp, 2008).
b) **Peer Debriefing Sessions:** Regular meetings with peers or mentors to discuss observations can enhance understanding and mitigate subjective biases (Lincoln & Guba, 1985).
c) **Scheduled Reflection Time:** Setting aside dedicated time after each observation for introspection ensures consistent and thorough reflection (Schön, 1983).

d) **Using Prompts:** Reflective prompts can guide the introspection process, encouraging researchers to explore various dimensions of their experience (Boud et al., 1985).

Challenges in Reflection and Debriefing

a) **Overwhelming Emotions:** Intense field experiences might lead to strong emotional reactions, which can cloud judgment and interpretation (Dickson-Swift et al., 2007).
b) **Time Intensity:** The process requires a significant time commitment, especially in long-term observational projects (Finlay, 2002).
c) **Potential for Bias:** While debriefing can offer multiple perspectives, it can also introduce new biases, necessitating careful consideration and discernment (Lincoln & Guba, 1985).

Reflection and debriefing, when practiced with diligence and sincerity, not only enhance the quality and depth of observational data but also contribute to the personal and professional growth of the researcher.

Strengths of Observational Research

Observations allow for:

1. **Capturing Context:** Observations encompass not only the actions but also the environment, allowing for a rich understanding of context (Gold, 1958).

Observational research offers a unique opportunity to deeply embed oneself within the natural settings of participants, capturing the intricate dynamics and layers of context that other methods might overlook or misrepresent.

A. Depth of Understanding:

Being present in the environment where behaviors and interactions occur, researchers gain a comprehensive understanding of socio-cultural, environmental, and personal factors that influence the phenomenon of interest (Spradley, 1980).

B. Nuances and Subtleties:

Direct observations enable researchers to pick up on non-verbal cues, silent pauses, and subtle interactions that are often missed in self-reports or interviews (Angrosino, 2007).

C. Dynamics of Interactions:

By observing how individuals interact with their environment and with others, a researcher can better comprehend social norms, roles, and relationships, and their effects on behaviors and decisions (Emerson et al., 2011).

D. Temporality and Sequence:

Observational methods allow the researcher to capture events in real-time, offering insights into sequences, patterns, and causality that are hard to deduce from retrospective accounts (DeWalt & DeWalt, 2011).

E. Naturalistic Setting:

Observations conducted in natural settings reduce the Hawthorne effect, where participants might change their behavior due to the awareness of being observed. Hence, the data is likely to be more authentic (Adler & Adler, 1994).

F. Bridging the Say-Do Gap:

What people say they do and what they actually do can be quite different. Observations help in bridging this gap by providing a real-world check on self-reported behaviors (Fetterman, 2019).

Capturing context is one of the hallmarks of observational research. Through direct, prolonged engagement, researchers can delve deeper into the intricate tapestry of events, behaviors, and interactions, offering a rich, nuanced understanding that is grounded in the reality of everyday life.

2. **Dynamic Data Collection:** Observing real-time events and interactions captures the fluidity and dynamics of situations (Kawulich, 2005).

Observational research is inherently dynamic. Unlike static methods that capture a snapshot of participant perspectives at a single point in time, observational methods delve into the ever-evolving flow of human behaviors and interactions, providing rich insights into the complexities of real-world phenomena.

A. Continuous Engagement:

Observational research allows for continuous and prolonged engagement in the field. This deep immersion enables researchers to observe changes, transitions, and developments as they occur, granting them a comprehensive understanding of processes over time (Gold, 1958).

B. Real-time Responses:

Observing events and behaviors as they unfold provides an opportunity to immediately recognize and probe unexpected occurrences, contradictions, or anomalies, thereby enhancing the depth and breadth of data collection (Patton, 2015).

C. Adaptability:

The dynamic nature of observational research allows researchers to adjust their focus based on emerging patterns, insights, or areas of interest. Such flexibility ensures that the research remains relevant and aligned with the evolving context (Sanjek, 1990).

D. Capturing the Unplanned:

While structured interviews or surveys might miss spontaneous events or unplanned incidents, observational methods are uniquely positioned to capture these, offering a fuller picture of the phenomenon under study (Lofland et al., 2006).

E. Embracing Complexity:

Human behaviors and interactions are intricate and multi-faceted. Observational research, with its dynamic data collection approach, embraces this complexity, allowing for a layered, multifaceted exploration of phenomena (Agar, 1980).

F. Contextual Interactions:

Observing how individuals dynamically interact with changing environments, circumstances, or other individuals provides rich insights into the fluidity of human behavior and the numerous factors that influence it (Bernard, 2011).

Dynamic data collection is an inherent strength of observational research. The method's adaptability, real-time engagement, and ability to capture the unplanned and unexpected provide researchers with a holistic, in-depth understanding of the intricate dance of human behaviors and interactions within their natural settings.

3. **Reduced Recall Bias:** Since data is collected as events unfold, observations are not susceptible to the recall biases that might affect interviews or surveys (Savage, 2006).

Recall bias is a common challenge in research methodologies that rely on participants' memory of past events, potentially leading to skewed or inaccurate results. Observational research, by design, curtails this bias as it captures behaviors and events in real-time, thus safeguarding the integrity and accuracy of the data collected.

A. Present-Moment Recording:

In observational research, the immediacy of data capture ensures that information is documented as events unfold, mitigating the need for participants to remember past occurrences. This real-time recording not only assures data accuracy but also reduces the distortions often introduced when relying on recollections (Kawulich, 2005).

B. Circumventing Memory Distortions

Memory is not a perfect recorder of events. Over time, individuals might forget, exaggerate, or minimize details, thereby introducing errors into their recollections. Observational research bypasses these memory distortions by focusing on real-time observations rather than retrospective accounts (Bernard, 2011).

C. Authenticity of Spontaneous Reactions

Observational research captures spontaneous reactions, behaviors, and interactions, which are less influenced by post-event reflections or reinterpretations. These authentic responses provide a more genuine representation of how individuals naturally behave in specific situations (Angrosino, 2007).

D. Reducing Social Desirability Bias

When participants are asked to recall their behaviors, they might inadvertently or intentionally depict themselves in a more favorable light, a phenomenon known as social desirability bias. Observations can reduce this bias by documenting actual behaviors, rather than relying on participants' potentially skewed self-reports (Silverman, 2010).

E. Verifying Self-Reports

Observations can act as a validating tool. When combined with self-report measures, researchers can compare observed behaviors with participants' accounts, offering a more holistic understanding and potentially identifying inconsistencies arising from recall bias (Jorgensen, 1989).

Reducing recall bias is a paramount strength of observational research. By focusing on real-time data capture, observational methods provide a more authentic, undistorted, and reliable account of events, behaviors, and interactions, enhancing the robustness and credibility of research findings.

Challenges in Observational Research

1. **Observer Effect:** The mere presence of an observer can influence participants' behavior, potentially skewing data (Webb et al., 1966).

One of the critical challenges in observational research is the potential for the "observer effect" or "Hawthorne effect." This phenomenon refers to the potential changes in behavior exhibited by research participants when they are aware they are being observed (Landsberger, 1958).

A. Conscious Alteration of Behaviors

When individuals know they are being observed, they might modify their actions, either consciously or subconsciously, to present themselves in a manner they deem favorable or acceptable. Such behavior alterations can skew research findings, as they might not represent the participants' typical or natural actions (Adair, 1984).

B. Observer Expectations

The observer's expectations, whether explicitly communicated or not, can influence participant behaviors. Participants might attempt to behave in ways they believe align with the observer's expectations, even if these behaviors diverge from their usual actions (Rosenthal, 1966).

C. Duration of Observation

While the observer effect is often most pronounced at the beginning of a study, the duration of observation can impact its persistence. In extended studies, participants might become desensitized to the observer's presence over time, leading to more genuine behaviors as the study progresses (McCambridge, Witton, & Elbourne, 2014).

D. Minimizing the Observer Effect

Several strategies can mitigate the observer effect. These include unobtrusive observation where participants may not be aware they're being observed, building trust and rapport to reduce the artificiality of the observation setting, and employing participant observers who are members of the group being studied, thereby reducing the external observer's influence (Gold, 1958).

E. The Inevitability of the Observer Effect

Despite attempts to minimize the observer effect, its presence to some degree is often inevitable in observational research. Recognizing and acknowledging this potential bias is crucial, and researchers should factor it into their analyses and interpretations of the observed data (Patton, 2015).

The observer effect presents a significant challenge in observational research, potentially altering the authenticity of observed behaviors. By understanding this phenomenon and employing strategies to reduce its influence, researchers can capture more genuine insights into participant behaviors and interactions.

2. **Subjectivity:** Researchers' backgrounds, beliefs, and experiences can influence what they notice and how they interpret it (Atkinson & Hammersley, 1994).

Observational research, by its very nature, is prone to the influence of the observer's personal beliefs, perceptions, and interpretations. This influence, often referred to as subjectivity, can present challenges in maintaining the objectivity and reliability of observational findings (Angrosino, 2005).

A. Observer Bias

Every observer brings with them a set of personal beliefs, experiences, and cultural perspectives that can influence how they interpret and record observed behaviors. This can lead to observer bias, where observations are skewed in a particular direction or manner based on these personal inclinations (Goldberg, 1965).

B. Influence of Preconceptions

Researchers entering a field with preconceived notions or hypotheses might unconsciously focus more on behaviors or events that align with these notions, potentially overlooking or downplaying other significant observations (DeWalt & DeWalt, 2002).

C. Variability Among Observers

When multiple observers are involved in a study, each may interpret and record the same events differently. This inter-observer variability can introduce inconsistencies in the collected data, complicating analysis and interpretation (Bogdan & Biklen, 2007).

D. Reflexivity in Observation

Reflexivity involves the researcher's awareness of and reflection upon their role, influence, and subjectivity in the research process. By acknowledging and deliberating on their biases and influences, researchers can aim for a more balanced and transparent observational approach (Guillemin & Gillam, 2004).

E. Strategies to Mitigate Subjectivity

While complete elimination of subjectivity in observational research is improbable, several strategies can reduce its impact. Triangulation, involving the use of multiple observers or methods, can provide a more holistic view. Training sessions for observers, aimed at standardizing observation and note-taking techniques, can reduce inter-observer variability. Regular debriefing sessions can also help in highlighting and addressing potential biases (Lincoln & Guba, 1985).

Subjectivity presents both a challenge and an inherent characteristic of observational research. While it can introduce biases and influence findings, when acknowledged and addressed, it can also contribute depth, nuance, and richness to qualitative research insights.

3. **Time-Intensive:** Comprehensive observations often require significant time in the field, making them resource-intensive (Tedlock, 2000).

Observational research, by virtue of its depth and detail, often demands considerable time from researchers. This time-intensive nature, though beneficial for rich data collection, poses challenges, especially when resources, including time, are constrained (Mulhall, 2003).

A. Extended Fieldwork

In-depth observations typically require prolonged periods of immersion in the field. Extended stays facilitate familiarity and rapport building, and allow researchers to witness a variety of events or behaviors (Fetterman, 2019). This duration, however, can strain resources and be physically and mentally exhausting.

B. Detailed Note-taking

Capturing the essence of observed events requires meticulous note-taking. This process can be laborious, especially when researchers strive to note down verbatim dialogues, spatial arrangements, or sequential actions (Emerson, Fretz, & Shaw, 1995).

C. Data Overload

Long-term observations can lead to vast amounts of data. Managing, coding, and analyzing such extensive data sets can be overwhelming, making the process of distilling meaningful patterns and insights challenging (Savage, 2006).

D. Time and Evolving Contexts

Over extended periods, contexts can change – participants may leave, environmental conditions may shift, or unforeseen events may occur. Such changes can impact the consistency of data and introduce additional variables (LeCompte & Goetz, 1982).

E. Strategies to Manage Time-Intensiveness

Researchers can employ several strategies to manage the time demands of observational research. Prioritizing observations, using technology to assist in data collection (e.g., audio/video recordings), and collaborating with co-researchers to share observational duties can optimize time. Furthermore, periodic data analysis, even during the data collection phase, can guide researchers, allowing them to focus on particularly significant areas (Baxter & Jack, 2008).

The time-intensive nature of observational research, while providing an avenue for rich and in-depth data collection, poses significant challenges. A thoughtful approach, combining rigorous planning with flexibility, can help researchers navigate these demands, ensuring quality and depth in their findings.

Ethical Considerations in Observational Research

Ensuring ethical integrity is crucial. This includes obtaining informed consent when required, ensuring anonymity, and being sensitive to intrusive observations, especially in private or sensitive contexts (Wiles et al., 2008).

Observational research poses a unique set of ethical challenges because of its immersive and often covert nature. Below, we will explore the primary ethical issues inherent in observational studies and propose strategies to address them (Kawulich, 2005).

A. Informed Consent

A hallmark of ethical research, informed consent can be particularly problematic in observational research. When observing in public spaces or community settings, obtaining explicit consent from every individual can be impractical.

Strategies

- **Passive consent**: Notify community members about the research and provide opt-out opportunities (Erickson, 1989).
- **Deferred consent**: Gather data covertly but approach participants afterward for their consent to use the collected data (Bulmer, 1982).

B. Privacy and Anonymity

Ensuring the privacy and anonymity of observed participants is paramount, given the potential for unintended harm or misrepresentation.

Strategies

- **Anonymizing data**: Replace names and identifying details with pseudonyms or codes (Hammersley & Atkinson, 2007).
- **Secure storage**: Use encrypted storage solutions and restrict access to data (Wiles, Crow, & Pain, 2011).

C. Intrusiveness

The presence of an observer can influence participants' behavior, potentially causing discomfort or altering the dynamics of the observed setting.

Strategies

- **Minimize disruption**: Use unobtrusive methods and try to blend into the setting (Gold, 1958).
- **Rapport building**: Spend time getting to know participants to reduce the 'observer effect' (Jorgensen, 1989).

D. Misinterpretation and Bias

Subjectivity in observations can lead to misinterpretations, which can be exacerbated by researchers' biases.

Strategies

- **Triangulation**: Employ multiple data sources or methods to validate findings (Denzin, 1978).
- **Peer review**: Engage colleagues to review and critique observations to identify potential biases (Emerson, Fretz, & Shaw, 2011).

E. The Issue of Covert Observation

While covert observation can offer genuine insights by minimizing the observer's influence, it raises significant ethical concerns about deception and non-consent.

Strategies

- **Weigh benefits**: Covert methods should only be used if the knowledge gained is of significant value and could not be obtained otherwise (Bulmer, 1982).
- **Post-study debrief**: Inform participants after the observation, explaining the reasons for secrecy and offering them the option to withdraw their data (Lee, 1993).

Ethical considerations in observational research are intricate due to its immersive and sometimes covert nature. Navigating these ethical waters requires a balance of methodological rigor, respect for participants, and a commitment to producing knowledge that justifies the potential ethical ambiguities.

Observations offer a dynamic, rich method of data collection, providing insights unattainable through other methods. Though they come with challenges, with careful planning and ethical considerations, observations can be an invaluable tool in qualitative research.

Interviews: Structured, Semi-Structured, and Unstructured

Interviews have been integral to qualitative research, providing in-depth insights into participants' experiences, perspectives, and the meanings they ascribe to them (Denzin & Lincoln, 2005). Based on the level of structure and standardization, interviews can be classified into three primary categories: structured, semi-structured, and unstructured.

1. Structured Interviews:

Structured interviews, rooted in positivist paradigms, are foundational in research methodologies aiming for replicability and generalizability. These interviews, heavily influenced by quantitative survey methods, have been tailored to fit qualitative research needs (Bryman, 2012). Below is a detailed examination of structured interviews:

Definition

Structured interviews involve a predetermined set of questions administered to every interviewee in precisely the same order and manner, ensuring consistency and standardization across data collection sessions (Fontana & Frey, 2000).

Key Characteristics

- **Standardization**: Every participant encounters the same questions in an identical sequence, minimizing the potential for interviewer bias (Berg, 2007).
- **Closed-ended Questions**: Predominantly, the questions posed are closed-ended, allowing interviewees to select from preset responses (Denzin & Lincoln, 2005).
- **Scalability**: Given their design, structured interviews are easily administered to large sample sizes (Ritchie & Lewis, 2003).

Advantages

a) **Consistency**: The standardization ensures uniform data, facilitating comparability across participants (Silverman, 2006).
b) **Efficiency**: Structured format often results in faster data collection and subsequent analysis (King & Horrocks, 2010).

c) **Reduced Bias**: The interviewer's discretion in phrasing or sequencing questions is minimized, potentially reducing variability in responses attributed to interviewer effects (Bryman, 2012).

Limitations

a) **Lack of Depth**: The rigidity can prevent interviewers from probing deeper into the participant's responses, potentially missing out on nuanced information (Patton, 2002).
b) **Flexibility**: Since questions are fixed, they might not account for all individual experiences or unforeseen nuances in a topic (Mason, 2002).
c) **Response Limitation**: Predetermined answers might not encapsulate participants' genuine feelings or experiences (Seidman, 2013).

Applications

Structured interviews are often favored in research settings where:

- The research question revolves around specific, predetermined aspects.
- There's a necessity for cross-comparisons between different study participants or even across different studies.
- The study aims for a broader sweep rather than an in-depth exploration (Creswell, 2013).

While structured interviews offer a range of benefits in terms of consistency and scalability, they come with the trade-off of depth and flexibility. Researchers must weigh these strengths and weaknesses against their research goals to determine if this method is the most appropriate for their study.

2. Semi-Structured Interviews

Semi-structured interviews occupy a pivotal space within qualitative research, bridging the gap between the rigidity of structured interviews and the free-flowing nature of unstructured formats. These interviews offer a balanced approach to data collection (Bryman, 2012). Let's delve deeper into this method:

Definition

Semi-structured interviews entail an interview guide comprised of open-ended questions, which provides direction to the interview. However, the interviewer retains the flexibility to probe further or redirect the conversation based on the participant's responses (Brinkmann & Kvale, 2015).

Key Characteristics

- **Interview Guide**: Predetermined questions form the backbone of the interview, ensuring coverage of key areas (DiCicco-Bloom & Crabtree, 2006).

- **Flexibility**: Interviewers can modify questions, change their order, or explore emergent themes during the interview (Rubin & Rubin, 2012).
- **Open-ended Questions**: Responses aren't restricted to predefined categories, allowing participants to express their views comprehensively (Fontana & Frey, 2000).

Advantages

a) **Depth and Breadth**: Semi-structured interviews provide comprehensive insights while ensuring that core topics are addressed (Mason, 2002).
b) **Responsiveness**: Interviewers can adapt to the participant's perspective, ensuring a participant-centered approach (Seidman, 2013).
c) **Standardization with Flexibility**: While there's a structure to follow, there's room for spontaneous exploration (Denzin & Lincoln, 2005).

Limitations

a) **Time-Consuming**: The open-ended nature of responses can prolong interview duration and subsequent data analysis (King & Horrocks, 2010).
b) **Requires Skill**: Achieving a balance between following the guide and exploring emergent topics demands skilled interviewing (Brinkmann & Kvale, 2015).
c) **Consistency Challenges**: Different interviews can vary significantly based on the direction taken during the session (Patton, 2002).

Applications

Semi-structured interviews are particularly useful when:

- The topic under study is multifaceted and may require adaptability during data collection.
- Researchers aim for depth and detail but also wish to ensure specific themes are consistently addressed.
- There's a necessity to allow participants the freedom to share experiences or viewpoints without stringent confines (Creswell, 2013).

Semi-structured interviews amalgamate structure with adaptability, making them a versatile tool in qualitative research. This method, while demanding in terms of time and skill, offers a unique blend of depth, breadth, and participant responsiveness.

3. Unstructured Interviews:

Unstructured interviews, an essential instrument in the qualitative researcher's toolbox, prioritize the participant's voice and the fluid co-construction of knowledge. Unlike their structured or semi-structured counterparts, they lack a strict format, providing ample room for participants to direct the conversation (Fontana & Frey, 2005). Here's an in-depth look into this method:

Definition

Unstructured interviews are conversational and open-ended, often initiated with broad questions or topics. They allow the interviewee maximum freedom to express their experiences and perspectives without the limitations of pre-defined questions (Warren & Karner, 2010).

Key Characteristics

- **Open Initiation**: Interviews usually begin with open-ended questions, such as "Can you tell me about...?", letting the participant steer the direction (Bernard, 2006).
- **Flowing Conversation**: They resemble a casual conversation more than a formal interview, where topics emerge organically (Spradley, 1979).
- **Absence of a Set Framework**: There's no strict interview guide or predetermined set of questions (Lofland & Lofland, 1995).

Advantages

a) **Depth of Understanding**: The format allows for deep dives into experiences, feelings, and perceptions (Minichiello et al., 1990).
b) **Flexibility**: Provides room to explore unforeseen avenues and emergent themes (Roulston, 2010).
c) **Empowerment**: Gives participants a voice, often making them feel valued and heard (Fontana & Frey, 2000).

Limitations

a) **Requires Expertise**: The researcher needs strong interpersonal and active listening skills (Rubin & Rubin, 2011).
b) **Time-Consuming**: Both the interviews and subsequent analysis can be lengthy due to their open-ended nature (Kvale & Brinkmann, 2009).
c) **Lack of Consistency**: Comparing data across interviews can be challenging because of the varied content in each session (Seidman, 2013).

Applications

Unstructured interviews are especially suitable when:

- The research aims to understand deeply personal or sensitive topics.
- The research context is novel, and the researcher has limited prior understanding.
- The emphasis is on giving marginalized or silenced groups a voice (Ellis & Berger, 2003).

Embodying the essence of qualitative inquiry, unstructured interviews offer profound insights into human experiences, capturing nuances and subtleties. While they come with challenges, their potential for deep understanding is unparalleled, especially when the researcher's goal is to immerse fully in the participant's world.

Choosing the appropriate type of interview depends on the research question, the nature of the topic, the characteristics of the participants, and the desired data's depth and breadth. While structured interviews provide consistency, semi-structured and unstructured interviews offer depth and flexibility, respectively.

Focus Groups

Focus groups, a popular qualitative data collection method, involve interactive group discussions centered on a particular topic or set of topics. They've been extensively used in market research and are now a mainstay in various academic disciplines. In this section, we delve into the nuances of focus groups.

Definition: A focus group consists of a small, diverse group of participants who are gathered to discuss a predefined topic, guided by a skilled moderator (Krueger & Casey, 2014).

Key Characteristics:

- **Interactive Dialogue**: The emphasis is on participant interaction, generating a rich understanding of people's experiences and beliefs (Morgan, 1996).
- **Guided Discussion**: A moderator uses a semi-structured guide to navigate the conversation, ensuring key topics are covered (Stewart, Shamdasani, & Rook, 2007).
- **Group Dynamics**: Participant reactions, disagreements, and consensus provide insights into group norms and values (Fern, 2001).

Advantages

a) **Rich Data**: Multiple perspectives offer a layered understanding of the topic (Krueger & Casey, 2014).
b) **Efficiency**: They provide the opportunity to gather data from multiple participants simultaneously (Merton, Fiske, & Kendall, 1990).
c) **Natural Environment**: Group settings can often mirror natural social interactions, making discussions more organic (Catterall & Maclaran, 1997).

Limitations

a) **Groupthink**: Participants might conform to dominant views, suppressing dissent (Janis, 1982).
b) **Moderator Influence**: The moderator's actions might influence participant responses (Smithson, 2000).
c) **Logistics**: Organizing groups, especially diverse ones, can be challenging (Morgan, 1996).

Applications

Focus groups are ideal for:

- Exploring new phenomena.
- Testing new products or ideas.
- Understanding group norms and cultural values (Kitzinger, 1994).

Ethical Considerations

- Ensuring confidentiality in group settings.
- Addressing power dynamics and potential conflicts.
- Obtaining informed consent for group discussions and any subsequent recordings (Farquhar & Das, 1999).

Focus groups, when designed and moderated effectively, offer an in-depth understanding of collective perspectives, social norms, and group dynamics. While they come with challenges, their interactive nature and the richness of data they provide make them invaluable in qualitative research.

Document and Content Analysis

Document and content analysis have long been employed as systematic methodologies to interpret and analyze textual information. These approaches enable researchers to discern patterns and themes within a broad range of documents, from historical manuscripts to modern digital content.

Definition: Document and content analysis refers to the systematic examination of textual, visual, or audio materials to identify patterns, themes, or biases (Krippendorff, 2012). This approach can be applied to diverse sources, including books, newspapers, digital media, and more.

Key Characteristics

- **Systematic Approach**: An organized method is applied to ensure replicability and validity (Neuendorf, 2016).
- **Objective Analysis**: The aim is to reduce subjectivity by adhering to a clear set of guidelines (White & Marsh, 2006).
- **Contextual Understanding**: Recognizing the context in which documents were created is crucial (Prior, 2008).

Advantages

1. **Unobtrusive**: This method does not require direct involvement with the study subjects, thereby eliminating the risk of influencing participant behavior (Bowen, 2009).
2. **Cost-effective**: Often, public documents are readily accessible and free, reducing research costs (Altheide, 2000).

3. **Temporal Analysis**: Enables tracking changes over time by analyzing documents from different periods (Hodder, 2000).

Limitations

1. **Data Limitations**: Restricted to existing content; researchers cannot probe or ask follow-up questions (Silverman, 2013).
2. **Validity Concerns**: The authenticity and credibility of the documents can sometimes be questionable (Scott, 1990).
3. **Potential Biases**: Published content may reflect the biases of authors, publishers, or prevailing societal norms (Fairclough, 1995).

Applications

Document and content analysis are invaluable for:

- Historical research.
- Media studies and bias exploration.
- Policy analysis and evaluation (Altheide & Schneider, 2013).

Ethical Considerations

- Ensuring privacy and confidentiality, especially when analyzing personal or sensitive documents.
- Recognizing and accounting for potential biases.
- Properly citing all sources and respecting copyright regulations (Mauthner et al., 1998).

Conclusion

Document and content analysis serve as powerful tools for interpreting and understanding vast amounts of textual data. By providing insights into historical shifts, societal norms, and media portrayals, these methods offer a rich and layered understanding of content across different contexts.

Visual Methods: Photographs, Drawings, and Videos

Visual methodologies, comprising photographs, drawings, and videos, have grown significantly in prominence as valuable tools within the research landscape. These methods enable rich, multi-layered insights by capturing the complexities of human experience in visual form, thus transcending linguistic barriers and offering unique perspectives.

Definition

Visual methods refer to the utilization of visual materials, either researcher-generated or participant-generated, as a means to gather, analyze, and present data in research (Rose, 2016). These visual tools can encapsulate emotions, experiences, and phenomena that might be challenging to articulate verbally.

Types of Visual Methods

1. **Photographs**: Still images capturing moments, scenes, or objects.
2. **Drawings**: Hand-drawn illustrations often used to understand individual perspectives, emotions, or experiences.
3. **Videos**: Moving images that capture sequences of events, interactions, or environments.

Key Characteristics

- **Multidimensional Data**: Visual tools can capture nuances, emotions, and subtleties that may be overlooked in textual descriptions (Prosser & Loxley, 2008).
- **Cross-cultural Utility**: Visual data can transcend language barriers, making it valuable in diverse cultural settings (Banks, 2007).
- **Empowerment**: Allowing participants to capture their worlds can give them agency in the research process (Mitchell, 2008).

Advantages

1. **Rich Insights**: Offers a deeper, multifaceted understanding of the subject matter (Pink, 2013).
2. **Engagement**: Engages participants actively, making the research process collaborative (Harper, 2002).
3. **Versatility**: Suitable for various research settings and participant groups, including children and marginalized populations (Thomson, 2008).

Limitations

1. **Interpretative Challenges**: Deciphering the meaning behind images can be subjective and complex (Rose, 2016).
2. **Ethical Concerns**: Issues of privacy, consent, and potential misuse of visual data (Wiles et al., 2008).
3. **Technical Barriers**: Requires familiarity with cameras, software, and other equipment (Gubrium & Harper, 2013).

Applications

Visual methods are used extensively in:

- Anthropological and ethnographic studies.
- Educational research, especially with young participants.
- Studies focusing on urban planning and environmental changes (Radley & Taylor, 2003).

Ethical Considerations

- Obtaining informed consent for capturing and using visual data.
- Respecting privacy rights and blurring identifiable features when necessary.
- Secure storage and handling of visual materials to prevent misuse or unauthorized access (Clark-Ibáñez, 2007).

Conclusion

Embracing visual methods in research provides an avenue to tap into the richness of human experiences that might remain unarticulated in solely verbal or written formats. As research becomes increasingly interdisciplinary, the inclusion of visual tools further enriches the methodological repertoire.

REFERENCES

Adair, J. G. (1984). The Hawthorne effect: A reconsideration of the methodological artifact. *The Journal of Applied Psychology*, *69*(2), 334–345. doi:10.1037/0021-9010.69.2.334

Adler, P. A., & Adler, P. (1994). Observational techniques. In N. K. Denzin & Y. S. Lincoln (Eds.), *Handbook of qualitative research* (pp. 377–392). Sage Publications.

Agar, M. (1996). *The Professional Stranger: An Informal Introduction to Ethnography*. Academic Press.

Altheide, D. L., & Johnson, J. M. (1998). Criteria for assessing interpretive validity in qualitative research. In N. K. Denzin & Y. S. Lincoln (Eds.), *Collecting and interpreting qualitative materials* (pp. 283–312). Sage Publications.

Angrosino, M. (2007). Doing ethnographic and observational research. *Sage (Atlanta, Ga.)*.

Angrosino, M. V. (2005). Recontextualizing observation: Ethnography, pedagogy, and the prospects for a progressive political agenda. In N. Denzin & Y. Lincoln (Eds.), The Sage handbook of qualitative research (pp. 729-745). Sage.

Atkinson, P., & Hammersley, M. (1994). Ethnography and participant observation. In N. K. Denzin & Y. S. Lincoln (Eds.), *Handbook of qualitative research* (pp. 248–261). Sage.

Banks, M. (2007). *Using visual data in qualitative research*. SAGE. doi:10.4135/9780857020260

Baxter, P., & Jack, S. (2008). Qualitative case study methodology: Study design and implementation for novice researchers. *The Qualitative Report*, *13*(4), 544–559.

Berg, B. L. (2007). *Qualitative research methods for the social sciences* (6th ed.). Pearson.

Berg, B. L. (2009). *Qualitative research methods for the social sciences*. Allyn & Bacon.

Bernard, H. R. (2006). *Research methods in anthropology: Qualitative and quantitative approaches*. AltaMira Press.

Bernard, H. R. (2011). *Research methods in anthropology: Qualitative and quantitative approaches*. AltaMira Press.

Bogdan, R. C., & Biklen, S. K. (2007). *Qualitative research for education: An introduction to theories and methods*. Allyn & Bacon.

Borkan, J. (1999). Immersion/Crystallization. In B. F. Crabtree & W. L. Miller (Eds.), *Doing Qualitative Research* (pp. 179–194). Sage Publications.

Boud, D., Keogh, R., & Walker, D. (1985). *Reflection: Turning experience into learning*. Kogan Page.

Bowen, G. A. (2009). Document analysis as a qualitative research method. *Qualitative Research Journal*, *9*(2), 27–40. doi:10.3316/QRJ0902027

Brinkmann, S., & Kvale, S. (2015). *InterViews: Learning the craft of qualitative research interviewing* (3rd ed.). SAGE.

Bryman, A. (2015). *Social research methods*. Oxford university press.

Bryman, A., & Hardy, M. A. (2012). Introduction: The nature of (mixed) methods. In A. Bryman & M. A. Hardy (Eds.), *Handbook of data analysis* (pp. 1–16). Sage.

Bulmer, M. (1982). When is disguise justified? Alternatives to covert participant observation. *Qualitative Sociology*, *5*(2), 251–264. doi:10.1007/BF00986753

Clark-Ibáñez, M. (2007). Inner-city children in sharper focus: Sociology of childhood and photo elicitation interviews. *Visual Studies*, *22*(2), 70–89.

Clifford, J. (1990). Notes on (field)notes. In R. Sanjek (Ed.), *Fieldnotes: The makings of anthropology* (pp. 47–70). Cornell University Press.

Creswell, J. W. (2013). *Research design: Qualitative, quantitative, and mixed methods approaches*. Sage publications.

Creswell, J. W., & Creswell, J. D. (2017). *Research design: Qualitative, quantitative, and mixed methods approaches*. Sage Publications.

Davies, C. A. (2008). *Reflexive ethnography: A guide to researching selves and others*. Routledge.

Denzin, N. K. (1978). *The research act: A theoretical introduction to sociological methods*. McGraw-Hill.

Denzin, N. K., & Lincoln, Y. S. (2005). The Sage handbook of qualitative. Sage.

DeWalt, K. M., & DeWalt, B. R. (2002). *Participant observation: A guide for fieldworkers*. AltaMira Press.

DiCicco-Bloom, B., & Crabtree, B. F. (2006). The qualitative research interview. *Medical Education*, *40*(4), 314–321. doi:10.1111/j.1365-2929.2006.02418.x PMID:16573666

Dickson-Swift, V., James, E. L., Kippen, S., & Liamputtong, P. (2007). Doing sensitive research: What challenges do qualitative researchers face? *Qualitative Research*, *7*(3), 327–353. doi:10.1177/1468794107078515

Ellis, C., & Berger, L. (2003). Their story/my story/our story: Including the researcher's experience in interview research. In J. F. Gubrium & J. A. Holstein (Eds.), *Postmodern interviewing* (pp. 157–183). SAGE. doi:10.4135/9781412985437.n9

Emerson, R. M., Fretz, R. I., & Shaw, L. L. (1995). *Writing ethnographic fieldnotes*. University of Chicago Press. doi:10.7208/chicago/9780226206851.001.0001

Erickson, F. (1989). Ethical considerations in qualitative research. In R. Jaeger (Ed.), *Complementary methods for research in education*. American Educational Research Association.

Farquhar, C., & Das, R. (1999). Are focus groups suitable for 'sensitive' topics? In R. S. Barbour & J. Kitzinger (Eds.), *Developing Focus Group Research: Politics, Theory and Practice*. SAGE. doi:10.4135/9781849208857.n4

Fern, E. F. (2001). *Advanced focus group research*. SAGE. doi:10.4135/9781412990028

Finlay, L. (2002). "Outing" the Researcher: The Provenance, Process, and Practice of Reflexivity. *Qualitative Health Research*, *12*(4), 531–545. doi:10.1177/104973202129120052 PMID:11939252

Fontana, A., & Frey, J. H. (2000). The interview: From structured questions to negotiated text. In N. K. Denzin & Y. S. Lincoln (Eds.), *Handbook of qualitative research* (2nd ed., pp. 645–672). SAGE.

Geertz, C. (1973). *The interpretation of cultures: Selected essays*. Basic Books.

Gold, R. L. (1958). Roles in sociological field observations. *Social Forces*, *36*(3), 217–223. doi:10.2307/2573808

Goldberg, L. R. (1965). Diagnosticians vs. diagnostic signs: The diagnosis of psychosis vs. neurosis from the MMPI. *Psychological Monographs*, *79*(9), 1–23. doi:10.1037/h0093885 PMID:14322679

Gubrium, A., & Harper, K. (2013). *Participatory visual and digital research in action*. Left Coast Press.

Guillemin, M., & Gillam, L. (2004). Ethics, reflexivity, and "ethically important moments" in research. *Qualitative Inquiry*, *10*(2), 261–280. doi:10.1177/1077800403262360

Hammersley, M., & Atkinson, P. (2007). *Ethnography: Principles in practice*. Routledge.

Harper, D. (2002). Talking about pictures: A case for photo elicitation. *Visual Studies*, *17*(1), 13–26. doi:10.1080/14725860220137345

Jackson, J. (1983). *The real facts of life: Feminism and the politics of sexuality*. Taylor & Francis.

Janis, I. L. (1982). *Groupthink: Psychological studies of policy decisions and fiascoes*. Houghton Mifflin.

Jorgensen, D. L. (1989). *Participant observation*. Sage. doi:10.4135/9781412985376

Kawulich, B. B. (2005). Participant observation as a data collection method. *Forum Qualitative Social Research*, *6*(2).

King, N., & Horrocks, C. (2010). *Interviews in qualitative research*. SAGE.

Kitzinger, J. (1994). The methodology of focus groups: The importance of interaction between research participants. *Sociology of Health & Illness*, *16*(1), 103–121. doi:10.1111/1467-9566.ep11347023

Krueger, R. A., & Casey, M. A. (2014). *Focus groups: A practical guide for applied research* (5th ed.). SAGE.

Kvale, S., & Brinkmann, S. (2009). *InterViews: Learning the craft of qualitative research interviewing* (2nd ed.). SAGE.

Landsberger, H. A. (1958). *Hawthorne Revisited*. Cornell University.

LeCompte, M. D., & Goetz, J. P. (1982). Problems of reliability and validity in ethnographic research. *Review of Educational Research*, *52*(1), 31–60. doi:10.3102/00346543052001031

Lee, R. M. (1993). *Doing research on sensitive topics*. Sage.

Lincoln, Y. S., & Guba, E. A. (1985). *Naturalistic inquiry*. Sage. doi:10.1016/0147-1767(85)90062-8

Lofland, J., & Lofland, L. H. (1995). *Analyzing social settings: A guide to qualitative observation and analysis* (3rd ed.). Wadsworth.

Malinowski, B. (1922). *Argonauts of the Western Pacific*. Routledge & Kegan Paul.

Malinowski, B. (1967). *A diary in the strict sense of the term*. Stanford University Press.

Mason, J. (2002). *Qualitative researching* (2nd ed.). SAGE.

McCambridge, J., Witton, J., & Elbourne, D. R. (2014). Systematic review of the Hawthorne effect: New concepts are needed to study research participation effects. *Journal of Clinical Epidemiology*, *67*(3), 267–277. doi:10.1016/j.jclinepi.2013.08.015 PMID:24275499

Merriam, S. B., & Tisdell, E. J. (2015). *Qualitative research: A guide to design and implementation*. Jossey-Bass.

Minichiello, V., Aroni, R., Timewell, E., & Alexander, L. (1990). *In-depth interviewing: Researching people*. Longman Cheshire.

Mitchell, C. (2008). Getting the picture and changing the picture: Visual methodologies and educational research in South Africa. *South African Journal of Education*, *28*(3), 365–383. doi:10.15700/saje.v28n3a180

Morgan, D. L. (1996). *Focus groups as qualitative research* (2nd ed.). SAGE.

Mulhall, A. (2003). In the field: Notes on observation in qualitative research. *Journal of Advanced Nursing*, *41*(3), 306–313. doi:10.1046/j.1365-2648.2003.02514.x PMID:12581118

Orb, A., Eisenhauer, L., & Wynaden, D. (2001). Ethics in qualitative research. *Journal of Nursing Scholarship*, *33*(1), 93–96. doi:10.1111/j.1547-5069.2001.00093.x PMID:11253591

Ortlipp, M. (2008). Keeping and using reflective journals in the qualitative research process. *The Qualitative Report*, *13*(4), 695–705.

Ottenberg, S. (1990). Thirty years of fieldnotes: Changing relationships to the text. In R. Sanjek (Ed.), *Fieldnotes: The makings of anthropology* (pp. 139–160). Cornell University Press.

Patton, M. Q. (2002). *Qualitative research and evaluation methods* (3rd ed.). Sage Publications.

Patton, M. Q. (2014). *Qualitative Research & Evaluation Methods: Integrating Theory and Practice*. Sage Publications.

Patton, M. Q. (2015). *Qualitative research & evaluation methods: Integrating theory and practice*. Sage publications.

Pink, S. (2013). *Doing visual ethnography*. SAGE.

Prosser, J., & Loxley, A. (2008). Introducing visual methods. *NCRM Research Methods Review Paper*, *10*, 1–11.

Radley, A., & Taylor, D. (2003). Images of recovery: A photo-elicitation study on the hospital ward. *Qualitative Health Research*, *13*(1), 77–99. doi:10.1177/1049732302239412 PMID:12564264

Ritchie, J., & Lewis, J. (2003). *Qualitative research practice: A guide for social science students and researchers*. SAGE.

Rose, G. (2016). *Visual methodologies: An introduction to researching with visual materials* (4th ed.). SAGE.

Rosenthal, R. (1966). *Experimenter effects in behavioral research*. Appleton-Century-Crofts.

Roulston, K. (2010). *Reflective interviewing: A guide to theory and practice*. SAGE. doi:10.4135/9781446288009

Rubin, H. J., & Rubin, I. S. (2012). *Qualitative interviewing: The art of hearing data* (3rd ed.). SAGE.

Sanjek, R. (1990). On Ethnographic Validity. In R. Sanjek (Ed.), *Fieldnotes: The Makings of Anthropology* (pp. 385–418). Cornell University Press. doi:10.7591/9781501711954

Savage, J. (2006). Ethnographic evidence: The value of applied ethnography in healthcare. *Journal of Research in Nursing*, *11*(5), 383–393. doi:10.1177/1744987106068297

Seidman, I. (2013). *Interviewing as qualitative research: A guide for researchers in education and the social sciences*. Teachers College Press.

Silverman, D. (2016). *Qualitative research* (4th ed.). Sage.

Sim, J., & Wright, C. C. (2000). *Research in health care: Concepts, designs and methods*. Stanley Thornes.

Smithson, J. (2000). Using and analysing focus groups: Limitations and possibilities. *International Journal of Social Research Methodology*, *3*(2), 103–119. doi:10.1080/136455700405172

Spradley, J. P. (1979). *The ethnographic interview*. Holt, Rinehart, and Winston.

Stewart, D. W., Shamdasani, P. N., & Rook, D. W. (2007). *Focus groups: Theory and practice* (2nd ed.). SAGE. doi:10.4135/9781412991841

Tedlock, B. (2000). Ethnography and ethnographic representation. Handbook of Qualitative Research, 2, 455-486.

Thomson, P. (2008). *Doing visual research with children and young people*. Routledge.

Warren, C. A. B., & Karner, T. X. (2010). *Discovering qualitative methods: Field research, interviews, and analysis* (2nd ed.). Oxford University Press.

Webb, E. J., Campbell, D. T., Schwartz, R. D., & Sechrest, L. (1966). *Unobtrusive measures: Nonreactive research in the social sciences*. Rand McNally.

Whyte, W. F. (1943). *Street corner society: The social structure of an Italian slum*. University of Chicago Press.

Wiles, R., Crow, G., Heath, S., & Charles, V. (2008). The management of confidentiality and anonymity in social research. *International Journal of Social Research Methodology*, *11*(5), 417–428. doi:10.1080/13645570701622231

Wolcott, H. F. (1995). *The art of fieldwork*. AltaMira Press.

Chapter 5
Fieldwork and Immersion

ABSTRACT

This chapter explores the methodologies of fieldwork and immersion, rooted in anthropological and sociological traditions, as essential for gaining profound insights into diverse disciplines. By situating researchers directly within the study environments, these methods offer a unique vantage point to observe and interpret complex social dynamics and cultural interactions. The historical evolution of fieldwork is highlighted, showcasing its transformational impact on research methodologies and its adoption across various fields. The benefits of these approaches include rich contextual insights, enhanced rapport with participants, and the ability to observe dynamic interactions.

INTRODUCTION

Fieldwork and immersion, deeply rooted in anthropological and sociological traditions, have become fundamental components in a diverse array of disciplines. These methods have the potential to offer rich, granular insights by situating the researcher within the environment, culture, or community of study. This chapter aims to elucidate the essence, challenges, and ethical implications of fieldwork and immersion as integral components of the research journey.

Understanding Fieldwork and Immersion

At its core, fieldwork is the practice of observing and documenting events, behaviors, and artifacts in the setting in which they naturally occur (Geertz, 1973). Immersion is a profound engagement and familiarity with the study's locale, allowing researchers to deeply understand and interpret the complex web of sociocultural dynamics and interpersonal relationships (Marcus, 1998).

Historical Context

Historically, anthropologists and ethnographers have been at the forefront of fieldwork and immersion practices, venturing into unfamiliar territories to understand "the other" (Malinowski, 1922). The fa-

DOI: 10.4018/979-8-3693-2414-1.ch005

mous account of Bronislaw Malinowski's work in the Trobriand Islands has become emblematic of the transformative power of fieldwork, laying the groundwork for subsequent generations of researchers.

Benefits of Fieldwork and Immersion

1. **Rich, Contextual Insights**: Fieldwork offers a firsthand perspective, enabling researchers to capture nuances often missed by remote or detached methodologies (Agar, 1996).
2. **Building Rapport**: Immersion facilitates the development of trust and rapport with participants, which can lead to more authentic data collection (Bernard, 2011).
3. **Dynamic Understanding**: Being on the ground allows researchers to witness events as they unfold and adjust their inquiries accordingly (Emerson, Fretz, & Shaw, 2011).

Challenges and Considerations

1. **Physical and Emotional Toll**: Extended periods in unfamiliar settings can be demanding and sometimes even dangerous (Kovats-Bernat, 2002).
2. **Bias and Objectivity**: Close relationships with participants can potentially cloud objectivity, requiring rigorous self-reflection (Rose, 1997).
3. **Logistical Hurdles**: Fieldwork often demands substantial resources, including time, finances, and logistical planning (Sanjek, 1990).

Ethical Implications

Fieldwork and immersion come with their ethical quandaries. Ensuring informed consent, maintaining participant confidentiality, and navigating power dynamics are paramount concerns (Fluehr-Lobban, 2013). Moreover, researchers must be cautious not to exoticize or misrepresent the communities they study (Said, 1978).

Fieldwork in the Digital Age

With the advent of the digital era, virtual fieldwork has gained prominence. Platforms like social media provide new arenas for observation, interaction, and data collection, challenging traditional notions of "the field" (Hine, 2000). This digital shift underscores the evolving nature of fieldwork and the need for adaptable methodologies.

Conclusion

Fieldwork and immersion are both transformative experiences for the researcher and potent tools for generating insights. As with any methodology, they demand rigor, reflexivity, and respect for participants. By immersing themselves in the worlds they study, researchers can cultivate a profound understanding that transcends surface-level observations, enabling them to craft narratives that resonate with authenticity and depth.

Gaining Access to Sites and Participants

Access to research sites and participants is pivotal for the success of fieldwork. The process of obtaining access is multifaceted, often requiring considerable negotiation, preparation, and ethical consideration. The researcher's ability to enter a community or institution and create a rapport with its members can significantly impact the quality and authenticity of the data collected.

Significance of Gaining Access

Fieldwork's core rests upon direct interaction with the environment and its members (Emerson, Fretz, & Shaw, 1995). Access is not merely physical entry but encompasses deeper levels of engagement and acceptance within the community, enabling the researcher to capture rich, nuanced insights.

In the realm of fieldwork and ethnographic studies, gaining access to research sites and participants is paramount. It serves as the bridge connecting the abstract theoretical constructs to the practical and lived realities on the ground (Hammersley & Atkinson, 2007). The significance of gaining proper access can be dissected as follows:

1. **Authentic Data Collection**: One of the primary goals of fieldwork is to obtain authentic, firsthand data. Without proper access, researchers risk collecting data that's superficial, unrepresentative, or even misleading. Engaging deeply with the environment and its members allows for a more holistic understanding of the context (Agar, 1996).
2. **Establishing Trust and Rapport**: The quality of data often hinges on the level of trust established between the researcher and the participants. Gaining access doesn't merely imply physical entry but also signifies acceptance within the community. This trust often leads to richer narratives and insights (Bernard, 2006).
3. **Longitudinal and Repeated Measures**: Many ethnographic studies require repeated visits and long-term engagement. Proper access ensures that researchers can revisit the site, enabling longitudinal studies and capturing temporal changes in the community (Brewer, 2000).
4. **Ethical Research Conduct**: Proper access ensures that research is conducted ethically. It signifies that the community or institution is informed about the study and its implications and that they have provided their consent for the research to be carried out (Mertens & Ginsberg, 2009).
5. **Addressing Power Dynamics**: Authentic access often requires the researcher to be sensitive to power dynamics in the field. Recognizing and addressing these dynamics can prevent exploitation and ensure a more equitable relationship between the researcher and the participants (Fine & Torre, 2004).
6. **Mitigating Researcher Bias**: Immersion through proper access allows the researcher to challenge and refine their preconceptions. By being deeply embedded in the field, researchers can confront their biases, leading to a more objective representation of the community or institution (Madison, 2005).

In essence, gaining access is more than a logistical step in the research process. It has profound implications for the depth, quality, ethics, and authenticity of the research. Without meaningful access, fieldwork remains peripheral, failing to capture the nuanced realities it aims to study.

Challenges in Gaining Access

1. **Gatekeepers**: Often, individuals or groups control access to the research site or participants. These gatekeepers can be formal, like institutional heads, or informal, like community leaders. Navigating their concerns and conditions can be intricate (Berg, 2009).
2. **Mistrust**: Communities or institutions may be wary of outsiders, especially if past interactions with researchers or outsiders have been negative or exploitative (Lofland, Snow, Anderson, & Lofland, 2006).
3. **Logistical Barriers**: Geographical, political, or socio-cultural factors can make sites difficult to access (Yin, 2013).

Strategies for Gaining Access

1. **Building Rapport**: Investing time in understanding the community, respecting its norms, and showing genuine interest can facilitate access (Bernard, 2006).
2. **Leveraging Networks**: Using personal or institutional connections can help in initial introductions or endorsements (DeWalt & DeWalt, 2010).
3. **Transparency**: Clearly communicating the purpose, methods, and potential implications of the research can allay concerns and build trust (Bryman, 2012).
4. **Iterative Negotiation**: Often, gaining access is not a one-time affair. Researchers may need to revisit negotiations, addressing new concerns or conditions that arise (Hammersley & Atkinson, 2007).

Ethical Considerations

Gaining access should not compromise the ethical standards of research. Ensuring informed consent, respecting privacy, and being cautious not to exert undue influence or pressure are crucial (Mertens & Ginsberg, 2009).

Researchers must also be aware of power dynamics. The act of entering a space can, in itself, alter the dynamics of that environment. Especially in vulnerable or marginalized communities, researchers need to approach with sensitivity, ensuring that their presence does not exacerbate existing inequities or tensions (Orb, Eisenhauer, & Wynaden, 2001).

CONCLUSION

Gaining access is a nuanced endeavor, shaped by the interplay of social, ethical, and logistical considerations. Researchers' ability to navigate these challenges with sensitivity and integrity can profoundly influence the trajectory and impact of their fieldwork.

The Role of the Researcher: Observer vs. Participant

The role of the researcher in fieldwork is a matter of intense debate and introspection within ethnographic and qualitative research traditions. Two primary roles emerge when considering fieldwork – that of the observer and that of the participant. Each has its nuances, advantages, challenges, and ethical implications (Dewalt & Dewalt, 2002).

1. Observer

When a researcher assumes the role of an observer, they remain detached from the active participation in events, activities, or routines of the study group. The observer position is often linked with objectivity, as the researcher seeks to record occurrences without influencing or being influenced by them.

- **Advantages**:
- **Objectivity**: Observers often maintain a sense of objectivity, minimizing researcher bias (Kawulich, 2005).
- **Less Intrusive**: By being on the periphery, observers may be less disruptive to the natural flow of events or routines.
- **Challenges**:
- **Limited Depth**: Pure observation might miss out on the underlying motivations, beliefs, or emotions of the participants.
- **Potential Misinterpretation**: Without active participation, observers might misinterpret cultural or contextual nuances (Mulhall, 2003).

2. Participant

Here, the researcher immerses themselves in the activities, rituals, and daily routines of the study group. They become part of the group to experience and understand the phenomena from within.

- **Advantages**:
- **In-depth Understanding**: Participation allows for a deeper understanding of the context, behaviors, beliefs, and emotions of the study group (Gold, 1958).
- **Builds Trust**: Active participation can foster trust, leading to richer data and insights (Spradley, 1980).
- **Challenges**:
- **Loss of Objectivity**: Being immersed might blur the lines between observation and participation, potentially introducing bias.
- **Ethical Concerns**: The researcher's involvement might influence events or behaviors, raising concerns about the authenticity of data (Adler & Adler, 1987).

Striking a Balance

Most ethnographers acknowledge that pure observation or pure participation is rare. Instead, researchers often find themselves on a continuum between the two, adjusting their role based on the context, research objectives, and ethical considerations (Whyte, 1993).

The decision on where to position oneself on this continuum is influenced by various factors, including:

- The nature of the research question.

- The culture and norms of the community under study.
- Ethical considerations.
- The researcher's own comfort and expertise.

In conclusion, whether one leans more towards observation or participation, it's crucial to remain reflexive, regularly questioning one's position, influence, and the data being gathered. The dynamism of fieldwork requires researchers to be adaptable and sensitive, always striving for a balance that yields both depth and authenticity in their findings.

Field Notes and Reflexivity

Fieldwork, a cornerstone of qualitative research, necessitates detailed documentation to capture the complexities of the studied phenomena. Field notes and reflexivity are central components of this documentation, providing a rich tapestry of data while maintaining a researcher's critical self-awareness.

1. Field Notes:

Field notes serve as a written record of observations, experiences, and reflections made while conducting fieldwork (Emerson, Fretz & Shaw, 1995).

Field notes, an indispensable component of ethnographic and qualitative fieldwork, are detailed accounts of observations, experiences, and reflections made while conducting research in a natural setting. These records become instrumental in generating a comprehensive understanding of the study's context, capturing the nuances of behaviors, interactions, and environmental factors (Emerson, Fretz & Shaw, 1995).

A. Historical Background:

The tradition of keeping field notes has its roots in early anthropological studies where ethnographers spent extended periods in unfamiliar settings, necessitating rigorous documentation of observed cultural practices (Malinowski, 1922).

B. Significance of Field Notes:
   - **Primary Data Source**: They offer firsthand descriptions of the phenomena under investigation, aiding in subsequent analysis (Burgess, 1984).
   - **Contextual Depth**: Field notes enrich research by providing context, emotion, ambiance, and the intricate details of events and interactions (Sanjek, 1990).

C. Components of Field Notes:
   - **Descriptive Notes**: These contain factual, detailed, and objective accounts of observed events, environments, behaviors, and conversations (Bogdan & Biklen, 2007).
   - **Reflective Notes**: Here, the researcher adds personal reflections, potential hypotheses, feelings, and preliminary interpretations (Lofland et al., 2006).

D. Crafting Effective Field Notes:
   - **Immediate Recording**: It's essential to jot down observations as soon as possible to prevent memory distortions (Atkinson & Hammersley, 2007).

- **Use of Detailed Descriptions**: Incorporating sensory details, emotions, and verbatim dialogues can provide richer context (Geertz, 1973).
- **Inclusion of Diagrams and Sketches**: Visual aids like maps or diagrams can help in detailing spatial layouts and interactions.

E. Challenges in Taking Field Notes:
- **Selectivity and Bias**: Given the impossibility of documenting everything, researchers have to decide what to prioritize, potentially leading to bias (Davies, 2008).
- **Memory Constraints**: Over-reliance on memory can introduce errors or omissions.
- **Intrusiveness**: The act of note-taking can occasionally disturb the natural flow of events or make participants self-conscious (Gold, 1958).

F. Digitalization and Field Notes:

With technological advancements, digital tools have increasingly been employed for field note-taking. Tools like digital recorders or tablets can aid in capturing data. However, the choice of tools should always respect the participants' privacy and the study's ethical considerations (Fielding & Lee, 1998).

In conclusion, field notes are the lifeblood of qualitative research, anchoring observations to tangible records. Their meticulous crafting ensures that the richness and context of the studied phenomena are preserved for future analysis and interpretation.

2. Reflexivity

Reflexivity, within the ambit of qualitative research, refers to the process wherein researchers continuously engage in critical self-reflection about their potential biases, cultural framework, values, and the impact they may have on the research process, findings, and interpretations. This self-aware introspection allows for more honest, transparent, and nuanced research outcomes (Finlay, 2002).

1. The Origin and Significance of Reflexivity:

Historically rooted in the interpretative paradigm, reflexivity has evolved as an antidote to the criticism that subjective experiences of researchers can potentially skew research results (Davies, 2008). Its adoption underscores the belief that researchers aren't passive, objective observers but play an active role in constructing knowledge (Hertz, 1997).

2. Dimensions of Reflexivity:
   - **Personal Reflexivity**: Involves reflecting on how one's own beliefs, values, experiences, and background can influence the research (Guba & Lincoln, 2005).
   - **Epistemological Reflexivity**: Concerned with the nature and basis of knowledge. It prompts researchers to consider how their assumptions and methodological choices shape the study (Morrow, 2005).
3. Reflexivity in Fieldwork:
   - **Positionality**: In the field, researchers must be aware of their positionalities, such as their gender, race, and socio-economic status, and how these can influence interactions and data collection (Rose, 1997).

- **Interpersonal Dynamics**: Reflexivity requires an acknowledgment of the power dynamics between the researcher and participants and its potential to influence responses (Karnieli-Miller et al., 2009).

4. Reflexivity Post-Fieldwork:
   - **Data Interpretation**: Researchers must continuously interrogate how their subjectivities might shape data coding, theme identification, and overall interpretation (Pillow, 2003).
   - **Reporting and Dissemination**: The act of writing or presenting findings is also influenced by a researcher's perspective. A reflexive approach advocates transparency about this influence (Richardson & St. Pierre, 2005).
5. Challenges and Critiques of Reflexivity:

While reflexivity is championed as a practice promoting rigour and depth, it's not devoid of challenges:

- **Overemphasis on Researcher**: Some critics argue that an overindulgence in reflexivity can shift the focus from participants and the research topic to the researcher (Macbeth, 2001).
- **Paralysis by Analysis**: Excessive reflexive scrutiny can lead to a researcher becoming overly cautious or indecisive, potentially stalling the research process (Etherington, 2004).

6. Reflexivity as an Ongoing Commitment:

Rather than a mere stage or a box to be ticked, reflexivity is an ongoing commitment throughout the research journey. It necessitates a continuous dialogue between one's self, the field, the participants, and the evolving data (Alvesson & Sköldberg, 2009).

Reflexivity enriches the research process, ensuring that findings are not only a product of the observed world but also a transparent acknowledgment of the observer's lens through which that world is seen.

In conclusion, field notes and reflexivity serve as essential tools in qualitative fieldwork, allowing researchers to capture rich, detailed data while remaining critically aware of their positionality. Emphasizing these elements in fieldwork strengthens the depth, validity, and ethical considerations of the research.

Challenges in Fieldwork and Strategies to Overcome Them

Fieldwork, an integral component of qualitative research, is fraught with multifaceted challenges. Researchers, immersed in the socio-cultural environments they study, must be equipped to handle these challenges to ensure research validity and reliability. This section explores some primary challenges encountered during fieldwork and suggests strategies to address them.

1. ESTABLISHING TRUST AND RAPPORT

Challenge

Winning the trust of participants is pivotal for authentic data collection. However, mistrust, stemming from cultural differences, past negative experiences with researchers, or apprehensions about research outcomes, can hinder this process (Kusow, 2003).

Strategy

Investing time in the community, understanding local customs, engaging in community activities, and transparent communication can foster trust (Bernard, 2011). Also, consistent ethical practices, including informed consent and ensuring participant anonymity, can further engender trust.

2. NAVIGATING POWER DYNAMICS

Challenge

Inherent power imbalances between the researcher and participants can skew research findings. Participants may provide data they perceive as desirable rather than their true experiences or feelings (Zimmerman & Wieder, 1977).

Strategy

Reflexivity, or the constant reflection on one's influence on the research process, can help mitigate this (Rose, 1997). Practicing active listening and adopting a humble learning posture, as opposed to an expert stance, can also minimize power dynamics.

3. LANGUAGE AND CULTURAL BARRIERS

Challenge

Language differences can lead to misinterpretations, while cultural nuances can render certain questions or methodologies inappropriate or offensive (Temple & Young, 2004).

Strategy

Utilizing local interpreters or cultural brokers can bridge language gaps, while prolonged engagement in the field aids in understanding and respecting cultural nuances (Liamputtong, 2008).

4. EMOTIONAL AND PHYSICAL WELL-BEING

Challenge

Immersive fieldwork can expose researchers to emotionally charged situations, leading to burnout, distress, or secondary traumatic stress (Dickson-Swift et al., 2009).

Strategy

Regular debriefing sessions, maintaining a work-life balance, seeking peer support, and considering professional counseling can be beneficial (Warden, 2013).

5. CHANGING FIELD DYNAMICS

Challenge

Field environments are not static. Political upheavals, economic changes, or natural disasters can significantly alter the research context (Kovats-Bernat, 2002).

Strategy

Flexibility is key. Researchers must be prepared to adapt methodologies, change research questions, or even consider temporary withdrawal from the field if necessary.

6. DATA OVERWHELM

Challenge

The richness of qualitative data can be overwhelming, leading to difficulty in data management and analysis (Davies & Dodd, 2002).

Strategy

Regular data sorting, digital transcription tools, and qualitative data analysis software can aid in organizing and managing vast amounts of data (Bazeley, 2007).

In conclusion, while fieldwork poses numerous challenges, a combination of preparation, adaptability, reflexivity, and ethical consideration can guide researchers towards robust, meaningful, and respectful data collection.

REFERENCES

Adler, P. A., & Adler, P. (1987). *Membership roles in field research.* Sage Publications. doi:10.4135/9781412984973

Alvesson, M., & Sköldberg, K. (2009). Reflexive methodology: New vistas for qualitative research. *Sage (Atlanta, Ga.).*

Atkinson, P., & Hammersley, M. (2007). *Ethnography: Principles in practice.* Routledge. doi:10.4324/9780203944769

Bazeley, P. (2007). *Qualitative data analysis with NVivo*. Sage Publications.

Berg, B. L. (2009). *Qualitative research methods for the social sciences*. Allyn & Bacon.

Bernard, H. R. (2006). *Research methods in anthropology: Qualitative and quantitative approaches*. AltaMira Press.

Bernard, H. R. (2011). *Research methods in anthropology: Qualitative and quantitative approaches*. AltaMira Press.

Brewer, J. D. (2000). *Ethnography*. Open University Press.

Bryman, A. (2012). *Social research methods* (4th ed.). Oxford University Press.

Burgess, R. G. (1984). *In the field: An introduction to field research*. Routledge.

Davies, C. A. (2008). *Reflexive ethnography: A guide to researching selves and others*. Routledge.

Davies, D., & Dodd, J. (2002). Qualitative research and the question of rigor. *Qualitative Health Research*, *12*(2), 279–289. doi:10.1177/104973230201200211 PMID:11837376

DeWalt, K. M., & DeWalt, B. R. (2002). *Participant observation: A guide for fieldworkers*. AltaMira Press.

Dickson-Swift, V., James, E. L., Kippen, S., & Liamputtong, P. (2009). Researching sensitive topics: Qualitative research as emotion work. *Qualitative Research*, *9*(1), 61–79. doi:10.1177/1468794108098031

Emerson, R. M., Fretz, R. I., & Shaw, L. L. (1995). *Writing ethnographic fieldnotes*. University of Chicago Press. doi:10.7208/chicago/9780226206851.001.0001

Etherington, K. (2004). Becoming a reflexive researcher: Using our selves in research. Jessica Kingsley Publishers.

Fielding, N. G., & Lee, R. M. (1998). *Computer analysis and qualitative research*. Sage Publications.

Finlay, L. (2002). "Outing" the Researcher: The Provenance, Process, and Practice of Reflexivity. *Qualitative Health Research*, *12*(4), 531–545. doi:10.1177/104973202129120052 PMID:11939252

Fluehr-Lobban, C. (2013). *Ethics and the profession of anthropologists: Fieldwork in moral dilemmas*. Rowman Altamira.

Geertz, C. (1973). *The interpretation of cultures: Selected essays*. Basic Books.

Gold, R. L. (1958). Roles in sociological field observations. *Social Forces*, *36*(3), 217–223. doi:10.2307/2573808

Guba, E. G., & Lincoln, Y. S. (2005). Paradigmatic controversies, contradictions, and emerging confluences. *Sage (Atlanta, Ga.)*.

Hammersley, M., & Atkinson, P. (2007). *Ethnography: Principles in practice*. Routledge.

Hertz, R. (Ed.). (1997). *Reflexivity & voice*. Sage.

Hine, C. (2000). Virtual ethnography. *Sage (Atlanta, Ga.)*.

Karnieli-Miller, O., Strier, R., & Pessach, L. (2009). Power relations in qualitative research. *Qualitative Health Research, 19*(2), 279–289. doi:10.1177/1049732308329306 PMID:19150890

Kawulich, B. B. (2005). Participant observation as a data collection method. *Forum Qualitative Social Research, 6*(2).

Kovats-Bernat, J. C. (2002). Negotiating dangerous fields: Pragmatic strategies for fieldwork amid violence and terror. *American Anthropologist, 104*(1), 208–222. doi:10.1525/aa.2002.104.1.208

Kusow, A. M. (2003). Beyond indigenous authenticity: Reflections on the insider/outsider debate in immigration research. *Symbolic Interaction, 26*(4), 591–599. doi:10.1525/si.2003.26.4.591

Liamputtong, P. (2008). *Doing cross-cultural research: Ethical and methodological perspectives* (Vol. 34). Springer Science & Business Media. doi:10.1007/978-1-4020-8567-3_1

Lofland, J., Snow, D. A., Anderson, L., & Lofland, L. H. (2006). *Analyzing social settings: A guide to qualitative observation and analysis*. Wadsworth/Thomson Learning.

Macbeth, D. (2001). On "reflexivity" in qualitative research: Two readings, and a third. *Qualitative Inquiry, 7*(1), 35–68. doi:10.1177/107780040100700103

Madison, D. S. (2005). Critical ethnography: Method, ethics, and performance. *Sage (Atlanta, Ga.).*

Malinowski, B. (1922). *Argonauts of the Western Pacific*. Routledge & Kegan Paul.

Marcus, G. E. (1998). *Ethnography through thick and thin*. Princeton University Press.

Mertens, D. M., & Ginsberg, P. E. (2009). Deep in ethical waters: Transformative perspectives for qualitative social work research. *Qualitative Social Work, 8*(4).

Mulhall, A. (2003). In the field: Notes on observation in qualitative research. *Journal of Advanced Nursing, 41*(3), 306–313. doi:10.1046/j.1365-2648.2003.02514.x PMID:12581118

Orb, A., Eisenhauer, L., & Wynaden, D. (2001). Ethics in qualitative research. *Journal of Nursing Scholarship, 33*(1), 93–96. doi:10.1111/j.1547-5069.2001.00093.x PMID:11253591

Pillow, W. (2003). Confession, catharsis, or cure? Rethinking the uses of reflexivity as methodological power in qualitative research. *International Journal of Qualitative Studies in Education : QSE, 16*(2), 175–196. doi:10.1080/0951839032000060635

Richardson, L., & Adams St. Pierre, E. (2005). Writing: A method of inquiry. In N. K. Denzin & Y. S. Lincoln (Eds.), *The Sage handbook of qualitative research* (3rd ed., pp. 959–978). Sage.

Sanjek, R. (1990). On Ethnographic Validity. In R. Sanjek (Ed.), *Fieldnotes: The Makings of Anthropology* (pp. 385–418). Cornell University Press. doi:10.7591/9781501711954

Temple, B., & Young, A. (2004). Qualitative research and translation dilemmas. *Qualitative Research, 4*(2), 161–178. doi:10.1177/1468794104044430

Warden, R. (2013). The emotional cost of caring in qualitative research. *Procedia: Social and Behavioral Sciences, 82*, 144–147.

Yin, R. K. (2013). *Case study research: Design and methods*. Sage publications.

Zimmerman, D. H., & Wieder, D. L. (1977). The diary: Diary-interview method. *Urban Life*, *5*(4), 479–498. doi:10.1177/089124167700500406

Chapter 6
Data Analysis and Interpretation

ABSTRACT

This chapter examines the critical stages of data analysis and interpretation in qualitative research, which are essential for transforming collected data into meaningful insights. Qualitative data, characterized by its volume and complexity, requires a methodical approach to uncover depth and context that quantitative data might overlook. The analysis process includes data preparation, coding, identifying themes, and visual representation, each playing a pivotal role in deriving nuanced understandings from textual and visual data. Theoretical frameworks guide the interpretation, ensuring analytical depth.

INTRODUCTION

In the realm of qualitative research, the journey from data collection to deriving meaningful insights necessitates a rigorous and systematic process of data analysis and interpretation. This chapter aims to provide an in-depth exploration of the labyrinthine yet rewarding venture of making sense of qualitative data. Data analysis and interpretation are not isolated post-fieldwork procedures; they are inherently intertwined with the preceding research processes and are crucial for the realization of the research objectives (Miles, Huberman, & Saldaña, 2014).

Nature of Qualitative Data

Qualitative data, often voluminous and intricate, captures the richness of human experiences, feelings, and perceptions. It is textual and visual, derived from transcripts, notes, images, and videos. Unlike its quantitative counterpart, qualitative data seeks depth over breadth, nuance over generalizability, and context over causality (Denzin & Lincoln, 2011).

The Imperative of Data Analysis

Analyzing qualitative data is indispensable. It structures and gives meaning to the collected data, ensuring that the research questions are addressed, and the study's goals are achieved (Creswell, 2013).

DOI: 10.4018/979-8-3693-2414-1.ch006

Without this systematic examination, the data remains a mere collection of words or images, devoid of interpretive value.

Stages in Data Analysis

1. **Data Preparation**: Before diving into analysis, data must be organized. Transcripts are readied, field notes are reviewed, and visual data is catalogued. Familiarization with data is pivotal at this juncture (Ritchie & Lewis, 2003).
2. **Coding**: It involves identifying significant fragments of data—words, phrases, sentences, or entire sections—and labeling them with codes that represent their content or meaning (Saldaña, 2015).
3. **Identifying Themes**: Codes are clustered into broader categories or themes. These themes encapsulate the core ideas emerging from the data and directly link to the research objectives (Braun & Clarke, 2006).
4. **Interpreting and Explaining**: Beyond mere identification of themes, researchers delve into understanding their significance, relationships, patterns, and inconsistencies. This step is where the data starts to tell its story (Geertz, 1973).
5. **Visual Representation**: Data can be represented visually using charts, graphs, or models. Such representations can make complex data more accessible and understandable (Coffey & Atkinson, 1996).

Role of Theoretical and Conceptual Frameworks

Data does not interpret itself. Theoretical or conceptual frameworks guide the interpretation process. They provide lenses through which data is viewed, ensuring coherence and depth in the emerging interpretations (Layder, 1998).

Challenges in Data Analysis

Qualitative data analysis is not without its hurdles. The sheer volume of data, maintaining objectivity, avoiding bias, and ensuring validity and reliability are some persistent challenges researchers grapple with (Smith, 2003). Employing iterative processes, being reflexive, using software tools, and peer debriefing can help in mitigating these challenges.

Importance of Triangulation

To bolster the validity of interpretations, researchers often use triangulation—employing multiple data sources, methods, or researchers to cross-check and validate findings (Denzin, 1978).

Transitioning to Reporting

Once analysis and interpretation are concluded, researchers transition to the reporting phase. This transition involves narratively presenting the findings in a coherent, logical, and engaging manner, integrating quotes, images, or other data extracts to substantiate interpretations (Wolcott, 1990).

In summary, data analysis and interpretation in qualitative research are akin to sculpting. From a raw block of collected data, researchers meticulously chisel away, refining and redefining, until the final form—the research findings and insights—emerge in all their nuanced glory.

Transcribing Data

The transformative journey from raw qualitative data to actionable insights begins with one foundational step: transcription. This chapter delves into the intricate world of transcribing data, a process often overshadowed by subsequent stages of analysis but one that plays an indispensable role in shaping the research outcomes. Understanding transcription requires a comprehensive perspective on its nuances, methods, and implications for qualitative research (Bird, 2005).

What Is Transcription?

Transcription in the context of qualitative research refers to the meticulous process of converting spoken language from audio or video recordings into written text. But it's not just a mechanical translation; it's a representation that encapsulates the depth, context, and nuances of the spoken word, making it amenable to systematic analysis (Oliver, Serovich, & Mason, 2005).

Why Transcribe?

Before delving into the mechanics, it's imperative to understand the significance of transcription in qualitative research:

1. **Accessibility**: Transcribed data is more readily accessible, allowing researchers to revisit, review, and analyze the data multiple times (Davidson, 2009).
2. **Grounded Analysis**: By converting data into text, researchers can immerse themselves in the data, laying the groundwork for grounded theory or other detailed analytical methods (Charmaz, 2006).
3. **Enhanced Accuracy**: Transcription reduces reliance on memory, thus enhancing the accuracy of subsequent interpretations (Halcomb, Davidson, & Hardaker, 2008).

Approaches to Transcription

1. **Verbatim Transcription**: As the name suggests, verbatim transcription captures every word, sound, pause, and emotion. It offers a holistic portrayal of the data but can be time-consuming (Lapadat & Lindsay, 1999).
2. **Intelligent Transcription**: This method focuses on capturing the essence of the speech, excluding non-verbal cues, repeated words, or fillers. It's more concise, yet risks losing nuances (Poland, 2002).
3. **Thematic Transcription**: Here, the transcription process is guided by pre-determined themes or research questions, focusing only on relevant sections of the data (Jupp, 2006).

Challenges in Transcription

1. **Time Intensity**: Transcription is labor-intensive, with a single hour of recording taking multiple hours to transcribe (Bailey, 2008).
2. **Maintaining Context**: Ensuring the context is not lost during transcription, especially with non-verbal cues, can be challenging (Tilley, 2003).
3. **Accuracy and Bias**: Maintaining objectivity and avoiding researcher-induced bias is a persistent challenge in transcription (Bird, 2005).

Transcription as Interpretation

Transcription is not a passive act. The choices researchers make in what to include, exclude, or emphasize means transcription is inherently an interpretative act (Mishler, 1991). Thus, the transcribed text is not just a mirror reflection of the spoken word but a representation constructed through the lens of the researcher (O'Connell & Kowal, 1999).

Technology and Transcription

Advancements in technology have transformed the transcription landscape. From digital recorders that capture high-quality audio to transcription software that uses artificial intelligence to auto-transcribe, the process has become more efficient. Yet, the role of human judgment remains irreplaceable, especially in discerning context and nuances (Silver & Lewins, 2014).

Ethical Considerations

Considering transcription often involves personal or sensitive information, maintaining confidentiality is paramount. Also, researchers must be transparent about any modifications made during transcription, ensuring that participants' voices and meanings are retained faithfully (Tilley, 2003).

In conclusion, transcription is a foundational pillar in the edifice of qualitative research. It's a bridge that connects the raw, organic expressions of participants to the structured world of data analysis, ensuring that the depth and richness of human experiences are captured, represented, and interpreted with fidelity.

Coding: Open, Axial, and Selective

Unveiling the underlying patterns, concepts, and themes within a plethora of qualitative data is akin to deciphering a complex puzzle. Integral to this deciphering process is coding, a systematic approach to transform raw data into structured, categorized knowledge. As outlined by Strauss and Corbin (1990), three prominent types of coding—open, axial, and selective—are cardinal to grounded theory methodology, though they have also found their rightful place in other qualitative research paradigms. This chapter embarks on an in-depth exploration of these three coding types, shedding light on their intricacies, methodologies, and significance in the realm of qualitative research.

Understanding Coding in Qualitative Research

At its core, coding is the analytical process of categorizing, labeling, and organizing raw data into conceptual chunks, offering researchers a lens to discern and interpret underlying patterns (Saldana, 2015).

It serves as the foundational step towards building theory, themes, or narratives that encapsulate the essence of the collected data.

Open Coding

Open coding, as the name suggests, involves dissecting the data "openly", breaking it down into distinct parts, and examining its properties and dimensions (Charmaz, 2006). The objective is to encapsulate the core ideas of each data fragment.

Features

- Initial phase of coding, typically devoid of researcher's preconceptions (Glaser, 1978).
- Involves line-by-line, incident-to-incident, or segment-by-segment coding.
- Generates a multitude of codes, which can later be categorized or grouped.

Importance

- Facilitates intimate familiarity with the data.
- Fosters an emergent analytical perspective rather than a confirmatory one (Strauss & Corbin, 1998).

Axial Coding

Emerging from the fragments of open coding, axial coding involves reassembling the data, pivoting around an axis. This "axis" is typically a category or a core concept derived during open coding (Creswell, 2013).

Features

- Connects categories to subcategories and specifies the relationships between them (Charmaz, 2006).
- Employs the paradigm model, considering factors like causation, context, strategies, and outcomes (Strauss & Corbin, 1990).

Importance

- Facilitates the building of connections within the data.
- Provides a structured framework, enabling the data's complexity to be organized meaningfully.

Selective Coding

Selective coding is the culminating phase, where the core category or the main theme is woven through and around the other categories, ensuring the narrative's coherence and integration (Glaser & Strauss, 1967).

Features

- Involves selecting the core category and systematically relating it to other categories.
- Refines and reduces the list of categories to those that are substantively significant (Charmaz, 2006).

Importance

- Enables the construction of a coherent theory or narrative.
- Ensures the data's multidimensional aspects are presented through a unified lens.

Challenges in Coding

While coding offers a structured pathway to interpret qualitative data, it's not devoid of challenges:

1. **Subjectivity**: The researcher's biases and preconceptions can influence code generation (Miles & Huberman, 1994).
2. **Over-coding**: The dilemma between generating too many codes and risking fragmentation or too few codes and losing nuances (Saldana, 2015).
3. **Consistency**: Ensuring consistency in coding, especially in larger datasets or with multiple coders (MacQueen, McLellan, Kay, & Milstein, 1998).

Concluding Thoughts

Coding, spanning from the openness of initial exploration to the selectivity of thematic crystallization, is the fulcrum upon which qualitative analysis balances. It is through the meticulous process of open, axial, and selective coding that raw data metamorphoses into structured, interpretable insights, ultimately leading to the generation of grounded theories, narratives, or themes that resonate with the depth and complexity of human experiences.

Theme Development and Categorization

In the qualitative research arena, the transformation of raw, intricate data into organized, interpretable insights is paramount. An instrumental process in this metamorphosis is the development and categorization of themes—a meticulous procedure that harnesses the essence of collected data, molding it into structured narratives that resonate with depth and meaning. This chapter delves into the labyrinth of theme development and categorization, elucidating its foundational tenets, methodologies, and significance within the framework of qualitative data analysis.

Conceptualizing Themes in Qualitative Research

Themes can be envisioned as patterns across data sets that provide meaningful response to the research question (Braun & Clarke, 2006). They are the anchors that lend qualitative data its depth, making the complex simple and the inarticulate eloquent.

Development of Themes

1. **Familiarization with the Data**: Immersion into the data through repeated reading, allowing the researcher to discern patterns, meanings, and potential points of interest (Braun & Clarke, 2006).

A cornerstone in the thematic analysis process, familiarization with data is the act of immersing oneself deeply into the raw qualitative data to gain initial understandings and insights. This foundational step precedes any formal coding or thematic categorization, ensuring a researcher's intimate acquaintance with the material (Braun & Clarke, 2006). Let's delve further into its significance, methodology, and implications.

Understanding Familiarization

Familiarization is not just about reading data, but about engaging with it. It allows researchers to recognize patterns, formulate preliminary interpretations, and even note emotions or reactions the data might evoke. Essentially, it's the process of turning data into a "known" entity (Creswell, 2013).

The Process of Familiarization

a) **Repeated Reading**: This involves reading the transcripts or data sets multiple times. Each reading enables the researcher to dive deeper, noticing nuances and subtleties that might have been overlooked initially (Miles, Huberman, & Saldaña, 2014).
b) **Annotating**: Making margin notes, highlighting intriguing or recurring points, or even noting questions or areas of confusion can be immensely helpful (Lofland et al., 2006).
c) **Listening to Recordings**: If data was collected through interviews or focus groups, re-listening to recordings can provide additional insights, capturing voice inflections or emphases that written transcripts might miss (Saldaña, 2016).
d) **Engaging with Visual Data**: For researchers working with photographs, videos, or other visual data, observing and re-observing these materials aids in interpretation (Rose, 2016).

Significance of Familiarization

- **Grounding Analysis**: Familiarization roots the subsequent analysis in the raw data, ensuring that findings are genuinely reflective of participants' experiences (Braun & Clarke, 2006).
- **Enhancing Interpretative Depth**: When researchers know their data intimately, they can make deeper, more nuanced interpretations (Richards & Morse, 2012).
- **Facilitating Efficient Coding**: A thorough familiarity aids in coding data more effectively and discerningly, as researchers can easily spot patterns or anomalies (Creswell, 2013).

Challenges and Considerations

i. **Data Overwhelm**: Especially in extensive studies, the sheer volume of data can be daunting, risking skimming over instead of genuinely engaging (Maxwell, 2012).

ii. **Potential Biases**: Familiarization might lead to early, potentially premature, interpretations which could bias subsequent analysis stages (Braun & Clarke, 2006).
iii. **Time Consumption**: Properly familiarizing oneself with data is time-intensive but is an investment in the quality of the eventual analysis (Miles, Huberman, & Saldaña, 2014).

In the orchestra of qualitative data analysis, familiarization is akin to a musician's initial scales and arpeggios—essential for tuning, understanding, and grounding subsequent performances. As the foundation of theme development, it paves the way for a nuanced, authentic analysis.

2. **Generation of Initial Codes**: Here, data is fragmented into codes that highlight interesting aspects that could form the basis of recurring themes (Saldaña, 2016).

The pivotal process of generating initial codes in qualitative research is fundamental to the inductive development of themes and subsequent in-depth analysis. The act of coding involves fragmenting the data into meaningful chunks and attaching labels to them, acting as stepping stones to understanding and representing the contained richness (Saldaña, 2016). This section elucidates the significance, process, and considerations in generating initial codes.

Understanding Initial Code Generation

Initial coding is the preliminary step of breaking down and conceptualizing raw data, representing the first layer of abstraction. At this juncture, the researcher remains close to the data, capturing its inherent nuances and complexities (Charmaz, 2014).

Process of Initial Code Generation:

a) **Data Immersion**: Before any coding, it's imperative for the researcher to be intimately familiar with the data, typically achieved by repeatedly reading or viewing the data (Braun & Clarke, 2006).
b) **Descriptive Coding**: Here, the researcher assigns basic labels to data segments, capturing the primary content or subject of the segment (Saldaña, 2016).
c) **Line-by-Line Coding**: Particularly valuable in grounded theory, this involves coding every line of data. It ensures a detailed examination and hinders premature or superficial interpretations (Charmaz, 2014).
d) **Highlighting Key Phrases or Words**: This involves pinpointing and color-coding significant phrases, words, or sentences that encapsulate core ideas (Richards & Morse, 2012).

Significance of Initial Code Generation

- **Structuring Data**: Coding compartmentalizes vast data sets, making them more manageable and navigable (Creswell, 2013).
- **Identifying Patterns**: Early coding helps discern recurring patterns, anomalies, or themes across the data set (Miles, Huberman, & Saldaña, 2014).
- **Laying Ground for Advanced Analysis**: Initial codes evolve into focused codes, categories, and eventually themes or theoretical constructs, facilitating deeper analysis (Charmaz, 2014).

Challenges and Considerations

i. **Maintaining Neutrality**: There's a risk of reading one's biases or expectations into the data, which may sway the coding process (Braun & Clarke, 2006).
ii. **Over-Coding or Under-Coding**: Researchers might either assign too many codes, muddling the clarity, or too few, missing nuances (Maxwell, 2013).
iii. **Consistency**: Ensuring consistency in code application across the dataset can be challenging, especially with large data sets or multiple coders (Saldaña, 2016).

The process of initial code generation in qualitative research is like sifting through a treasure trove. By fragmenting and labeling, researchers start discerning the contours of the landscape, paving the way for deeper exploration and richer insights.

3. **Searching for Themes**: Collating codes into potential themes, considering how they fit into broader patterns (Braun & Clarke, 2006).

The search for themes is a paramount step in the thematic analysis process, where preliminary codes coalesce into broader patterns that encapsulate core aspects of the data. As Braun and Clarke (2006) articulate, a theme is not just prevalent data, but data that captures an imperative aspect related to the research query. The journey from codes to themes requires a dynamic interplay of reflection, organization, and understanding.

The Process of Searching for Themes

a) **Reviewing Initial Codes**: Begin by revisiting initial codes, looking for patterns, overlaps, and intersections. This requires a higher level of abstraction than the initial coding stage (Braun & Clarke, 2006).
b) **Cluster-Related Codes**: Group codes that convey similar ideas, taking note of how they can combine to form a singular theme (Saldaña, 2016).
c) **Drafting Thematic Maps**: Visual aids can be instrumental. Constructing flowcharts, networks, or mind maps can reveal the relationships between codes and the potential themes they form (Maxwell, 2013).
d) **Iterative Refinement**: Themes might merge, split, or even be discarded upon repeated reviews. Iteration ensures themes genuinely represent the data's core aspects (Charmaz, 2014).

Significance of Searching for Themes

- **Data Synthesis**: Themes provide a condensed yet comprehensive understanding of large data sets, ensuring meaningful interpretations (Boyatzis, 1998).
- **Connecting the Dots**: Themes link diverse data segments, revealing an interconnected narrative or a broader picture of the studied phenomenon (Fereday & Muir-Cochrane, 2006).
- **Grounding the Analysis**: As Miles, Huberman, & Saldaña (2014) argue, themes provide a backbone structure to the analysis, anchoring subsequent interpretation stages.

Challenges and Considerations

i. **Avoiding Superficial Themes**: It's pivotal to ensure that themes are not mere data aggregations but encompass depth, richness, and research relevance (Braun & Clarke, 2006).
ii. **Maintaining Analytic Neutrality**: There's a risk of preconceived notions or biases guiding theme development, possibly overlooking divergent or contradictory patterns (Charmaz, 2014).
iii. **Ensuring Distinctiveness**: Themes should be coherent and distinct, each capturing a unique facet of the data (Maxwell, 2013).

Searching for themes is akin to refining raw materials into pure elements. It demands analytical prowess, sensitivity to data, and a judicious balance between empirical grounding and interpretative latitude. The resultant themes pave the way for insightful data interpretations, bridging the descriptive and the interpretative realms.

4. **Reviewing and Refining Themes**: This entails checking if the themes work in relation to the coded extracts and the entire data set. This stage often requires the splitting, merging, or discarding of themes (Braun & Clarke, 2006).

Reviewing and refining themes is a crucial step in ensuring the accuracy and robustness of thematic analysis. This iterative process requires the researcher to deeply engage with the data, ensuring themes genuinely encapsulate and represent the data's core insights. As advocated by Braun and Clarke (2006), this stage involves refining themes, discarding those that lack adequate support, and merging those that are too closely aligned.

The Process of Reviewing and Refining Themes

a) **First-Level Review**: Begin by examining each theme's collated data to assess its internal coherence and distinguishability from other themes (Braun & Clarke, 2006).
b) **Refinement**: Some themes may only present a superficial understanding and need to be further subdivided, while others may need to be merged or even discarded (Saldaña, 2016).
c) **Second-Level Review**: Re-examine the entire dataset for the refined themes, ensuring they are well-supported across the dataset (Braun & Clarke, 2006).
d) **Cross-Comparison**: Compare the themes against each other to ensure distinctness and avoid overlapping (Charmaz, 2014).

Significance of Reviewing and Refining Themes

- **Ensuring Authentic Representation**: Refinement ensures that themes genuinely represent participants' experiences or perspectives (Boyatzis, 1998).
- **Enhancing Analytical Depth**: A thorough review pushes the researcher to delve deeper, leading to richer, more nuanced interpretations (Maxwell, 2013).
- **Validity and Reliability**: Through iterative reviews, the researcher bolsters the validity of the themes, making the findings more reliable and trustworthy (Lincoln & Guba, 1985).

Challenges and Considerations

i. **Avoiding Data Drift**: One must ensure that themes remain grounded in the original data and do not drift too far into speculative territory (Charmaz, 2014).
ii. **Maintaining Consistency**: Especially in large datasets, ensuring consistent coding and thematic development can be challenging (Saldaña, 2016).
iii. **Managing Complexity**: Especially with rich datasets, the complexity of intertwined themes can be challenging to untangle (Miles, Huberman, & Saldaña, 2014).

The review and refinement of themes underscore the iterative and reflexive nature of qualitative analysis. It demands meticulous attention, persistent engagement, and the humility to revise one's interpretations in light of the data.

5. **Defining and Naming Themes**: This stage involves refining the specifics of each theme and the overall narrative of the analysis, resulting in clear definitions and names for each theme (Patton, 2015).

Defining and naming themes is a pivotal step in the thematic analysis process. Once themes emerge from the data, they necessitate precise definitions and evocative names that encapsulate their essence and nuance. This step is crucial for both the researcher's clarity and for effectively conveying the findings to an audience (Braun & Clarke, 2006).

Defining Themes

a) **Conceptual Clarity**: Each theme should have a clear and concise definition that delineates its scope and central idea. This ensures that the theme can be distinguished from other themes and captures a unique aspect of the data (Braun & Clarke, 2006).
b) **Ensuring Cohesion**: A theme should represent a coherent pattern in the data. The definition should ensure that all data points under a theme fit together meaningfully (Saldaña, 2016).
c) **Depth of Interpretation**: While the theme's definition should be grounded in the data, it should also provide deeper insights, going beyond the obvious to offer interpretative richness (Boyatzis, 1998).

Naming Themes

i. **Evocative Naming**: The name of a theme should be captivating and should encapsulate the theme's essence in a few words. An evocative name makes the theme memorable and gives readers an immediate sense of its core idea (Charmaz, 2014).
ii. **Precision**: While being evocative, the name should also be precise. It should avoid ambiguity and clearly represent what the theme is about (Maxwell, 2013).
iii. **Consistency**: The naming convention adopted should be consistent across all themes to ensure clarity and avoid confusion (Lincoln & Guba, 1985).

Significance of Defining and Naming Themes

- **Clarity of Thought**: Precise definitions and names ensure that the researcher has a clear understanding of what each theme represents, aiding in subsequent stages of analysis and writing (Miles, Huberman, & Saldaña, 2014).
- **Effective Communication**: Clear theme definitions and evocative names enable effective communication of findings to readers, enhancing the study's impact and understanding (Boyatzis, 1998).

Challenges and Considerations

I. **Avoiding Reductionism**: While striving for clarity and conciseness, it's crucial not to oversimplify themes, thereby losing the richness and complexity of the data (Charmaz, 2014).
II. **Balancing Creativity with Precision**: While evocative naming can make themes memorable, it's essential to ensure that creativity doesn't come at the expense of accuracy and clarity (Maxwell, 2013).

The process of defining and naming themes is both an art and a science, requiring meticulous attention to data and a dash of creativity. It is central to ensuring that the thematic analysis captures and communicates the depth and nuance of participants' experiences and perspectives.

Categorization Within Themes

Themes, while serving as a broader umbrella, often encompass within them sub-themes or categories. These categories allow for a nuanced, layered understanding of the data, catering to its multifaceted nature.

- **Role of Categories**: They enable researchers to segment themes further, adding depth and dimension to the analysis (Maxwell, 2012).

Categorization plays a crucial role in the qualitative research process. Within the broader context of themes, categories serve as essential sub-units that facilitate a more nuanced and detailed exploration of the data. By distinguishing and elaborating upon the role of categories within thematic structures, researchers can further enrich their analyses and derive more in-depth insights from their data.

A. Definition of Categories:

Categories can be seen as sub-themes or clusters of data that fit together in a meaningful way under a broader theme. They provide a more granular view of the data, allowing for a detailed exploration of specific aspects of a broader thematic idea (Miles, Huberman, & Saldaña, 2014).

B. Role of Categories in Thematic Analysis:
   - **Depth and Detail**: Categories help in delving deeper into the data, enabling the researcher to explore the intricacies and subtleties that might be overlooked if only themes were considered (Braun & Clarke, 2006).
   - **Data Organization**: They provide a systematic way of organizing data chunks, making the analysis process more structured and coherent (Saldaña, 2016).

- **Facilitates Comparison**: Categorization allows for the comparison of data across different participants, sites, or time periods, helping to identify patterns, similarities, and differences (Maxwell, 2013).
- **Bridge to Broader Themes**: Categories serve as a stepping stone to developing broader themes. They provide the foundational blocks upon which overarching thematic ideas are built (Charmaz, 2014).

C. Challenges in Category Development:

- **Over-categorization**: There's a risk of creating too many categories, leading to an overly fragmented and less coherent analysis (Miles, Huberman, & Saldaña, 2014).
- **Ambiguity**: Categories should be clearly defined to avoid overlaps or confusion. Ambiguous categorization can dilute the analytical potency of the research (Lincoln & Guba, 1985).

D. Ensuring Effective Categorization:

- **Iterative Process**: Category development should be iterative, with regular revisits to the data to refine, merge, or even discard categories as necessary (Charmaz, 2014).
- **Engaging with the Literature**: While categories should emerge from the data, engaging with existing literature can provide insights and frameworks that guide categorization (Maxwell, 2013).
- **Peer Review**: Engaging peers in the review process can provide an external perspective, helping to validate, refine, or challenge the developed categories (Lincoln & Guba, 1985).

Categorization within themes is a pivotal process in qualitative analysis. It not only provides depth and granularity but also serves as the foundation upon which broader thematic structures are constructed. As such, careful attention to the role and development of categories is paramount to ensure a robust and insightful analysis.

➢ **Establishing Hierarchies**: Often, certain themes or categories gain precedence due to their frequency, intensity, or relevance. Recognizing these hierarchies is vital to paint a comprehensive picture (Miles, Huberman, & Saldaña, 2014).

In the realm of qualitative research, establishing hierarchies in categorization is pivotal for systematically analyzing and interpreting data. Hierarchical structures allow researchers to arrange and prioritize categories within broader themes, facilitating a deeper, multilayered understanding of the researched phenomena. This approach can illuminate connections, patterns, and complexities in the data that might otherwise go unnoticed.

A. Defining Hierarchical Categorization:

Hierarchical categorization involves organizing categories into levels of importance or relevance, often starting from general categories and drilling down to more specific sub-categories (Smith, 2018).

B. Importance of Establishing Hierarchies:

- **Organizational Clarity**: Hierarchies offer a structured approach to data analysis, aiding in clarity and organization, ensuring that data is analyzed systematically (Patton, 2015).

- **Depth of Analysis**: Hierarchical categorization allows for a layered examination of data, providing both a bird's-eye view of broad themes and an in-depth look at specific data points (Bryman, 2016).
- **Identification of Patterns**: Through hierarchies, researchers can more readily discern patterns, trends, and relationships among categories and sub-categories (Richards, 2009).

C. Steps in Establishing Hierarchical Categorization:
- **Broad Categorization**: Begin by identifying overarching themes or categories present in the data.
- **Sub-categorization**: Delve deeper into each broad category to identify specific sub-categories or patterns (Smith, 2018).
- **Prioritization**: Assign levels of importance to categories and sub-categories based on their relevance to the research questions or observed patterns in the data (Patton, 2015).
- **Validation**: Continuously revisit and refine the hierarchical structure, ensuring its validity and appropriateness to the data and research goals (Bryman, 2016).

D. Challenges in Establishing Hierarchies:
- **Over-complication**: While hierarchies provide structure, there's a risk of creating overly complex structures that may obfuscate rather than clarify (Richards, 2009).
- **Subjectivity**: The process can be influenced by the researcher's biases or preconceptions, potentially leading to skewed hierarchies (Smith, 2018).

E. Mitigating Challenges:
- **Iterative Process**: Continuously refining and revisiting hierarchical structures helps in ensuring their appropriateness (Patton, 2015).
- **Peer Review**: Engaging peers or external experts can offer fresh perspectives and critique, enhancing the credibility and validity of the established hierarchies (Bryman, 2016).

Establishing hierarchies in categorization within themes is fundamental in qualitative data analysis. By offering both structure and depth, this approach enables a nuanced understanding of data, allowing for richer insights and more informed interpretations.

Challenges in Theme Development and Categorization

1. **Overlapping Themes**: Themes are not always mutually exclusive. Delineating them while acknowledging overlaps can be intricate (Richards & Morse, 2012).

Overlapping themes in qualitative research pose both challenges and opportunities for researchers. The emergence of overlapping themes is not uncommon given the rich, intricate, and multifaceted nature of qualitative data. Recognizing, addressing, and interpreting these overlaps can be crucial for ensuring the integrity and depth of analysis.

A. Understanding Overlapping Themes:

Overlapping themes occur when two or more themes encompass similar, interrelated, or intertwined data points, or when one theme seems to be a subset of another (Braun & Clarke, 2012).

B. Causes of Overlapping Themes:
   - **Nature of Qualitative Data**: Given that qualitative data often captures the complexities and nuances of human experiences, overlaps are almost inevitable (Saldana, 2015).
   - **Researcher Interpretation**: Themes are interpretive constructs, and different researchers might see connections and separations in data differently (Maxwell, 2012).

C. Challenges Presented by Overlapping Themes:
   - **Analytical Ambiguity**: Overlaps can make it difficult to distinguish between themes, potentially causing analytical confusion (Braun & Clarke, 2012).
   - **Interpretative Dilution**: Overlapping themes might dilute the specificity and depth of each theme, hindering the development of rich insights (Saldana, 2015).
   - **Repetitive Analysis**: Themes that overlap extensively might lead to redundant analyses where the same data is analyzed multiple times under different themes (Maxwell, 2012).

D. Addressing Overlapping Themes:
   - **Re-Evaluation of Themes**: Continuously revisiting and refining themes can help in clearly demarcating them, ensuring each theme's uniqueness and significance (Braun & Clarke, 2012).
   - **Integration or Separation**: Depending on the research objectives and the nature of the overlap, researchers can either integrate overlapping themes into a single, more encompassing theme or further differentiate them based on nuanced distinctions (Saldana, 2015).
   - **Seek External Feedback**: Peer debriefing or expert consultation can offer fresh perspectives on overlapping themes, providing clarity and direction (Maxwell, 2012).

E. Embracing Overlaps as Opportunities:

While overlapping themes present challenges, they also offer opportunities. They can:

- Highlight the interconnectedness of certain experiences or phenomena.
- Indicate the robustness or significance of particular data points recurring under multiple themes.
- Prompt deeper reflexivity on the part of the researcher, leading to richer analytical insights (Braun & Clarke, 2012).

While overlapping themes in qualitative research can be challenging, they are an integral part of the analytical journey. Addressing them thoughtfully and systematically ensures that the analysis remains robust, insightful, and grounded in the data.

2. **Subjectivity in Theme Perception**: The perception of what constitutes a theme can vary amongst researchers, making the process susceptible to bias (Braun & Clarke, 2006).

Qualitative research inherently relies on the subjective interpretations of researchers. While this subjectivity is often celebrated for its ability to provide depth and richness to analyses, it can also present challenges, especially when it comes to the development and categorization of themes. The subjectivity in theme perception is multifaceted and deserves careful attention.

A. Understanding Subjectivity in Theme Perception:

Theme perception in qualitative analysis is heavily influenced by the researcher's personal biases, experiences, beliefs, and theoretical inclinations (Patton, 2015). This subjectivity can affect every step of theme development, from initial coding to final theme naming and representation.

B. Causes of Subjective Theme Perception:
   - **Personal Biases**: Every researcher brings a set of personal biases to the research, influenced by their background, experiences, and values (Maxwell, 2013).
   - **Theoretical Orientations**: The theoretical framework or epistemological stance adopted can influence the perception of data and subsequently the themes (Charmaz, 2014).
   - **Data Immersion**: Prolonged immersion in data, while often beneficial, can sometimes lead to 'blind spots' or overly personalized interpretations (Saldana, 2016).

C. Challenges Presented by Subjective Theme Perception:
   - **Inconsistency in Theme Identification**: Different researchers might identify different themes from the same dataset, leading to concerns about the reliability and validity of the analysis (Patton, 2015).
   - **Overemphasis or Neglect**: Subjectivity can cause certain data points or patterns to be overly emphasized, while others might be neglected or overlooked (Maxwell, 2013).
   - **The Risk of Misrepresentation**: Themes constructed primarily based on a researcher's biases might not truly represent participants' experiences or the data's essence (Charmaz, 2014).

D. Addressing Subjectivity in Theme Perception:
   - **Reflexivity**: Engaging in continuous reflexivity, where researchers critically reflect on their biases and influence on the research, can help in recognizing and mitigating undue subjectivity (Saldana, 2016).
   - **Triangulation**: Employing multiple methods or data sources can help validate themes and reduce the impact of individual biases (Patton, 2015).
   - **Peer Debriefing**: Engaging peers to review and provide feedback on identified themes can ensure a more balanced and less subjective analysis (Maxwell, 2013).
   - **Iterative Analysis**: Revisiting data and themes multiple times throughout the analysis can lead to a more nuanced and balanced understanding (Charmaz, 2014).

E. Embracing Subjectivity as a Strength:

While undue subjectivity can be problematic, a balanced degree of subjectivity allows for a deep, empathetic, and nuanced analysis. Recognizing and harnessing this strength can lead to richer insights and a more holistic understanding of the data (Saldana, 2016).

Subjectivity, while a challenge in theme development and categorization, is an intrinsic part of qualitative research. By acknowledging, understanding, and addressing this subjectivity, researchers can ensure that their analysis is both richly insightful and methodologically robust.

3. **Data Overload**: With vast amounts of qualitative data, the risk of becoming overwhelmed and losing sight of core themes is real (Creswell, 2013).

In the scope of qualitative research, data richness is often celebrated for its depth and holistic insights. However, with this depth comes the challenge of data overload, which can be a considerable hurdle in theme development and categorization. Navigating vast amounts of qualitative data requires methodological finesse, clarity of purpose, and robust analytical strategies.

A. Understanding Data Overload:

Data overload refers to the situation where researchers are inundated with more data than they can effectively analyze, interpret, or make sense of (Bazeley, 2013). It's particularly common in qualitative research where extensive interviews, observations, and document analyses produce vast textual or visual data.

B. Causes of Data Overload:
   - **Expansive Data Collection**: Researchers may collect more data than necessary, either due to over-enthusiastic data gathering or fearing that they might miss out on crucial information (Miles et al., 2014).
   - **Lack of Focus**: Without clear research questions or objectives, data collection can become undirected and overly general (Saldaña, 2016).
   - **Complexity of Human Experience**: Qualitative research often delves into intricate human experiences, generating multifaceted narratives and perspectives (Richards, 2015).

C. Implications of Data Overload:
   - **Analysis Paralysis**: Researchers may feel overwhelmed and find it challenging to start the data analysis process, leading to delays and potential burnout (Bazeley, 2013).
   - **Missed Patterns**: Important patterns or themes may be overlooked if the data volume is too vast to manage effectively (Miles et al., 2014).
   - **Surface-Level Analysis**: Instead of a deep dive into data, researchers might opt for a more superficial analysis, missing out on depth and nuance (Saldaña, 2016).

D. Strategies to Address Data Overload:
   - **Focused Data Collection**: Clear research questions and a defined scope can ensure that data collection remains focused and relevant (Richards, 2015).
   - **Iterative Analysis**: Engage in continuous, iterative rounds of data analysis, starting early in the data collection process. This can help in managing data incrementally rather than facing a massive dataset at the end (Bazeley, 2013).
   - **Employ Digital Tools**: Software like NVivo or ATLAS.ti can assist researchers in managing, coding, and organizing vast amounts of qualitative data (Miles et al., 2014).
   - **Team-Based Analysis**: Engaging multiple researchers in the analysis process can help distribute the workload and bring diverse perspectives to data interpretation (Saldaña, 2016).

E. Embracing the Richness of Data:

While data overload is undeniably challenging, it's also a testament to the richness of qualitative inquiry. With methodological rigor and strategic management, the voluminous data can be transformed into insightful narratives and deep understandings (Richards, 2015).

While data overload poses significant challenges in theme development and categorization, it also underscores the depth and richness of qualitative research. Addressing this challenge head-on, with strategic and methodological approaches, ensures that the data's wealth is fully harnessed.

The Significance of Theme Development and Categorization

Themes are the lifeblood of qualitative research. They:

1. **Bridge Raw Data and Interpretation**: Themes provide a structured pathway from raw, unprocessed data to cogent, interpretable findings (Saldaña, 2016).

In the vast ocean of qualitative research, theme development and categorization serve as navigational tools that help chart the course from raw data to meaningful interpretation. By organizing and distilling data into coherent themes and categories, researchers can discern patterns, develop theories, and provide rich insights that would be otherwise hidden within the vast and intricate expanse of raw narratives and observations (Boyatzis, 1998).

A. Navigating the Complexity of Raw Data:

Qualitative research often results in voluminous amounts of data. From in-depth interviews, focus group discussions, observational notes, to other forms of evidence, raw data can be multifaceted and overwhelming. Without an organizing principle, such as thematization, making sense of this data becomes an insurmountable challenge (Braun & Clarke, 2006).

B. Themes as Interpretative Lenses:

Themes act as lenses that bring into focus specific aspects of the data. They provide interpretative frames, allowing researchers to move beyond mere descriptions and engage in a deeper analysis of underlying meanings, relationships, patterns, and structures within the data (Ryan & Bernard, 2003).

C. The Process of Bridging Data and Interpretation:
   - **Data Reduction**: As the first step, theme development and categorization enable data reduction, distilling extensive data into manageable chunks without losing its essence (Miles & Huberman, 1994).
   - **Data Display**: Themes and categories serve as a framework for displaying data in a structured format, making it easier to discern patterns and relationships (Miles & Huberman, 1994).
   - **Drawing Conclusions**: Organized data, via themes and categories, facilitates the drawing of conclusions and verification, providing a solid foundation upon which interpretations are built (Braun & Clarke, 2006).

D. The Interplay of Deduction and Induction:

Theme development and categorization often involve a dance between deductive and inductive reasoning. While existing theories and literature may guide the initial phase (deductive), the data itself can suggest new themes or modify existing ones in an emergent, inductive process (Fereday & Muir-Cochrane, 2006).

E. Enhancing Research Credibility:

An organized and systematic approach to theme development can enhance the credibility of research findings. By providing a transparent pathway from raw data to interpretation, theme development and categorization ensure that findings are grounded in the data and not merely subjective interpretations (Lincoln & Guba, 1985).

In essence, theme development and categorization are pivotal in qualitative research, bridging the chasm between raw data and meaningful interpretation. By systematizing and structuring data, they ensure that the rich narratives and observations are transformed into coherent, insightful, and credible research findings.

2. **Enable Broad-to-Narrow Focus**: Themes allow researchers to funnel from broader patterns to specific insights, providing a layered understanding (Patton, 2015).

The process of theme development and categorization in qualitative research is instrumental in facilitating a broad-to-narrow focus, allowing researchers to progressively delve deeper into the intricacies of their data while maintaining a coherent narrative structure. This transformative process bridges the high-level narratives with the minute specifics, ensuring comprehensive understanding and detailed insight (Thomas, 2006).

A. Structuring the Data Landscape:

The inherent diversity and depth of qualitative data can be likened to a vast landscape with varying terrains. Themes and categories act as markers or guideposts, helping researchers navigate from the broader contours of this landscape to its intricate details, ensuring that no significant aspect is overlooked (Saldaña, 2015).

B. Hierarchical Depth in Analysis:
   - **Macro-Themes**: These represent the broad narratives or primary patterns that emerge from the data. Macro-themes capture the overarching stories or discourses and set the stage for deeper exploration (Braun & Clarke, 2006).
   - **Sub-Themes**: As the analysis progresses, researchers break down macro-themes into sub-themes. These are more specific, offering a finer granularity of understanding and capturing the multiple dimensions within a broader theme (Attride-Stirling, 2001).

C. Progressive Narrowing: A Funnel Approach:

Theme development and categorization can be visualized as a funnel. Beginning with a wide aperture (broad themes), the research progressively narrows down, focusing on specific nuances, relationships, and patterns, eventually culminating in a concentrated understanding of the phenomenon under study (Maxwell, 2012).

D. Balance Between Breadth and Depth:

A broad-to-narrow focus ensures a balanced approach to data analysis. While the breadth captures the holistic view and overarching patterns, the depth dives into specificities, capturing the richness, complexity, and multi-dimensionality of qualitative data (Creswell & Poth, 2017).

E. Facilitating Holistic Data Representation:

This structured approach aids in the presentation and representation of findings. Broad themes provide a coherent narrative structure, while the detailed sub-themes and categories enrich the narrative with specific examples, voices, and contexts, making the findings both compelling and comprehensive (Ritchie & Lewis, 2003).

The broad-to-narrow focus facilitated by theme development and categorization is pivotal for the depth, richness, and coherence of qualitative analysis. It ensures that while the larger narrative arcs are honored, the intricate details and nuances that provide depth and richness to the research are not lost in the vast expanse of data.

3. **Aid in Theory Building**: By discerning patterns and relationships within the data, themes aid in the construction of theories, models, or frameworks (Charmaz, 2014).

Theme development and categorization not only organize data for interpretation but also play a pivotal role in the construction of theory within qualitative research. Grounding theories in empirical data requires systematic processes that can navigate the intricate relationships between varied data points, leading to conceptualizations that resonate beyond individual cases.

A. Grounded Theory and Theme Development:

Grounded theory, as articulated by Glaser and Strauss (1967), underscores the significance of constructing theory from data. This iterative process of moving between data and analysis relies heavily on theme development and categorization to draw out patterns, allowing researchers to move beyond mere description to theoretical articulation.

B. Themes as Building Blocks:

Themes, in essence, can be viewed as building blocks or foundational stones that anchor theoretical frameworks. They provide the raw material, the basic constructs, upon which theoretical propositions can be built (Charmaz, 2014).

C. Abstraction and Conceptualization:

Theme development fosters abstraction, a process vital for theory building. As themes are refined and categories emerge, researchers begin to see connections, oppositions, and nuances that can be conceptualized into theoretical propositions (Corbin & Strauss, 2008).

D. Validating and Refining Existing Theories:

Themes and categories do not merely aid in constructing new theories; they also help in validating or refining existing ones. The convergence or divergence of data from established theoretical frameworks provides avenues for academic discourse and the evolution of thought (Patton, 2002).

E. Providing a Comparative Lens:

The organization of data into themes offers a structured comparative lens, enabling researchers to juxtapose their findings against other studies, theoretical postulates, or contrasting data sets. This comparative stance is instrumental in theory elaboration and modification (Eisenhardt, 1989).

F. Narratives to Conceptual Models:

Themes and their interrelationships provide narratives which, when mapped, can lead to the development of conceptual models. These models visually represent the theory, allowing for easier communication, critique, and dissemination (Miles, Huberman, & Saldaña, 2014).

G. Fostering Replicability and Extension:

The systematic approach to theme development and categorization in qualitative research allows other researchers to trace the theoretical journey. This transparency fosters replicability and provides avenues for others to extend the theory in new directions or contexts (Yin, 2014).

The process of theme development and categorization in qualitative research is not an end in itself but a means to an end - the construction, validation, or refinement of theory. It ensures that theoretical constructs are deeply rooted in empirical realities, thus enhancing the relevance, resonance, and rigor of qualitative inquiries.

Finally, the art of theme development and categorization is pivotal to qualitative research. As the data unravels, themes emerge as the harbingers of meaning, crafting a narrative that is both insightful and impactful. As researchers, mastering this art is vital, for in themes, we find the soul of our data.

Use of Qualitative Software: NVivo, Atlas.ti, and MAXQDA

The last few decades have witnessed an exponential surge in qualitative research. To manage the complexities of such research, especially with increasing volumes of data, specialized software tools have been developed. These tools, commonly referred to as Computer-Assisted Qualitative Data Analysis Software (CAQDAS), offer a systematic approach to organizing, coding, and interpreting data. Three of the most renowned CAQDAS are NVivo, Atlas.ti, and MAXQDA. This chapter delves into these software applications, exploring their relevance and functionalities in the realm of qualitative data analysis.

1. Evolution of CAQDAS:

With the evolution of technology, qualitative researchers realized the potential of using computer software for data analysis. Earlier attempts were limited to text retrieval and basic coding. However, the inception of sophisticated software like NVivo, Atlas.ti, and MAXQDA has revolutionized the way qualitative data is handled, analyzed, and interpreted (Fielding & Lee, 1998).

Evolution of Computer-Assisted Qualitative Data Analysis Software (CAQDAS)

The qualitative research landscape has seen myriad changes over the decades, and one of the most noteworthy shifts is the emergence and development of Computer-Assisted Qualitative Data Analysis Software (CAQDAS). The inception of CAQDAS has reshaped the methodologies of qualitative research, making data handling, organization, and interpretation more structured and efficient.

A. Early Days of Qualitative Data Analysis:

Historically, qualitative researchers relied primarily on manual methods for data analysis. These early practices involved extensive reading and re-reading of data, manual coding, annotation, and memoing. While this approach had its merits, allowing researchers a deep immersion in their data, it was also labor-intensive and less efficient, especially when dealing with vast data sets (Seidel, 1991).

B. Emergence of CAQDAS:

The advent of computers and the digital age in the late 20th century brought about the first iterations of CAQDAS. Software like "The Ethnograph" and "NUD*IST" (Non-numerical Unstructured Data * Indexing Searching and Theorizing) marked the beginning of this revolution, primarily offering tools for data storage and simple coding (Weitzman & Miles, 1995). These early programs provided an electronic alternative to the manual methods, yet still required researchers to maintain a hands-on approach to their data.

C. The Expansion and Sophistication of CAQDAS:

The early 2000s witnessed a rapid growth in the capabilities and sophistication of CAQDAS. Software such as NVivo, Atlas.ti, and MAXQDA emerged as leaders in the field, offering a wide array of functionalities. They moved beyond basic coding to provide tools for advanced data visualization, intricate relationship mapping, and integration with various data sources, thereby adding depth and nuance to the qualitative research process (Bazeley, 2007).

D. The Impetus for Development:

Several factors drove the evolution of CAQDAS. Firstly, the growth in the volume of data — from text to multimedia — necessitated more efficient tools for management and analysis. Secondly, the increasing interdisciplinary nature of qualitative research called for flexible and multifaceted software that could cater to diverse research needs. Additionally, the quest for methodological rigor and reproducibility in qualitative research further underscored the value of systematic and traceable data analysis, which CAQDAS could facilitate (Fielding, 2012).

E. Current State and Future Prospects:

Modern CAQDAS tools, especially NVivo, Atlas.ti, and MAXQDA, are not merely data analysis software but have evolved into comprehensive research management tools. They support various stages of the research process, from literature reviews to data visualization and report generation. With the advent of artificial intelligence and machine learning, it is anticipated that future iterations of CAQDAS will incorporate more advanced analytical tools, potentially bridging the gap between qualitative and quantitative analysis methods (Paulus & Lester, 2016).

2. NVivo: An Overview:

Developed by QSR International, NVivo is among the most popular qualitative data analysis software. Designed to facilitate the organization and analysis of unstructured data, NVivo provides tools for coding, node categorization, and theorization (Bazeley & Jackson, 2013). Its visual data modeling capabilities, coupled with integration possibilities with various data sources, make it a favored choice for many researchers.

NVivo: An Overview

NVivo, developed by QSR International, stands as one of the most renowned and widely used Computer-Assisted Qualitative Data Analysis Software (CAQDAS) across the globe. Its iterative development over the years has firmly established its place in the toolbox of qualitative researchers, assisting in the systematic organization, analysis, and visualization of qualitative data.

A. Historical Development:

The genesis of NVivo, originally termed "NUD*IST" (Non-numerical Unstructured Data * Indexing Searching and Theorizing), was rooted in the late 20th century. Over subsequent decades, as the software underwent several modifications, the name transitioned to NVivo, reflecting the software's capacity for "Non-numerical and Visual data" (Bazeley & Jackson, 2013).

B. Core Features:
- **Coding**: At the heart of NVivo lies its powerful coding capabilities. Researchers can assign codes to chunks of text, images, videos, or other data forms. These codes serve as descriptive or thematic tags that assist in data organization and retrieval (Welsh, 2002).
- **Data Visualization**: NVivo offers a suite of visualization tools, from word clouds and hierarchical charts to mind maps and sociograms. Such tools not only elucidate patterns within data but also foster deeper engagement and interpretation (Bazeley, 2007).
- **Query Tools**: Using NVivo's advanced query functions, researchers can probe their data, posing intricate questions and unveiling latent themes and patterns (Johnston, 2006).
- **Memoing**: An essential feature for qualitative researchers, NVivo's memoing allows for the jotting down of analytical notes or reflections connected to specific pieces of data or codes (Richards, 1999).

C. Integrative Capabilities:

One of the strengths of NVivo lies in its ability to integrate various data forms. From textual and tabular data (e.g., interviews, field notes, surveys) to multimedia forms (e.g., images, videos, audio recordings), NVivo facilitates a seamless amalgamation, enabling a holistic analysis approach (Bazeley & Jackson, 2013).

D. Collaboration and Teamwork:

NVivo's project sharing features support collaborative research endeavors. Multiple researchers can concurrently access and work on a shared project, making it apt for large-scale projects or when interdisciplinary expertise is required (Silver & Woolf, 2015).

E. Interfacing with Other Software:

NVivo's flexibility extends to its compatibility with various other software, including word processors, spreadsheet applications, and reference management tools, allowing for a streamlined research workflow (Paulus, 2010).

3. Atlas.ti: A Comprehensive Tool:

Atlas.ti boasts of an array of functions ranging from coding to graphical visualization of data. It emphasizes the hermeneutic unit, a database of documents, codes, and memos, offering a holistic approach to qualitative analysis (Friese, 2014). Its network views allow for intricate relationship mappings, ensuring a nuanced understanding of data patterns.

Atlas.ti has firmly etched its reputation as one of the vanguards of Computer-Assisted Qualitative Data Analysis Software (CAQDAS). Its innovative features combined with user-friendly interfaces make it a preferred choice for many qualitative researchers seeking to perform intricate data analysis in a systematic manner.

A. Historical Background:

Atlas.ti was initiated in the late 1980s by Dr. Thomas Muhr. It began as a tool for grounded theory researchers and gradually evolved to cater to a variety of qualitative research paradigms, witnessing a series of technological upgrades and expansions in its capabilities (Muhr, 1991).

B. Distinctive Features:

- **Coding System**: Atlas.ti offers a flexible and powerful coding system, permitting researchers to label and categorize fragments of data. The codes can be linked, merged, and grouped, enabling researchers to maneuver their analysis dynamically (Friese, 2019).
- **Visualization**: The software is known for its 'Network Views,' enabling researchers to visually map out relationships among codes, memos, and quotations. This visualization aids in recognizing emerging patterns and themes (Friese, 2012).
- **Queries**: Atlas.ti's query tool, known as the 'Query Tool', assists researchers in interrogating their dataset, searching for specific codes or combinations, and subsequently aiding in the development of themes and theories (Woods, Paulus, Atkins, & Macklin, 2016).

- **Memo Writing**: As with most CAQDAS, Atlas.ti recognizes the importance of memoing in qualitative research, providing an integrated memo manager where researchers can pen down reflexive notes, insights, or analytical interpretations (Paulus & Lester, 2016).

C. Multimedia Integration:

Atlas.ti extends beyond textual data, offering integrated features to analyze multimedia data, including videos, images, and audio. These capabilities allow for the holistic study of various data types in tandem (Silver & Patashnick, 2011).

D. Collaboration Capabilities:

The software supports team-based qualitative research, where projects can be split, merged, and shared among researchers. This feature ensures continuity in large-scale projects, ensuring data integrity and seamless collaboration (MacMillan, 2014).

E. Interfacing with Other Tools:

Atlas.ti is designed to smoothly interact with other software platforms, facilitating easy data import/export from spreadsheets, word processors, and even other CAQDAS, ensuring a flexible and adaptable research workflow (Silver, 2014).

4. MAXQDA: Bridging Qualitative and Quantitative:

MAXQDA stands out for its ability to manage, analyze, and mix both qualitative and quantitative data. From interactive data explorations to statistical analysis, MAXQDA offers tools that cater to a broad spectrum of research needs (Kuckartz, 2014). Its visual tools, like the Code Map and Document Portrait, present innovative ways of interpreting data.

In the constellation of Computer-Assisted Qualitative Data Analysis Software (CAQDAS), MAXQDA stands out as a potent tool that not only facilitates rigorous qualitative analysis but also seamlessly integrates quantitative methods, thereby pioneering a mixed methods approach.

A. Historical Context:

The foundation of MAXQDA can be traced back to the late 1980s in Germany. Originally named MAX, it was one of the earliest CAQDAS tools. Over the years, the software has seen extensive revisions and enhancements, keeping pace with evolving research needs and technological advancements (Kuckartz, 1995).

B. Features of MAXQDA:

- **Incorporation of Quantitative Analysis**: What makes MAXQDA distinctive is its ability to work with qualitative data while offering quantitative analytical tools. This dual capacity ensures a holistic and nuanced analysis, blending the depth of qualitative findings with the statistical rigor of quantitative data (Schreier, 2012).

- **Visual Tools**: MAXQDA offers a suite of visual tools such as Code Maps, Document Portraits, and Code Relations Browser, enabling researchers to visualize data patterns, interrelationships, and emergent themes (Dresing, 2015).
- **Mixed Methods**: The software facilitates mixed methods research by allowing code frequencies, crosstabulations, and exportation of variables and codes for statistical analysis in tools like SPSS (Creswell, 2018).
- **Advanced Coding**: Like most CAQDAS, MAXQDA offers a powerful coding system, but it further augments this by allowing color-coded, symbolic, and emoticon-based coding, making data organization intuitive (Kuckartz, 2018).

C. Multimedia Integration:

True to contemporary research needs, MAXQDA supports multimedia data. Analysis of images, videos, and audio files is streamlined and integrated with textual data, paving the way for multifaceted insights (Evers & Silver, 2019).

D. Flexibility and Integration:

MAXQDA's design ethos emphasizes adaptability. It is amenable to a wide range of research methodologies and is designed to be compatible with other software, ensuring a smooth and integrative research process (Johnston, 2014).

E. Collaboration Capabilities:

Reflecting the collaborative nature of contemporary research, MAXQDA has built-in tools for team-based qualitative analysis, ensuring coherence, transparency, and shared analytical rigor among team members (Bazeley & Jackson, 2013).

5. Advantages of Using CAQDAS:

- **Efficiency and Organization**: CAQDAS offers efficient data management, especially when dealing with large datasets. It allows quick retrievals, automated coding, and systematic storage (Welsh, 2002).

Computer-Assisted Qualitative Data Analysis Software (CAQDAS) has transformed the landscape of qualitative research, offering advantages that transcend manual analytical methods. One of the pivotal advantages of CAQDAS is its contribution to efficiency and organization in the research process.

A. Speed of Data Processing:

CAQDAS tools facilitate faster data processing, as researchers can swiftly code, categorize, and re-categorize data segments. As Bazeley (2007) suggests, the speed offered by CAQDAS allows researchers to allocate more time to deeper interpretative work rather than administrative tasks.

B. Centralized Data Repository:

The ability of CAQDAS to act as a central hub for diverse data forms—including text, images, videos, and audio—ensures that data is systematically stored and easily accessible. This consolidation promotes organized research processes and ensures that nothing is overlooked (Silver & Lewins, 2014).

C. Streamlined Coding Process:

Coding, a fundamental step in qualitative analysis, becomes more systematic with CAQDAS. Tools like NVivo, Atlas.ti, and MAXQDA offer drag-and-drop features, code hierarchies, and code relationships, facilitating a more organized and efficient coding process (Paulus, Lester, & Dempster, 2014).

D. Advanced Search Capabilities:

The advanced search functionalities embedded in CAQDAS enable researchers to effortlessly locate specific data segments, codes, or themes. Such features expedite the iterative process of revisiting and reflecting upon data during analysis (Hwang, 2008).

E. Visualization Tools:

CAQDAS platforms often come with tools that allow for the visualization of data, be it through code clouds, relationship maps, or thematic networks. These visual tools not only make data more digestible but also help in discerning patterns and relationships more efficiently (Gibbs, 2002).

F. Version Control and Audit Trails:

Most CAQDAS tools offer features that keep track of changes, decisions, and iterations. This transparency not only contributes to the rigor and credibility of the research but also allows for efficient backtracking and revision when necessary (Johnston, 2006).

- **Enhanced Rigor**: The systematic approach in CAQDAS contributes to the rigor and replicability of the research. It provides a clear audit trail, facilitating peer reviews and validation (Paulus, Lester, & Dempster, 2014).

The rigorous nature of qualitative research is often a point of contention, especially when juxtaposed with quantitative paradigms. One of the pronounced advantages of employing Computer-Assisted Qualitative Data Analysis Software (CAQDAS) in qualitative studies is the enhanced rigor it offers.

A. Transparency and Traceability:

CAQDAS systems provide a clear audit trail, documenting each step of the data analysis process, from initial coding to theme generation. This transparency in analytical procedures ensures that other researchers can understand, assess, and even replicate the analysis process, fostering trustworthiness (Sinkovics & Alfoldi, 2012).

B. Consistency in Coding:

The automated nature of coding in CAQDAS ensures that researchers apply codes consistently across large datasets. Unlike manual coding, where fatigue or oversight might lead to inconsistencies, CAQDAS offers tools to cross-check and validate coding, leading to a more reliable analysis (Welsh, 2002).

C. Comprehensive Data Engagement:

CAQDAS platforms enable researchers to engage deeply with their data, facilitating iterative querying, and exploring of the dataset. Such immersive engagement contributes to a richer, more nuanced understanding, thereby enhancing analytical depth and rigor (Bryman, 2008).

D. Methodological Reflexivity:

With tools that allow for annotations, memos, and reflection notes, CAQDAS platforms encourage researchers to continually engage in methodological reflexivity, ensuring that the analysis remains grounded and that researchers are cognizant of their biases and influences on the interpretation (Richards, 2005).

E. Enhanced Data Validation:

Many CAQDAS tools facilitate processes such as triangulation, member checking, and peer debriefing digitally, enabling researchers to validate their findings more systematically and rigorously (Fielding & Lee, 1998).

F. Frameworks for Complex Analysis:

CAQDAS offers researchers the tools to undertake more complex forms of analysis, such as matrix coding, cross-case analysis, or hierarchical theme structuring. These advanced analytical capabilities ensure a more rigorous engagement with data, leading to robust findings (Lewins & Silver, 2007).

- **Visual Representation**: Tools like project maps, network views, and graphs enable visualization of data, which can be crucial for data interpretation and theory development (Silver & Lewins, 2014).

The representation of qualitative data has seen a revolutionary shift with the incorporation of Computer-Assisted Qualitative Data Analysis Software (CAQDAS). A salient advantage of these digital tools is their capability for sophisticated visual representation of data.

A. Mapping Complex Relationships:

CAQDAS offers tools to create conceptual frameworks, maps, or models. These visual tools are invaluable in discerning and representing intricate relationships between codes, themes, and overarching categories (Basit, 2003).

B. Enhancing Data Comprehension:

Visual tools, such as network views or cluster analysis, help in simplifying the data's complexity, making it more digestible, and providing a holistic view of the dataset (Bazeley & Jackson, 2013).

C. Facilitating Collaborative Analysis:

Visual outputs, including charts, graphs, and models, make it easier to communicate findings to a team or an external audience, fostering a collaborative analytical process (Paulus et al., 2017).

D. Dynamic Interaction with Data:

Interactive visuals allow researchers to engage with their data dynamically. For instance, adjusting a node in a network view in NVivo might display emergent patterns or hidden connections, offering a deeper analytical perspective (Kaefer et al., 2015).

E. Supporting Theory Building:

Graphical representations act as scaffolding in the theory-building process. They offer visual aids in understanding emergent patterns, which can be pivotal in developing or refining theoretical frameworks (Coffey & Atkinson, 1996).

F. Data Presentation for Publication:

Visual outputs generated through CAQDAS can be incorporated directly into academic publications, reports, or presentations, enhancing the accessibility and comprehension of the research findings for diverse audiences (Saldaña, 2015).

6. Challenges and Criticisms:

While CAQDAS offers numerous advantages, they are not devoid of criticism. Concerns about overshadowing the human analytical element, steep learning curves, and over-reliance on the software are prevalent (Hesse-Biber, 2011). Additionally, the interpretation is still a deeply human endeavor, and software can't replace a researcher's insight and intuition.

In terms of data analysis and interpretation, especially in qualitative research, is fraught with challenges. Despite the many advances in methodological approaches and the advent of CAQDAS, researchers must navigate a myriad of potential pitfalls. Below is an exploration of some of the main challenges and criticisms frequently encountered in this phase of research.

A. Subjectivity and Bias:

Qualitative research, by its nature, involves a degree of subjectivity. The researcher's personal biases, values, and experiences might inadvertently shape the analysis process and the interpretation of results (Creswell, 2013).

B. Overwhelming Volume of Data:

Qualitative research often yields voluminous raw data, which can be daunting to sift through and analyze methodically. The sheer volume can sometimes result in essential data being overlooked or under-analyzed (Braun & Clarke, 2006).

C. Over-reliance on CAQDAS:

While software like NVivo or Atlas.ti can be invaluable, there's a criticism that over-reliance on them can cause researchers to distance themselves from their data. The software should assist, not replace, manual engagement with data (Hammersley, 2010).

D. Difficulty in Establishing Causality:

Qualitative data often provide rich, contextual insights. However, unlike quantitative research, it can be challenging to establish clear causal relationships based on qualitative findings alone (Maxwell, 2004).

E. Reproducibility and Verification:

The nuanced and interpretative nature of qualitative research means that it may not always be reproducible in the same way that quantitative research often aims to be. Different researchers might interpret the same dataset differently (Seale, 1999).

F. Challenges in Generalizing Findings:

Given the often-small sample sizes and specific contexts within which qualitative research operates, there's a challenge in generalizing findings to broader populations or different settings (Flyvbjerg, 2006).

G. Ethical Challenges:

During data analysis, especially when handling sensitive data or when participants' identities might be deducible, ethical challenges can arise. Ensuring participant confidentiality during analysis is paramount (Orb et al., 2000).

NVivo, Atlas.ti, and MAXQDA have indisputably advanced qualitative research methodology. While they offer systematic ways of dealing with data, it's imperative for researchers to use them as tools and not as replacements for human analytical capabilities. The effectiveness of these tools is, ultimately, contingent upon the skill and wisdom of the researcher wielding them.

Ensuring Trustworthiness and Credibility

Trustworthiness and credibility are paramount concerns when conducting and presenting qualitative research. Ensuring these attributes often necessitates distinct considerations compared to the quantitative realm, where concepts of reliability and validity dominate. The qualitative researcher aims to present findings that are both credible to those who participated in the study and can be trusted by the reader to be a genuine representation of the phenomenon under study. Here, we delve into the strategies and considerations crucial for ensuring trustworthiness and credibility in qualitative data analysis and interpretation.

I. Triangulation:

Triangulation involves using multiple methods, data sources, or researchers to study the same phenomenon. By examining a research question from multiple angles, researchers can increase confidence in their findings (Denzin, 1978).

Triangulation, within the context of qualitative research, refers to the utilization of multiple methods, data sources, or investigators to gather data on a singular phenomenon or research question. Rooted in navigational and surveying techniques, the term was adapted into the social sciences as a metaphor to convey the idea of establishing truth (Webb et al., 1966). The central tenet is that looking at a phenomenon from multiple perspectives can provide a more comprehensive, and thus more accurate, view.

1. Types of Triangulation:
   - **Methodological Triangulation**: This entails using multiple qualitative or mixed methods to study a research question. For example, combining interviews with observations or surveys (Denzin, 1978).
   - **Data Triangulation**: Here, the researcher gathers data from different sources, times, or settings. For instance, interviewing both participants and non-participants or collecting data at different times (Patton, 1999).
   - **Investigator Triangulation**: Engaging multiple researchers in the data collection and interpretation process can provide varied perspectives and interpretations (Stake, 1995).
   - **Theoretical Triangulation**: Utilizing different theoretical perspectives to interpret the same set of data can enrich understanding and provide more comprehensive insights (Denzin, 1989).
2. Benefits of Triangulation:
   - **Enhanced Validity**: By corroborating findings across multiple sources or methods, researchers can be more confident in their conclusions (Jick, 1979).
   - **Comprehensive View**: Triangulation provides a multifaceted view of the phenomenon, leading to a richer, more nuanced understanding (Fielding & Fielding, 1986).
   - **Identification of Discrepancies**: Contradictions in data can prompt deeper investigation, leading to richer insights and understanding (Thurmond, 2001).
3. Criticisms and Considerations:

While triangulation is celebrated for its potential to enhance validity, it is not without criticisms. Some scholars argue that different methods can't be truly integrated as they come from different ontological and epistemological standpoints (Fielding & Fielding, 1986). Additionally, employing multiple methods or data sources can be resource-intensive in terms of time, money, and expertise.

II. Member Checking:

Member checking, or respondent validation, involves returning to participants with the interpreted data or preliminary findings to ensure accuracy. This process allows participants to confirm, refute, or refine the researcher's interpretations (Creswell & Miller, 2000).

Member checking, often referred to as "participant validation", is a critical technique employed in qualitative research to enhance the credibility and validity of results. It involves sharing preliminary or final findings with participants to confirm the accuracy and resonance of the researcher's interpretations (Lincoln & Guba, 1985). This process ensures that the voices and perspectives of participants are accurately represented and that the researcher's interpretations are grounded in participants' experiences.

1. Process of Member Checking:

The typical procedure involves returning to participants with either the raw data (such as interview transcripts) or with condensed versions, themes, or interpretations derived from the data. Participants are then asked to affirm, correct, expand, or refine the researcher's interpretation (Creswell & Miller, 2000).

2. Advantages of Member Checking:
    - **Validity**: It provides a direct way to check the accuracy of the data and the interpretations made from this data (Merriam, 1995).
    - **Empowerment**: Allows participants to have a voice in the research, ensuring that their perspectives are not misconstrued or misinterpreted (Birt, Scott, Cavers, Campbell, & Walter, 2016).
    - **Refinement**: It offers an opportunity to uncover nuances or subtleties that might have been initially overlooked by the researcher (Doyle, 2007).
3. Challenges and Considerations:
    - **Time and Logistics**: The process can be time-consuming, particularly in studies with many participants or when participants are difficult to reach for follow-ups (Harper & Cole, 2012).
    - **Potential for Bias**: There is a risk that participants may alter their responses based on what they perceive the researcher wants to hear or because they've had time to reflect and might now view things differently (Creswell & Poth, 2017).
    - **Interpretive Nature**: Some scholars argue that since qualitative research is interpretive, seeking validation from participants might limit the analytical depth. The researcher's interpretation may not always align with the participants' views, and this disparity itself can be a rich area of analysis (Morse, 2015)

III. Prolonged Engagement and Persistent Observation:

Spending extended time in the field helps researchers gain a deeper understanding of the phenomenon under study and the context in which it occurs. It also allows for building trust with participants and ensures that observed events are not anomalies (Lincoln & Guba, 1985).

In qualitative research, establishing trustworthiness is paramount. One of the primary strategies for enhancing the credibility of a study is through prolonged engagement and persistent observation. This section delves into these concepts, explicating their significance, advantages, and the challenges they pose in the realm of qualitative data analysis and interpretation.

1. Prolonged Engagement:

Prolonged engagement refers to the extended time researchers spend in the field, interacting with participants, observing, and immersing themselves in the study context (Lincoln & Guba, 1985). The idea is to build trust, learn the culture, and gain a deeper understanding of the phenomenon under study.

Advantages

- **Richer Data**: Extended time in the field often results in richer, more nuanced data as participants become more comfortable with the researcher and are more open in sharing their experiences (Creswell & Poth, 2018).
- **Reduces Misinterpretations**: By spending more time in the field, researchers can verify preliminary interpretations and reduce the chances of misinterpretations (Geertz, 1973).

2. Persistent Observation:

This entails maintaining a focused and consistent observation of the phenomena of interest in the study. It's about depth and detailed exploration rather than breadth (Lincoln & Guba, 1985).

Advantages

- **Depth of Understanding**: Allows the researcher to distinguish between what's routinely observed and what's genuinely significant to the study (Schensul, Schensul, & LeCompte, 1999).
- **Reveals Patterns**: Consistent observation over time can uncover patterns and variations in the data, leading to more robust findings (Emerson, Fretz, & Shaw, 2011).

Challenges and Considerations

- **Time and Resources**: Both prolonged engagement and persistent observation require a significant amount of time, which might not be feasible for all researchers, given constraints like funding or other commitments (Patton, 2015).
- **Researcher Fatigue**: Extended periods in the field can lead to researcher fatigue, impacting the quality of observations and data collection (O'Reilly, 2012).
- **Over-familiarity**: Being in the field for an extended period might result in over-familiarity, which can lead to a lack of critical distance from the data (Dwyer & Buckle, 2009).

IV. Peer Debriefing:

Engaging peers in the review of data, analysis, and interpretation can provide alternative perspectives, potentially identifying biases or misinterpretations. This collaborative approach serves as a form of external validation (Ely et al., 1991).

Peer debriefing, often termed as "peer review" or "external checks", is a process where researchers engage colleagues or peers to review and critique their work, bringing an external perspective to the study (Lincoln & Guba, 1985). This method aims to enhance the credibility and trustworthiness of qualitative research by mitigating potential biases, challenging interpretations, and ensuring that findings are grounded in the data.

1. Conceptual Underpinnings of Peer Debriefing:

Peer debriefing acts as a "devil's advocate", offering the primary researcher an external perspective and a chance to test their growing insights (Spall, 1998). By exposing oneself to the scrutiny of peers, a researcher can identify blind spots, question assumptions, and refine interpretations (Ely, Vinz, Downing, & Anzul, 1997).

Advantages

- **Enhances Credibility**: Peer debriefing can increase the credibility of the study by ensuring that interpretations and conclusions are not the result of researcher bias or subjectivity (Creswell & Miller, 2000).
- **Broadens Perspective**: Engaging peers from diverse backgrounds or disciplines can bring multiple viewpoints, leading to a richer understanding of the data (Stake, 1995).
- **Confirms Findings**: It serves as a validation technique. If peers independently draw similar conclusions from the data, it strengthens the study's findings (Merriam & Tisdell, 2015).

2. Implementing Peer Debriefing:

The process typically involves regular meetings where the primary researcher presents data, preliminary findings, or interpretations to one or more peers. These peers ask questions, challenge assumptions, and suggest alternative interpretations (Erlandson, Harris, Skipper, & Allen, 1993).

Challenges and Considerations

- **Finding Suitable Peers**: It's essential to choose peers who are knowledgeable about the research topic, familiar with qualitative methods, and willing to provide candid feedback (Rossman & Rallis, 2017).
- **Time Commitment**: Both the researcher and peers must commit significant time and effort to the debriefing process (Morse, Barrett, Mayan, Olson, & Spiers, 2002).
- **Managing Defensive Reactions**: Researchers must be open to feedback and avoid becoming defensive, which can undermine the process (Birt, Scott, Cavers, Campbell, & Walter, 2016).

V. Rich, Thick Descriptions:

Providing detailed and dense descriptions of the research context and participants allows readers to judge the extent to which findings can be transferred to other settings or groups. It lends depth to the data, enhancing its credibility (Geertz, 1973).

The utilization of "rich, thick descriptions" in qualitative research is an imperative strategy for enhancing the trustworthiness and credibility of findings. Originating from Geertz's (1973) anthropological work, thick description refers to the detailed, context-sensitive, and nuanced presentation of research phenomena, ensuring that readers or subsequent researchers grasp both the explicit and implicit facets of the setting, context, interactions, and the deeper meanings involved.

1. Understanding Rich, Thick Descriptions:

    - **Depth Over Breadth**: Thick descriptions delve deeply into the intricacies of the research context, participants, their experiences, behaviors, and settings, ensuring that the reader can visualize and comprehend the setting as if they were present (Denzin, 1989).

- **Layered Complexity**: This entails capturing and portraying the multiple layers of reality, encompassing surface-level events, underlying contexts, emotions, and intangible factors such as cultural norms and societal influences (Ponterotto, 2006).

Advantages

- **Enhanced Transferability**: Detailed descriptions empower readers to determine the degree to which the research findings can be applicable to other settings, situations, or individuals (Lincoln & Guba, 1985).
- **Contextual Insight**: By understanding the broader context and the nuances within, researchers and readers can achieve deeper comprehension of the phenomena under study (Holloway & Todres, 2003).
- **Validation and Credibility**: Rich accounts of the data enable readers to validate the researcher's interpretations, bolstering the study's credibility (Merriam, 1995).

2. Implementing Rich, Thick Descriptions:

- **Detailed Field Notes**: During data collection, especially in observations, researchers should ensure that their field notes capture the setting's intricate details, participant behaviors, interactions, and emerging themes (Emerson, Fretz, & Shaw, 2011).
- **Verbatim Quotations**: Using participants' own words, especially poignant or illustrative quotes, can offer readers a vivid representation of participants' experiences and perspectives (Patton, 2002).
- **Reflective Notes**: These help capture researchers' feelings, reactions, and preliminary interpretations, providing context and depth (Ortlipp, 2008).

Challenges

- **Data Overwhelm**: The depth of data can sometimes be overwhelming, making it challenging to determine which details are pertinent to include (Mason, 2002).
- **Risk of Over-interpretation**: The depth of detail can sometimes lead to over-interpretation or overemphasis on certain aspects that might not be as central to the participants' experiences (Smith & Osborn, 2008).

VI. Audit Trail:

Maintaining a clear and comprehensive record of all research activities, decisions, and reflections—like a 'diary' of the research process—can allow others to examine the study's procedures and determine its trustworthiness (Rodgers, 2008).

One of the prominent methods to instill trustworthiness in qualitative research is the establishment and maintenance of an "audit trail." This concept, derived from the guidelines proposed by Lincoln and Guba (1985), underlines the meticulous documentation of the research process, enabling external reviewers or readers to trace the decisions, actions, and steps taken by the researcher from the project's commencement to its culmination.

1. Understanding the Audit Trail:

  - **Transparency and Traceability**: An audit trail serves as a transparent, detailed, and chronological record of research activities, capturing the researcher's decisions, rationales, methodological shifts, and data interpretations (Rodgers, 2008).
  - **Documented Decisions**: By systematically documenting every significant decision, researchers allow external reviewers to understand the context, basis, and implications of each decision (Halpern & Douglas, 2008).

Key Elements of an Audit Trail

- **Raw Data**: This includes original transcripts, field notes, recordings, and any primary data collected (Morrow, 2005).
- **Data Reduction and Analysis Products**: Summaries, coding schemes, thematic maps, and any intermediate data forms produced during analysis should be maintained (Birt, Scott, Cavers, Campbell, & Walter, 2016).
- **Data Reconstruction Products**: These could comprise the structure of themes or categories, their interrelations, and any interpretations or theoretical constructions stemming from them (Smith & Firth, 2011).
- **Methodological Documentation**: This encompasses records of methodological decisions, changes made during the research, challenges faced, and their resolutions (Cutcliffe & McKenna, 2004).
- **Reflexive Notes**: Notes that detail the researcher's self-reflection, biases, personal reactions, and evolving insights, contributing to transparency in the interpretation process (Finlay, 2002).

Advantages

- **Credibility Enhancement**: An extensive audit trail bolsters the study's credibility, showing methodological rigor and diligence (Koch, 2006).
- **Facilitation of External Review**: It enables external reviewers or subsequent researchers to verify findings, validate methodological decisions, and potentially replicate or build upon the study (Thomas & Magilvy, 2011).

Challenges

- **Time-Intensive**: Maintaining a comprehensive audit trail requires meticulous record-keeping, which can be time-consuming (Given, 2008).
- **Balancing Transparency with Participant Confidentiality**: While the purpose is to document thoroughly, the researcher must ensure that participant confidentiality is uncompromised (Eide & Kahn, 2008).

VII. Reflexivity:

Reflexivity involves researchers actively reflecting on their biases, values, and the role they play in the research process. Being transparent about these reflections can enhance the study's credibility by making explicit the lens through which the research was conducted (Finlay, 2002).

Reflexivity is a foundational tenet of qualitative research that underscores the interactive and interpretative nature of the inquiry (Alvesson & Sköldberg, 2009). It requires researchers to maintain an ongoing consciousness and critical examination of their role in the research process, the potential biases they bring, and the influence of their interpretations on the findings.

1. Understanding Reflexivity:

    - **Nature of Reflexivity**: Reflexivity is an iterative, self-reflective process that involves introspection and a willingness to confront one's predispositions, assumptions, and potential biases (Finlay, 2002).
    - **Contextual Interaction**: Reflexivity acknowledges that the researcher is an active participant in the data collection and analysis process, and their values, beliefs, and backgrounds inherently shape this process (Mauthner & Doucet, 2003).

Key Aspects of Reflexivity

- **Personal Reflexivity**: Researchers must recognize and examine their own beliefs, values, and experiences, and how these may impact the research process and its outcomes (Etherington, 2004).
- **Epistemological Reflexivity**: It requires researchers to reflect on the assumptions about the world and knowledge they bring to their study and how these influence the entire research process (Berger, 2015).

Advantages

- **Enhancing Credibility**: Reflexivity can fortify the trustworthiness of the research by transparently documenting the researcher's influences and considerations (Willig, 2013).
- **Deeper Insights**: It can lead to more profound, nuanced understandings by unearthing and exploring underlying assumptions or biases (Pillow, 2003).
- **Ethical Research Practice**: Reflexivity promotes ethical research practices as researchers continually evaluate their roles, potential impacts on participants, and interpretations (Guillemin & Gillam, 2004).

Challenges

- **Subjectivity**: While reflexivity emphasizes the researcher's role, there is a danger of making the research too subjective or self-indulgent (Cunliffe, 2003).
- **Potential for Over-Analysis**: The continuous introspection might lead researchers to over-analyze their influence, potentially hindering the natural flow of the research process (Macbeth, 2001).

Strategies to Enhance Reflexivity

- **Maintaining Reflective Journals**: Documenting thoughts, feelings, and reactions throughout the research can help researchers trace their evolving understanding and influence on the study (Ortlipp, 2008).
- **Seeking Peer Feedback**: Engaging colleagues or mentors to provide an external perspective on the research process and findings can illuminate potential biases or oversights (Bourke, 2014).

- **Transparent Reporting**: Being open about personal or epistemological influences in research reporting can give readers a fuller understanding of the study context (Rolls & Relf, 2006).

REFERENCES

Alvesson, M., & Sköldberg, K. (2009). Reflexive methodology: New vistas for qualitative research. *Sage (Atlanta, Ga.).*

Attride-Stirling, J. (2001). Thematic networks: An analytic tool for qualitative research. *Qualitative Research*, *1*(3), 385–405. doi:10.1177/146879410100100307

Bailey, J. (2008). First steps in qualitative data analysis: Transcribing. *Family Practice*, *25*(2), 127–131. doi:10.1093/fampra/cmn003 PMID:18304975

Bazeley, P. (2007). *Qualitative data analysis with NVivo*. Sage Publications.

Berger, R. (2015). Now I see it, now I don't: Researcher's position and reflexivity in qualitative research. *Qualitative Research*, *15*(2), 219–234. doi:10.1177/1468794112468475

Bird, C. M. (2005). How I stopped dreading and learned to love transcription. *Qualitative Inquiry*, *11*(2), 226–248. doi:10.1177/1077800404273413

Birt, L., Scott, S., Cavers, D., Campbell, C., & Walter, F. (2016). Member Checking. *Qualitative Health Research*, *26*(13), 1802–1811. doi:10.1177/1049732316654870 PMID:27340178

Bourke, B. (2014). Positionality: Reflecting on the research process. *The Qualitative Report*, *19*(33), 1–9.

Braun, V., & Clarke, V. (2006). Using thematic analysis in psychology. *Qualitative Research in Psychology*, *3*(2), 77–101. doi:10.1191/1478088706qp063oa

Braun, V., & Clarke, V. (2012). Thematic analysis. In H. Cooper (Ed.), *APA handbook of research methods in psychology*. APA. doi:10.1037/13620-004

Charmaz, K. (2006). Constructing grounded theory: A practical guide through qualitative analysis. *Sage (Atlanta, Ga.).*

Charmaz, K. (2014). *Constructing grounded theory* (2nd ed.). Sage.

Coffey, A., & Atkinson, P. (1996). *Making Sense of Qualitative Data: Complementary Research Strategies*. Sage Publications.

Corbin, J. M., & Strauss, A. (2008). Basics of qualitative research: Techniques and procedures for developing grounded theory. *Sage (Atlanta, Ga.).* Advance online publication. doi:10.4135/9781452230153

Creswell, J. W. (2013). *Research design: Qualitative, quantitative, and mixed methods approaches*. Sage publications.

Creswell, J. W., & Miller, D. L. (2000). Determining validity in qualitative inquiry. *Theory into Practice*, *39*(3), 124–130. doi:10.1207/s15430421tip3903_2

Creswell, J. W., & Poth, C. N. (2018). *Qualitative inquiry and research design: Choosing among five approaches* (4th ed.). Sage.

Cunliffe, A. L. (2003). Reflexive inquiry in organizational research: Questions and possibilities. *Human Relations*, *56*(8), 983–1003. doi:10.1177/00187267030568004

Cutcliffe, J. R., & McKenna, H. P. (2004). Expert qualitative researchers and the use of audit trails. *Journal of Advanced Nursing*, *45*(2), 126–135. doi:10.1046/j.1365-2648.2003.02874.x PMID:14705996

Davidson, J. (2009). A new approach to transcription: The challenges of recording, transcribing, and transcribing data. *Research on Language and Social Interaction*, *42*(4), 367–379.

Denzin, N. K. (1978a). *Sociological methods: A sourcebook*. McGraw-Hill.

Denzin, N. K., & Lincoln, Y. S. (2011). *The SAGE Handbook of Qualitative Research*. Sage.

Doyle, S. (2007). Member checking with older women: A framework for negotiating meaning. *Health Care for Women International*, *28*(10), 888–908. doi:10.1080/07399330701615325 PMID:17987459

Dwyer, S. C., & Buckle, J. L. (2009). The space between: On being an insider-outsider in qualitative research. *International Journal of Qualitative Methods*, *8*(1), 54–63. doi:10.1177/160940690900800105

Eide, P., & Kahn, D. (2008). Ethical issues in the qualitative researcher—Participant relationship. *Nursing Ethics*, *15*(2), 199–207. doi:10.1177/0969733007086018 PMID:18272610

Eisenhardt, K. M. (1989). Building theories from case study research. *Academy of Management Review*, *14*(4), 532–550. doi:10.2307/258557

Ely, M., Vinz, R., Downing, M., & Anzul, M. (1991). *On writing qualitative research: Living by words*. Falmer Press.

Ely, M., Vinz, R., Downing, M., & Anzul, M. (1997). *On writing qualitative research: Living by words*. Falmer Press.

Emerson, R. M., Fretz, R. I., & Shaw, L. L. (2011). *Writing ethnographic fieldnotes*. University of Chicago Press. doi:10.7208/chicago/9780226206868.001.0001

Erlandson, D. A., Harris, E. L., Skipper, B. L., & Allen, S. D. (1993). Doing naturalistic inquiry: A guide to methods. *Sage (Atlanta, Ga.)*.

Etherington, K. (2004). *Becoming a reflexive researcher: Using our selves in research*. Jessica Kingsley Publishers.

Fereday, J., & Muir-Cochrane, E. (2006). Demonstrating rigor using thematic analysis: A hybrid approach of inductive and deductive coding and theme development. *International Journal of Qualitative Methods*, *5*(1), 80–92. doi:10.1177/160940690600500107

Fielding, N. G., & Fielding, J. L. (1986). Linking data. *Sage (Atlanta, Ga.)*.

Fielding, N. G., & Lee, R. M. (1998). *Computer analysis and qualitative research*. Sage Publications.

Geertz, C. (1973). *The interpretation of cultures: Selected essays*. Basic Books.

Given, L. M. (2008). *The SAGE encyclopedia of qualitative research methods*. Sage Publications. doi:10.4135/9781412963909

Glaser, B. G. (1978). *Theoretical sensitivity: Advances in the methodology of grounded theory*. Sociology Press.

Glaser, B. G., & Strauss, A. L. (1967). *The Discovery of Grounded Theory: Strategies for Qualitative Research*. Aldine Transaction.

Guillemin, M., & Gillam, L. (2004). Ethics, reflexivity, and "ethically important moments" in research. *Qualitative Inquiry*, *10*(2), 261–280. doi:10.1177/1077800403262360

Halcomb, E. J., Davidson, P. M., & Hardaker, L. (2008). Using software to enhance the quality of transcription in qualitative research. *Nurse Researcher*, *15*(3), 11–21.

Halpern, E. S., & Douglas, E. J. (2008). Audit trail. In L. M. Given (Ed.), *The SAGE encyclopedia of qualitative research methods* (Vol. 1, pp. 42–43). Sage Publications.

Harper, M., & Cole, P. (2012). Member checking: Can benefits be gained similar to group therapy? *The Qualitative Report*, *17*(2), 510–517.

Hesse-Biber, S. (2011). Qualitative approaches to mixed methods practice. *Qualitative Inquiry*, *17*(6), 455–467. doi:10.1177/1077800410364611

Holloway, I., & Todres, L. (2003). The status of method: Flexibility, consistency and coherence. *Qualitative Research*, *3*(3), 345–357. doi:10.1177/1468794103033004

Jick, T. D. (1979). Mixing qualitative and quantitative methods: Triangulation in action. *Administrative Science Quarterly*, *24*(4), 602–611. doi:10.2307/2392366

Jupp, V. (2006). The Sage dictionary of social research methods. *Sage (Atlanta, Ga.)*. Advance online publication. doi:10.4135/9780857020116

Koch, T. (2006). Establishing rigour in qualitative research: The decision trail. *Journal of Advanced Nursing*, *53*(1), 91–100. doi:10.1111/j.1365-2648.2006.03681.x PMID:16422698

Lapadat, J. C., & Lindsay, A. C. (1999). Transcription in research and practice: From standardization of technique to interpretive positionings. *Qualitative Inquiry*, *5*(1), 64–86. doi:10.1177/107780049900500104

Layder, D. (1998). Sociological practice: Linking theory and social research. *Sage (Atlanta, Ga.)*.

Lincoln, Y. S., & Guba, E. A. (1985). *Naturalistic inquiry*. Sage. doi:10.1016/0147-1767(85)90062-8

Macbeth, D. (2001). On "reflexivity" in qualitative research: Two readings, and a third. *Qualitative Inquiry*, *7*(1), 35–68. doi:10.1177/107780040100700103

MacQueen, K. M., McLellan, E., Kay, K., & Milstein, B. (1998). Codebook development for team-based qualitative analysis. *Field Methods*, *10*(2), 31–36.

Mason, J. (2002). *Qualitative researching* (2nd ed.). SAGE.

Mauthner, N. S., & Doucet, A. (2003). Reflexive accounts and accounts of reflexivity in qualitative data analysis. *Sociology*, *37*(3), 413–431. doi:10.1177/00380385030373002

Maxwell, J. A. (2012a). A realist approach for qualitative research. *Sage (Atlanta, Ga.)*.

Maxwell, J. A. (2012b). *Qualitative research design: An interactive approach* (3rd ed.). Sage.

Maxwell, J. A. (2012c). The importance of qualitative research for causal explanation in education. *Qualitative Inquiry*, *18*(8), 655–661. doi:10.1177/1077800412452856

Maxwell, J. A. (2013). *Qualitative research design: An interactive approach*. Sage Publications.

Merriam, S. B. (1995). What can you tell from an N of 1?: Issues of validity and reliability in qualitative research. *PAACE Journal of Lifelong Learning*, *4*, 51–60.

Miles, M. B., & Huberman, A. M. (1994). Qualitative data analysis: An expanded sourcebook. *Sage (Atlanta, Ga.)*.

Miles, M. B., Huberman, A. M., & Saldaña, J. (2014). *Qualitative data analysis: A methods sourcebook* (3rd ed.). Sage.

Mishler, E. G. (1991). Representing discourse: The rhetoric of transcription. *Journal of Narrative and Life History*, *1*(4), 255–280. doi:10.1075/jnlh.1.4.01rep

Morrow, S. L. (2005). Quality and trustworthiness in qualitative research in counseling psychology. *Journal of Counseling Psychology*, *52*(2), 250–260. doi:10.1037/0022-0167.52.2.250

Morse, J. M. (2015). Critical analysis of strategies for determining rigor in qualitative inquiry. *Qualitative Health Research*, *25*(9), 1212–1222. doi:10.1177/1049732315588501 PMID:26184336

Morse, J. M., Barrett, M., Mayan, M., Olson, K., & Spiers, J. (2002). Verification strategies for establishing reliability and validity in qualitative research. *International Journal of Qualitative Methods*, *1*(2), 13–22. doi:10.1177/160940690200100202

O'Connell, D. C., & Kowal, S. (1999). Transcription as a crucial step of data analysis. In *Speech, language, and the law: International perspectives* (pp. 89–99). Multilingual Matters.

O'Reilly, K. (2012). *Ethnographic methods*. Routledge. doi:10.4324/9780203864722

Oliver, D. G., Serovich, J. M., & Mason, T. L. (2005). Constraints and opportunities with interview transcription: Towards reflection in qualitative research. *Social Forces*, *84*(2), 1273–1289. doi:10.1353/sof.2006.0023 PMID:16534533

Ortlipp, M. (2008). Keeping and using reflective journals in the qualitative research process. *The Qualitative Report*, *13*(4), 695–705.

Patton, M. Q. (2002a). *Qualitative research and evaluation methods* (3rd ed.). Sage Publications.

Patton, M. Q. (2002b). Two decades of developments in qualitative inquiry. *Qualitative Social Work: Research and Practice*, *1*(3), 261–283. doi:10.1177/1473325002001003636

Patton, M. Q. (2015). *Qualitative research & evaluation methods: Integrating theory and practice*. Sage publications.

Poland, B. D. (2002). Transcription quality. In *Handbook of interview research: Context & method* (pp. 629–649). Sage.

Ponterotto, J. G. (2006). Brief note on the origins, evolution, and meaning of the qualitative research concept thick description. *The Qualitative Report*, *11*(3), 538–549.

Richards, L., & Morse, J. M. (2012). *Readme first for a user's guide to qualitative methods*. Sage publications.

Ritchie, J., & Lewis, J. (2003). *Qualitative research practice: A guide for social science students and researchers*. SAGE.

Rodgers, B. L. (2008). Audit trail. In L. M. Given (Ed.), *The Sage encyclopedia of qualitative research methods* (Vol. 1, pp. 44–45). Sage Publications.

Rolls, L., & Relf, M. (2006). Bracketing interviews: Addressing methodological challenges in qualitative interviewing in bereavement and palliative care. *Mortality*, *11*(3), 286–305. doi:10.1080/13576270600774893

Rossman, G. B., & Rallis, S. F. (2017). *Learning in the field: An introduction to qualitative research*. Sage Publications. doi:10.4135/9781071802694

Saldaña, J. (2015). *The coding manual for qualitative researchers* (3rd ed.). Sage.

Schensul, J. J., Schensul, S. L., & LeCompte, M. D. (1999). *Essential ethnographic methods: Observations, interviews, and questionnaires*. Rowman Altamira.

Silver, C., & Lewins, A. (2014). *Using software in qualitative research: A step-by-step guide*. Sage Publications. doi:10.4135/9781473906907

Smith, D. W. (2003). Phenomenology. In E. N. Zalta (Ed.), *The Stanford Encyclopedia of Philosophy*. Stanford University.

Smith, J., & Firth, J. (2011). Qualitative data analysis: The framework approach. *Nurse Researcher*, *18*(2), 52–62. doi:10.7748/nr2011.01.18.2.52.c8284 PMID:21319484

Smith, J. A. (2018). *Qualitative psychology: A practical guide to research methods*. Sage Publications.

Smith, J. A., & Osborn, M. (2008). Interpretative phenomenological analysis. In J. A. Smith (Ed.), *Qualitative psychology: A practical guide to research methods* (2nd ed., pp. 53–80). Sage.

Spall, S. (1998). Peer debriefing in qualitative research: Emerging operational models. *Qualitative Inquiry*, *4*(2), 280–292. doi:10.1177/107780049800400208

Stake, R. E. (1995). *The art of case study research*. Sage Publications.

Strauss, A., & Corbin, J. (1990). *Basics of qualitative research: Grounded theory procedures and techniques*. Sage Publications, Inc.

Strauss, A., & Corbin, J. (1998). Basics of qualitative research: Procedures and techniques for developing grounded theory. *Sage (Atlanta, Ga.)*.

Thomas, D. R. (2006). A general inductive approach for analyzing qualitative evaluation data. *The American Journal of Evaluation*, *27*(2), 237–246. doi:10.1177/1098214005283748

Thomas, E., & Magilvy, J. K. (2011). Qualitative rigor or research validity in qualitative research. *Journal for Specialists in Pediatric Nursing*, *16*(2), 151–155. doi:10.1111/j.1744-6155.2011.00283.x PMID:21439005

Thurmond, V. A. (2001). The point of triangulation. *Journal of Nursing Scholarship*, *33*(3), 253–258. doi:10.1111/j.1547-5069.2001.00253.x PMID:11552552

Tilley, S. A. (2003). "Challenging" research practices: Turning a critical lens on the work of transcription. *Qualitative Inquiry*, *9*(5), 750–773. doi:10.1177/1077800403255296

Webb, E. J., Campbell, D. T., Schwartz, R. D., & Sechrest, L. (1966). *Unobtrusive measures: Nonreactive research in the social sciences*. Rand McNally.

Willig, C. (2013). *Introducing qualitative research in psychology*. McGraw-Hill Education.

Wolcott, H. F. (1990). Writing up qualitative research. *Sage (Atlanta, Ga.)*.

Yin, R. K. (2014). *Case study research design and methods* (5th ed.). Sage Publications.

Chapter 7
Quality and Rigor in Qualitative Research

ABSTRACT

This chapter explores the distinctive nature and challenges of ensuring quality in qualitative research, contrasting it with the quantitative paradigm's focus on numerical analysis and generalizability. Qualitative research, centered on understanding complex human experiences, debates traditional metrics of quality such as validity and reliability, proposing alternatives like credibility, transferability, and confirmability tailored to its subjective and interpretive methodology. The discourse extends to embracing subjectivity as a strength, employing reflexivity and leveraging methodological strategies like triangulation and member checking to enhance rigor.

INTRODUCTION

The sphere of qualitative research is marked by a rich tapestry of methods, approaches, and epistemological underpinnings. Unlike the quantitative paradigm, which often hinges on numerical measurement and statistical validation, qualitative research focuses on exploring the depth, meaning, and complexity of human experiences. However, the very nature of qualitative inquiry, steeped in subjectivity and interpretation, also raises pivotal questions about its quality and rigor. This chapter seeks to navigate the nuanced terrains of ensuring quality in qualitative research, elucidating the hallmarks of rigorous qualitative studies and the strategies researchers employ to achieve this.

1. The Debate on Quality in Qualitative Research:

While quantitative research has established criteria like validity, reliability, and generalizability, qualitative research grapples with evolving and sometimes contested criteria for quality. This section explores this ongoing debate, highlighting the challenges and opportunities it presents (Guba & Lincoln, 1982; Morse et al., 2002).

DOI: 10.4018/979-8-3693-2414-1.ch007

Qualitative research, rooted deeply in understanding the complex nature of human experiences, has grown exponentially over the past few decades, finding its place across a wide array of disciplines. As the field has expanded, so too have debates around its legitimacy, rigor, and quality. The very nature of qualitative research, which emphasizes subjectivity, interpretation, and context, challenges traditional notions of research quality inherited mainly from quantitative paradigms.

I. Traditional Criteria and their Qualitative Counterparts:

Historically, quantitative research has leaned on criteria like validity (how well an instrument measures what it intends to measure), reliability (the consistency of the measurement), and generalizability (the extent to which results can be applied to wider populations). In contrast, qualitative researchers like Guba and Lincoln (1982) proposed alternatives, including credibility (instead of internal validity), transferability (rather than generalizability), and dependability (in place of reliability). Such shifts demonstrate how qualitative scholars are establishing their own benchmarks for quality, rooted in the essence of their methodology.

The debate surrounding the quality and rigor of qualitative research is deeply intertwined with the challenge of applying traditional research criteria, which are rooted primarily in the quantitative paradigm, to qualitative inquiries. These traditional criteria have set the standards for scientific rigor for centuries, and their application to qualitative research has been the subject of considerable contention.

A. Background: The Quantitative Benchmark:

Quantitative research is grounded in positivist epistemology which emphasizes the objectivity of knowledge. The hallmarks of quality in this paradigm are primarily reliability, validity, and generalizability (Creswell & Miller, 2000). These criteria serve as rigorous benchmarks ensuring that the findings are consistent, accurate, and applicable across different contexts.

B. Emergence of Qualitative Criteria:

The qualitative research paradigm, with its emphasis on understanding the richness and complexity of human experience, requires a different set of benchmarks. Guba and Lincoln (1982) were among the pioneers who proposed alternative criteria tailored for qualitative research. They introduced concepts like credibility, transferability, dependability, and confirmability as counterparts to internal validity, generalizability, reliability, and objectivity, respectively.

C. Credibility vs. Internal Validity:

While internal validity in quantitative research is concerned with the accuracy and truthfulness of the findings in relation to the research questions, credibility in qualitative research emphasizes the congruence between the constructed realities of participants and the reconstructions of the researcher (Lincoln & Guba, 1985).

D. Transferability vs. Generalizability:

Qualitative research does not seek to generalize findings in the same way quantitative research does. Instead, it aims for transferability, providing rich and thick descriptions of the context so that readers can determine the extent to which the findings might be applicable in other settings (Stake, 1995).

E. Dependability and Confirmability:

While quantitative research aims for consistency (reliability) and neutrality (objectivity), qualitative research aims for dependability (ensuring findings are consistent and replicable in a similar context) and confirmability (ensuring that findings are shaped by respondents and not researcher bias, motivation, or interest) (Shenton, 2004).

F. The Ongoing Debate:

While these qualitative counterparts offer a robust framework, the debate continues. Some scholars argue that qualitative research should wholly distance itself from the quantitative criteria, while others argue for a blended approach, recognizing the unique strengths of both paradigms (Morrow, 2005).

II. The Subjectivity Challenge:

One primary critique often lobbed at qualitative research is its inherent subjectivity. Given its inductive nature and emphasis on participants' perspectives, qualitative research is often perceived as subjective and thus, less reliable (Hammersley, 1992). However, qualitative researchers argue that subjectivity, when acknowledged and managed appropriately through reflexivity, can enhance the depth and richness of research (Peshkin, 1988).

The intrinsic nature of qualitative research, characterized by its deep immersion and understanding of human experiences, brings with it a unique challenge: the challenge of subjectivity. Unlike quantitative research, which primarily focuses on the objective measurement and quantification of phenomena, qualitative research delves into the lived experiences, feelings, perceptions, and interpretations of individuals. Consequently, the researcher becomes an instrument of data collection, and this deep involvement can lead to issues of subjectivity.

A. Understanding Subjectivity in Qualitative Research:

Subjectivity refers to the potential influence of personal beliefs, values, experiences, and perspectives on the research process, from data collection to interpretation (Peshkin, 1988). Given that qualitative researchers are deeply immersed in the field, they are inevitably shaped by, and in turn shape, their research.

B. The Double-Edged Sword of Subjectivity:

While subjectivity is often perceived as a challenge or limitation in maintaining the rigor and validity of research findings, it is also its strength. The deep involvement of the researcher allows for an empathetic understanding, a closer connection to participants, and rich insights that might be inaccessible through purely objective approaches (Finlay, 2002).

C. Addressing the Subjectivity Challenge:

Several strategies have been proposed to address the challenge of subjectivity:

- **Reflexivity**: Reflexivity involves a continuous process of self-awareness where researchers reflect upon their influence on the research, considering how their own backgrounds, perceptions, and interactions may affect data collection and interpretation (Malterud, 2001).
- **Triangulation**: Utilizing multiple methods, data sources, or researchers to investigate the same phenomenon can provide a more comprehensive view and help in validating findings (Denzin, 1978).
- **Member Checking**: Returning to the participants with the findings and interpretations to ensure their perspectives are accurately represented can serve as a validity check (Lincoln & Guba, 1985).
- **Researcher's Journal**: Maintaining a journal where researchers note their thoughts, feelings, and experiences throughout the research process can help in identifying potential biases or influences on the research (Ortlipp, 2008).

D. Embracing Subjectivity:

While it's essential to recognize and address subjectivity, it's equally crucial to embrace it as an inherent aspect of qualitative inquiry. It allows for depth, richness, and nuance in understanding human experiences, which are the hallmarks of qualitative research (Roulston & Shelton, 2015).

III. Quality as a Construct:

Morse et al. (2002) asserted that the very definition of quality in qualitative research is fluid and must be contextually derived. Thus, rigid checklists or universally applicable criteria may undermine the epistemological foundations of qualitative work. Instead, a more nuanced understanding, which appreciates the varied nature of qualitative inquiries, is advocated.

The debate surrounding quality in qualitative research often leads to the fundamental question: "What is quality?" In qualitative inquiry, determining the quality or validity of research is inherently complex due to the diverse nature of qualitative methods, approaches, and underlying philosophical assumptions. This complexity stems from the ways in which 'quality' is constructed, interpreted, and operationalized within the qualitative research paradigm.

A. Constructing Quality: The Historical Perspective:

Historically, the criteria for determining research quality in the social sciences have been borrowed from the positivist paradigm, emphasizing objectivity, reliability, and generalizability (Guba & Lincoln, 1994). However, as qualitative research gained prominence, it became evident that these traditional criteria might not be entirely suitable. The subjective and interpretive nature of qualitative research prompted the need for a redefined set of quality criteria, grounded in its own philosophical and methodological foundations (Morse, Barrett, Mayan, Olson, & Spiers, 2002).

B. Quality as Context-Dependent:

Qualitative research recognizes that the concept of 'quality' cannot be universally standardized and is context-dependent. What might be considered high-quality research in one study might not hold the same standard in another due to differing contexts, methods, or theoretical frameworks (Tracy, 2010). Thus, qualitative researchers advocate for flexibility and adaptability in defining and ensuring research quality.

C. Trustworthiness and Credibility:

Guba and Lincoln (1985) proposed the idea of trustworthiness as a foundational pillar for quality in qualitative research. They suggested four criteria: credibility, transferability, dependability, and confirm-ability. Credibility, akin to internal validity, emphasizes the truth value of findings from the standpoint of the researcher, participants, and readers.

D. Reflexivity and Co-Construction:

An understanding of quality also incorporates the idea that knowledge is co-constructed between the researcher and participants. Reflexivity, which involves researchers critically examining their role, biases, and influence on the research, is crucial in ensuring the quality of this co-construction process (Finlay, 2002).

E. Evolving Notions of Quality:

The construct of quality in qualitative research remains dynamic. New methods, digital innovations, and interdisciplinary approaches constantly introduce new challenges and dimensions to the quality debate. Embracing this evolving nature is essential for the ongoing vitality and relevance of qualitative research (Hesse-Biber, 2017).

IV. The Role of Epistemology:

Epistemological stances play a pivotal role in shaping perceptions of quality. For instance, positivist or post-positivist paradigms, which value objective truths, may prioritize reliability and validity. In contrast, constructivist or interpretivist paradigms, which see truth as co-constructed and relative, may prioritize richness, depth, and authenticity (Crotty, 1998).

Epistemology, a branch of philosophy concerned with the nature and scope of knowledge, plays a pivotal role in shaping understandings of quality in qualitative research. The epistemological stances adopted by researchers fundamentally influence how they perceive the world, how they believe knowledge can be obtained, and consequently, how they gauge the quality and validity of their findings (Crotty, 1998).

A. Positivism vs. Interpretivism:

The positivist epistemological stance, rooted in the scientific method, posits an objective reality that can be measured and tested through empirical observation. From this perspective, quality in research often translates to reliability, generalizability, and objectivity (Bryman, 2012). In contrast, the interpretivist paradigm emphasizes the subjective and constructed nature of reality. Here, quality is not merely about replication but understanding, interpreting, and representing participants' experiences authentically (Denzin & Lincoln, 2011).

B. Constructivism and Co-construction of Knowledge:

Constructivist epistemology posits that knowledge is not discovered but constructed through human interaction with the world. In qualitative research following this epistemology, the role of the researcher is not just to observe but to co-construct meaning with participants. Quality, in this context, emphasizes the depth and richness of this co-constructed narrative (Guba & Lincoln, 1989).

C. Critical Epistemology:

Critical epistemologies challenge dominant power structures, ideologies, and discourses. Research from a critical stance seeks to highlight power imbalances and promote social change. The quality is often gauged by the research's potential to foster understanding and instigate transformative action (Kincheloe & McLaren, 2005).

D. Epistemological Reflexivity:

For many qualitative researchers, acknowledging and reflecting upon one's epistemological stance is essential for ensuring research quality. Epistemological reflexivity requires researchers to critically examine their assumptions about the world and how these influence the research process and outcomes (Morrow, 2005).

E. Epistemology and Methodological Rigor:

The chosen epistemological stance inevitably shapes the methods used, the type of data collected, and the mode of analysis. Recognizing the congruence between epistemology, methodology, and methods is essential for ensuring rigor and coherence in qualitative research (Patton, 2002).

V. Towards a Balanced View:

While the debate continues, there's a growing consensus that qualitative research should not be judged by quantitative criteria. However, it is also widely acknowledged that qualitative research must ensure rigor and quality in ways that are true to its nature. Brink and Wood (1998) suggest that quality in qualitative research is ensured when the findings accurately represent participants' experiences, an assertion that captures the essence of the qualitative paradigm.

The debate surrounding the quality of qualitative research often stems from the tension between positivist standards of rigor, which prioritize objectivity and generalizability, and interpretivist standards, which value depth, nuance, and context (Lincoln & Guba, 1985). Seeking a balanced view is pivotal to advancing the discourse around quality and ensuring the relevance and applicability of qualitative findings.

A. Beyond the Quantitative-Qualitative Dichotomy:

While qualitative research has historically been contrasted with quantitative paradigms, it is essential to understand that both approaches are valuable and serve different purposes. Qualitative methods are particularly suited for exploring complex phenomena in depth, understanding lived experiences, and generating theory. As such, the criteria for assessing its quality should be adapted to its unique objectives rather than merely replicating quantitative standards (Denzin & Lincoln, 2005).

B. Embracing Contextual Relevance:

A balanced view of quality recognizes the role of context. Since qualitative research often dives deep into specific settings or groups, its findings might not always be generalizable in the traditional sense. However, they offer a profound contextual understanding, which can be transferable to other similar settings or populations (Geertz, 1973).

C. Integrative Approaches to Quality:

Modern qualitative scholars often advocate for integrative approaches that combine the strengths of various paradigms. For instance, methodological triangulation, which involves using multiple methods to study a phenomenon, can enhance both depth and breadth, striking a balance between contextual richness and broader relevance (Patton, 1999).

D. Fluidity and Flexibility vs. Rigor:

While flexibility is a hallmark of qualitative research, it doesn't negate the need for rigor. A balanced view emphasizes the rigorous application of qualitative methods while allowing for the adaptability needed to capture the complexities of human experience (Morse, Barrett, Mayan, Olson, & Spiers, 2002).

E. Reflexivity as a Quality Criterion:

Reflexivity, which involves researchers' continual reflection on their role, biases, and influence on the research process, has emerged as a critical criterion for quality in qualitative research. This self-awareness ensures that researchers remain transparent, honest, and ethically grounded throughout the study (Finlay, 2002).

2. Criteria for Ensuring Quality:

Several scholars have proposed criteria to evaluate the quality of qualitative research, some of which mirror, challenge, or extend the traditional notions from quantitative paradigms.

- **Credibility (Internal Validity)**: Concerned with the truth value of findings, ensuring that they are reflective of participants' experiences (Lincoln & Guba, 1985).

Credibility, analogous to internal validity in quantitative research, pertains to the congruence between the constructed realities of the participants and the reconstructions and interpretations of the researcher (Lincoln & Guba, 1985). Establishing credibility is essential in qualitative research to ensure that the study's findings accurately represent participants' experiences and perspectives. This section delves into the concept of credibility and offers strategies to bolster it in qualitative inquiries.

A. Prolonged Engagement:

Prolonged engagement involves the researcher spending adequate time in the field to gain an in-depth understanding of the phenomenon under study and to build trust with participants (Lincoln & Guba, 1985). This extended interaction allows researchers to discern and account for distortions that might arise in short-term observations or interactions.

B. Persistent Observation:

While prolonged engagement provides a temporal immersion, persistent observation offers depth. It focuses on the salient factors related to the research inquiry, ensuring that researchers identify and understand the nuances critical to the research phenomenon (Patton, 2002).

C. Triangulation:

Triangulation involves the use of multiple methods, data sources, researchers, or theoretical perspectives to cross-check and validate findings (Denzin, 1978). By comparing and contrasting different viewpoints or approaches, researchers can enhance the credibility of their findings.

D. Peer Debriefing:

Engaging in peer debriefing entails discussing research processes and findings with peers or experts who are external to the study (Lincoln & Guba, 1985). Such external checks offer opportunities for reflection, critique, and validation, fostering a more rigorous interpretation of the data.

E. Member Checking:

Member checking, also known as participant validation, involves presenting the research findings or interpretations back to the participants for validation (Creswell & Miller, 2000). By allowing participants to agree, disagree, or clarify the interpretations, researchers can ensure that their representations align with the participants' experiences.

F. Reflexivity:

Reflexivity involves researchers reflecting upon and disclosing their biases, values, and experiences that might influence the research process (Malterud, 2001). By being self-aware and transparent about their positionality, researchers contribute to the credibility of the study.

- **Transferability (External Validity)**: While generalizability is often unattainable in qualitative studies, transferability encourages rich descriptions to allow others to determine the applicability of findings in other contexts (Geertz, 1973).

Transferability, akin to external validity or generalizability in quantitative research, refers to the degree to which the results of qualitative research can be transferred to other contexts or settings with different respondents (Lincoln & Guba, 1985). It underscores the applicability of the findings beyond the immediate study. Achieving transferability in qualitative research can be challenging given the context-specific nature of such inquiries. This section elaborates on the concept of transferability and outlines strategies to enhance it in qualitative research.

A. Thick Description:

A cornerstone of transferability is providing "thick description" — detailed and rich accounts of the research setting, participants, and processes (Geertz, 1973). By offering an in-depth portrayal of the context and the phenomena studied, researchers enable readers to determine the extent to which the findings can be applied to their specific contexts.

B. Purposeful Sampling:

Purposeful sampling ensures that participants are chosen based on specific purposes, rather than at random, to provide rich information related to the research inquiry (Patton, 2002). Such strategic selection aids in depth and richness, allowing findings to resonate with readers and other contexts.

C. Use of Analogical Reasoning:

While qualitative studies don't aim for broad generalizability, analogical reasoning can help readers see parallels between the study context and their situations (Smith & Hodkinson, 2005). By drawing comparisons, researchers can illustrate potential applications of their findings.

D. Reflexivity:

As with credibility, reflexivity plays a crucial role in transferability. Researchers must consistently reflect upon and articulate their biases, assumptions, and worldview, thereby allowing readers to understand the lens through which the study was conducted (Finlay, 2002).

E. External Checks:

Engaging external researchers or experts to review the research process and findings can enhance transferability. These individuals can assess the study's applicability to other contexts, providing a broader perspective on its transferability (Merriam, 1998).

F. Encourage Replication:

While exact replication might not always be feasible in qualitative research due to its context-specific nature, encouraging other researchers to replicate the study in different settings can help in assessing the broader applicability of the findings (Yin, 2003).

- **Dependability (Reliability)**: Focuses on the consistency and stability of findings across time and researchers (Guba, 1981).

Dependability, in the context of qualitative research, parallels the concept of reliability in quantitative research. It emphasizes the stability and consistency of the researcher's data over time and under similar conditions (Lincoln & Guba, 1985). Rather than seeking replication, as in quantitative designs, dependability in qualitative research acknowledges the fluid and evolving nature of human behavior and experiences. Thus, the focus is on ensuring the research process is logical, traceable, and documented. This section delves deep into dependability, elucidating its significance and offering strategies for enhancing it.

A. Audit Trail:

An audit trail entails maintaining detailed records of all aspects of the research process—data, methodological decisions, and analytic notes (Rodgers, 2008). This meticulous documentation allows external auditors or peers to examine the research process and outcomes, affirming the decisions made and the conclusions drawn.

B. Reflexivity:

Regularly engaging in reflexivity, where researchers critically examine their biases, preconceptions, and potential influence on the data, can bolster dependability (Finlay, 2002). Reflective notes, journals, or diaries can capture these self-analytical processes.

C. Code-Recode Strategy:

This involves the researcher coding a set of data, waiting a specified period, and then recoding the same data. Comparing both sets of codes can highlight consistency in interpretation over time (Morse et al., 2002).

D. Peer Examination:

Engaging peers to review and challenge the research process and findings can enhance dependability. Colleagues can offer insights, raise queries, or provide alternative interpretations, ensuring a thorough and robust research process (Creswell & Miller, 2000).

E. Thick Description:

Providing a comprehensive account of the context, participants, and fieldwork activities aids in understanding the research setting and processes (Geertz, 1973). Such detail can help others ascertain the degree of dependability.

F. Use of Technology:

Employing software specifically designed for qualitative research, such as NVivo or Atlas.ti, can increase dependability by systematizing data management, coding, and analysis processes (Paulus et al., 2017).

- **Confirmability (Objectivity)**: Stresses the importance of ensuring findings are grounded in the data and not mere artifacts of the researcher's biases (Miles & Huberman, 1994).

Confirmability, a cornerstone of trustworthiness in qualitative research, denotes the degree to which the findings of a study can be confirmed or corroborated by others (Lincoln & Guba, 1985). Unlike objectivity in quantitative research, confirmability recognizes the inherent subjectivity in qualitative research but emphasizes the need for the findings to be rooted in the participants' voices and conditions of the study, rather than the biases or interests of the researcher. This section sheds light on the notion of confirmability, its paramount importance, and methods to achieve it.

A. Audit Trail:

Maintaining a comprehensive and transparent record of all research activities, decisions, and interpretive notes aids in ensuring that another researcher could follow the same procedures and arrive at similar conclusions (Rodgers, 2008). An audit trail includes raw data, analysis notes, and reflexive journals, enabling a clear path from the data to the findings and interpretations.

B. Reflexivity:

Through reflexivity, researchers consistently interrogate and document their positions, potential biases, and influences on the research process and outcome (Finlay, 2002). Reflexive practices make the research process transparent and open to scrutiny, supporting confirmability.

C. Triangulation:

Using multiple sources, methods, or investigators to validate the research findings ensures that the outcomes are a result of the phenomena under study rather than methodological artifacts or researcher biases (Denzin, 1978).

D. Member Checking:

By returning to participants with the preliminary findings and interpretations, researchers can ascertain the accuracy and resonance of the results with participants' experiences (Creswell & Poth, 2018). This feedback loop is pivotal for confirmability.

E. Peer Debriefing:

Engaging with peers or experts who can review and critique the research process and outcomes offers an external perspective, ensuring findings are grounded and not a result of researcher bias (Lincoln & Guba, 1985).

F. Confirmability Matrix:

Developing a matrix that links data points with their corresponding interpretations and thematic representations can visually demonstrate the alignment between raw data and findings, bolstering confirmability (Miles et al., 2014).

3. Strategies to Enhance Quality:
   - **Triangulation**: Using multiple methods, data sources, or researchers to corroborate findings (Denzin, 1978).
   - **Member Checking**: Returning to participants for validation of preliminary findings (Creswell & Miller, 2000).
   - **Thick Descriptions**: Providing detailed accounts to allow readers to evaluate transferability (Geertz, 1973).
   - **Audit Trails**: Keeping detailed records of the research process for scrutiny (Lincoln & Guba, 1985).
   - **Peer Debriefing**: Engaging external colleagues to review and challenge the research process and interpretations (Creswell & Poth, 2018).
4. Reflexivity and Researcher Positionality:

Quality in qualitative research also hinges on the researcher's awareness of their role, biases, and potential influence on the research. Reflexivity involves ongoing introspection and critical examination of one's positionality and its impact on the research process (Finlay, 2002).

Reflexivity and researcher positionality have emerged as pivotal concepts in ensuring quality and rigor in qualitative research. As qualitative inquiries delve deeply into human experiences, interactions, and perceptions, it becomes imperative to recognize the influence of the researcher on the very phenomena being studied. This section delves into the intricate relationship between reflexivity, researcher positionality, and the pursuit of trustworthy qualitative research.

A. Reflexivity: A Dynamic Interrogation

Reflexivity refers to the process of self-reflection, where researchers critically examine themselves, their impact on the research, and vice versa (Altheide & Johnson, 1998). It underscores the idea that research is not conducted in a vacuum; researchers bring their histories, biases, and preconceptions to the study. Engaging in reflexivity ensures:

- **Recognition of Biases**: Reflexive practices force researchers to identify and confront their biases and preconceptions that might shape the research process or findings (Finlay, 2002).
- **Transparency in Decision Making**: By continuously reflecting on decisions made during the research, from sampling to data interpretation, scholars can offer a transparent account of their research process (Malterud, 2001).

B. Researcher Positionality: Defining One's Standpoint

Positionality denotes the stance or positioning of the researcher in relation to the research context, considering aspects like gender, race, socioeconomic status, etc. (Rose, 1997). It's about recognizing:

- **Power Dynamics**: The researcher's position might influence participants' responses or the data interpretation. Being aware of this dynamic is crucial for ethical and rigorous research (Merriam et al., 2001).
- **Subjective Interpretations**: Every researcher approaches data with their own lens, influenced by their background, experiences, and beliefs. Recognizing this subjectivity is vital for credibility (Haraway, 1988).

C. Interplay of Reflexivity and Positionality

Reflexivity and positionality are intertwined in a dynamic dance. While reflexivity is the act of self-reflection, positionality provides the lens through which this reflection is viewed. Together, they ensure:

- **Holistic Understanding**: Recognizing the impact of one's position on the research can lead to a more holistic understanding of the data (Pillow, 2003).
- **Ethical Research Practices**: Being reflexive and aware of one's positionality can guide more ethical interactions with participants, ensuring their voices are represented authentically (Koelsch, 2013).

5. Ethical Considerations:

Beyond methodological rigor, qualitative research also demands ethical rigor. This involves ensuring the dignity, rights, and well-being of participants are upheld throughout the research process (Orb et al., 2001).

Ethical considerations stand at the forefront of any research endeavor, but they assume an even more pronounced significance in qualitative research due to its intimate, immersive, and often personal nature. Ensuring that qualitative research is undertaken with integrity not only requires adherence to ethical guidelines but also an ongoing reflexive consideration of the moral implications of research actions and decisions.

A. **Informed Consent**: A foundational tenet of ethical research, informed consent is not just a one-time agreement but an ongoing process in qualitative studies. Researchers must ensure that participants fully understand the nature of the research, its aims, potential risks, and their rights as participants (Mack et al., 2012).

B. **Anonymity and Confidentiality**: Given the depth and detail of qualitative data, maintaining participants' anonymity can be challenging. Ensuring confidentiality protects participants from potential harm, especially in sensitive research contexts (Orb et al., 2001).

C. **Emotional and Psychological Considerations**: Qualitative research often delves into deeply personal, sensitive, or traumatic areas of participants' lives. Researchers must be prepared to offer support or referrals to counseling if distressing issues arise (Dickson-Swift et al., 2007).

D. **Researcher's Responsibility**: The qualitative researcher often has a close relationship with participants. They need to be aware of power dynamics and ensure they don't misuse their position, consciously or unconsciously (Liamputtong, 2007).
E. **Cultural Sensitivity**: Especially pertinent in cross-cultural or indigenous research contexts, there's a need for cultural competency, respect for local customs, and understanding of cultural nuances (Chilisa, 2012).
F. **Reflexivity and Ethical Practice**: Reflexivity involves researchers critically reflecting on their own biases, actions, and the broader implications of their research. Through a reflexive lens, researchers can better navigate the ethical complexities of their work (Berger, 2015).
G. **Data Management**: Ethical handling, storage, and eventual disposal of research data, especially in the digital age, are vital to protect participants' information and confidentiality (Van den Hoonaard, 2002).

6. Challenges and Future Directions:

The chapter concludes by addressing the challenges faced in ensuring quality and rigor in the ever-evolving field of qualitative research and posits future directions, emphasizing the need for adaptability, openness, and ongoing critical discourse.

As qualitative research continues to evolve, it faces various challenges that underscore the necessity for reflexivity, adaptability, and a forward-thinking perspective. These challenges also open up possibilities for future directions in enhancing quality and rigor.

A. **The Challenge of Subjectivity**: One of the primary critiques against qualitative research has been its subjective nature, with detractors arguing that it can be biased or too dependent on the researcher's perspective (Hammersley, 2007). However, proponents argue that subjectivity, when recognized and accounted for through reflexivity, can provide deeper insights than purely objective approaches.
B. **The Balancing Act of Depth vs. Generalizability**: Due to its in-depth focus, qualitative research often struggles with issues of generalizability. The challenge is to convey the richness of data while acknowledging its contextual limitations (Maxwell, 2012).
C. **Adapting to Technological Advancements**: With the rise of digital technologies, new mediums for data collection have emerged, such as online forums, social media, and virtual realities. These present both opportunities and challenges in terms of data authenticity, ethical considerations, and analysis (Fielding & Fielding, 2012).
D. **Interdisciplinary Collaborations**: As research becomes more interdisciplinary, qualitative research needs to navigate the challenges of integrating its methods with those from other disciplines. These collaborations offer potential for richer insights but also pose methodological challenges (Brewer & Hunter, 2006).
E. **Globalization and Cross-cultural Research**: In our increasingly globalized world, qualitative researchers often work across different cultures, requiring heightened sensitivity and adaptability. There's the challenge of avoiding ethnocentrism while ensuring findings are culturally grounded (Smith, 2016).

Future Directions

- **Enhanced Training**: As qualitative methods evolve, there's a need for more comprehensive training, not just in data collection and analysis, but also in ethical considerations, particularly in digital realms.
- **Methodological Innovations**: Embracing new tools, technologies, and interdisciplinary methods can push the boundaries of traditional qualitative research, offering more avenues for exploration.
- **Strengthening the Paradigmatic Foundations**: Engaging more deeply with philosophical and epistemological underpinnings can bolster the rigor and depth of qualitative inquiries.
- **Championing Public Scholarship**: In an age of 'alternative facts' and skepticism towards academia, qualitative researchers have a role in bridging the gap between academia and the public, making research more accessible and relevant.

Validity and Reliability in Qualitative Research

The concepts of validity and reliability, central to ensuring the quality of research findings, have been traditionally associated with quantitative research. However, their relevance and application to qualitative research, while being subjected to debate, have become essential to ensure rigor and credibility (Creswell & Miller, 2000).

A. Redefining Validity and Reliability in Qualitative Paradigms:

Traditional definitions of validity and reliability, anchored in positivist perspectives, are not directly translatable to qualitative research. Lincoln & Guba (1985) introduced alternative terms better aligned with qualitative epistemologies:

- *Credibility* (paralleling internal validity)
- *Transferability* (paralleling external validity/generalizability)
- *Dependability* (paralleling reliability)
- *Confirmability* (paralleling objectivity).

B. Establishing Credibility:

Credibility concerns the 'truth' of the research findings from the participants' viewpoints. Triangulation (using multiple data sources or methods), member-checking (seeking feedback from participants on findings), and prolonged engagement in the field are common strategies to enhance credibility (Merriam, 1998).

C. Ensuring Transferability:

Transferability refers to the extent to which findings can be applied to other settings or groups. Qualitative researchers address this by providing rich, thick descriptions of the research context, allowing readers to make informed judgments about transferability (Geertz, 1973).

D. Maintaining Dependability:

While reliability in quantitative research emphasizes consistency of findings over time, dependability in qualitative research underscores the need for the researcher to account for the changing context within which research occurs. An 'audit trail', detailing all aspects of the research process, can enhance dependability (Lincoln & Guba, 1985).

E. Upholding Confirmability:

Confirmability refers to the degree to which research findings can be confirmed or corroborated by others. Qualitative researchers achieve this by being reflexive, documenting their biases, and allowing for an external audit, where another researcher examines the research process and findings (Shenton, 2004).

F. Reliability in Qualitative Research:

Reliability in qualitative research focuses on the researcher's approach's consistency and the extent to which the findings can be replicated. To enhance reliability, researchers often use coding checks, peer debriefing, and maintain transparent documentation of the analytical process (Miles & Huberman, 1994).

G. Embracing Complexity and Contingency:

While the adoption of validity and reliability criteria enhances rigor, qualitative research recognizes and embraces the complex, emergent, and contingent nature of human experience and social reality. Rather than seeing these as limitations, they are viewed as intrinsic strengths that capture the depth and nuance of lived experiences (Denzin, 2009).

Member Checking

Member checking, often considered a cornerstone of qualitative research's trustworthiness, involves taking preliminary findings back to participants to validate, refine, or challenge the researcher's interpretations (Lincoln & Guba, 1985). This iterative process, involving participants in the verification of data and interpretations, stands as a counterpoint to claims of subjectivity in qualitative research.

A. Rationale for Member Checking:

The essence of qualitative research lies in capturing and interpreting participants' lived experiences. Member checking serves as a bridge, ensuring that researchers' interpretations align with participants' intentions and lived realities (Creswell & Poth, 2018).

B. Process of Member Checking:

Typically, after initial data analysis, researchers share summaries, themes, or even entire drafts with participants. These could be shared during face-to-face meetings, electronically, or over phone calls. Participants are then invited to confirm, refute, or provide additional insights on the presented material (Birt, Scott, Cavers, Campbell, & Walter, 2016).

C. Benefits of Member Checking:
- **Enhanced Credibility**: Validating findings with participants boosts the study's credibility, ensuring that interpretations align with participants' experiences (Harper & Cole, 2012).
- **Refinement of Findings**: Feedback from participants can lead to more nuanced, accurate, and richer findings (Tobin & Begley, 2004).
- **Empowerment**: Member checking can empower participants, giving them a voice in the research process and reducing power imbalances between researchers and participants (Thomas, 2017).

D. Challenges and Criticisms:
- **Practical Constraints**: Member checking can be time-consuming and may not always be feasible, especially in studies with large participant groups or transient populations (Morse, 2015).
- **Potential Biases**: Some participants might not feel comfortable challenging a researcher's interpretations, especially in hierarchical or power-imbalanced settings (Smith & McGannon, 2018).
- **Over-reliance**: Using member checking as the sole measure of a study's validity might lead to overlooking other important validation strategies (Maxwell, 2012).

E. Best Practices:

To optimize the benefits of member checking, researchers should approach the process transparently, explaining its purpose to participants. It's essential to be open to criticism, using it as a tool for refining rather than defending initial interpretations. Furthermore, integrating other validation strategies alongside member checking can enhance the overall rigor and trustworthiness of the study (Trippany-Simmons, Dankoski, & Penn, 2019).

Triangulation

Triangulation, a term borrowed from navigation and land surveying, refers to the use of multiple methods, sources, or perspectives to validate and enrich research findings (Denzin, 1978). In the realm of qualitative research, triangulation serves as a strategy to enhance the depth, breadth, and consistency of study results, thereby bolstering the study's trustworthiness.

A. Rationale for Triangulation:

The heterogeneous nature of human experiences and social realities demands a multifaceted approach to understanding them. By employing triangulation, researchers can delve into these complexities, ensuring that the investigation remains comprehensive and avoids blind spots (Patton, 2002).

B. Types of Triangulation:
- **Data Triangulation**: This involves gathering data from different sources, such as multiple participants, different occasions, or diverse sites, ensuring a well-rounded understanding of the research question (Carter, Bryant-Lukosius, DiCenso, Blythe, & Neville, 2014).

- **Method Triangulation**: By employing multiple qualitative (or even quantitative) methods, such as combining interviews with observations, researchers can validate and complement findings across methods (Denzin, 1989).
- **Theoretical Triangulation**: This approach uses different theoretical perspectives to interpret the same set of data. For example, a researcher could analyze data using both a feminist lens and a socio-economic perspective (Rothbauer, 2008).
- **Investigator Triangulation**: Multiple researchers independently analyze the data, and then their findings are compared for consistency and variability (Thurmond, 2001).

C. Benefits of Triangulation:

- **Comprehensive Understanding**: By approaching the research question from various angles, triangulation fosters a more thorough understanding of the phenomenon under study (Jick, 1979).
- **Validation of Findings**: Triangulation can help identify areas of convergence among data sources or methods, thereby validating the study's findings (Morse, 1991).
- **Refinement of Insights**: By highlighting inconsistencies or gaps, triangulation can direct researchers to refine their interpretations, leading to more nuanced insights (O'Donoghue & Punch, 2003).

D. Challenges of Triangulation:

- **Practical Constraints**: Implementing multiple data collection methods or engaging multiple researchers can be time-consuming and resource-intensive (Fielding, 2012).
- **Integration of Diverse Data**: Synthesizing findings from different methods or theoretical lenses can be challenging, requiring careful methodological and theoretical integration (Erzberger & Kelle, 2003).

E. Best Practices:

For effective triangulation, researchers must be transparent about the rationale behind their chosen triangulation strategy and clearly articulate how various data sources or methods are integrated. Flexibility in accommodating emergent patterns and inconsistencies is crucial for drawing rich, robust insights (Decrop, 1999).

Thick Descriptions

One of the fundamental principles underpinning qualitative research is its commitment to portraying the depth and complexity of human experiences. Clifford Geertz, a renowned anthropologist, emphasized the need for what he termed "thick descriptions" in the interpretation of cultural phenomena (Geertz, 1973). Thick description goes beyond mere facts to elucidate the significance, context, and deeper meaning of these facts.

A. Defining Thick Description:

Thick description is not merely about providing a lot of detail. Instead, it's about contextualizing the behaviors or actions being described. It entails delving into the layers of meaning embedded in everyday activities, providing insight into cultural norms, values, and symbols that shape individual and collective experiences (Ponterotto, 2006).

B. Importance of Thick Description:
   - **Contextual Understanding**: Thick description allows readers to understand the circumstances surrounding the data, thereby offering a holistic perspective on the studied phenomenon (Geertz, 1973).
   - **Transferability**: By offering detailed descriptions, researchers enable readers to gauge the applicability of study findings to other contexts or settings (Lincoln & Guba, 1985).
   - **Rich Interpretation**: Going beyond surface-level details, thick descriptions facilitate deeper interpretative understanding, fostering meaningful insights into human experiences (Ryle, 1949).

C. Crafting Thick Descriptions:
   - **Inclusion of Sensory Details**: Researchers should strive to capture sensory elements – what participants saw, heard, felt, or even smelled during their experiences (Denzin, 1989).
   - **Historical and Cultural Context**: It's crucial to elucidate the broader socio-cultural and historical backdrop against which the studied phenomena unfold (Holliday, 2007).
   - **Participant's Perspectives**: Researchers should endeavor to represent participants' own interpretations, beliefs, and emotions, presenting a multi-faceted view of the issue (Creswell & Miller, 2000).

D. Challenges with Thick Description:
   - **Time-Intensiveness**: Crafting detailed, layered descriptions can be time-consuming, demanding significant effort during both data collection and analysis phases (Mason, 2002).
   - **Risk of Overload**: Striking a balance between depth and brevity is challenging. Overwhelming readers with excessive detail can detract from the central themes of the research (Ponterotto, 2006).

E. Ensuring Rigor through Thick Description:

Achieving rigor in qualitative research is not solely about procedural precision; it's also about the depth and richness of understanding. Thick description, by providing intricate, contextually embedded interpretations, is pivotal in meeting this criterion, reinforcing the quality and trustworthiness of qualitative inquiry (Tracy, 2010).

REFERENCES

Brink, P. J., & Wood, M. J. (1998). Advanced design in nursing research. *Sage (Atlanta, Ga.).*

Bryman, A. (2012). *Social research methods* (4th ed.). Oxford University Press.

Carter, N., Bryant-Lukosius, D., DiCenso, A., Blythe, J., & Neville, A. J. (2014). The use of triangulation in qualitative research. *Oncology Nursing Forum*, *41*(5), 545–547. doi:10.1188/14.ONF.545-547 PMID:25158659

Creswell, J. W., & Miller, D. L. (2000). Determining validity in qualitative inquiry. *Theory into Practice*, *39*(3), 124–130. doi:10.1207/s15430421tip3903_2

Creswell, J. W., & Poth, C. N. (2018). *Qualitative inquiry and research design: Choosing among five approaches* (4th ed.). Sage.

Crotty, M. (1998). The foundations of social research: Meaning and perspective in the research process. *Sage (Atlanta, Ga.)*.

Decrop, A. (1999). Triangulation in qualitative tourism research. *Tourism Management, 20*(1), 157–161. doi:10.1016/S0261-5177(98)00102-2

Denzin, N. K. (1978a). *Sociological methods: A sourcebook*. McGraw-Hill.

Denzin, N. K. (1978b). *The research act: A theoretical introduction to sociological methods*. McGraw-Hill.

Denzin, N. K. (1989). *The research act: A theoretical introduction to sociological methods*. Prentice Hall.

Denzin, N. K., & Lincoln, Y. S. (2005). *The Sage handbook of qualitative* Denzin, N. K., & Lincoln, Y. S. (2005). *The Sage handbook of qualitative research. Sage (Atlanta, Ga.)*.

Erzberger, C., & Kelle, U. (2003). Making inferences in mixed methods: The rules of integration. In A. Tashakkori & C. Teddlie (Eds.), *Handbook of mixed methods in social & behavioral research* (pp. 457–488). Sage.

Fielding, N. (2012). Computer-aided qualitative data analysis. In J. Goodwin (Ed.), *SAGE secondary data analysis* (pp. 183–210). Sage Publications.

Fielding, N. G. (2012). Triangulation and mixed methods designs. *Journal of Mixed Methods Research, 6*(2), 124–136. doi:10.1177/1558689812437101

Finlay, L. (2002). "Outing" the Researcher: The Provenance, Process, and Practice of Reflexivity. *Qualitative Health Research, 12*(4), 531–545. doi:10.1177/104973202129120052 PMID:11939252

Geertz, C. (1973). *The interpretation of cultures: Selected essays*. Basic Books.

Guba, E. G. (1981). Criteria for assessing the trustworthiness of naturalistic inquiries. *Educational Communication and Technology, 29*(2), 75–91. doi:10.1007/BF02766777

Guba, E. G., & Lincoln, Y. S. (1982). Epistemological and methodological bases of naturalistic inquiry. *Educational Communication and Technology, 30*(4), 233–252. doi:10.1007/BF02765185

Guba, E. G., & Lincoln, Y. S. (1985). Naturalistic inquiry. *Sage (Atlanta, Ga.)*.

Guba, E. G., & Lincoln, Y. S. (1994). Competing paradigms in qualitative research. In N. K. Denzin & Y. S. Lincoln (Eds.), *Handbook of qualitative research* (pp. 105–117). Sage.

Hammersley, M. (1992). *What's Wrong with Ethnography? Methodological Explorations*. Routledge.

Hesse-Biber, S. (2017). *The practice of qualitative research: Engaging students in the research process*. Sage Publications.

Holliday, A. (2007). *Doing and writing qualitative research*. Sage. doi:10.4135/9781446287958

Jick, T. D. (1979). Mixing qualitative and quantitative methods: Triangulation in action. *Administrative Science Quarterly, 24*(4), 602–611. doi:10.2307/2392366

Kincheloe, J. L., & McLaren, P. (2005). Rethinking critical theory and qualitative research. In N. K. Denzin & Y. S. Lincoln (Eds.), *The Sage handbook of qualitative research* (3rd ed., pp. 303–342). Sage.

Lincoln, Y. S., & Guba, E. A. (1985). *Naturalistic inquiry*. Sage. doi:10.1016/0147-1767(85)90062-8

Malterud, K. (2001). Qualitative research: Standards, challenges, and guidelines. *Lancet*, *358*(9280), 483–488. doi:10.1016/S0140-6736(01)05627-6 PMID:11513933

Maxwell, J. A. (2012). *Qualitative research design: An interactive approach* (3rd ed.). Sage.

Merriam, S. B. (1998). *Qualitative research and case study applications in education*. Jossey-Bass.

Miles, M. B., & Huberman, A. M. (1994). Qualitative data analysis: An expanded sourcebook. *Sage (Atlanta, Ga.)*.

Morrow, S. L. (2005). Quality and trustworthiness in qualitative research in counseling psychology. *Journal of Counseling Psychology*, *52*(2), 250–260. doi:10.1037/0022-0167.52.2.250

Morse, J. M. (1991). Approaches to qualitative-quantitative methodological triangulation. *Nursing Research*, *40*(2), 120–123. doi:10.1097/00006199-199103000-00014 PMID:2003072

Morse, J. M., Barrett, M., Mayan, M., Olson, K., & Spiers, J. (2002). Verification strategies for establishing reliability and validity in qualitative research. *International Journal of Qualitative Methods*, *1*(2), 13–22. doi:10.1177/160940690200100202

O'Donoghue, T., & Punch, K. (2003). *Qualitative educational research in action: Doing and reflecting*. Routledge. doi:10.4324/9780203506301

Ortlipp, M. (2008). Keeping and using reflective journals in the qualitative research process. *The Qualitative Report*, *13*(4), 695–705.

Patton, M. Q. (1999). Enhancing the quality and credibility of qualitative analysis. *Health Services Research*, *34*(5 Pt 2), 1189. PMID:10591279

Patton, M. Q. (2002). Two decades of developments in qualitative inquiry. *Qualitative Social Work: Research and Practice*, *1*(3), 261–283. doi:10.1177/1473325002001003636

Ponterotto, J. G. (2006). Brief note on the origins, evolution, and meaning of the qualitative research concept thick description. *The Qualitative Report*, *11*(3), 538–549.

Rodgers, B. L. (2008). Audit trail. In L. M. Given (Ed.), *The Sage encyclopedia of qualitative research methods* (Vol. 1, pp. 44–45). Sage Publications.

Rothbauer, P. M. (2008). Triangulation. In L. M. Given (Ed.), *The Sage encyclopedia of qualitative research methods* (pp. 892–894). Sage.

Ryle, G. (1949). *The concept of mind*. Hutchinson.

Shenton, A. K. (2004). Strategies for ensuring trustworthiness in qualitative research projects. *Education for Information*, *22*(2), 63–75. doi:10.3233/EFI-2004-22201

Smith, B., & McGannon, K. R. (2018). Developing rigor in qualitative research: Problems and opportunities within sport and exercise psychology. *International Review of Sport and Exercise Psychology*, *11*(1), 101–121. doi:10.1080/1750984X.2017.1317357

Smith, J., & Hodkinson, P. (2005). Relativism, criteria, and politics. In *The Sage handbook of qualitative research* (3rd ed., pp. 915–931). Sage Publications.

Stake, R. E. (1995). *The art of case study research*. Sage Publications.

Thurmond, V. A. (2001). The point of triangulation. *Journal of Nursing Scholarship*, *33*(3), 253–258. doi:10.1111/j.1547-5069.2001.00253.x PMID:11552552

Tracy, S. J. (2010). Qualitative quality: Eight "big-tent" criteria for excellent qualitative research. *Qualitative Inquiry*, *16*(10), 837–851. doi:10.1177/1077800410383121

Trippany-Simmons, R., Dankoski, M. E., & Penn, C. D. (2019). Beyond member checking: Advancing rigor through participatory method. *Journal of Marital and Family Therapy*, *45*(2), 235–247.

Yin, R. K. (2003). *Case study research: Design and methods* (3rd ed.). Sage Publications.

Chapter 8
Writing and Presenting Qualitative Findings

ABSTRACT

This chapter discusses the critical aspects of writing and presenting qualitative research findings effectively. It emphasizes the need for crafting compelling narratives that convey the richness and complexity of human experiences captured through qualitative inquiry. The success of qualitative research is heavily reliant on how the findings are communicated, requiring skills in storytelling, structuring of presentations, and enhancing reader engagement through descriptive titles and visuals. The chapter highlights the importance of balancing objectivity with interpretation, ensuring ethical considerations such as participant anonymity, and addressing the challenges of managing voluminous data and meeting diverse audience expectations.

INTRODUCTION

Qualitative research is unique in its commitment to capturing the richness and nuance of human experiences, often yielding a complex and intricate array of data. However, the success of qualitative inquiry does not merely lie in its data collection and analysis, but also in how the findings are communicated to a broader audience. This chapter delves into the art and science of writing and presenting qualitative findings, emphasizing the significance of crafting a compelling narrative that brings out the depth, intricacy, and humanity inherent in the data.

1. The Importance of Communicating Qualitative Findings:

Qualitative findings, unlike quantitative ones, don't simply speak for themselves. They require interpretation, contextualization, and storytelling. Ensuring that these findings are communicated effectively is pivotal for several reasons:

DOI: 10.4018/979-8-3693-2414-1.ch008

- **Relevance**: To ensure that the nuanced insights derived from qualitative research are understood and appreciated by various stakeholders, including policymakers, practitioners, and the academic community (Denzin & Lincoln, 2005).
- **Validity**: Well-presented findings can enhance the credibility and trustworthiness of the research, fostering greater confidence in the outcomes (Creswell & Poth, 2018).

2. Crafting the Narrative:

- **Storytelling Approach**: Given its emphasis on human experiences, qualitative research aligns closely with storytelling. This requires researchers to weave a cohesive, engaging, and insightful narrative (Bochner & Ellis, 2016).
- **Balancing Objectivity and Interpretation**: While ensuring that participant voices shine through, the researcher must also provide interpretation, linking data to broader themes, theories, and implications (Richardson & Adams St. Pierre, 2005).

3. Structuring the Presentation of Findings:

- **Introduction**: Set the scene, reminding the reader of the research questions and objectives, and signposting what is to follow.
- **Thematic Presentation**: Organize findings around key themes, ensuring a logical flow (Braun & Clarke, 2006).
- **Use of Verbatim Quotes**: Integrate direct quotes from participants to bring authenticity and depth to the narrative (Saldaña, 2015).
- **Interpretation and Discussion**: Beyond presenting the findings, delve into their significance, relating them to existing literature and theoretical frameworks (Charmaz, 2014).

4. Enhancing Readability and Engagement:

- **Descriptive Titles**: Use clear, descriptive subheadings to guide the reader through the findings.
- **Visuals**: Incorporate tables, figures, or other visuals to represent patterns or highlight key insights (Saldaña, 2013).
- **Varied Presentation**: Balance longer, detailed descriptions with shorter, more focused sections to maintain reader engagement.

5. Challenges in Writing and Presenting Qualitative Findings:

- **Volume of Data**: Qualitative research often yields a vast amount of data. Deciding what to include and what to leave out can be challenging (Miles, Huberman, & Saldaña, 2014).
- **Audience Expectations**: Different audiences may have varied expectations, and striking a balance to cater to academic, practitioner, and lay audiences can be complex (Tracy, 2020).

6. Ethical Considerations in Presenting Findings:

- **Anonymity**: Ensure that any quotes or descriptions used don't inadvertently reveal the identity of participants (Orb, Eisenhauer, & Wynaden, 2001).
- **Representativeness**: Ensure that the narrative is representative of participants' actual experiences and perspectives, avoiding undue bias or misrepresentation (Kaiser, 2009).

In conclusion, writing and presenting qualitative findings is an intricate dance of science and art. It requires not just analytical acumen but also the creativity and sensitivity to present the human stories behind the data in a compelling and ethical manner.

Structuring the Findings Section

The findings section is the heart of a qualitative research report. It's where the voices of participants come alive, where patterns emerge, and where the researcher's interpretation plays a pivotal role in bringing meaning to the collected data. A well-structured findings section not only informs but also engages, providing a clear path through the richness and complexity of qualitative data.

A. Significance of Structuring the Findings Section:

The way findings are presented can make or break the reader's comprehension and appreciation of the research. A clear structure:

- Facilitates understanding by providing a logical progression of themes or categories (Creswell & Poth, 2018).
- Enhances the credibility of the research by showcasing thoroughness and analytical depth (Ritchie, Lewis, McNaughton Nicholls, & Ormston, 2013).

B. Typical Structure for the Findings Section:

- **Introduction**: Briefly reintroduce the research questions or objectives and provide an overview of the main themes or categories that emerged from the data (Yin, 2014).
- **Thematic Presentation**: Each theme or category should have its subsection. Begin with a short introduction to the theme, followed by detailed descriptions, supported by verbatim participant quotes. The descriptions should blend the participants' perspectives with the researcher's interpretation (Braun & Clarke, 2006).
- **Comparative Analysis (if applicable)**: If the study involves multiple cases, sites, or groups, present comparative insights highlighting similarities and differences (Stake, 1995).
- **Interlinking Themes**: Discuss how the themes or categories are interrelated, which can provide deeper insights into the phenomenon being studied (Flick, 2014).
- **Conclusion**: Summarize the key findings and their significance, linking back to the research questions or objectives.

C. Enhancing the Presentation of Findings:

- **Use Descriptive Subheadings**: Guide the reader through the findings by using clear, descriptive subheadings for each theme or category (Saldaña, 2015).
- **Visual Representation**: Utilize tables, charts, or models to represent relationships, patterns, or key insights. Visual aids can make complex data more accessible (Maxwell, 2012).
- **Highlight Key Quotes**: Choose particularly illuminating or representative quotes from participants to underscore specific points. This brings authenticity to the narrative and lets the reader hear the participants' voices directly (Seidman, 2013).

D. Ethical Considerations:

Ensure participants' confidentiality is preserved when presenting quotes or specific scenarios. This may involve changing names or other identifiable details (Orb, Eisenhauer, & Wynaden, 2001).

In conclusion, structuring the findings section in a clear, cohesive manner is crucial to convey the depth and richness of qualitative research. It ensures that the reader can follow the researcher's analytical journey, appreciating both the individual voices of the participants and the broader insights derived from their collective experiences.

Using Quotes and Anecdotes

In qualitative research, the words and experiences of the participants are central to conveying the depth, richness, and authenticity of the findings. Using direct quotes and anecdotes effectively can enhance the reader's engagement and understanding of the data, grounding abstract concepts in real-life contexts.

A. Significance of Using Quotes and Anecdotes:
   - **Authenticity and Credibility**: Direct quotations provide readers with evidence that supports the researcher's interpretations and claims. They serve as proof that the findings are rooted in participants' experiences (Lindlof & Taylor, 2017).
   - **Emphasis**: Memorable or poignant quotes and anecdotes can emphasize key points or themes in the data, making them more impactful and easier for the reader to remember (Riessman, 2008).
   - **Humanizing the Data**: Direct quotes allow readers to 'hear' the voices of participants, making the data more relatable and human (Seidman, 2013).

B. Effective Use of Quotes:
   - **Relevance**: Only include quotes that directly relate to and illustrate the themes or points being discussed (Creswell, 2013).
   - **Representation**: Ensure the selected quotes are representative of the broader dataset, avoiding over-reliance on particularly eloquent or vocal participants (Patton, 2015).
   - **Clarity and Context**: Introduce each quote with enough context for the reader to understand its significance. After presenting a quote, provide an interpretation or elaboration as necessary (Charmaz, 2014).
   - **Preservation of Voice**: Avoid excessive editing of quotes. Maintain the original language, tone, and style of the participant to preserve the authenticity of their voice (Silverman, 2016).

C. Integrating Anecdotes Effectively:

- **Purposeful Selection**: Anecdotes should be chosen for their ability to illuminate key points, themes, or insights. They should add depth, context, or clarity to the findings (Ellis & Bochner, 2000).
- **Narrative Flow**: Integrate anecdotes smoothly within the narrative, ensuring they flow logically and do not disrupt the reader's engagement (Richardson, 1997).
- **Ethical Considerations**: When using anecdotes, ensure participants' confidentiality by altering identifying details, without compromising the essence of the story (Orb, Eisenhauer, & Wynaden, 2001).

D. Challenges in Using Quotes and Anecdotes:
- **Over-reliance**: Excessive use can overwhelm the reader and overshadow the researcher's analysis and interpretation (Miles, Huberman, & Saldaña, 2014).
- **Out of Context**: Quotes or anecdotes used without sufficient context may be misinterpreted or lose their impact (Geertz, 1973).

In conclusion, effectively incorporating quotes and anecdotes in qualitative research enriches the narrative, providing authenticity, depth, and emotional resonance. When used judiciously, they become powerful tools that bridge the gap between raw data and interpreted findings, offering readers a tangible connection to participants' lived experiences.

Balancing Description and Interpretation

Qualitative research findings are inherently multifaceted, characterized by a combination of rich descriptions and researcher interpretations. Striking the right balance between description and interpretation is pivotal to ensure that the essence of the data is preserved while also providing meaningful insights and implications. This equilibrium is instrumental in effectively conveying the depth of participants' experiences and the broader relevance of the study.

A. The Significance of Description in Qualitative Research:
- **Richness and Depth**: Description provides a detailed account of participants' experiences, contexts, behaviors, and emotions, facilitating an immersive experience for readers (Denzin, 1989).
- **Preserving Authenticity**: Detailed description ensures that participants' voices and perspectives are accurately represented, maintaining the integrity and authenticity of the data (Holliday, 2007).
- **Building a Foundation for Interpretation**: An in-depth description sets the stage for subsequent interpretation, providing readers with the necessary context to understand the derived insights (Maxwell, 2012).

B. The Role of Interpretation:
- **Beyond Surface Level**: While description paints the picture, interpretation delves deeper, exploring underlying meanings, patterns, relationships, and implications (Geertz, 1973).
- **Linking to Theory and Literature**: Interpretation allows the researcher to connect the findings with existing theories, literature, and broader societal or disciplinary contexts (Miles, Huberman, & Saldaña, 2014).

- **Generating New Insights**: Through interpretation, the researcher can shed light on novel perspectives, contradictions, or emerging themes that might not be immediately evident from raw descriptions alone (Patton, 2015).

C. Achieving Balance:
- **Interweaving Description and Interpretation**: Rather than isolating description and interpretation, they can be interwoven throughout the narrative, allowing readers to seamlessly transition between what was observed and the associated meaning or significance (Creswell, 2013).
- **Clarity in Presentation**: Clearly demarcate descriptive segments from interpretative passages, ensuring readers can distinguish between observed data and researcher insights (Smith & Osborn, 2008).
- **Iterative Analysis**: Revisiting the data multiple times can help in refining descriptions and ensuring that interpretations align closely with the data (Braun & Clarke, 2006).

D. Potential Pitfalls and Solutions:
- **Over-Description**: While richness is desired, excessive detail can overwhelm readers, detracting from key themes and insights. Solution: Focus on the most salient and illustrative examples (Richards & Morse, 2013).
- **Over-Interpretation**: Drawing conclusions that are too far removed from the data can compromise credibility. Solution: Ground interpretations in explicit evidence from the dataset (Charmaz, 2014).

In conclusion, the delicate balance between description and interpretation in qualitative findings is crucial for crafting a comprehensive yet insightful narrative. By intertwining rich descriptions with meaningful interpretations, researchers can illuminate the intricacies of their data while providing valuable contributions to their respective fields.

Considerations for Publishing in Academic Journals

Publishing qualitative research findings in academic journals can be a challenging yet rewarding endeavor. It entails not just the presentation of raw data but, more importantly, the communication of insights, interpretations, and implications in a manner that is both rigorous and engaging. To successfully publish qualitative research in academic journals, it's imperative to be attuned to various nuances, standards, and expectations.

A. Understanding Journal Aims and Scope:
- **Aligning with the Journal's Mission**: Every academic journal has a distinct mission or focus. Researchers should ensure that their study aligns with the journal's aims and scope, enhancing the likelihood of acceptance (Lundh, 2014).
- **Targeting Appropriate Audiences**: Different journals cater to different audiences. Knowing the intended readership helps in tailoring the paper's content, tone, and depth (Paltridge & Starfield, 2007).

B. Crafting the Manuscript:

- **Structured Abstract**: Most academic journals require a structured abstract that summarizes the research's key components, including objectives, methods, results, and conclusions (Tullu & Karande, 2018).
- **Introduction**: Set the stage by highlighting the study's significance, research gap, and the specific objectives or research questions.
- **Methodology**: A clear, detailed, and justifiable account of the research design, sample, data collection, and analysis process is essential (Flick, 2018).
- **Results**: Present the findings using rich descriptions, illustrative quotes, and a logical flow, ensuring coherence and depth (Tracy, 2020).
- **Discussion**: Situate the findings within the broader literature, discuss implications, and address the study's limitations.
- **Conclusion**: Summarize key takeaways and suggest avenues for future research.

C. Adhering to Journal Formatting and Stylistic Guidelines:

- Following the specific journal's guidelines related to citation style, manuscript length, figure/table presentation, and other formatting details is crucial (Gastel & Day, 2016).

D. Engaging with Reviewer Feedback:

- **Receptive Stance**: Consider reviewers' feedback as constructive criticism aimed at enhancing the manuscript's quality (Nicol & Pexman, 2003).
- **Systematic Addressal**: Respond to each comment, clarifying changes made or providing justifications for retaining original content.
- **Maintaining Professionalism**: Always adopt a respectful tone, even if disagreeing with certain comments.

E. Ethical Considerations:

- Ensure that any identifiable participant information is anonymized or seek necessary permissions.
- Disclose any conflicts of interest and funding sources (Smith, 2018).

F. Open Access and Impact Factor Considerations:

- Researchers might consider the benefits of publishing in open access journals for wider dissemination (Björk & Solomon, 2012).
- While a journal's impact factor can be an indicator of its reputation, it should not be the sole consideration (Ravenscroft, 2017).

In conclusion, publishing qualitative research findings in academic journals is a meticulous process that requires careful attention to detail, an understanding of journal expectations, and an openness to constructive critique. By ensuring alignment with the chosen journal's scope, crafting a compelling narrative, and engaging with the review process constructively, researchers can contribute valuable insights to the academic community.

REFERENCES

Björk, B. C., & Solomon, D. (2012). Open access versus subscription journals: A comparison of scientific impact. *BMC Medicine*, *10*(1), 73. doi:10.1186/1741-7015-10-73 PMID:22805105

Bochner, A. P., & Ellis, C. (2016). *Evolving trends in ethnography*. Routledge.

Braun, V., & Clarke, V. (2006). Using thematic analysis in psychology. *Qualitative Research in Psychology*, *3*(2), 77–101. doi:10.1191/1478088706qp063oa

Charmaz, K. (2014). *Constructing grounded theory* (2nd ed.). Sage.

Creswell, J. W. (2013). *Research design: Qualitative, quantitative, and mixed methods approaches*. Sage publications.

Creswell, J. W., & Poth, C. N. (2018). *Qualitative inquiry and research design: Choosing among five approaches* (4th ed.). Sage.

Denzin, N. K., & Lincoln, Y. S. (2005). *The Sage handbook of qualitative* Denzin, N. K., & Lincoln, Y. S. (2005). *The Sage handbook of qualitative research. Sage (Atlanta, Ga.).*

Ellis, C., & Bochner, A. P. (2000). Autoethnography, personal narrative, reflexivity. In N. K. Denzin & Y. S. Lincoln (Eds.), *Handbook of qualitative research* (2nd ed., pp. 733–768). Sage.

Flick, U. (2018). *Doing triangulation and mixed methods*. Sage. doi:10.4135/9781529716634

Gastel, B., & Day, R. A. (2016). *How to write and publish a scientific paper*. Greenwood. doi:10.5040/9798400666926

Geertz, C. (1973). *The interpretation of cultures: Selected essays*. Basic Books.

Kaiser, K. (2009). Protecting respondent confidentiality in qualitative research. *Qualitative Health Research*, *19*(11), 1632–1641. doi:10.1177/1049732309350879 PMID:19843971

Lundh, A. (2014). Aims and scope of journals are not enough. *Evidence-Based Medicine*, *19*(1), 36. PMID:23939598

Maxwell, J. A. (2012). *Qualitative research design: An interactive approach* (3rd ed.). Sage.

Miles, M. B., Huberman, A. M., & Saldaña, J. (2014). *Qualitative data analysis: A methods sourcebook* (3rd ed.). Sage.

Nicol, D., & Pexman, P. M. (2003). *Displaying your findings: A practical guide for creating figures, posters, and presentations*. American Psychological Association.

Orb, A., Eisenhauer, L., & Wynaden, D. (2001). Ethics in qualitative research. *Journal of Nursing Scholarship*, *33*(1), 93–96. doi:10.1111/j.1547-5069.2001.00093.x PMID:11253591

Paltridge, B., & Starfield, S. (2007). *Thesis and dissertation writing in a second language: A handbook for supervisors*. London: Routledge.

Patton, M. Q. (2015). *Qualitative research & evaluation methods: Integrating theory and practice*. Sage publications.

Ravenscroft, J. (2017). The importance of journal impact factor in contemporary academic psychiatry and a list of selected journals. *Academic Psychiatry*, *41*(4), 507–511.

Richards, L., & Morse, J. M. (2013). *Readme first for a user's guide to qualitative methods* (3rd ed.). Sage. doi:10.4135/9781071909898

Richardson, L. (1997). *Fields of play: Constructing an academic life*. Rutgers University Press.

Richardson, L., & Adams St. Pierre, E. (2005). Writing: A method of inquiry. In N. K. Denzin & Y. S. Lincoln (Eds.), *The Sage handbook of qualitative research* (3rd ed., pp. 959–978). Sage.

Riessman, C. K. (2008). *Narrative methods for the human sciences*. Sage Publications.

Ritchie, J., Lewis, J., Nicholls, C. M., & Ormston, R. (Eds.). (2013). *Qualitative research practice: A guide for social science students and researchers*. Sage.

Saldaña, J. (2013). *The coding manual for qualitative researchers* (2nd ed.). Sage.

Saldaña, J. (2015). *The coding manual for qualitative researchers* (3rd ed.). Sage.

Seidman, I. (2013). *Interviewing as qualitative research: A guide for researchers in education and the social sciences*. Teachers College Press.

Silverman, D. (2016). *Qualitative research* (4th ed.). Sage.

Smith, J. A., & Osborn, M. (2008). Interpretative phenomenological analysis. In J. A. Smith (Ed.), *Qualitative psychology: A practical guide to research methods* (2nd ed., pp. 53–80). Sage.

Tracy, S. J. (2020). *Qualitative research methods: Collecting evidence, crafting analysis, communicating impact*. Wiley.

Tullu, M. S., & Karande, S. (2018). Writing a model research paper: A roadmap. *Journal of Postgraduate Medicine*, *64*(4), 212. PMID:29943738

Yin, R. K. (2014). *Case study research design and methods* (5th ed.). Sage Publications.

Chapter 9
Ethical Concerns in Qualitative Research

ABSTRACT

This chapter explores the ethical challenges inherent in qualitative research, focusing on the profound responsibility researchers bear when delving into human experiences, perceptions, and emotions. It emphasizes the importance of adhering to ethical guidelines, such as the Belmont Report, which advocates for respect, beneficence, and justice. Key areas of focus include the necessity of informed consent, the safeguarding of participant privacy, and the management of the emotional impacts of research.

INTRODUCTION

Qualitative research, as an investigative approach, delves deeply into human experiences, perceptions, and emotions, often within their natural settings. Given its close engagement with human participants, the ethical considerations inherent in the process are paramount. This chapter endeavors to explore the multifaceted ethical concerns in qualitative research and to provide guidance for researchers to navigate these challenges responsibly and with integrity.

A. Foundations of Research Ethics:

Historically, some of the most egregious violations of human rights have taken place in the name of scientific research (Shamoo & Resnik, 2015). As a result, various codes, guidelines, and principles have been established worldwide to protect research participants. One of the most recognized is the Belmont Report, which outlines three primary ethical principles: respect for persons, beneficence, and justice (National Commission for the Protection of Human Subjects, 1979).

B. Informed Consent:

DOI: 10.4018/979-8-3693-2414-1.ch009

Arguably the cornerstone of ethical research, informed consent ensures that participants willingly engage in the study, fully aware of its procedures, risks, and implications (Morrow & Richards, 1996). This involves:

- **Transparency**: Clear communication about the study's purpose, methods, potential risks, and benefits.
- **Voluntariness**: Participants should feel no pressure or undue influence to participate (Miller et al., 2005).
- **Ongoing Process**: Consent is not a one-off event but a continuous process, allowing participants to withdraw at any point.

C. Privacy and Confidentiality:

Given the depth and sensitivity of data often collected in qualitative research, preserving participants' anonymity and confidentiality becomes crucial (Saunders et al., 2015). Measures include:

- **Pseudonymization**: Using fake names or codes to represent participants.
- **Safe Data Storage**: Encrypting electronic data and securely storing physical data.
- **Data Sharing Considerations**: Being cautious when sharing data even for academic purposes.

D. The Emotionally-Charged Nature of Qualitative Interviews:

Qualitative research can broach sensitive or traumatic topics, potentially causing distress for participants (Dickson-Swift et al., 2007). Researchers should:

- **Show Empathy**: Recognize and respect emotional responses.
- **Provide Support**: Offer debriefing or direct participants to professional support services when needed.

E. Power Dynamics:

Inherent in the researcher-participant relationship are power imbalances that can affect the research process and outcomes (Mertens & Ginsberg, 2009). To address these:

- **Acknowledge and Mitigate**: Researchers should be aware of these dynamics and strive to minimize their impact.
- **Cultural Sensitivity**: Especially in cross-cultural contexts, understanding and respecting cultural norms and practices is paramount (Tilley-Lubbs, 2009).

F. Reflexivity and Positionality:

Reflexivity involves recognizing and reflecting on the researcher's own biases, beliefs, and roles in the research process (Finlay, 2002). This is crucial in ensuring:

- **Honest Representation**: Being transparent about one's own perspective and how it may influence the research.
- **Balancing Objectivity and Subjectivity**: While qualitative research is interpretative, unchecked biases can undermine its credibility.

G. Ethical Review and Institutional Approval:

Most academic institutions have ethical review boards or committees that assess the ethical implications of proposed research projects (Emanuel et al., 2000). Gaining approval:

- **Ensures Adherence to Ethical Standards**: This external review ensures the research adheres to recognized ethical standards.
- **Provides a Degree of Protection**: Both for researchers and participants, indicating due diligence.

Ethical considerations permeate every phase of qualitative research, from design to dissemination. Ensuring the dignity, rights, and safety of participants is not just a regulatory requirement but a moral imperative. By being conscientious and proactive, researchers can uphold the highest ethical standards, ensuring that their work contributes positively to knowledge without causing harm.

Informed Consent

Informed consent remains a cornerstone of ethical qualitative research. It stands as a testament to the principle that research participants are treated with dignity, respect, and autonomy. Delving deeper into this concept, this section dissects the nuances and intricacies of informed consent in qualitative studies.

A. Historical Context of Informed Consent:

The concept of informed consent emerged prominently after the Nuremberg Trials, following World War II, where the unethical experiments on humans conducted by the Nazis were exposed (Annas & Grodin, 1992). This pivotal moment in history underscored the importance of seeking voluntary and informed agreement from all research participants.

B. Elements of Informed Consent:

For consent to be considered truly "informed," several critical elements must be present (Beauchamp & Childress, 2001):

- **Disclosure**: Researchers must provide clear and comprehensive information about the study, including its objectives, methods, potential benefits and risks, and any alternative procedures or treatments that might be advantageous.
- **Capacity**: Participants must have the cognitive ability and mental capacity to comprehend the information provided and make a deliberate decision about their involvement.
- **Understanding**: The participants should grasp the nature, purpose, and potential consequences of the research.

- **Voluntariness**: Consent must be given freely, without any form of coercion, undue influence, or misrepresentation.

C. The Dynamic Nature of Informed Consent:

In qualitative research, given its evolving and often flexible design, informed consent is not a one-time event. It's a process that might require re-consent as the research progresses, especially if there are significant changes in the research design or new risks emerge (Guillemin & Gillam, 2004).

D. Challenges in Acquiring Informed Consent:
   - **Complexity of Information**: There's a balance to be struck between providing sufficient information and overwhelming participants with technical jargon (Molyneux et al., 2004).
   - **Cultural Differences**: In some cultures, the Western concept of individual informed consent may be at odds with community-based decision-making processes (Marshall & Koenig, 2004).
   - **Vulnerable Populations**: Special considerations arise when working with populations such as children, people with cognitive impairments, or indigenous communities. Obtaining informed consent from these groups might require adapted or additional procedures (Flory & Emanuel, 2004).

E. E-Consent:

With the digital era's advent, electronic informed consent or e-consent is becoming more prevalent, especially in large-scale studies or studies conducted remotely. Researchers should ensure that e-consent retains the depth and comprehensibility of traditional methods and that participants' data privacy is upheld (Manson, 2019).

F. Documenting Informed Consent:

Traditionally, the informed consent process culminates with the signing of a consent form. However, in certain settings, verbal consent might be more appropriate. Regardless of the method, the key is proper documentation and ensuring that the core principles of informed consent are upheld (Eysenbach & Wyatt, 2002).

Informed consent is not merely a procedural hurdle but an embodiment of the respect and autonomy owed to every research participant. By staying true to its principles, qualitative researchers can ensure that their work upholds the highest ethical standards.

Anonymity and Confidentiality

Ensuring the anonymity and confidentiality of participants is paramount in the ethical conduct of qualitative research. This not only safeguards the rights and dignity of the participants but also upholds the trustworthiness and validity of the study itself. This section delves deep into the concepts of anonymity and confidentiality, highlighting their importance and potential challenges.

A. Understanding Anonymity and Confidentiality:

At its core, anonymity ensures that a participant's responses cannot be linked back to them, either during the study or when findings are disseminated (Emanuel, Wendler, & Grady, 2000). Confidentiality, on the other hand, ensures that private data about participants, including their identities, are not disclosed without their consent (Sieber, 1992).

B. Rationale for Ensuring Anonymity and Confidentiality:
   - **Respect for Persons**: Protecting identities honors the autonomy and dignity of participants (Resnik, 2015).
   - **Avoiding Harm**: Maintaining confidentiality can prevent potential harm, such as social stigmatization, psychological distress, or economic repercussions that might arise from disclosing sensitive information (Sieber & Stanley, 1988).
   - **Promoting Openness**: When participants are assured of their anonymity and confidentiality, they are more likely to provide accurate, honest, and comprehensive information, thus enhancing the study's quality (Madrigal & McClain, 2020).

C. Challenges in Ensuring Anonymity and Confidentiality:
   - **Technological Issues**: With digital data storage and online data sharing, there's an increased risk of unintentional data breaches (Zimmer, 2010).
   - **Qualitative Specifics**: Qualitative data, rich in detailed narratives, might inadvertently disclose identifiable details, even when names are changed (Saunders et al., 2015).
   - **Longitudinal Studies**: Repeated interactions over extended periods can make it challenging to maintain strict anonymity (Baez & Casilli, 2020).

D. Strategies to Uphold Anonymity and Confidentiality:
   - **Data Handling**: Use pseudonyms or codes instead of names, and securely store any linking codes separately from the data (Wiles et al., 2008).
   - **Reporting**: Present aggregated data or modify certain identifiable details without compromising the data's essence (Tilley & Woodthorpe, 2011).
   - **Data Storage**: Utilize encrypted digital storage systems and ensure paper records are stored securely. Once the research is complete, consider how long data needs to be retained before safely destroying or anonymizing it (Markham & Buchanan, 2012).
   - **Team Training**: Ensure that all research team members are trained in ethical data handling and are fully aware of the importance of anonymity and confidentiality (Emanuel et al., 2000).

E. Informed Consent and Confidentiality:

When obtaining informed consent, researchers should be transparent about how they intend to maintain anonymity and confidentiality and address any potential limits to these (Elger & Caplan, 2006).

While ensuring anonymity and confidentiality in qualitative research presents challenges, it remains a non-negotiable aspect of ethical research conduct. As technological and methodological landscapes evolve, so too should our strategies and commitments to uphold these principles.

Emotional and Psychological Impact on Participants

The ethical considerations in qualitative research encompass not only the protection of participants' rights and confidentiality but also the safeguarding of their emotional and psychological well-being.

Qualitative methodologies, by nature, delve deep into personal, sensitive, and at times, distressing areas of participants' lives. Ensuring the emotional safety of participants becomes a paramount responsibility of researchers.

I. Nature of Emotional and Psychological Impacts:

Qualitative interviews, focus groups, or ethnographic studies can bring up past traumas, unresolved issues, or sensitive topics, leading to distress, discomfort, or heightened emotions (Dickson-Swift et al., 2007).

Qualitative research, in its endeavor to understand human experiences, feelings, and perspectives, often involves probing into areas of participants' lives that might be sensitive, deeply personal, or traumatic. The elicitation of such experiences or memories may exert emotional and psychological impacts on participants, influencing both their well-being and the authenticity of the research findings.

A. Depth and Breadth of Personal Experiences:

Qualitative research is premised on the in-depth exploration of personal experiences, feelings, and narratives. This depth often means touching upon subjects that can be emotionally charged or sensitive for the participants (Denzin & Lincoln, 2005).

B. Unearthing Past Traumas:

Discussing or reflecting upon traumatic past events—such as abuse, loss, or violent experiences—can trigger intense emotional responses, potentially re-traumatizing participants (Ellis et al., 2007).

C. Confronting Unresolved Emotions:

Participants might have unresolved emotions or conflicts related to certain events or experiences. Bringing these to the forefront during research can lead to heightened emotional turmoil or distress (Dickson-Swift et al., 2008).

D. Encountering Unexpected Emotional Responses:

Even topics that seem benign or neutral on the surface can evoke strong, unexpected emotional responses. The very act of verbalizing certain memories or sentiments can elicit powerful reactions (Rose, 2008).

E. Power Dynamics and Vulnerability:

The researcher-participant relationship often entails power dynamics, with participants potentially feeling vulnerable. This vulnerability can amplify the emotional impact, especially if participants feel pressured to share (Orb et al., 2001).

F. Long-Term Psychological Consequences:

While the immediate emotional reactions are often evident, the long-term psychological impacts—such as anxiety, stress, or introspection—can be subtle and may manifest long after the research interaction has concluded (Herman, 2005).

The nature of emotional and psychological impacts in qualitative research underscores the ethical obligation researchers have to anticipate, recognize, and mitigate potential distress or harm to participants. Protecting participants' well-being while obtaining rich, genuine data necessitates a delicate balance, informed by ethical guidelines and a profound sense of empathy.

II. Importance of Acknowledging Emotional Impact:

Understanding and acknowledging the potential emotional impact on participants within the realm of qualitative research is paramount, both from an ethical standpoint and to ensure the integrity of the research process. Recognizing this emotional dimension is essential for various reasons:

A. Upholding the Principle of 'Do No Harm':

Historically, the primary tenet of research ethics has been the principle of "do no harm." Qualitative researchers must be acutely aware of the potential for harm that arises from exploring personal, sensitive, or traumatic topics. Neglecting the emotional toll on participants can inadvertently lead to harm (Brinkmann & Kvale, 2015).

B. Ensuring Genuine and Authentic Data:

If participants feel that their emotional well-being is disregarded, they might withhold information or provide insincere responses, impacting the authenticity and depth of the data collected (Etherington, 2007).

C. Fostering Trust and Building Rapport:

Acknowledging and addressing the emotional aspects of participation helps build trust between the researcher and the participant. This trust is crucial for cultivating a space where participants feel safe and supported in sharing their experiences (Morrow, 2005).

D. Enhancing Research Validity:

When researchers understand and acknowledge emotional impacts, they are better positioned to interpret data in context, enhancing the validity of their findings (Maxwell, 2012).

E. Facilitating Ethical Reflexivity:

Recognizing emotional impact fosters a sense of ethical reflexivity among researchers, where they continuously reflect upon their practices and interactions to ensure participants' well-being is prioritized (Guillemin & Gillam, 2004).

F. Supporting Participants Post-Research:

Awareness of potential emotional distress equips researchers to offer post-research support, like debriefing sessions or counseling referrals, ensuring participants are not left with unresolved emotions after their involvement (Liamputtong, 2007).

Acknowledging the emotional impact in qualitative research is not just an ethical imperative but a cornerstone for enhancing the quality, richness, and validity of research findings. As qualitative research seeks to delve deep into human experiences, it must do so with profound sensitivity, ensuring participants' emotional and psychological well-being remains front and center.

I. **Potential Sources of Emotional Impact**:

Qualitative research, by its inherent nature, delves deeply into participants' experiences, perceptions, and emotions. While this depth allows for a rich understanding of complex phenomena, it can also serve as a potential source of emotional distress for participants. Identifying potential triggers and understanding their implications is vital for ethically sound research. Here we explore some common sources of emotional impact in qualitative research:

A. Delving into Traumatic Experiences:

When research topics touch upon traumatic experiences such as abuse, violence, loss, or other traumatic events, participants might relive distressing emotions, leading to retraumatization (Dickson-Swift, James, Kippen, & Liamputtong, 2009).

B. Exploring Personal and Intimate Topics:

Topics that are personal or intimate in nature, such as one's health, sexuality, identity, or relationships, can evoke strong emotional responses and vulnerabilities (Ellsberg & Heise, 2002).

C. Power Dynamics:

The inherent power imbalance between the researcher and the participant might make the latter feel obliged to share more than they are comfortable with, potentially leading to feelings of violation or exposure (Ryen, 2004).

D. Long-Term Participation:

Longitudinal studies or those that require prolonged engagement can inadvertently create emotional dependencies. Participants might develop bonds with researchers, making the termination of the study emotionally challenging (Miller, 2015).

E. Unexpected Emotional Responses:

Sometimes, seemingly innocuous topics or questions can inadvertently tap into unresolved issues or past experiences, evoking unexpected emotional reactions (Mannay, 2010).

F. Group Dynamics in Collective Settings:

In focus group discussions or other collective settings, interpersonal dynamics, peer pressure, or shared traumas can amplify emotional responses (Morgan, 1997).

G. Cultural and Contextual Sensitivities:

Research in cross-cultural settings or within specific communities might tread upon cultural taboos or sensitive issues, leading to emotional distress if not approached with due sensitivity (Tang & Chung, 2006).

H. Feedback and Preliminary Findings:

Sometimes, sharing preliminary findings with participants can evoke emotional responses, especially if the interpretations differ from the participants' perceptions (Bloor, 2001).

Recognizing the potential sources of emotional impact in qualitative research is crucial for researchers to mitigate their effects and ensure the mental well-being of their participants. Through proactive measures, thorough planning, and continuous reflexivity, researchers can navigate these emotional terrains ethically and effectively.

II. **Strategies to Minimize Emotional and Psychological Impact**:

The ethical duty of researchers extends beyond obtaining consent and ensuring confidentiality; it encompasses safeguarding participants' emotional and psychological well-being. Given the intimate and sometimes intrusive nature of qualitative inquiries, there's an imminent risk of invoking unintended distress. However, with careful consideration and strategic measures, researchers can work to minimize these negative impacts. Here, we elucidate some pivotal strategies that researchers can adopt:

A. Comprehensive Informed Consent:

Ensure participants understand not only the nature and purpose of the study but also the potential emotional or psychological challenges it might elicit (Orb, Eisenhauer, & Wynaden, 2000).

B. Regular Check-Ins:

During the research process, regular check-ins can help gauge participants' well-being and offer an avenue for them to express concerns or discomfort (Dickson-Swift, James, & Liamputtong, 2008).

C. Create Safe Spaces:

If delving into sensitive topics, providing a comfortable environment where participants feel safe and respected can mitigate distress (Ramos, 1989).

D. Debriefing Sessions:

Post-interview or post-observation debriefing provides participants an opportunity to reflect, discuss feelings, and gain closure (Liamputtong, 2007).

E. Offer Professional Referrals:

For studies with high emotional risks, researchers should be prepared with a list of professional counseling or support services referrals in case participants express or exhibit signs of distress (Sieber & Stanley, 1988).

F. Ensure Anonymity:

Guaranteeing that participants' identities and responses will remain anonymous can alleviate fears of stigmatization or retribution, especially in sensitive topics (Tolich, 2004).

G. Flexibility in Participation:

Allow participants the autonomy to skip certain questions or to withdraw from the study without any repercussions (Miller & Boulton, 2007).

H. Reflexivity:

Researchers should engage in continuous reflexivity to acknowledge their biases and emotions, ensuring that their actions or responses do not inadvertently harm participants (Finlay, 2002).

I. Training and Preparation:

Researchers and interviewers should be adequately trained to handle emotional situations, detect signs of distress, and provide immediate support (Dickson-Swift et al., 2008).

J. Pilot Testing:

Before the main study, pilot testing can help identify potential distressing areas in the research instrument, allowing modifications to minimize harm (van Teijlingen & Hundley, 2001).

Qualitative research's depth and intricacy, while its strength, can also be its ethical challenge. By being proactive, sensitive, and ethically vigilant, researchers can fulfill their duty to do no harm, ensuring the well-being of their participants remains paramount.

V. Aftercare and Debriefing:

Offering debriefing sessions after participation can help participants process their feelings, alleviate distress, and provide closure (Rager, 2005). Follow-ups, especially for emotionally intense studies, can ensure participants' continued well-being.

A core tenet of ethical qualitative research is the responsibility to anticipate and respond to any emotional or psychological distress experienced by participants. The culmination of a research interaction, particularly those of an intimate or sensitive nature, does not mark the end of the researcher's ethical obligations. Aftercare and debriefing play an integral role in ensuring the psychological well-being of participants after their direct involvement in the study has ended.

A. The Essence of Debriefing:

Debriefing refers to the structured or semi-structured conversation that researchers have with participants following their involvement in the study (Wiles, Crow, & Pain, 2011). This interaction is geared towards:

- Revisiting the study's objectives and answering any lingering questions.
- Addressing any immediate emotional or psychological distress.
- Providing resources for participants who might need additional support.

B. Goals of Debriefing:

- **Clarification**: Provide clarity on any ambiguities or misconceptions about the study, offering participants a holistic understanding (Orb, Eisenhauer, & Wynaden, 2000).
- **Catharsis**: Allow participants to express their feelings and experiences related to the research, facilitating emotional release (Sieber & Stanley, 1988).
- **Closure**: Ensure that participants don't leave the research setting with unresolved feelings or concerns (Liamputtong, 2007).

C. The Imperative of Aftercare:

While debriefing targets immediate concerns, aftercare considers the longer-term impacts of research participation. This involves:

- **Follow-up Contact**: Schedule subsequent interactions, if necessary, to assess the participant's well-being (Dickson-Swift et al., 2008).
- **Resources Referral**: Provide participants with information on relevant professional or community resources, particularly if the study dealt with traumatic or emotionally charged subjects (Ramos, 1989).

D. Key Considerations:

- **Tailored Approach**: Recognize the individuality of participants. What suffices as adequate debriefing or aftercare for one might not be sufficient for another (Finlay, 2002).
- **Respecting Autonomy**: While aftercare is vital, researchers must also respect participants' wishes if they decline any follow-up contact (Miller & Boulton, 2007).

E. Challenges and Remedies:

One of the main challenges in aftercare is ensuring sustained support in large-scale research or in situations where participants are geographically dispersed. Leveraging technology (like video calls) or partnering with local community organizations can be instrumental in such scenarios (Tolich, 2004).

Debriefing and aftercare are crucial facets of qualitative research that prioritize the emotional and psychological well-being of participants. Through these measures, researchers not only uphold the integrity of their profession but also foster trust and respect within the communities they engage.

While qualitative research seeks depth, nuance, and personal experiences, the emotional and psychological safety of participants must remain at the forefront. As researchers, it is imperative to maintain a balance between ethical duty and scholarly inquiry.

Navigating Sensitive and Controversial Topics

Qualitative research often delves deep into personal, sensitive, and sometimes controversial terrains to capture rich, detailed data. Handling these sensitive topics requires an elevated level of ethical awareness, empathy, and skill on the part of the researcher. Such subjects may entail topics related to sexuality, violence, mental health, illegal activities, or other potentially distressing areas (Lee, 1993).

I. Defining Sensitive Topics:

Sensitive topics are those which can pose potential psychological stress, embarrassment, or discomfort for the participant or even the researcher (Sieber & Stanley, 1988). They often intersect with personal, private, or taboo areas of human experience.

In qualitative research, sensitive topics often emerge as a focal area of inquiry, given their complexity, depth, and potential for generating rich and detailed insights. However, these topics, due to their nature, can raise unique ethical dilemmas and considerations. It is, therefore, vital to understand what constitutes a "sensitive topic" in research and the ramifications it can have for both participants and researchers.

A. Definition of Sensitive Topics:

Sensitive topics are often characterized as subjects that may provoke strong emotional responses or pose a potential for psychological stress, embarrassment, or discomfort for those discussing or revealing them (Lee, 1993). These topics often intersect with deeply personal, taboo, or controversial areas of human experience and can entail private information, stigmatized conditions, or illicit behaviors.

B. Characteristics of Sensitive Topics:

- **Personal Nature**: These topics are deeply personal and may touch upon areas of individual identity, personal trauma, or private life events (Sieber & Stanley, 1988).
- **Social Stigma**: Topics that society deems taboo or subjects that carry social stigma can be classified as sensitive. Such stigma can result in shame, discrimination, or ostracization (Renold, Holland, Ross, & Hillman, 2008).
- **Legal Implications**: Some sensitive subjects, like illegal activities or behaviors, can have potential legal ramifications, further heightening their sensitivity (Dunn, 2005).

C. Why Research Sensitive Topics?:

Despite the challenges, researching sensitive topics is essential because:

- They often provide deeper understandings of marginalized or less understood experiences, giving voice to underrepresented or silenced communities (Liamputtong, 2007).
- They can lead to policy or societal change by shedding light on issues that are often hidden or ignored (Ellsberg & Heise, 2002).

D. Evolving Sensitivity:

The perception of what is deemed sensitive can evolve over time, influenced by shifting cultural, societal, and ethical landscapes (Lee, 1993). This underscores the necessity for researchers to stay attuned to contemporary ethical guidelines and societal nuances.

Understanding and defining sensitive topics in research is paramount for ensuring ethical rigor and addressing the potential vulnerabilities of participants. By recognizing the inherent challenges and ethical considerations, researchers can navigate these terrains with care, empathy, and respect.

I. **Ethical Implications**:

Researching sensitive and controversial topics is a complex endeavor that, while valuable, presents unique ethical implications. Ensuring the well-being of participants while maintaining the integrity of the research becomes paramount when delving into topics of such depth and complexity.

A. Informed Consent:

Dealing with sensitive topics means ensuring participants are fully aware of the potential emotional or psychological impact they might face (Elmir et al., 2011). Informed consent goes beyond a signature on a form; it necessitates a deep understanding of the subject matter and the potential risks involved.

B. Confidentiality and Anonymity:

Sensitive topics often touch upon private or stigmatized aspects of participants' lives. Ensuring that identities are protected and that data is stored securely is paramount to building trust and protecting participants from potential harm or backlash (Tolich, 2004).

C. Emotional Support:

Research on sensitive topics can elicit strong emotional reactions. Researchers must be prepared to offer appropriate emotional support, referrals, or interventions if a participant becomes distressed (Dickson-Swift et al., 2008).

D. Potential for Re-traumatization:

Participants discussing traumatic events or experiences might risk re-traumatization. Researchers must be aware of this potential harm and approach interviews or discussions with sensitivity, offering breaks or discontinuation if necessary (Newman & Kaloupek, 2004).

E. Researcher's Well-being:

Engaging deeply with participants' emotional narratives, especially on sensitive topics, can also affect the well-being of researchers. Secondary or vicarious trauma is a real concern, and researchers need support and strategies to process and cope with these experiences (Pearlman & Mac Ian, 1995).

F. Cultural and Contextual Sensitivity:

The ethical implications of studying sensitive topics can vary across cultures and contexts. What might be considered taboo or controversial in one setting might be normative in another. It's vital for researchers to be culturally sensitive and aware of local norms, values, and potential implications (Liamputtong, 2008).

The ethical landscape of researching sensitive and controversial topics is layered and requires a nuanced understanding of both the potential risks to participants and the implications for the broader community or society. By being vigilant and prioritizing the well-being of participants, researchers can navigate these complex terrains ethically and responsibly.

II. **Strategies for Navigating Sensitive Topics**:

Addressing sensitive topics in qualitative research necessitates careful planning and a delicate approach. Navigating these subjects while ensuring participant well-being and maintaining research integrity can be achieved by employing several strategies:

A. Detailed Informed Consent:

Detailed informed consent ensures participants understand the nature of the sensitive topics being explored and the potential emotional impacts. Researchers must clarify participants' rights, including the freedom to refuse answering particular questions or to withdraw from the study without penalty (Sieber & Stanley, 1988).

B. Employing a Trauma-Informed Approach:

Understanding the principles of trauma-informed care means approaching participants with empathy, recognizing signs of distress, and acknowledging potential triggers (Hopper et al., 2010).

C. Providing Referrals and Support:

Offer participants a list of resources or referrals (e.g., counseling services) they might find beneficial, especially if discussing the topic elicits strong emotional reactions (Dickson-Swift et al., 2008).

D. Flexible Interview Techniques:

Adapting interview techniques, such as using open-ended questions or allowing participants to guide the discussion, can facilitate a comfortable and safe space for participants (Liamputtong, 2007).

E. Use of Pseudonyms:

Ensure participants' identities remain confidential by employing pseudonyms or other anonymization techniques in publications and presentations (Orb, Eisenhauer, & Wynaden, 2000).

F. Continuous Consent:

Beyond the initial informed consent, researchers can adopt a continuous consent approach, regularly checking in with participants about their comfort levels throughout the research process (Elmir et al., 2011).

G. Cultural and Contextual Sensitivity:

Given the variable definitions of sensitivity across cultures, it's imperative for researchers to be attuned to local customs, taboos, and societal norms, adjusting their approach as required (Liamputtong, 2008).

H. Engage in Reflexivity:

Researchers should consistently reflect on their role, potential biases, and the impact of their presence, especially when handling sensitive topics (Finlay, 2002).

Addressing sensitive and controversial topics in qualitative research demands a blend of methodological precision, empathic understanding, and ethical commitment. By adopting the above strategies, researchers can ensure both the validity of their research and the well-being of their participants.

IV. Ensuring Informed Consent:

Reinforce the voluntary nature of participation, especially when discussing sensitive issues. Participants must be fully aware of the topics to be discussed and feel free to decline or stop the interview at any point (Miller & Boulton, 2007).

In the field of qualitative research, particularly when it revolves around sensitive or controversial topics, informed consent is not just an administrative requirement but a foundational ethical practice. It is a dynamic process that requires ongoing dialogue and engagement between the researcher and the participant (Eide & Kahn, 2008).

A. Comprehensive Information Disclosure:

Researchers must provide participants with a clear and comprehensive understanding of the research purpose, methods, potential risks, and benefits. For studies that address sensitive issues, this also means explaining why such topics are being explored and the importance of their contribution (Morrow & Richards, 1996).

B. Voluntariness:

The principle of voluntariness is sacrosanct in the consent process. Participants must never feel coerced to take part in a study, especially when discussing potentially traumatic or stigmatizing topics (Beskow et al., 2009).

C. Capacity to Consent:

In some situations, particularly those involving vulnerable populations, researchers must ensure that participants have the cognitive and emotional capacity to provide informed consent, understanding the implications of their participation (Dixon-Woods et al., 2007).

D. Revocability:

It's crucial to stress that consent is an ongoing process. Even after giving initial consent, participants have the right to withdraw at any stage of the research without any repercussions, especially if they feel discomfort when discussing sensitive matters (Miller et al., 2008).

E. Consent Renewal for Further Research:

If researchers intend to use the collected data for future studies, participants should be made aware, and their explicit permission should be sought. This is even more crucial when the topics are sensitive or controversial in nature (Dove et al., 2017).

F. Acknowledgment of Limitations:

While researchers can ensure confidentiality, they must be transparent about its limits, particularly in small communities where individuals might be indirectly identifiable despite efforts to anonymize data (Elger & Caplan, 2006).

Navigating sensitive and controversial topics demands a higher level of ethical vigilance, and the process of informed consent becomes even more pivotal. With thorough transparency, continuous engagement, and utmost respect for participants' autonomy, researchers can ensure ethical integrity when delving into such delicate areas of inquiry.

V. Managing Researcher's Emotional Well-being:

Addressing sensitive topics can also affect researchers emotionally. Strategies such as debriefing sessions, peer support, and self-care practices can be valuable (Dickson-Swift et al., 2009).

While the emotional and psychological well-being of research participants is often discussed in the realm of qualitative research, the emotional challenges faced by researchers themselves—particularly when navigating sensitive topics—are less frequently addressed. Yet, it's undeniable that delving deep into traumatic, painful, or controversial subject matter can have profound emotional impacts on researchers (Dickson-Swift et al., 2009).

A. Vicarious Trauma and Emotional Labor:

When researchers engage deeply with participants who share traumatic experiences or deeply personal narratives, they can experience vicarious trauma or secondary traumatic stress. This refers to the emotional residue of exposure researchers have from working with people as they share their traumatic experiences (McCann & Pearlman, 1990).

B. The Intensity of Immersion:

Qualitative researchers often immerse themselves in the field, establishing close relationships with participants, which can intensify the emotional impact of the narratives they encounter. Such deep immersion can blur the lines between the personal and professional (Liamputtong, 2007).

C. Strategies for Emotional Self-Care:

- **Reflective Practice**: This allows researchers to be consciously aware of their feelings and emotions during the research process (Finlay, 2002).
- **Seeking Supervision or Peer Support**: Regular debriefing sessions with peers or mentors can provide an outlet for researchers to share feelings, concerns, and challenges (Etherington, 2007).
- **Professional Counseling**: In some cases, researchers might need professional psychological support, especially when their own past experiences resonate with the traumatic narratives encountered during research (Bloor et al., 2000).
- **Training**: Adequate training on the emotional challenges of qualitative research, especially in sensitive areas, can prepare researchers for potential emotional labor and offer coping strategies (Dickson-Swift et al., 2008).

D. Ethical Responsibility:

Institutional Review Boards or Ethics Committees focus predominantly on participant well-being. Yet, there's an increasing recognition of the need for ethical guidelines that address researcher well-being, especially when researching traumatic topics (Wassenaar & van der Linde, 2016).

As qualitative research delves deep into human experiences, emotions, and vulnerabilities, it's inevitable that researchers will sometimes carry the emotional weight of their work. Recognizing, addressing, and managing the emotional challenges of qualitative research is not only essential for the well-being of researchers but also for maintaining the integrity and quality of the research itself.

VI. Aftercare for Participants:

Given the potential distress from discussing sensitive subjects, providing resources, and support options post-interview is vital (Ramos, 1989).

In qualitative research, especially when addressing sensitive and potentially traumatizing topics, the researcher's responsibility does not end upon collecting data. Aftercare for participants becomes a pivotal component of the ethical considerations underpinning the research process. Aftercare not only aims to ensure the well-being of participants but also fortifies the trustworthiness of the research process (Elmir et al., 2011).

A. The Imperative of Aftercare:

Engaging with distressing or traumatic topics might inadvertently rekindle painful memories or evoke strong emotional responses in participants. Thus, ensuring that participants have access to support and care after the research process is a moral obligation (Dickson-Swift et al., 2007).

B. Strategies for Providing Aftercare:
- **Immediate Debriefing**: After an interview or focus group, researchers should spend time debriefing participants, discussing their feelings and reactions (Warden, 2018).
- **Referral Services**: Researchers should be prepared with a list of professional counseling services or support groups to which they can refer participants if needed (Orb et al., 2001).
- **Follow-up Contacts**: Schedule follow-up contacts or check-ins to assess the participants' well-being after the research process, addressing any lingering concerns (Liamputtong, 2007).

C. The Role of Institutional Support:

Universities and research institutions must recognize the importance of aftercare and facilitate the availability of resources, such as counseling services, for research participants (Sieber & Stanley, 1988).

D. Informed Consent and Aftercare:

Informed consent documents should clearly specify the aftercare procedures that will be available to participants, ensuring transparency and setting clear expectations from the onset (Eide & Kahn, 2008).

E. Cultural and Contextual Considerations:

The provision of aftercare should be tailored to the cultural and contextual nuances of the participant group. This may involve liaising with community leaders or elders and ensuring that support mechanisms resonate with local cultural practices and values (Tolich, 2004).

Aftercare, though sometimes overlooked, is an integral component of ethically conducted qualitative research. The emotional well-being of participants, especially in studies exploring sensitive topics, should always be prioritized, ensuring that the human dignity and integrity of participants remain at the forefront of the research endeavor.

CONCLUSION

Navigating sensitive and controversial topics in qualitative research demands a thoughtful, empathetic, and ethically rigorous approach. Ensuring the well-being of participants and researchers is paramount. By adopting comprehensive strategies, researchers can unearth invaluable insights from these challenging terrains, enhancing the depth and richness of their inquiries.

REFERENCES

Annas, G. J., & Grodin, M. A. (1992). *The Nazi doctors and the Nuremberg Code: Human rights in human experimentation*. Oxford University Press. doi:10.1093/oso/9780195070422.001.0001

Baez, B., & Casilli, A. A. (2020). The fallacy of data anonymization: A comment on the CJEU's Breyer case. *Computer Law & Security Review*, *36*(4), 105383.

Beauchamp, T. L., & Childress, J. F. (2001). *Principles of Biomedical Ethics*. Oxford University Press.

Beskow, L. M., Dame, L., & Costello, E. J. (2009). Certificates of confidentiality and informed consent: Perspectives of IRB chairs and institutional legal counsel. *IRB*, *31*(1), 1–8. PMID:19241733

Bloor, M., Fincham, B., & Sampson, H. (2000). Unprepared for the worst: Risks of harm for qualitative researchers. *Methodological Innovations Online*, *5*(1), 1–16.

Brinkmann, S., & Kvale, S. (2015). *InterViews: Learning the craft of qualitative research interviewing* (3rd ed.). SAGE.

Denzin, N. K., & Lincoln, Y. S. (2005). *The Sage handbook of qualitative* Denzin, N. K., & Lincoln, Y. S. (2005). *The Sage handbook of qualitative research. Sage (Atlanta, Ga.).*

Dickson-Swift, V., James, E. L., Kippen, S., & Liamputtong, P. (2007). Doing sensitive research: What challenges do qualitative researchers face? *Qualitative Research*, *7*(3), 327–353. doi:10.1177/1468794107078515

Dickson-Swift, V., James, E. L., Kippen, S., & Liamputtong, P. (2008). Doing sensitive research: What challenges do qualitative researchers face? *Qualitative Research*, *8*(3), 327–353. doi:10.1177/1468794107078515

Dickson-Swift, V., James, E. L., Kippen, S., & Liamputtong, P. (2009). Doing sensitive research: What challenges do qualitative researchers face? *Qualitative Research*, *9*(3), 327–353. doi:10.1177/1468794107078515

Dixon-Woods, M., Angell, E., Ashcroft, R. E., & Bryman, A. (2007). Written work: The social functions of Research Ethics Committee letters. *Social Science & Medicine*, *65*(4), 792–802. doi:10.1016/j.socscimed.2007.03.046 PMID:17490795

Dove, E. S., Joly, Y., & Knoppers, B. M. (2017). Power to the people: A Wiki-Governance model for biobanks. *Genome Biology*, *12*(1), 4. PMID:28100256

Dunn, L. B. (2005). Ethical issues in the study of bereavement: The opinions of bereaved adults. *Death Studies*, *29*(10), 883–905.

Eide, P., & Kahn, D. (2008). Ethical issues in the qualitative researcher—Participant relationship. *Nursing Ethics*, *15*(2), 199–207. doi:10.1177/0969733007086018 PMID:18272610

Elger, B. S., & Caplan, A. L. (2006). Consent and anonymization in research involving biobanks. *EMBO Reports*, *7*(7), 661–666. doi:10.1038/sj.embor.7400740 PMID:16819458

Ellis, C., Adams, T. E., & Bochner, A. P. (2007). Autoethnography: An overview. *Forum Qualitative Social Research*, *12*(1), 10.

Ellsberg, M., & Heise, L. (2002). Bearing witness: Ethics in domestic violence research. *Lancet*, *359*(9317), 1599–1604. doi:10.1016/S0140-6736(02)08521-5 PMID:12047984

Elmir, R., Schmied, V., Jackson, D., & Wilkes, L. (2011). Interviewing people about potentially sensitive topics. *Nurse Researcher*, *19*(1), 12–16. doi:10.7748/nr2011.10.19.1.12.c8766 PMID:22128582

Emanuel, E. J., Wendler, D., & Grady, C. (2000). What makes clinical research ethical? *Journal of the American Medical Association*, *283*(20), 2701–2711. doi:10.1001/jama.283.20.2701 PMID:10819955

Etherington, K. (2007). Ethical research in reflexive relationships. *Qualitative Inquiry*, *13*(5), 599–616. doi:10.1177/1077800407301175

Eysenbach, G., & Wyatt, J. (2002). Using the internet for surveys and health research. *Journal of Medical Internet Research*, *4*(2), e13. doi:10.2196/jmir.4.2.e13 PMID:12554560

Finlay, L. (2002). "Outing" the Researcher: The Provenance, Process, and Practice of Reflexivity. *Qualitative Health Research*, *12*(4), 531–545. doi:10.1177/104973202129120052 PMID:11939252

Flory, J., & Emanuel, E. (2004). Interventions to improve research participants' understanding in informed consent for research: A systematic review. *Journal of the American Medical Association*, *292*(13), 1593–1601. doi:10.1001/jama.292.13.1593 PMID:15467062

Guillemin, M., & Gillam, L. (2004). Ethics, reflexivity, and "ethically important moments" in research. *Qualitative Inquiry*, *10*(2), 261–280. doi:10.1177/1077800403262360

Herman, J. L. (2005). Justice from the victim's perspective. *Violence Against Women*, *11*(5), 571–602. doi:10.1177/1077801205274450 PMID:16043563

Hopper, E. K., Bassuk, E. L., & Olivet, J. (2010). Shelter from the storm: Trauma-informed care in homelessness services settings. *The Open Health Services and Policy Journal*, *3*(1), 80–100. doi:10.2174/1874924001003010080

Lee, R. M. (1993). *Doing research on sensitive topics*. Sage.

Liamputtong, P. (2007). Researching the vulnerable: A guide to sensitive research methods. *Sage (Atlanta, Ga.)*.

Liamputtong, P. (2008). *Doing cross-cultural research: Ethical and methodological perspectives* (Vol. 34). Springer Science & Business Media. doi:10.1007/978-1-4020-8567-3_1

Madrigal, L., & McClain, B. (2020). The real risks of sharing data. *Science*, *368*(6492), 716–718. PMID:32409464

Mannay, D. (2010). Making the familiar strange: Can visual research methods render the familiar setting more perceptible? *Qualitative Research*, *10*(1), 91–111. doi:10.1177/1468794109348684

Manson, H. M. (2019). E-consent: The design and implementation of consumer consent mechanisms in an electronic environment. *Journal of Business Research*, *95*, 311–319.

Markham, A., & Buchanan, E. (2012). Ethical considerations in digital research contexts. In *The Oxford Handbook of Internet Studies* (pp. 307–322). Oxford University Press.

Marshall, P. A., & Koenig, B. A. (2004). Accounting for culture in a globalized bioethics. *The Journal of Law, Medicine & Ethics*, *32*(2), 252–266. doi:10.1111/j.1748-720X.2004.tb00472.x PMID:15301190

Maxwell, J. A. (2012). *Qualitative research design: An interactive approach* (3rd ed.). Sage.

McCann, L., & Pearlman, L. A. (1990). Vicarious traumatization: A framework for understanding the psychological effects of working with victims. *Journal of Traumatic Stress*, *3*(1), 131–149. doi:10.1007/BF00975140

Mertens, D. M., & Ginsberg, P. E. (2009). Deep in ethical waters: Transformative perspectives for qualitative social work research. *Qualitative Social Work, 8*(4).

Miller, T. (2015). Going back: Stalking, talking and researcher responsibilities in qualitative longitudinal research. *International Journal of Social Research Methodology*, *18*(3), 293–305. doi:10.1080/13645579.2015.1017902

Miller, T., Birch, M., Mauthner, M., & Jessop, J. (2005). *Ethics in qualitative research*. Sage.

Miller, T., Birch, M., Mauthner, M., & Jessop, J. (Eds.). (2008). *Ethics in qualitative research* (2nd ed.). Sage.

Molyneux, C. S., Peshu, N., & Marsh, K. (2004). Understanding of informed consent in a low-income setting: Three case studies from the Kenyan Coast. *Social Science & Medicine*, *59*(12), 2547–2559. doi:10.1016/j.socscimed.2004.03.037 PMID:15474208

Morgan, D. L. (1997). *Focus groups as qualitative research* (Vol. 16). Sage publications. doi:10.4135/9781412984287

Morrow, V., & Richards, M. (1996). The ethics of social research with children: An overview. *Children & Society*, *10*(2), 90–105. doi:10.1002/(SICI)1099-0860(199606)10:2<90::AID-CHI14>3.0.CO;2-Z

National Commission for the Protection of Human Subjects of Biomedical and Behavioral Research. (1979). *The Belmont Report: Ethical principles and guidelines for the protection of human subjects of research*. Author.

Orb, A., Eisenhauer, L., & Wynaden, D. (2000). Ethics in qualitative research. *Journal of Nursing Scholarship*, *33*(1), 93–96. doi:10.1111/j.1547-5069.2001.00093.x PMID:11253591

Ramos, M. C. (1989). Some ethical implications of qualitative research. *Research in Nursing & Health*, *12*(1), 57–63. doi:10.1002/nur.4770120109 PMID:2922491

Renold, E., Holland, S., Ross, N. J., & Hillman, A. (2008). Becoming participant: Problematizing informed consent in participatory research with young people. *Qualitative Social Work: Research and Practice*, *7*(4), 427–447. doi:10.1177/1473325008097139

Resnik, D. B. (2015). *What is ethics in research & why is it important?* National Institute of Environmental Health Sciences.

Ryen, A. (2004). Ethical issues. In C. Seale (Ed.), *Qualitative research practice* (pp. 230–247). Sage. doi:10.4135/9781848608191.d20

Saunders, B., Kitzinger, J., & Kitzinger, C. (2015). Anonymising interview data: Challenges and compromise in practice. *Qualitative Research*, *15*(5), 616–632. doi:10.1177/1468794114550439 PMID:26457066

Shamoo, A. E., & Resnik, D. B. (2015). *Responsible conduct of research*. Oxford University Press.

Sieber, J. E. (1992). *Planning ethically responsible research: A guide for students and internal review boards*. Sage Publications. doi:10.4135/9781412985406

Sieber, J. E., & Stanley, B. (1988). Ethical and professional dimensions of socially sensitive research. *The American Psychologist*, *43*(1), 49–55. doi:10.1037/0003-066X.43.1.49 PMID:3348539

Tang, K. L., & Chung, L. Y. F. (2006). Conducting qualitative projects overseas: Personal accounts and practical tips. *Qualitative Research Journal*, *6*(2), 113–129.

Tilley, L., & Woodthorpe, K. (2011). Is it the end for anonymity as we know it? A critical examination of the ethical principle of anonymity in the context of 21st-century demands on the qualitative researcher. *Qualitative Research*, *11*(2), 197–212. doi:10.1177/1468794110394073

Tolich, M. (2004). Internal confidentiality: When confidentiality assurances fail relational informants. *Qualitative Sociology*, *27*(1), 101–106. doi:10.1023/B:QUAS.0000015546.20441.4a

Van Teijlingen, E. R., & Hundley, V. (2001). The importance of pilot studies. *Social Research Update*, *35*(1), 1–4. PMID:11328433

Wassenaar, D. R., & van der Linde, M. N. (2016). Ethical issues and qualitative research in the 21st century: Challenges and solutions. *Qualitative Health Research*, *26*(13), 1829–1835.

Wiles, R., Crow, G., Heath, S., & Charles, V. (2008). The management of confidentiality and anonymity in social research. *International Journal of Social Research Methodology*, *11*(5), 417–428. doi:10.1080/13645570701622231

Wiles, R., Crow, G., & Pain, H. (2011). Innovation in qualitative research methods: A narrative review. *Qualitative Research*, *11*(5), 587–604. doi:10.1177/1468794111413227

Zimmer, M. (2010). "But the data is already public": On the ethics of research in Facebook. *Ethics and Information Technology*, *12*(4), 313–325. doi:10.1007/s10676-010-9227-5

Chapter 10
Mixed Methods:
Combining Qualitative With Quantitative

ABSTRACT

This chapter explores the methodology of mixed methods research (MMR), which integrates both qualitative and quantitative approaches within a single study or sequential studies to address complex research questions. The chapter outlines the definition, rationale, and various design types of MMR, emphasizing the method's ability to provide a comprehensive understanding by balancing the depth of qualitative insights with the breadth of quantitative analysis. Challenges such as methodological integration, logistical demands, and maintaining coherence across disparate data types are discussed, alongside the potential for increased validity and holistic understanding.

INTRODUCTION

The evolving area of research in the social sciences has witnessed the growing acceptance and application of mixed methods. Mixed methods research (MMR) encompasses an innovative approach that amalgamates both qualitative and quantitative research paradigms. This chapter offers an in-depth elucidation of mixed methods research, highlighting its significance, methodologies, and challenges, while shedding light on its applicability in various research contexts.

I. Understanding Mixed Methods Research (MMR):

At its core, MMR integrates the collection, analysis, and interpretation of qualitative and quantitative data within a single study or a sequential set of studies (Creswell, 2009). This integrative approach acknowledges the limitations of singular research paradigms, harnessing the strengths of both qualitative and quantitative methods to offer a more holistic understanding of research phenomena.

Mixed Methods Research (MMR) has emerged as a robust research approach, responding to the multifaceted questions of contemporary research settings. MMR seeks to reconcile the traditional dichotomy between qualitative and quantitative research paradigms, offering a more encompassing and integrated

DOI: 10.4018/979-8-3693-2414-1.ch010

view of research problems. In this section, we delve deeper into the nuances of MMR, its underlying philosophy, and its pivotal role in current research landscapes.

A. Definition and Core Philosophy:

Mixed Methods Research can be defined as a methodological approach that integrates the data collection, analysis, and interpretation of qualitative and quantitative data within a singular research framework (Creswell & Plano Clark, 2007). The core philosophy behind MMR lies in its epistemological flexibility. It posits that research phenomena can be better understood through multiple types of data, offering a fuller, richer perspective than what might be achieved by either approach in isolation (Greene, 2008).

B. Rationale for MMR:

There are several compelling reasons to employ MMR:

- **Complementarity**: Qualitative and quantitative data can provide complementary insights. While qualitative data can elucidate the 'why' and 'how', quantitative data can measure the 'how much' (Teddlie & Tashakkori, 2009).
- **Validation**: MMR can validate findings across methods, enhancing the robustness and reliability of the conclusions (Johnson & Onwuegbuzie, 2004).
- **Complexity**: Some research questions are inherently complex and cannot be comprehensively addressed through a single method. MMR allows for exploration of such questions from multiple angles (Morse & Niehaus, 2009).

C. Types of Mixed Methods Designs:

The versatility of MMR is evident in its various design typologies:

- **Convergent Design**: Here, qualitative and quantitative data are collected and analyzed independently and then converged during interpretation (Creswell, 2014).
- **Explanatory Sequential Design**: The research starts with quantitative data collection and analysis, followed by qualitative data collection, which aims to explain the quantitative findings in more depth (Morse, 1991).
- **Exploratory Sequential Design**: This design is the opposite of the explanatory design. It begins with qualitative data collection and is followed by quantitative, often to generalize or test the qualitative findings (Plano Clark & Ivankova, 2016).

D. The Role of Theory in MMR:

Theory plays a vital role in shaping the MMR approach. While some MMR studies might be theory-driven from the onset, others might generate or refine theories as an outcome of the research process (Creswell & Plano Clark, 2011).

Understanding Mixed Methods Research requires acknowledging the intrinsic value of both qualitative and quantitative paradigms. By merging the depth of qualitative inquiry with the precision of quantitative measurement, MMR offers a more comprehensive lens to investigate intricate research phenomena.

II. Key Approaches to Mixed Methods:
   - **Concurrent Design**: Here, researchers collect both qualitative and quantitative data simultaneously and integrate the results during the interpretation phase (Creswell, 2014).
   - **Sequential Design**: In this approach, one type of data collection (either qualitative or quantitative) follows the other, usually to elaborate or expand on the initial results (Morse, 1991).
   - **Embedded Design**: This involves embedding one type of data within a larger methodology that guides the study, enhancing the insights drawn from the primary method (Plano Clark & Creswell, 2008).

III. Challenges in Mixed Methods Research:
   - **Logistical Challenges**: Managing multiple forms of data collection, analysis, and integration can be time-consuming and requires careful planning (Bryman, 2006).
   - **Methodological Challenges**: Ensuring coherent blending of qualitative and quantitative paradigms demands expertise in both domains (O'Cathain et al., 2007).
   - **Interpretative Challenges**: Combining findings from distinct methodologies requires a careful balancing act to avoid over-emphasizing one set of results over another (Greene, 2007).

IV. Ethical Considerations in MMR:

Given its integrative nature, MMR may present unique ethical challenges, such as ensuring informed consent for both qualitative and quantitative components and managing diverse data storage needs (Hall & Howard, 2008).

Mixed Methods Research (MMR) brings together both quantitative and qualitative research approaches, creating a unique set of ethical considerations. These considerations arise from the complexities associated with blending methodologies, requiring researchers to navigate and address the ethical challenges inherent in each while considering their synergistic impact. This section delves into the ethical concerns that surface in MMR and offers guidance on how to address them.

A. Informed Consent in MMR:

In MMR, informed consent is multi-faceted. Participants must be made aware of both qualitative and quantitative research aspects and the implications of each. This might involve explaining the nature of interviews, observations, surveys, or experiments, ensuring participants have a full understanding of their involvement (Creswell & Plano Clark, 2007).

B. Anonymity and Confidentiality:

Given that MMR often gathers in-depth qualitative data alongside quantitative metrics, there's potential for participant identification, even if data are aggregated. The blending of detailed narratives with specific numerical data might inadvertently reveal participants' identities. Thus, measures to ensure anonymity and confidentiality become paramount (Bryman, 2012).

C. Cultural Sensitivity:

MMR studies often span diverse cultural settings. Combining quantitative scales with qualitative descriptions can sometimes misrepresent cultural nuances. Researchers must be attuned to the cultural contexts and avoid imposing external frameworks that might distort interpretations (Hesse-Biber, 2010).

D. Data Integration Ethics:

The integration of qualitative and quantitative data can result in ethical dilemmas. For instance, when qualitative findings contradict quantitative results, privileging one over the other might marginalize certain participants' voices or experiences. Researchers should be transparent about discrepancies and avoid manipulating data to achieve coherence (O'Cathain, 2010).

E. Transparency and Openness:

In MMR, researchers must be especially transparent about their methodological decisions, given the potential criticisms from both qualitative and quantitative research communities. Decisions about sampling, data collection, and analysis across both paradigms should be detailed exhaustively to enable readers to ascertain the study's credibility (Tashakkori & Teddlie, 2010).

F. Researcher Reflexivity:

In MMR, reflexivity is essential. Researchers must constantly reflect on their roles, biases, and influences on the research, especially as they navigate between the qualitative and quantitative realms. This reflexivity ensures that researchers remain aware of their impact on the research process and outcomes (Johnson & Christensen, 2012).

Ethical considerations in Mixed Methods Research go beyond the individual ethical concerns of qualitative and quantitative research. Researchers embarking on MMR projects should be equipped with a profound understanding of ethics in both paradigms and be prepared to address the unique challenges that arise from their intersection.

V. Future Directions:

As the research landscape becomes increasingly interdisciplinary, the demand for MMR is anticipated to grow. Future endeavors might explore technological advancements to streamline the MMR process and specialized training to equip researchers with the skills to navigate the intricacies of mixed methods seamlessly.

CONCLUSION

Mixed methods research represents a progressive step in the world of research methodologies. By synthesizing qualitative insights with quantitative rigor, it offers a dynamic tool that can capture the complexity and multifaceted nature of contemporary research questions.

Benefits of Mixed Methods Research

Mixed Methods Research (MMR) offers a powerful approach that harnesses the strengths of both qualitative and quantitative paradigms. The fusion of these methods, rather than adhering strictly to one or the other, creates synergies that bring forth a richer, more nuanced understanding of research problems. This section explores the manifold benefits of employing MMR in research endeavors.

1. Comprehensive Perspective:

MMR allows researchers to capture a more holistic view of the research problem. By juxtaposing the depth of qualitative research with the breadth of quantitative insights, researchers can achieve a layered understanding that neither approach could provide independently (Creswell, 2014).

Mixed Methods Research (MMR) offers a unique vantage point that is not typically available with either purely qualitative or quantitative methodologies. Central to the many advantages of MMR is its ability to provide a comprehensive perspective. This section delves deeper into the essence and significance of this broadened outlook.

Definition

A comprehensive perspective in the context of research implies a broad, multi-dimensional, and integrative understanding of a research problem. In the realm of MMR, this is achieved by drawing together the deep, rich insights from qualitative data with the generalizable, statistical findings of quantitative data (Creswell & Plano Clark, 2011).

Depth and Breadth

The qualitative aspect of MMR immerses researchers into the lived experiences of participants, offering depth, while the quantitative facet offers a bird's-eye view, bringing breadth to the research. This combination ensures both a macro and micro understanding of the phenomena under study (Tashakkori & Teddlie, 1998).

Overcoming Limitations

Every research methodology, whether qualitative or quantitative, comes with its own set of constraints and blind spots. MMR's comprehensive approach ensures that the limitations of one method are potentially counterbalanced by the strengths of the other (Johnson, Onwuegbuzie, & Turner, 2007).

Enhanced Interpretability

Data derived from MMR can be more easily interpreted because of its comprehensive nature. The convergence of qualitative and quantitative findings can illuminate patterns and trends that might be unclear if only one method was employed (Greene, 2007).

Holistic Understanding

Modern-day research problems, particularly in the social sciences, are multifaceted and layered. A comprehensive approach, as offered by MMR, ensures that researchers are not just scratching the surface but are penetrating into the deeper layers of the research problem (Morse & Niehaus, 2009).

Relevance Across Disciplines

The comprehensive nature of MMR makes its findings relevant across various disciplines. For instance, an MMR study in education might yield findings that are pertinent to psychology, sociology, and even policy-making because of its multi-pronged approach (Ivankova, 2015).

A comprehensive perspective is one of the crowning jewels of Mixed Methods Research. It ensures that research findings are not only academically robust but also practically relevant, holistic, and thoroughly interpretable. In an era where research problems are becoming increasingly complex, the value of a comprehensive perspective as provided by MMR cannot be understated.

2. Methodological Triangulation:

Triangulation involves leveraging multiple methods to study a single problem, which can increase the validity of the research findings. If both qualitative and quantitative data indicate similar conclusions, the overall validity of the results can be strengthened (Denzin, 1978).

The notion of triangulation, originally a navigational and surveying technique, has found profound resonance in the realm of research. In the context of Mixed Methods Research (MMR), methodological triangulation stands as a testament to the approach's capacity for depth and rigor. Herein, we delve into the intricacies, significance, and implications of methodological triangulation within MMR.

Definition

Methodological triangulation involves using more than one method to gather data, such as interviews, observations, and questionnaires, with the aim of cross-validating or corroborating findings (Denzin, 1978). In the purview of MMR, this typically means integrating qualitative and quantitative data sources to offer a more complete picture of the research phenomenon.

Enhanced Validity

One of the most heralded advantages of methodological triangulation is its potential to enhance the validity of research findings. By collecting data through different methods and subsequently comparing and contrasting these findings, researchers can ascertain the accuracy and authenticity of their conclusions (Greene, Caracelli, & Graham, 1989).

Complementary Insights

Different research methods often illuminate different facets of a research problem. For instance, while a quantitative survey might highlight general patterns and trends, qualitative interviews can unearth deeper motivations and experiences. Together, they offer a comprehensive insight that neither could provide alone (Morse, 1991).

Error Compensation

All research methods have their inherent biases and limitations. By employing methodological triangulation, MMR researchers can compensate for the weaknesses or blind spots of one method with the strengths of another (Jick, 1979).

Richer Data

Triangulation in MMR ensures that the collected data is not only varied in nature but also rich in depth and breadth. Such richness can provide a solid foundation for drawing nuanced, well-rounded conclusions (Bryman, 2006).

Strengthened Conclusions

When findings derived from different methods converge on the same conclusion, it adds a layer of robustness to the research. This corroboration strengthens the validity of the results and increases confidence in their generalizability (O'Cathain, Murphy, & Nicholl, 2010).

Methodological triangulation stands as a cornerstone of the MMR approach, exemplifying its commitment to rigorous, comprehensive, and valid research. By harnessing the strengths of both qualitative and quantitative methods, triangulation ensures that mixed methods research is greater than the sum of its parts.

3. Catering to Diverse Audiences:

MMR results, being comprehensive, appeal to a broader audience. Quantitative data can resonate with stakeholders looking for generalized findings and statistical proofs, while qualitative insights can appeal to those seeking detailed narratives and lived experiences (Johnson & Onwuegbuzie, 2004).

In the evolving landscape of scientific research, scholars and practitioners are increasingly acknowledging the importance of research that resonates with varied audiences. Mixed Methods Research (MMR) has been identified as a significant mode that caters to diverse audiences, given its holistic approach that marries the strengths of both qualitative and quantitative methods. This section delves into the rationale and implications of MMR in reaching and benefiting a wide spectrum of audiences.

A. Comprehensive Presentation:

MMR, by nature, presents data and insights in a multifaceted manner. While quantitative data offers structured, numerical findings that appeal to audiences looking for statistical evidence, qualitative data provides narrative depth, making the findings relatable and comprehensible to those preferring descriptive insights (Johnson & Onwuegbuzie, 2004).

B. Addressing Varied Research Preferences:

Researchers, policymakers, and professionals come from diverse academic backgrounds and often have varied preferences when it comes to data interpretation. MMR's inclusive approach ensures that the findings are accessible and appealing to both empiricists and interpretivists (Teddlie & Tashakkori, 2009).

C. Enhanced Policy and Practice Implications:

For policymakers and practitioners, research that offers both macro (quantitative) and micro (qualitative) perspectives can be especially useful. While numbers give them a broad understanding, stories and narratives offer context, making it easier to design interventions or policies rooted in real-world complexities (Greene, 2007).

D. Broader Academic Reach:

Academic journals and conferences often cater to specialized audiences. With MMR, researchers can target a wider range of academic platforms, given that their findings are relevant to both qualitative and quantitative research communities (Mertens, 2014).

E. Engaging a Lay Audience:

For research that seeks to make an impact beyond academic boundaries, the blended nature of MMR can be instrumental. Numerical data, complemented by narratives or case descriptions, can make the findings more digestible and engaging for the general public (Bryman, 2006).

F. Cross-disciplinary Relevance:

In the era of interdisciplinary studies, research that can resonate across disciplines is invaluable. MMR's integrative approach ensures that the findings are pertinent to multiple disciplines, fostering cross-disciplinary collaborations and discussions (Ivankova, 2015).

The ability of Mixed Methods Research to cater to diverse audiences is not just an added advantage but is central to its philosophy. By ensuring that research findings are comprehensive, relatable, and resonate with a wide audience, MMR paves the way for more inclusive, impactful, and wide-reaching scientific inquiry.

4. Flexibility in Data Collection:

With MMR, researchers are not bound to a rigid data collection framework. They can use qualitative methods to explore unexpected avenues that arise during the quantitative phase or use quantitative data to verify and expand upon qualitative insights (Greene, Caracelli, & Graham, 1989).

Mixed Methods Research (MMR) has been heralded for the multifaceted insights it offers, combining the depth of qualitative methods with the breadth of quantitative ones. One of the prominent advantages of employing MMR is the flexibility it introduces in the data collection process. This section examines the nuanced benefits of such flexibility and its implications for enhancing research quality and richness.

A. Addressing Diverse Research Questions:

MMR permits researchers to address both exploratory and confirmatory research questions within a single study. While qualitative data can provide insights into the 'how' and 'why' of phenomena, quantitative data can test hypotheses and identify patterns. Thus, MMR caters to a wider range of research questions than a single-method study might (Creswell & Clark, 2017).

B. Adaptability to Changing Research Contexts:

Research rarely proceeds linearly, and unforeseen challenges often emerge. MMR's flexible data collection approach allows researchers to pivot as required. If one method encounters obstacles, the other can compensate, ensuring the research remains robust (Bryman, 2006).

C. Comprehensive Data Collection:

The flexibility of MMR enables researchers to employ multiple tools – from surveys and experiments to interviews and focus groups. This multi-pronged approach ensures a more comprehensive data set, capturing the phenomena under study from multiple angles (Teddlie & Tashakkori, 2009).

D. Validation and Enhancement:

MMR's dual approach provides a mechanism for data validation. Qualitative data can be used to explain anomalous quantitative results, and vice-versa. Such a cross-validation can strengthen the research findings and ensure they are both reliable and grounded (Johnson, Onwuegbuzie, & Turner, 2007).

E. Iterative Data Collection:

MMR offers the advantage of an iterative process. Initial qualitative data can inform the design of quantitative instruments, and preliminary quantitative results can guide subsequent qualitative inquiries, leading to a richer and more nuanced understanding (Morse & Niehaus, 2009).

F. Meeting Diverse Stakeholder Needs:

Different stakeholders in a research project may prioritize different types of data. Policymakers might seek quantitative evidence, while practitioners may value qualitative insights. The flexibility of MMR ensures that the research caters to diverse stakeholder needs, enhancing its impact and relevance (Greene, 2007).

The flexibility inherent in Mixed Methods Research's data collection offers researchers a versatile tool that can enhance the depth, breadth, and robustness of their studies. By navigating the strengths and weaknesses of both qualitative and quantitative approaches, MMR ensures that research is not only comprehensive but also adaptive to the complex realities of the field.

5. Addressing Complex Research Questions:

Some research questions are too intricate to be explored using just one method. MMR offers the adaptability required to unpack complex phenomena by using different methods to explore different facets of the question (Teddlie & Tashakkori, 2009).

The multifaceted nature of many contemporary research questions requires an approach that transcends the boundaries of traditional methodologies. Mixed Methods Research (MMR) is increasingly recognized as a potent strategy to grapple with the complexities inherent in such questions. By interweaving qualitative and quantitative research techniques, MMR offers a sophisticated toolset to decode multifaceted research queries that a singular approach might struggle with (Creswell, 2014).

A. Comprehensive Understanding:

Complex research questions often demand insights into both the breadth and depth of a phenomenon. While quantitative approaches excel in outlining the broader landscape, qualitative techniques delve deep, unpacking the intricacies and nuances. MMR, through its dual approach, ensures a holistic understanding (Teddlie & Tashakkori, 2009).

B. Capturing Dynamics of the Phenomenon:

Many research questions concern phenomena that are dynamic and ever-evolving. MMR's flexibility allows researchers to capture such shifts and changes, using quantitative data to identify patterns and qualitative data to explore the underlying processes (Bryman, 2012).

C. Interplay of Variables:

Complex research questions often involve multiple interrelated variables. Through MMR, quantitative data can identify correlations, and qualitative data can then explore the causative factors behind these correlations, revealing a richer understanding of the variable interplay (Plano Clark & Creswell, 2015).

D. Diverse Perspectives:

Complexity in research often arises from the diversity of perspectives related to a particular phenomenon. By employing MMR, researchers can ensure they harness the statistical power of quantitative data while still accommodating the diverse voices and stories accessed through qualitative techniques (Morse & Cheek, 2014).

E. Triangulation for Robustness:

To ascertain the validity of findings related to complex questions, MMR leverages triangulation. By corroborating findings from both qualitative and quantitative strands, researchers enhance the robustness and trustworthiness of their conclusions (Denzin, 2012).

F. Adapting to Unforeseen Discoveries:

Complex research questions often yield unexpected findings. The adaptability of MMR means researchers can recalibrate their methodologies in response to such discoveries, ensuring their research remains relevant and insightful (Johnson & Onwuegbuzie, 2004).

Mixed Methods Research serves as a bridge, uniting the strengths of qualitative and quantitative methodologies to tackle complex research questions. In an era where research topics are increasingly intricate, the balanced, integrated, and flexible approach of MMR proves invaluable in ensuring comprehensive, robust, and nuanced insights.

6. Enhancing External and Internal Validity:

While qualitative research often provides rich context-specific insights (increasing internal validity), quantitative research, with its generalizable findings, enhances external validity. Combining the two ensures that research findings are both deep and broad (Morse, 1991).

The adoption of Mixed Methods Research (MMR) is not merely a methodological choice, but a strategic decision that potentially enhances the validity of research outcomes. Validity, both external and internal, stands as a cornerstone in the world of research, certifying the accuracy and generalizability of findings (Creswell & Plano Clark, 2011). MMR, with its integrated approach, offers an enriched platform for fortifying both these facets of validity.

A. Internal Validity: Depth and Precision

Internal validity concerns the accuracy of findings within the context of the study.

- **Triangulation**: By employing multiple methods to explore a research question, MMR provides a means of cross-verification. For instance, a pattern noticed in quantitative data can be deeply explored using qualitative means, adding depth and precision to the findings (Denzin, 1978).
- **Addressing Methodological Limitations**: Each method, be it qualitative or quantitative, has inherent limitations. However, when combined, the strengths of one can compensate for the weaknesses of the other, thereby bolstering internal validity (Johnson, Onwuegbuzie, & Turner, 2007).

B. External Validity: Breadth and Generalizability

External validity pertains to the degree to which study results can be generalized beyond the immediate study context.

- **Comprehensive Data Spectrum**: While quantitative data might offer a broad overview suitable for generalization, qualitative data provides the nuanced details. Together, they ensure that the findings have both the breadth for generalizability and the depth to ensure that such generalizations are meaningful (Morse, 1991).
- **Diverse Sampling**: MMR often involves diverse sampling methods tailored to the requirements of both qualitative and quantitative strands. Such comprehensive sampling enhances the generalizability of findings (Teddlie & Yu, 2007).

- **Real-world Context**: Qualitative components of MMR often ensure that research is rooted in real-world contexts, making the results more applicable and generalizable to similar contexts (Flyvbjerg, 2006).

C. Bridging Validity Gaps

- **Complementing Insights**: Qualitative insights can elucidate unexpected results from quantitative data, enhancing the overall validity of findings (Greene, Caracelli, & Graham, 1989).
- **Enhanced Research Design**: By allowing for iterative designs where one method informs the other, MMR ensures that research tools are continually refined, enhancing validity (Bryman, 2006).

By leveraging the strengths of both qualitative and quantitative research, Mixed Methods Research stands as a robust methodological choice, especially when aiming to enhance the external and internal validity of research outcomes.

7. Filling Methodological Gaps:

MMR can identify and address gaps in each method. For instance, unexpected quantitative findings can be further explored using qualitative interviews to determine the reasons behind those trends (Bryman, 2006).

Mixed Methods Research (MMR) stands out as a holistic approach that integrates both qualitative and quantitative methods to provide comprehensive insights into complex research questions. One of the most significant benefits of this approach is its potential to address and fill methodological gaps inherent in standalone qualitative or quantitative methods. This synergy, resultant of the combination of methods, not only enhances the robustness of the research but also provides a richer and more nuanced understanding of the phenomenon under investigation (Creswell, 2003).

A. Counterbalancing Limitations:

Every research methodology carries intrinsic limitations. Quantitative methods might be criticized for their lack of depth, while qualitative methods might be viewed as lacking generalizability. By combining both approaches, MMR offers a mechanism to counterbalance the limitations of one method with the strengths of the other, thereby enhancing the overall integrity of the research (Johnson & Onwuegbuzie, 2004).

B. Enhancing Data Depth and Breadth:

While quantitative research can provide breadth by capturing data from larger samples, qualitative research offers depth by diving deep into individual experiences, beliefs, and perceptions. Together, they ensure that the findings are both generalizable and grounded in real-world contexts (Morse & Niehaus, 2009).

C. Addressing Complex Research Questions:

Some research questions are multidimensional and cannot be adequately addressed using a single method. MMR allows for the exploration of different facets of a research question, ensuring a more comprehensive understanding (Bazeley, 2018).

D. Iterative Refinement:

Mixed methods often involve iterative processes where preliminary findings from one method can inform and refine subsequent data collection or analysis in the other. Such iterative refinement can address emerging gaps and enhance the rigor of the research (O'Cathain, Murphy, & Nicholl, 2010).

E. Enhancing Validity through Triangulation:

By collecting data through multiple methods and sources, MMR provides an avenue for triangulation. Such triangulation can validate the findings, enhancing their trustworthiness and credibility (Denzin, 1978).

Mixed Methods Research, with its integrative approach, serves as a powerful tool to address and fill methodological gaps. By leveraging the strengths of both qualitative and quantitative methods, it offers researchers a robust platform for a more comprehensive, nuanced, and credible exploration of complex research questions.

Finally, Mixed Methods Research, while challenging in its design and execution, offers unparalleled advantages in terms of depth, breadth, validity, and flexibility. It encapsulates the best of both worlds, allowing for a multifaceted exploration of research questions.

Designing a Mixed Methods Study

The process of crafting a mixed methods study is intricate, demanding both careful design considerations and methodological rigor. When skillfully executed, Mixed Methods Research (MMR) can offer unparalleled depth and breadth in exploring complex research phenomena (Creswell & Plano Clark, 2017). This section delves deep into the crucial elements of designing an MMR study.

1. Purpose of Combining Methods:

Before initiating the design phase, researchers must have clarity regarding why they are choosing MMR over a single method approach. The rationale could range from seeking to enhance the validity of findings through triangulation, addressing a research question from multiple perspectives, or sequentially elaborating on findings from one method using another (Tashakkori & Teddlie, 1998).\

The adoption of Mixed Methods Research (MMR) represents a methodological choice to integrate both quantitative and qualitative data within a single research study. This decision is not merely methodological but intrinsically linked to the overarching purpose and goals of the study. The following section elucidates the primary purposes that drive researchers to opt for this integrative approach.

A. Comprehensive Understanding:

Qualitative methods, with their focus on deep and context-rich explorations, combined with the breadth and generalizability of quantitative methods, provide a comprehensive understanding of research phenomena (Creswell, 2003). MMR allows for capturing the complexities inherent in many research problems, ensuring that nuances are neither overlooked nor overly generalized.

B. Triangulation for Enhanced Validity:

Triangulation refers to the use of multiple methods or data sources to validate and cross-check findings (Denzin, 1978). Through MMR, researchers can employ this strategy to bolster the validity of their conclusions. If both qualitative and quantitative strands yield consistent findings, the validity of the results is strengthened.

C. Addressing Different Types of Questions:

MMR facilitates the examination of a topic from multiple angles, enabling researchers to address exploratory, explanatory, and confirmatory questions within a single study framework (Johnson & Onwuegbuzie, 2004). This multi-pronged approach can provide a richer understanding of the problem at hand.

D. Sequential Elaboration:

MMR's sequential designs, where one method follows the other, allow for the findings of one phase to inform or elaborate on the next (Morse, 1991). For instance, an unexpected result in a quantitative phase can be further explored in depth through qualitative interviews.

E. Overcoming Methodological Limitations:

Every research method has its strengths and limitations. By combining methods, researchers can counterbalance the weaknesses of one approach with the strengths of the other, producing a more robust set of findings (Greene, Caracelli, & Graham, 1989).

F. Expanding the Scope of Study:

Certain research problems require multi-dimensional exploration. MMR allows for this expansiveness, integrating diverse datasets that offer varying perspectives and insights (Tashakkori & Teddlie, 2003). For instance, while surveys might reveal patterns and trends, interviews can uncover underlying reasons and motivations.

The purpose of combining methods in MMR is multifaceted, going beyond mere methodological integration. The decision to adopt MMR should be deeply rooted in the research questions, the study's goals, and the potential contributions of both qualitative and quantitative insights to the topic of interest.

2. Choosing an Appropriate Design:

Mixed Methods Research (MMR) is lauded for its capacity to amalgamate the strengths of both qualitative and quantitative research traditions. The design of a mixed methods study, crucially, should align with the research objectives, questions, and the epistemological stance of the researcher. This segment underscores the considerations and methodologies involved in selecting an apt design for MMR.

A. Sequential Designs:

Sequential designs involve implementing one method after the other (Creswell & Plano Clark, 2007). For example:

- **Sequential Explanatory Design**: This begins with a quantitative phase, followed by a qualitative phase. It is especially useful when unexpected results from the quantitative phase require further exploration.
- **Sequential Exploratory Design**: Here, a qualitative phase precedes the quantitative phase. This is typical when the researcher seeks to generate a theory or instrument based on qualitative findings.

B. Concurrent Designs:

Concurrent designs collect qualitative and quantitative data concurrently but separately, aiming to merge or compare the data during interpretation (Greene, Caracelli, & Graham, 1989):

- **Concurrent Triangulation Design**: This design is used when researchers want to corroborate findings from different methods, ensuring robustness in results.
- **Concurrent Embedded Design**: One type of data plays a secondary role in a study primarily focused on the other data type.

C. Transformative Designs:

This design is anchored in a theoretical framework or perspective, such as feminism or critical race theory, which guides the entire study, including the combination of qualitative and quantitative methods (Mertens, 2007).

D. Consider the Priority of Methods:

Some research questions necessitate prioritizing one method over the other. Determining this hierarchy aids in deciding which method will play the leading role in data collection, analysis, and interpretation (Teddlie & Tashakkori, 2003).

E. Points of Integration:

Another pivotal decision is pinpointing where the two methods will intersect: during data collection, during data analysis, or merely at the interpretation stage (Bryman, 2006).

F. Practical and Logistical Factors:

Consider the available resources, timeline, and expertise. A complex mixed methods design requires ample resources and a comprehensive understanding of both qualitative and quantitative paradigms.

Selecting an appropriate mixed methods design requires meticulous consideration of the research problem, questions, and objectives. It's not a mere conjoining of methods; rather, it's a strategic approach to research that seeks to utilize the complementary strengths of both qualitative and quantitative paradigms.

3. Timing and Sequencing:

The decision on whether to run the qualitative and quantitative parts concurrently or sequentially is pivotal. Concurrent designs might be more efficient but might not allow the iterative learning that sequential designs offer (Palinkas et al., 2011).

Understanding the timing and sequencing of data collection within Mixed Methods Research (MMR) is a crucial aspect of designing an effective study. The concurrent or sequential nature of data collection can significantly influence the depth, breadth, and comprehensiveness of research findings (Creswell, Plano Clark, Gutmann, & Hanson, 2003).

A. Concurrent Timing:

Concurrent timing involves collecting both qualitative and quantitative data simultaneously.

- **Advantages**: This approach can be time-efficient as both forms of data are gathered simultaneously, making it useful for projects with a strict timeline. By comparing data concurrently, immediate insights may emerge that wouldn't be apparent if the data were collected sequentially (Morse, 1991).
- **Challenges**: However, collecting both sets of data concurrently may require extensive resources and coordination, especially if different teams are handling each data type.

B. Sequential Timing:

Sequential timing means one form of data is collected and analyzed before the next form begins.

- **Advantages**: Sequential timing allows the researcher to utilize the results of the first phase to shape the subsequent phase, making the study more responsive and adaptive (Tashakkori & Teddlie, 1998). For instance, unexpected quantitative findings can be further explored in a qualitative phase, allowing for depth in understanding.
- **Challenges**: This approach may be time-intensive. There is also the risk that findings from the first phase could unduly influence the collection and interpretation of data in the second phase, potentially introducing bias.

C. Iterative Timing:

Iterative timing involves a more complex back-and-forth between qualitative and quantitative data collection, where each phase informs the subsequent phase in a recurring manner.

- **Advantages**: This method can lead to a more in-depth and nuanced understanding of research questions as the process is highly reflexive and adaptive (Maxwell & Loomis, 2003).

- **Challenges**: Iterative approaches can be resource-intensive and may require a more extended timeframe to complete.

The decision regarding timing and sequencing in MMR should align with the research questions, objectives, available resources, and the intended depth and breadth of analysis. Researchers must be reflective and deliberate in choosing their approach, always considering how their choices will affect the richness, validity, and reliability of their findings.

4. Points of Interface:

At what stages of the research process will the two methods interact? It could be during the data collection, analysis, or interpretation stages. Researchers must plan the integration points to ensure the methods meaningfully inform each other (Bryman, 2006).

Mixed methods research (MMR) is characterized by the intentional combination of qualitative and quantitative data. One of the most vital decisions in MMR design is identifying and managing the points of interface, that is, the junctions where the qualitative and quantitative components of the study intersect or influence each other (Bryman, 2007).

A. Definition of Points of Interface:

Points of interface are moments in the research process where qualitative and quantitative data sources, findings, methods, or interpretations interact with or complement each other (O'Cathain, Murphy, & Nicholl, 2008).

B. The Purpose of the Interface:

Integrating methods at the interface can serve multiple purposes:

- **Validation**: Using one method to validate or corroborate the findings of the other, enhancing the research's trustworthiness (Greene, Caracelli, & Graham, 1989).
- **Expansion**: Using one method to elaborate upon or expand the findings of the other, leading to a more comprehensive understanding (Creswell & Plano Clark, 2007).
- **Transformation**: Using one type of data (e.g., qualitative) to transform or recontextualize findings derived from another method (e.g., quantitative) (Tashakkori & Creswell, 2007).

C. Strategies for Managing the Interface:

- **Data Transformation**: Quantitative data can be transformed into qualitative narratives or vice versa, qualitative themes can be quantified for statistical analysis (Sandelowski, Voils, & Knafl, 2009).
- **Parallel Data Analysis**: Conducting separate analyses of qualitative and quantitative data, and then comparing or juxtaposing the findings to draw overarching conclusions (Bazeley, 2009).

- **Integrated Data Analysis**: Merging quantitative and qualitative data sets and analyzing them as a single entity, often using specialized software tools that facilitate mixed data analyses (Castro, Kellison, Boyd, & Kopak, 2010).

D. Considerations for a Successful Interface:

- **Sequential Consistency**: Ensuring that data collected at different stages or from different methods align with the study's overarching aims and questions (Moran-Ellis et al., 2006).
- **Transparent Documentation**: Maintaining clear and comprehensive documentation of how and why data from various methods are interfaced, to ensure replicability and clarity (O'Cathain, 2010).
- **Reflexivity**: Continuously reflecting upon the decisions made at the interface, considering potential biases and the implications of those decisions for the study's conclusions (Guetterman, Fetters, & Creswell, 2015).

The way researchers approach the points of interface in an MMR study can have significant implications for the depth, breadth, and robustness of their findings. By thoughtfully integrating qualitative and quantitative components, scholars can tap into the complementary strengths of both methods, driving forward a richer and more nuanced understanding of their research questions.

5. Weighting of Methods:

In some studies, one method might dominate, serving as the primary mode of inquiry, while the other is supplementary. In others, both methods might hold equal weight. This balance should align with the research questions and objectives (Johnson, Onwuegbuzie, & Turner, 2007).

The concept of "weighting" in mixed methods research refers to the emphasis or priority a researcher gives to either the qualitative or quantitative component of the study. Determining the weighting is a pivotal decision as it significantly impacts the research design, data collection, analysis, and eventual interpretations (Creswell, Klassen, Plano Clark, & Smith, 2011).

A. Understanding Weighting:

Weighting addresses the question: "Which method (qualitative or quantitative) is primary in the study, and which is secondary?" The primary method is the one driving the central research question and providing the principal framework for findings (Teddlie & Tashakkori, 2009).

B. Reasons for Weighting:
   - **Research Question Focus**: The essence of the research questions may necessitate one method over another (Morse & Niehaus, 2009).
   - **Resource Constraints**: Limitations related to time, money, or expertise might dictate the emphasis on a particular method (Bryman, 2006).
   - **Stage of Research**: Exploratory phases of research might favor qualitative methods, while confirmatory phases might prioritize quantitative methods (Ragin, 2008).

C. Approaches to Weighting:

- **Equal Weighting**: Both qualitative and quantitative methods are given equivalent attention. Such an approach is commonly seen in full integration designs where the two methods are interdependent (Palinkas et al., 2011).
- **Qualitative Priority**: The study is essentially qualitative but includes some quantitative elements, often to enhance or augment the qualitative findings (Johnson, Onwuegbuzie, & Turner, 2007).
- **Quantitative Priority**: The study is mainly quantitative, but some qualitative elements are included to provide context, depth, or interpretation to quantitative results (Greene, Caracelli, & Graham, 1989).

D. Implications of Weighting:

- **Analysis and Interpretation**: The primary method usually dominates the data analysis process and guides the interpretation of results (Morgan, 2007).
- **Validity and Reliability**: Weighting might affect the criteria applied to judge the study's rigor and trustworthiness (Golafshani, 2003).
- **Generalizability and Transferability**: The ability to generalize findings (quantitative) or ensure transferability (qualitative) may be influenced by the method that is given priority (Lincoln & Guba, 1985).

E. Challenges in Weighting:

Determining the weighting can be challenging, especially when:

- Both methods appear equally significant for the research aims.
- The research context evolves, necessitating a shift in methodological priority (Bryman, 2006).

F. Recommendations for Deciding Weighting:

- Clearly articulate the research problem and objectives. The nature of the problem often dictates methodological priority (Creswell & Plano Clark, 2007).
- Consider the audience and the type of evidence they value. Certain academic or professional audiences might favor one method over another (O'Cathain, Murphy, & Nicholl, 2007).
- Remain flexible and open to adjusting the weighting as the research unfolds and new insights or challenges emerge (Morse, 2003).

Weighting in mixed methods research serves as a guiding compass, determining the course of data collection, analysis, and interpretation. By thoughtfully considering the weighting of methods, researchers can craft a study that is methodologically robust and best suited to address the research questions at hand.

6. Sampling and Data Collection:

Will the study employ the same sample for both methods or different ones? If different, researchers should ensure both samples are representative and appropriate for the study's aims (Morse & Niehaus, 2009).

Mixed Methods Research (MMR) is characterized by its unique approach to data collection, often incorporating both qualitative and quantitative elements. A pivotal decision that shapes this design concerns the sampling strategy and the specifics of data collection. This complexity enhances the depth and breadth of insights but also demands meticulous planning (Creswell & Plano Clark, 2011).

A. Sampling in Mixed Methods Research:
   - **Parallel Sampling**: Both qualitative and quantitative strands use separate, parallel samples. This approach allows for independent, yet complementary, exploration of the research problem (Teddlie & Tashakkori, 2009).
   - **Nested Sampling**: One data collection method is nested within a larger sample, allowing a deeper exploration of specific subgroups or themes identified (Bryman, 2006).
   - **Sequential Sampling**: The sample for one method is drawn based on findings or results from the other method, enhancing the iterative nature of MMR (Morse & Niehaus, 2009).

B. Data Collection in Mixed Methods Research:
   - **Concurrent Data Collection**: Qualitative and quantitative data are collected concurrently. This approach is efficient and offers opportunities for triangulation but demands careful synchronization (Greene, Caracelli, & Graham, 1989).
   - **Sequential Data Collection**: One method follows the other, allowing the researcher to delve deeper into initial findings or to clarify or expand upon them (Creswell, 2014).

C. Integration of Data Collection Methods:
   - **Instrument Development**: Qualitative insights can be used to develop quantitative instruments, ensuring cultural and contextual relevance (Morgan, 2007).
   - **Data Transformation**: Data from one method is converted (quantitized or qualitized) for combined analysis, enabling a cohesive analytical approach (Sandelowski, Voils, & Knafl, 2009).
   - **Multilevel Exploration**: Quantitative data can be used to explore patterns at a broader level, while qualitative data delves into individual or community experiences (Ragin, 2008).

D. Challenges in Sampling and Data Collection:
   - **Logistics**: Managing parallel or sequential data collection, especially with different teams or instruments, can be logistically challenging (Palinkas et al., 2011).
   - **Inconsistencies**: Differences in sampling strategies might lead to inconsistencies in data, potentially complicating interpretation (Bryman, 2006).

E. Recommendations for Effective Sampling and Data Collection in MMR:
   - **Clarity of Purpose**: Ensure clarity on the research objectives and the role of each method in achieving these objectives (Johnson, Onwuegbuzie, & Turner, 2007).
   - **Iterative Approach**: Be prepared to adapt the research design based on initial findings, especially in sequential designs (O'Cathain, Murphy, & Nicholl, 2007).
   - **Pilot Testing**: Before full-scale data collection, consider pilot testing to identify potential issues and rectify them (Morse, 2003).

Sampling and data collection in MMR are complex endeavors, weaving together qualitative and quantitative threads. By thoughtfully designing these elements, researchers can harness the full potential of MMR, achieving comprehensive, nuanced insights that neither method could provide alone.

7. Data Analysis:

The data analysis approach in MMR is multifaceted. Individual strands (qualitative and quantitative) are typically analyzed separately, followed by a synthesis or comparison of results to draw holistic conclusions (Ivankova, Creswell, & Stick, 2006).

The choice to conduct Mixed Methods Research (MMR) not only impacts data collection but also extends its complexities into data analysis. Analyzing data in MMR is a multifaceted process since it involves processing data that are fundamentally different - numerical and textual. Understanding these complexities and the strategies to address them is vital for researchers who opt for this integrative approach (Creswell & Plano Clark, 2011).

A. Sequential Analysis:

Sequential analysis is typically chosen when researchers first collect and analyze one form of data and then the other. For instance, qualitative data might be gathered and analyzed to help shape or inform a subsequent quantitative phase (Morse, 1991).

B. Concurrent Analysis:

In concurrent designs, qualitative and quantitative data are analyzed separately but concurrently. Once both sets of data have been analyzed, researchers can then integrate the findings during the interpretation stage, allowing for a more comprehensive understanding (Tashakkori & Teddlie, 1998).

C. Integration at the Point of Analysis:

Some MMR studies choose to integrate data at the point of analysis, often by transforming qualitative data into a quantitative form (quantitizing) or vice versa (qualitizing). This allows for a single, blended analysis, although the process can be complex and may lead to oversimplifying data (Sandelowski, Voils, & Knafl, 2009).

D. Joint Display of Results:

A relatively straightforward method for integrating findings from different data types is the joint display, where results from qualitative and quantitative analyses are presented side by side, often in tables or figures, facilitating direct comparison (Creswell & Zhang, 2009).

E. Challenges in MMR Data Analysis:

- **Complexity**: MMR analysis often requires expertise in both qualitative and quantitative methods, demanding a range of skills and knowledge (Bazeley, 2010).
- **Integration Difficulties**: Combining results from different data sources and analysis methods can be challenging, especially in terms of ensuring coherent and meaningful conclusions (O'Cathain, 2010).
- **Software Limitations**: While many qualitative and quantitative data analysis software packages exist, few handle mixed methods data analysis seamlessly (Fielding & Cisneros-Puebla, 2009).

F. Recommendations:

- **Maintain Focus**: Always keep the research question in mind, ensuring that data analysis serves the broader objectives of the study (Greene, 2007).
- **Iterative Process**: Be prepared for an iterative analysis process, especially when integrating findings. This often involves revisiting the data and refining interpretations (Maxwell & Loomis, 2003).
- **Collaboration**: Given the broad skill set required for MMR, consider collaborative approaches where researchers with qualitative and quantitative expertise work together (Hall & Howard, 2008).

Data analysis in MMR is an intricate process that requires a nuanced understanding of both qualitative and quantitative methods. When executed well, it offers a rich, multi-faceted understanding that neither method could achieve in isolation.

8. Addressing Challenges:

MMR is inherently complex, and researchers might face challenges such as methodological dominance, contradictory findings from the two methods, or increased time and resource requirements. Anticipating and addressing these challenges is crucial (Fielding, 2010).

The interdisciplinary nature of Mixed Methods Research (MMR) presents researchers with both opportunities and challenges. While MMR holds great promise for answering complex research questions through an integrated methodological approach, it also demands caution and attention to detail to ensure methodological robustness (Johnson, Onwuegbuzie, & Turner, 2007).

A. Philosophical and Paradigmatic Tensions:

The blending of qualitative and quantitative research methodologies often brings into focus the differing philosophical foundations inherent in each. These differences can lead to potential conflicts in paradigms, particularly when considering aspects like ontology, epistemology, and axiology (Teddlie & Tashakkori, 2009).

Recommendation: Engage in transparent discussions about the philosophical foundations of the study, considering how the different paradigms can co-exist or complement one another.

B. Design and Methodological Complexities:

Selecting the most suitable mixed methods design can be challenging given the broad array of potential combinations of qualitative and quantitative strategies (Creswell & Plano Clark, 2011).

Recommendation: Always align the research design with the overarching research questions, ensuring that the chosen design adequately addresses the objectives.

C. Integration Difficulties:

One of the central challenges of MMR is integrating findings in a way that adds value beyond what each method would provide in isolation (Bryman, 2006).

Recommendation: Pre-plan points of integration, whether it's during the data collection, analysis, or interpretation stages. This can ensure a more seamless and coherent integration process.

D. Sampling and Data Collection Disparities:

Different data types often require different sampling techniques, which can pose logistical and methodological challenges (Collins, Onwuegbuzie, & Jiao, 2007).

Recommendation: Develop clear criteria for sampling that consider the needs of both qualitative and quantitative data collection, and be transparent in justifying any compromises made.

E. Skill and Training Gaps:

Conducting high-quality MMR typically requires expertise in both qualitative and quantitative methodologies, which may require interdisciplinary collaboration or additional training (O'Cathain, 2010).

Recommendation: Consider team-based approaches, engaging experts from both qualitative and quantitative backgrounds, or seeking further training where gaps exist.

F. Time and Resource Intensity:

Given the complexities involved, MMR can be more time-consuming and resource-intensive than mono-method studies (Morse & Niehaus, 2009).

Recommendation: Adequately budget time and resources, ensuring stakeholders or funding bodies understand the rationale behind the extended timeframe or increased costs.

While MMR offers profound advantages in research depth and breadth, it also demands a nuanced understanding of both its strengths and challenges. Addressing these challenges proactively can aid in ensuring the success and robustness of a mixed methods study.

In conclusion, designing a mixed methods study requires methodological acumen, clarity of purpose, and strategic planning. By giving due attention to the design's nuanced facets, researchers can harness MMR's full potential, providing rich, robust, and comprehensive insights into their research questions.

Integration of Data: Sequential, Concurrent, and Transformative Approaches

Mixed methods research is distinguished by its commitment to integrate quantitative and qualitative datasets to provide a comprehensive understanding of the research problem. The nature, timing, and processes of this integration can vary, giving rise to several approaches (Creswell & Plano Clark, 2011). This section delves deeper into three primary integration strategies: sequential, concurrent, and transformative approaches.

1. Sequential Integration:

Sequential integration involves collecting and analyzing one type of data first, followed by the other type. The sequence can start with either quantitative or qualitative data, but the essence lies in the clear phases that define the research process.

Strengths: This approach offers clarity and can help the researcher to use the results from the first phase to inform the second phase (Morse, 1991).

Challenges: The phased approach might elongate the research duration and may lead to the risk of prioritizing one dataset over the other.

Recommendation: Use sequential integration when the research questions demand a two-phase inquiry or when findings from one method can significantly inform the next phase.

2. Concurrent Integration:

Here, qualitative and quantitative data are collected and analyzed simultaneously. This approach seeks to compare or relate both datasets to discern convergences, divergences, or relationships (Greene, Caracelli, & Graham, 1989).

Strengths: Concurrent integration can save time compared to sequential approaches and allows simultaneous exploration of different dimensions of the research problem.

Challenges: Requires skillful management of both datasets simultaneously, and the need for a clear conceptual framework to guide integration.

Recommendation: Employ concurrent integration when the research objective is to understand different facets of a phenomenon in tandem or to validate one set of findings against another.

3. Transformative Integration:

The transformative approach is distinguished by its explicit philosophical or theoretical framework guiding the research. It usually focuses on issues of power, justice, and marginalized voices, using the integration of datasets to foster change (Mertens, 2007).

Strengths: It can lead to research outcomes that promote social justice and change.

Challenges: Requires deep grounding in the guiding theoretical framework and a commitment to ethical rigor.

Recommendation: Use transformative integration when the research is rooted in advocacy, social justice, or when aiming to provide a voice to marginalized groups.

Sequential Integration

Mixed methods research often employs diverse strategies to ensure that data is collected and analyzed in a manner that best aligns with the research objectives. One of these strategies is sequential integration, which involves the systematic collection and analysis of one type of data followed by another, either qualitative or quantitative (Creswell, 2014). This form of integration aims to leverage the strengths of each dataset to offer a comprehensive understanding of the research problem.

A. Phases of Sequential Integration:

Sequential integration can be bifurcated into two main phases:

Qualitative followed by Quantitative (QUAL → QUAN): Here, qualitative data is collected and analyzed first, and its findings subsequently inform the second quantitative phase. This order is often adopted when researchers want to generate a theory or hypothesis from qualitative data and then test it quantitatively (Morse & Niehaus, 2009).

Quantitative followed by Qualitative (QUAN → QUAL): In this structure, the quantitative phase precedes the qualitative one. This approach is used when researchers want to elaborate, enhance, or illustrate quantitative results with qualitative data (Teddlie & Tashakkori, 2009).

B. Strengths of Sequential Integration:
- *Clarity and Structure*: Sequential integration provides a clear structure, ensuring each method remains distinct, preserving the integrity and validity of each dataset (Ivankova, Creswell, & Stick, 2006).
- *Comprehensive Exploration*: By using two methods in succession, researchers can investigate a phenomenon deeply and broadly, ensuring a robust exploration (Greene, Caracelli, & Graham, 1989).

C. Challenges of Sequential Integration:
- *Time Intensive*: The phased approach requires two separate periods for data collection and analysis, extending the research duration (Creswell, Plano Clark, Gutmann, & Hanson, 2003).
- *Potential Bias*: There's a risk of the first phase unduly influencing or narrowing the scope of the second phase, potentially introducing biases (Bryman, 2006).

D. Recommendations for Successful Sequential Integration:
- *Clear Objectives*: Researchers must have clear objectives for each phase and ensure that the insights from the first phase genuinely inform the second (Creswell & Plano Clark, 2011).
- *Integration at Interpretation*: While data collection and analysis occur separately, integration should occur during the interpretation phase, drawing from both datasets to derive comprehensive conclusions (O'Cathain, Murphy, & Nicholl, 2010).

Sequential integration in mixed methods research is an invaluable strategy for researchers aiming to explore a phenomenon in depth. By harnessing the strengths of both qualitative and quantitative approaches in a structured manner, researchers can provide rich insights and a holistic understanding of their subject matter.

Concurrent Integration

Concurrent integration in mixed methods research involves the simultaneous collection and analysis of qualitative and quantitative data, with the intention of drawing on the strengths of both methods to achieve a holistic understanding of the research problem (Creswell, 2014). This approach allows researchers to validate and cross-check findings, leading to a more in-depth and nuanced comprehension of the research topic.

A. Characteristics of Concurrent Integration:
- **Parallel Data Collection**: Both qualitative and quantitative data are collected during the same timeframe, without one being contingent on the other (Tashakkori & Teddlie, 2003).
- **Separate yet Interactive Analyses**: While data is analyzed separately, findings often intersect, complement, or contrast, resulting in richer insights (Bryman, 2006).

B. Types of Concurrent Integration:
- **Equal Status Design**: Here, both qualitative and quantitative components have equal priority and are used to answer separate but related research questions (Greene, Caracelli, & Graham, 1989).
- **Nested Design**: One dataset plays a supportive, subsidiary role to the other primary dataset. This design is utilized when a secondary method is needed to clarify or enrich the primary data (Morse & Niehaus, 2009).

C. Strengths of Concurrent Integration:
- **Complementarity**: By juxtaposing the two types of data, researchers can gain more comprehensive insights, with each method illuminating unique aspects of the phenomenon (Johnson, Onwuegbuzie, & Turner, 2007).
- **Enhanced Validity**: The simultaneous use of qualitative and quantitative approaches can facilitate cross-validation, improving the study's internal validity (Creswell & Plano Clark, 2011).

D. Challenges of Concurrent Integration:
- **Methodological Complexity**: Handling both datasets can be methodologically demanding, requiring expertise in both qualitative and quantitative realms (Onwuegbuzie & Leech, 2005).
- **Resource Intensity**: Given the simultaneous nature, concurrent integration may demand more time, cost, and personnel, especially when data needs to be collected from large or diverse samples (Teddlie & Tashakkori, 2009).

E. Recommendations for Effective Concurrent Integration:
- **Clear Research Questions**: Ensure that the research questions are well-defined and aligned with the chosen mixed methods design, determining which aspects will be explored qualitatively and which quantitatively (Bazeley, 2012).
- **Integration at Interpretation**: Despite separate analyses, integration should occur at the interpretation stage, with conclusions derived from a synthesis of both data sets (O'Cathain, Murphy, & Nicholl, 2010).

Concurrent integration stands as a testament to the versatility of mixed methods research, allowing researchers to harness the simultaneous power of qualitative depth and quantitative breadth. It reaffirms the belief that diverse methodological approaches can coexist and complement, providing a more profound, multifaceted understanding of complex research topics.

Transformative Integration

Transformative integration in mixed methods research represents a paradigmatic shift in the way researchers approach the combined use of qualitative and quantitative methods. Rooted in transformative-emancipatory methodologies, this approach places special emphasis on the social and political context in which research is conducted, with the aim of advocating for social change or addressing issues of social justice (Mertens, 2007).

A. Characteristics of Transformative Integration:

- **Emphasis on Sociopolitical Activism**: The approach is deeply rooted in advocating for social change or highlighting issues of power, discrimination, and marginalization (Creswell & Plano Clark, 2011).
- **Incorporation of Participatory Methods**: Often, transformative mixed methods studies involve the community or the marginalized groups under study in the research process (Hesse-Biber, 2010).

B. Paradigmatic Roots:

The transformative paradigm arises from critical theory, feminist theory, and other frameworks emphasizing empowerment, collaboration, social justice, and change (Mertens, 2015). It challenges traditional paradigms by centering on the experiences of marginalized communities and groups.

C. Benefits of Transformative Integration:
- **Advocacy and Action**: It offers a platform for researchers to advocate for social change and action based on the findings (Mertens, 2007).
- **Holistic Understanding**: By integrating both qualitative and quantitative data, it provides a more holistic understanding of complex social issues, taking into account both the lived experiences of individuals and wider societal structures (Creswell, 2014).

D. Challenges of Transformative Integration:
- **Potential for Bias**: Given the activist stance, there's potential for researcher bias, or the findings being perceived as biased (Bryman, 2006).
- **Complexity**: It requires a nuanced understanding of both methodological approaches and the sociopolitical contexts under investigation (Teddlie & Tashakkori, 2009).

E. Recommendations for Effective Transformative Integration:
- **Collaborative Partnerships**: Engage the community or marginalized group in the research process, from defining the research questions to interpreting the findings (Hesse-Biber, 2010).
- **Continuous Reflexivity**: Researchers should engage in continuous reflexivity to acknowledge their positionality and potential biases in the research process (Lincoln & Guba, 1985).
- **Transparent Reporting**: Clearly articulate the rationale for adopting a transformative approach and ensure transparent reporting of methods, findings, and advocacy efforts (Creswell & Plano Clark, 2011).

Transformative integration in mixed methods research offers an avenue for researchers to advocate for meaningful change, highlighting and addressing the concerns of marginalized and disenfranchised communities. While it presents unique challenges, the potential for positive societal impact makes it a valuable approach in the arsenal of mixed-methods researchers.

CONCLUSION

The choice between sequential, concurrent, and transformative integration should be guided by the research question, objectives, and practical constraints of the study. Properly integrated mixed methods research can offer insights and depth that might not be achievable through a singular methodological lens.

Challenges in Mixed Methods Research

The adoption of mixed methods research (MMR) has seen significant growth over the past few decades. Its inherent ability to draw upon the strengths of both qualitative and quantitative research methodologies lends it a unique advantage. Yet, navigating the complexities of MMR is not without its challenges. In this section, we delve deep into the intricacies and potential pitfalls researchers might encounter in the realm of MMR.

1. Methodological Understanding:

To effectively employ MMR, a comprehensive understanding of both qualitative and quantitative research methods is imperative (Johnson & Onwuegbuzie, 2004). However, attaining proficiency in both methodologies can be demanding and time-consuming, often leading to a superficial engagement with one or the other.

Mixed methods research (MMR) holds the allure of amalgamating the strengths of both qualitative and quantitative approaches, providing a more holistic insight into research phenomena. However, one of the foundational challenges that often confronts researchers in this domain is the depth of methodological understanding required to effectively combine and integrate these distinct paradigms (Creswell, 2003). This section delves into the complexities and nuances of this challenge.

A. Dual Proficiency Requirement:

For a robust application of MMR, researchers must possess a comprehensive grasp of both qualitative and quantitative methodologies. This dual proficiency ensures that neither approach is superficially or inappropriately applied, thereby safeguarding the integrity of the research (Johnson, Onwuegbuzie, & Turner, 2007).

B. Research Design Complexity:

Crafting a research design that effectively integrates qualitative and quantitative methods necessitates not just an understanding of each method's individual intricacies but also how they can be synergized for complementarity (Bryman, 2006).

C. Analysis Integration:

The analytical techniques employed in qualitative and quantitative research vary considerably. A deep methodological understanding is crucial to ensure that data from both methods is integrated seamlessly during the analysis phase, without compromising the depth or integrity of findings (Teddlie & Tashakkori, 2009).

D. Navigating Criticisms:

Due to the blended nature of MMR, researchers often face criticisms from purists in both qualitative and quantitative camps. A solid methodological understanding equips researchers to defend their design choices and address potential critiques more effectively (Morse, 2010).

Deepening one's methodological understanding is not merely an academic exercise in the context of MMR. It is a foundational requisite that determines the quality, credibility, and validity of mixed methods studies. Researchers aspiring to leverage the power of MMR should invest in building a solid methodological foundation, drawing upon comprehensive training, mentorship, and continuous engagement with evolving methodological debates.

2. Paradigmatic Differences:

Historically, qualitative and quantitative research approaches stem from different epistemological traditions. The integration of two, sometimes conflicting paradigms, can pose philosophical and theoretical challenges (Morgan, 2007).

The endeavor to integrate qualitative and quantitative research paradigms in a Mixed Methods Research (MMR) often confronts paradigmatic challenges. Paradigms, in the research context, refer to a set of beliefs and feelings about the world and how it should be understood and studied (Denzin & Lincoln, 2000). The integration of differing paradigms inherent in qualitative and quantitative research methods can be fraught with complexities and tensions. This section seeks to elucidate the nature, implications, and potential resolutions of these paradigmatic differences in MMR.

A. Philosophical Foundations:
   - **Quantitative Research**: Typically grounded in positivism or post-positivism, this paradigm operates on the belief that reality is objective and can be measured and known. It emphasizes empirical observation and statistical analysis (Creswell, 2003).
   - **Qualitative Research**: Often rooted in constructivism or interpretivism, qualitative research assumes that reality is socially constructed and subjective. It emphasizes understanding and interpretation of participants' experiences and meanings (Denzin & Lincoln, 2011).

B. Points of Tension:
   - **Objectivity vs. Subjectivity**: While quantitative research strives for objectivity and generalizability, qualitative research values subjectivity and depth of understanding (Mertens, 2007).
   - **Generalizability vs. Contextuality**: Quantitative research seeks findings that can be generalized across populations, while qualitative research prioritizes context-specific insights (Bryman, 2006).
   - **Reductionism vs. Holism**: Quantitative approaches may reduce phenomena to measurable variables, while qualitative methods tend to capture the holistic essence of experiences (Teddlie & Tashakkori, 2009).

C. Bridging the Gap:
   - **Pragmatism**: One philosophical stance that has been advocated as a bridge between paradigms is pragmatism. Pragmatism posits that research questions should drive the methodological choices rather than vice versa. Thus, MMR guided by pragmatism prioritizes the research problem and employs both quantitative and qualitative methods synergistically to address it (Morgan, 2007).

- **Dialectical Stance**: Another perspective suggests that researchers should engage in a dialectic process where both paradigms are viewed as contradictory yet complementary, eventually leading to a richer understanding (Johnson & Onwuegbuzie, 2004).

D. Implications for Research Practice:

Researchers need to be transparent about their epistemological stance, making clear the paradigmatic basis for their methodological choices. Such transparency aids in addressing potential critiques and establishing the credibility of the MMR study (Greene, 2008).

While the paradigmatic differences between qualitative and quantitative research present challenges in MMR, they also offer opportunities for richer, more comprehensive insights. By embracing a pluralistic stance and leveraging the strengths of both paradigms, researchers can deliver holistic and robust findings that address complex research questions.

3. Integration Difficulties:

Simply employing both methods does not automatically yield the benefits of MMR. Effective integration of qualitative and quantitative data during the analysis phase can be challenging, requiring rigorous methodological design and clear research questions (Creswell & Plano Clark, 2011).

Mixed Methods Research (MMR) promises a comprehensive approach by combining the strengths of both qualitative and quantitative methods. However, one of the critical challenges faced by many researchers in MMR is the difficulty of effectively integrating the two data forms. Integration difficulties pertain to the challenges of meaningfully synthesizing, interpreting, and presenting data from both qualitative and quantitative sources in a coherent manner.

This section elucidates the nature of these challenges, their implications, and potential strategies to address them.

A. Nature of Integration Difficulties:

- **Diverse Data Types**: Qualitative and quantitative data inherently differ. While quantitative data is numerical and amenable to statistical analysis, qualitative data is textual or visual, requiring thematic interpretation (Creswell & Plano Clark, 2017).
- **Different Scales of Analysis**: Quantitative data often operates at a macro level, looking for broad patterns or trends, whereas qualitative data provides a micro perspective, diving deep into individual experiences or specific contexts (Teddlie & Tashakkori, 2009).
- **Differing Interpretations**: Findings from the two methods might not always converge or align, leading to potential contradictions (Bryman, 2006).

B. Implications for Research:

- **Validity Concerns**: Incoherent integration might lead to questions about the study's validity or the accuracy of its findings (Morse, 1991).
- **Complexity in Presentation**: Disparate findings can complicate the representation of results, making it challenging to communicate to audiences, especially those preferring a singular narrative (Greene, 2007).
- **Increased Time and Resource Requirements**: Effective integration often demands additional time for synthesis and can strain limited research resources.

C. Strategies for Addressing Integration Difficulties:

- **Sequential Integration**: Here, one method follows the other, allowing the results of the first method to inform the subsequent one. This step-by-step approach can ease integration challenges (Morse & Niehaus, 2009).
- **Joint Displays**: Creating tables or figures that juxtapose or combine both quantitative and qualitative results can be an effective visual method to facilitate integration (Creswell & Plano Clark, 2017).
- **Narrative Approach**: Researchers can weave together the findings from both methods into a cohesive narrative, ensuring that the story of the data is coherent and comprehensive (O'Cathain, Murphy, & Nicholl, 2008).
- **Engaging in Reflexivity**: Researchers should continually reflect on their integrative processes, acknowledging potential biases and being transparent about how decisions regarding integration were made (Rossman & Wilson, 1985).

While integration difficulties in MMR present substantial challenges, they also offer an opportunity for innovative thinking and methodological growth. With careful planning, reflection, and utilization of effective strategies, researchers can navigate these complexities to produce rich, multi-dimensional insights.

4. Time and Resource Intensiveness:

By virtue of its scope, MMR often demands more time and resources compared to mono-method research. This has implications for project timelines, funding, and logistics (Bryman, 2006).

Mixed Methods Research (MMR) undoubtedly offers a rich tapestry of insights by integrating both qualitative and quantitative methods. However, the fusion of these methodologies also inherently means an increased demand on time and resources. This section delves into the nuances of the time and resource challenges associated with MMR, exploring the reasons for these challenges, their implications for research, and potential strategies for managing them.

A. Underlying Reasons for Time and Resource Intensiveness:
    - **Dual Data Collection**: Implementing both qualitative and quantitative research methods requires designing, collecting, and processing two sets of data, each demanding its unique resources and time (Greene, Caracelli, & Graham, 1989).
    - **Data Analysis**: Qualitative data requires thematic analysis, coding, and interpretation, while quantitative data demands statistical evaluations. Both processes are time-consuming and require specialized software and expertise (Teddlie & Tashakkori, 2003).
    - **Integration Process**: Synthesizing and integrating findings from both data sets to generate coherent results adds an additional layer of complexity and time requirement (Bryman, 2006).

B. Implications for Research:
    - **Extended Research Timeline**: The multiple phases of MMR, from data collection to analysis and integration, naturally elongate the research timeline, potentially delaying publication or application of results (Johnson, Onwuegbuzie, & Turner, 2007).

- **Increased Costs**: The prolonged timeline combined with the need for specialized tools, software, and possibly more personnel can escalate the overall cost of the research project (Creswell, Klassen, Plano Clark, & Smith, 2011).
- **Potential for Research Fatigue**: Researchers and participants might experience fatigue given the extensive data collection and analytical procedures in MMR (Morse & Niehaus, 2009).

C. Strategies to Manage Time and Resource Challenges:

- **Phased Approach**: Breaking down the research into distinct phases with clear objectives can aid in efficient allocation and utilization of resources and time (Creswell & Plano Clark, 2017).
- **Use of Technology**: Leveraging specialized software for mixed methods (e.g., NVivo, MAXQDA) can streamline data management, analysis, and integration processes (Fielding & Fielding, 1986).
- **Collaborative Teams**: Forming interdisciplinary teams can help distribute the workload, with experts handling specific aspects of the qualitative and quantitative segments (Palinkas et al., 2011).
- **Pilot Testing**: Running pilot tests can refine the methodology, potentially reducing unnecessary time and resource expenditures in the main study (Bazeley, 2004).

The intricate weave of MMR, blending qualitative depth with quantitative breadth, inherently comes with increased demands on time and resources. However, with careful planning, the use of technology, and strategic team collaboration, researchers can navigate these challenges to harness the profound insights that MMR promises.

5. Interpretation Complexities:

When the quantitative and qualitative findings diverge or contradict each other, researchers are faced with the challenge of interpreting and presenting these discrepancies in a meaningful way (O'Cathain, 2010).

Mixed Methods Research (MMR) inherently combines two distinct methodologies: qualitative and quantitative. While this convergence can lead to a comprehensive understanding of the research question, it also introduces complexities, particularly in the realm of data interpretation. Here, we delve deeper into the challenges posed by interpretation complexities in MMR and ways to navigate them.

A. Nature of Interpretation Complexities in MMR:

- **Dual Data Sources**: In MMR, researchers confront dual datasets, each producing its set of results and interpretations. This dualism necessitates weaving together varied insights into a cohesive narrative, a task not always straightforward (Creswell & Plano Clark, 2007).
- **Differing Data Characteristics**: Qualitative data, with its depth and context, offers nuanced insights, while quantitative data provides empirical breadth. These differences can sometimes produce seemingly contradictory findings (Tashakkori & Teddlie, 1998).

B. Implications of Interpretation Complexities:

- **Risk of Misalignment**: If not adeptly managed, the integration of qualitative and quantitative data can lead to a fragmented or disjointed understanding of the research phenomena (O'Cathain, Murphy, & Nicholl, 2008).
- **Increased Cognitive Demand**: The need to interpret, juxtapose, and synthesize two types of data demands significant cognitive effort, increasing the potential for oversight or misinterpretation (Bazeley, 2010).

C. Strategies to Address Interpretation Complexities:

- **Sequential Interpretation**: One approach is to interpret the data sequentially, starting with one dataset and then moving to the other. This step-by-step process can make integration more manageable (Morse, 1991).
- **Joint Display**: Utilizing joint displays, where results from both methods are displayed side-by-side, can assist researchers in visualizing connections or disparities between the data sets (Guetterman, Fetters, & Creswell, 2015).
- **Iterative Process**: Instead of viewing interpretation as a singular step, approaching it iteratively allows for continuous refinement and cross-referencing between the datasets, leading to more nuanced results (Palinkas et al., 2011).
- **Collaborative Teams**: Engaging a multidisciplinary team, where members bring expertise in either qualitative or quantitative research, can assist in ensuring a balanced and thorough interpretation process (Saldaña, 2015).

While Mixed Methods Research promises comprehensive insights, it also presents inherent challenges, especially in the domain of data interpretation. Yet, with strategic approaches, careful planning, and a willingness to engage deeply with the data, researchers can navigate these complexities to produce rich, integrated findings that honor the essence of both qualitative and quantitative methodologies.

6. Publication Hurdles:

Given its intricate nature, presenting MMR findings comprehensively within the confines of word limits set by academic journals can be challenging. Moreover, the interdisciplinary nature of MMR might not fit neatly into the scope of discipline-specific journals, sometimes making publication more challenging (Bazeley, 2009).

Mixed Methods Research (MMR) has, over the decades, gained acceptance and credibility in the academic community. However, MMR researchers frequently encounter challenges related to publishing their studies. The integration of qualitative and quantitative data can sometimes lead to skepticism, misinterpretations, or apprehensions from reviewers and editors. This section provides an in-depth exploration of the publication hurdles faced by MMR researchers.

A. Complexity of MMR Publications:

- **Length and Depth**: Given the dual nature of MMR—combining qualitative with quantitative data—the resulting manuscripts can be lengthier and denser than those focusing on a single methodology (O'Cathain, 2010). This poses challenges in adhering to word limits set by many academic journals.

- **Diverse Audience**: MMR papers target a broad audience, spanning those with expertise in qualitative, quantitative, or both methodologies. This diversity can pose challenges in striking the right balance in terms of depth, jargon, and methodological detail (Brannen, 2005).

B. Paradigmatic Concerns:

- **Skepticism**: Some purists in either the qualitative or quantitative camps might view MMR with skepticism, questioning its rigor or validity (Morse, 2003).
- **Methodological Bias**: Journals and reviewers might have a predisposed inclination towards either qualitative or quantitative research, making it challenging for MMR studies to fit neatly within the established paradigms (Bryman, 2006).

C. Integration Difficulties:

- **Cohesiveness**: Ensuring that qualitative and quantitative components form a cohesive narrative can be challenging. Reviewers might criticize perceived imbalances or lack of integration in the manuscript (Johnson & Onwuegbuzie, 2004).

D. Navigating the Peer Review Process:

- **Reviewer Expertise**: Finding reviewers proficient in both qualitative and quantitative methods can be challenging. Lack of expertise in one of the methods might lead to inadequate or biased reviews (Fielding, 2012).
- **Iterative Revisions**: Given the complexities of MMR, manuscripts might undergo several rounds of revisions before acceptance, extending the publication timeline (Fetters, Curry, & Creswell, 2013).

E. Strategies to Overcome Publication Hurdles:

- **Targeted Journal Selection**: Opt for journals known for their openness to MMR or those that specialize in mixed methods (Fetters & Freshwater, 2015).
- **Clear Rationale**: Ensure that the manuscript clearly articulates the rationale for adopting an MMR approach and its relevance to the research question (Creswell & Tashakkori, 2007).
- **Seek Feedback**: Before submission, seek feedback from colleagues proficient in both methodologies to ensure balance, clarity, and rigor.

While MMR offers comprehensive insights and enriched findings, navigating the publication process poses distinct challenges. Awareness of these hurdles and a proactive approach to addressing them can ease the path to publication.

7. Training and Expertise:

Adequate training programs in MMR are still emerging. Researchers might find it challenging to access quality training resources or mentors with expertise in both qualitative and quantitative domains (Teddlie & Tashakkori, 2009).

Mixed Methods Research (MMR) is an intricate research approach that necessitates proficiency in both qualitative and quantitative methods. Researchers embarking on MMR encounters several challenges concerning training and expertise, which can significantly impact the quality of the research. This section delves deeper into these challenges and offers insights into potential solutions.

A. Dual Expertise Requirement:

- **Quantitative and Qualitative Competence**: MMR researchers need to be well-versed in both qualitative and quantitative methodologies. This requirement demands extensive training and expertise, which can be time-consuming (Tashakkori & Teddlie, 2008).

B. Interdisciplinary Collaboration:

- **Collaborative Skills**: MMR often involves collaboration among researchers with diverse methodological backgrounds. Effective communication and collaboration skills are crucial to ensure synergy (Caracelli & Greene, 1997).
- **Resource Allocation**: Collaboration might necessitate resource allocation, including funding and time, to facilitate cross-training and skill development (Teddlie & Tashakkori, 2009).

C. Training and Resources:

- **Limited Training Programs**: Many academic programs offer training in either qualitative or quantitative research but lack integrated MMR training (Onwuegbuzie & Johnson, 2006).
- **Resource Availability**: Access to resources such as software, data analysis tools, and mentorship in both methodologies is essential but may not always be readily available (Bazeley & Kemp, 2012).

D. The Need for Ongoing Learning:

- **Methodological Advancements**: Both qualitative and quantitative methodologies continuously evolve. MMR researchers must stay updated with the latest developments in both fields (Plano Clark & Ivankova, 2016).

E. Strategies to Overcome Training and Expertise Challenges:

- **Interdisciplinary Workshops**: Organize workshops that bring together experts from both qualitative and quantitative disciplines to facilitate knowledge exchange and skill development (Bryman & Hardy, 2012).
- **Mixed Methods Courses**: Academic institutions can develop dedicated MMR courses that cover the intricacies of combining qualitative and quantitative approaches (Teddlie & Tashakkori, 2009).
- **Mentorship Programs**: Establish mentorship programs wherein experienced MMR researchers guide novices in developing proficiency in both methodologies (Sandelowski et al., 2016).
- **Invest in Resources**: Academic institutions and research organizations should invest in resources such as software licenses and data analysis tools to support MMR (O'Cathain, Murphy, & Nicholl, 2007).
- **Stay Updated**: Regularly attend conferences, seminars, and webinars in both qualitative and quantitative research to stay informed about methodological advancements (Creswell & Plano Clark, 2011).

Training and expertise are fundamental challenges in MMR, requiring continuous learning and interdisciplinary collaboration. Researchers, institutions, and organizations must recognize the importance of addressing these challenges to conduct high-quality mixed methods research effectively.

8. Software Limitations:

While there are numerous software tools designed for qualitative or quantitative analysis, there are fewer options that effectively cater to the integrated analysis needs of MMR, potentially complicating the analysis process (Fielding & Fielding, 2008).

Mixed Methods Research (MMR) offers a comprehensive approach to research by combining qualitative and quantitative methods. However, MMR faces several challenges, one of which is software limitations. In this section, we will delve into the complexities associated with software use in MMR and propose strategies to mitigate these challenges.

A. Software Diversity and Compatibility:
   - **Variety of Tools**: MMR often requires the use of different software tools for qualitative and quantitative data analysis. Ensuring compatibility between these tools can be challenging (Teddlie & Tashakkori, 2009).
   - **Data Transfer**: Transferring data between software platforms may lead to data loss or formatting issues, complicating the research process (Fetters, Curry, & Creswell, 2013).

B. Learning Curve:
   - **Steep Learning**: Proficiency in software for qualitative and quantitative analysis requires substantial training. Researchers need to invest time and effort to become adept at using these tools (Plano Clark & Ivankova, 2016).

C. Limited Integration Tools:
   - **Few Seamless Integration Options**: The availability of tools that seamlessly integrate qualitative and quantitative data is limited. Researchers often need to employ workarounds to combine data effectively (O'Cathain, Murphy, & Nicholl, 2007).

D. Data Management:
   - **Data Overload**: MMR projects can generate large volumes of data. Managing and organizing this data across different software platforms can be overwhelming (Creswell & Plano Clark, 2011).

E. Strategies to Overcome Software Limitations:
   - **Interoperable Software**: Invest in software tools that are designed to work together. Some software packages are specifically developed for MMR, ensuring better compatibility (Teddlie & Tashakkori, 2009).
   - **Training and Support**: Institutions and organizations should provide training and support for researchers to become proficient in the software tools they need (Creswell & Creswell, 2017).
   - **Data Conversion Protocols**: Develop standardized data conversion protocols to minimize data loss or corruption when transferring data between different software (Fetters et al., 2013).
   - **Data Management Plans**: Establish robust data management plans from the outset to ensure that data organization and storage are well-structured (Creswell & Creswell, 2017).
   - **Integration Tool Development**: Advocate for the development of tools that facilitate the seamless integration of qualitative and quantitative data (O'Cathain et al., 2007).

Software limitations are a significant challenge in MMR. Researchers and institutions must recognize the importance of addressing these issues to streamline the process of combining qualitative and quantitative data effectively.

CONCLUSION

While MMR offers a robust approach to address complex research questions, it comes with its own set of challenges. Being forewarned of these challenges allows researchers to navigate them more effectively, ensuring the rigour and credibility of their mixed methods studies.

REFERENCES

Bryman, A. (2006). Paradigm peace and the implications for quality. *International Journal of Social Research Methodology*, *9*(2), 111–126. doi:10.1080/13645570600595280

Bryman, A. (2012). *Social research methods* (4th ed.). Oxford University Press.

Creswell, J. W. (2009). *Research design: Qualitative, quantitative, and mixed methods approaches.* Sage Publications.

Creswell, J. W. (2014). *A concise introduction to mixed methods research.* Sage.

Creswell, J. W., Klassen, A. C., Plano Clark, V. L., & Smith, K. C. (2011). *Best practices for mixed methods research in the health sciences.* National Institutes of Health. doi:10.1037/e566732013-001

Creswell, J. W., & Plano Clark, V. L. (2007). *Designing and conducting mixed methods research.* Sage Publications.

Greene, J. C. (2007). *Mixed methods in social inquiry.* Jossey-Bass.

Greene, J. C. (2008). Is mixed methods social inquiry a distinctive methodology? *Journal of Mixed Methods Research*, *2*(1), 7–22. doi:10.1177/1558689807309969

Hall, B., & Howard, K. (2008). A synergistic approach: Conducting mixed methods research with typological and systemic design considerations. *Journal of Mixed Methods Research*, *2*(3), 248–269. doi:10.1177/1558689808314622

Hesse-Biber, S. (2010). *Mixed methods research: Merging theory with practice.* Guilford Press.

Johnson, R. B., & Christensen, L. B. (2012). *Educational research: Quantitative, qualitative, and mixed approaches* (4th ed.). Sage Publications.

Johnson, R. B., & Onwuegbuzie, A. J. (2004). Mixed methods research: A research paradigm whose time has come. *Educational Researcher*, *33*(7), 14–26. doi:10.3102/0013189X033007014

Morse, J. M. (1991). Approaches to qualitative-quantitative methodological triangulation. *Nursing Research*, *40*(2), 120–123. doi:10.1097/00006199-199103000-00014 PMID:2003072

Morse, J. M., & Niehaus, L. (2009). *Mixed method design: Principles and procedures.* Left Coast Press.

O'Cathain, A., Murphy, E., & Nicholl, J. (2007). Why, and how, mixed methods research is undertaken in health services research in England: A mixed methods study. *BMC Health Services Research*, *7*(1), 1–11. doi:10.1186/1472-6963-7-85 PMID:17570838

O'Cathain, A., Murphy, E., & Nicholl, J. (2010). Three techniques for integrating data in mixed methods studies. *BMJ (Clinical Research Ed.)*, *341*(sep17 1), c4587. doi:10.1136/bmj.c4587 PMID:20851841

Plano Clark, V. L., & Creswell, J. W. (2008). *The mixed methods reader*. Sage.

Plano Clark, V. L., & Ivankova, N. V. (2016). *Mixed methods research: A guide to the field*. Sage Publications. doi:10.4135/9781483398341

Tashakkori, A., & Teddlie, C. (2010). *Handbook of mixed methods in social & behavioral research* (2nd ed.). Sage. doi:10.4135/9781506335193

Teddlie, C., & Tashakkori, A. (2009). Foundations of mixed methods research: Integrating quantitative and qualitative approaches in the social and behavioral sciences. *Sage (Atlanta, Ga.)*.

Chapter 11
Recent Trends and Future Directions

ABSTRACT

This chapter explores the evolution and future directions of qualitative research, focusing on its adaptation to societal changes and technological advancements. Recent trends include digital ethnography, which allows for the study of online communities, and an increased focus on intersectionality, enhancing the understanding of complex social identities. Narrative and visual methods have gained prominence, reflecting a shift towards more expressive forms of data presentation. Looking forward, qualitative research is expected to increasingly incorporate global perspectives and address ethical considerations with a growing awareness of the impacts of research on both subjects and researchers.

INTRODUCTION

Qualitative research has evolved significantly over the years, adapting to the changing landscape of the research world and the broader societal context. In this chapter, we will explore recent trends and future directions in qualitative research, shedding light on the exciting developments and challenges that researchers are likely to encounter in the years to come. As we delve into this exploration, we will be guided by scholarly insights and supported by relevant citations and references.

Introduction

Qualitative research, characterized by its emphasis on understanding the complexities of human experiences, has a rich history dating back to the early 20th century (Denzin & Lincoln, 2018). Over time, it has become a robust and multifaceted approach to inquiry, offering valuable insights into areas such as social sciences, healthcare, education, and more. Recent trends and emerging directions in qualitative research reflect both the dynamic nature of this field and the ever-evolving needs of researchers and society at large.

DOI: 10.4018/979-8-3693-2414-1.ch011

Recent Trends in Qualitative Research

1. Digital Ethnography and Online Research

The advent of the digital age has opened up new avenues for qualitative research. Digital ethnography, in particular, has gained prominence, allowing researchers to study online communities, social media, and virtual environments (Hine, 2015). This trend aligns with the growing importance of understanding digital cultures and the impact of technology on human behavior.

2. Intersectionality and Inclusivity

Qualitative research has become increasingly attuned to issues of intersectionality, acknowledging the complex interplay of various social identities such as race, gender, sexuality, and class (Cho, Crenshaw, & McCall, 2013). Researchers are now more inclined to adopt an intersectional lens to explore how multiple dimensions of identity shape lived experiences and social phenomena.

3. Narrative and Visual Methods

Narrative inquiry and visual methods have gained prominence as researchers recognize the power of storytelling and visual representation in conveying experiences and insights (Clandinin & Connelly, 2000). These approaches offer innovative ways to explore and present qualitative data.

4. Mixed Methods Research

The integration of qualitative and quantitative methods, known as mixed methods research, continues to evolve (Creswell & Creswell, 2017). Researchers are exploring more sophisticated ways of combining these approaches to gain a more comprehensive understanding of research questions.

Future Directions in Qualitative Research

A. Globalization and Cultural Context

As the world becomes increasingly interconnected, qualitative research is likely to delve deeper into the nuances of globalization and its impact on local cultures (Denzin & Lincoln, 2018). Researchers will need to grapple with questions of cultural sensitivity, power dynamics, and ethics in a globalized research landscape.

B. Technological Advancements

The rapid pace of technological advancements will continue to shape qualitative research. Virtual reality, artificial intelligence, and big data analytics are just a few examples of technologies that may influence data collection, analysis, and representation (Saldaña, 2015).

C. Critical and Decolonial Approaches

Qualitative researchers are increasingly adopting critical and decolonial approaches to challenge traditional power structures and Eurocentric perspectives (Tuhiwai Smith, 2012). This trend reflects a broader commitment to social justice and equity in research.

D. Ethical Considerations and Researcher Well-being

With growing awareness of the ethical complexities of qualitative research, future directions will likely involve more robust ethical guidelines and considerations (Hesse-Biber, 2012). Additionally, researchers' own well-being and emotional resilience may receive greater attention to mitigate potential emotional and psychological impacts.

The landscape of qualitative research is continually evolving, influenced by both external factors and the innovative thinking of researchers. Recent trends and future directions highlight the field's adaptability and commitment to addressing complex societal issues. As qualitative research continues to push boundaries, it remains a vital and dynamic approach to understanding the human experience.

In the subsequent sections of this chapter, we will delve into each of these trends and future directions, providing a comprehensive understanding of how qualitative research is poised to evolve and respond to the challenges and opportunities of the 21st century.

Advancements in Digital Qualitative Research

Digital qualitative research has emerged as a prominent trend in the field of qualitative inquiry, driven by the rapid proliferation of digital technologies and their influence on various aspects of human life. This section explores the recent advancements in digital qualitative research, shedding light on how technology is shaping the qualitative research landscape. Scholarly insights, citations, and references are provided to support the discussion.

The Digital Turn in Qualitative Research

The "digital turn" in qualitative research signifies the transformative impact of digital technologies on how researchers collect, analyze, and disseminate data. This shift has been facilitated by the widespread use of smartphones, social media, online communities, and other digital platforms that have become integral to people's daily lives (Pink, Horst, Postill, Hjorth, Lewis, & Tacchi, 2016).

Online Data Collection

One of the key advancements in digital qualitative research is the use of online platforms for data collection. Researchers can now engage with participants in virtual spaces, conducting interviews, surveys, and observations through video conferencing, email, or social media (Hesse-Biber & Griffin, 2019). This approach allows for broader and more diverse participant pools.

Social Media Analysis

The analysis of social media content has gained traction as a valuable source of qualitative data (Kozinets, 2010). Researchers can explore online communities, forums, and social networking sites to investigate a wide range of phenomena, from online activism to consumer behavior.

Digital Ethnography

Digital ethnography has become a prominent method for studying online communities and virtual environments (Pink et al., 2016). Researchers immerse themselves in digital spaces, observing interactions and behaviors in a manner analogous to traditional ethnography. This approach enables the exploration of digital cultures and online subcultures.

Data Visualization and Analysis Software

The development of sophisticated data visualization and analysis software has empowered researchers to manage and analyze large volumes of qualitative data efficiently (Charmaz & Belgrave, 2019). Tools like NVivo and ATLAS.ti provide researchers with the means to code, categorize, and visualize qualitative data in novel ways.

Ethical Considerations in Digital Qualitative Research

While digital technologies offer numerous advantages, they also raise ethical concerns that researchers must navigate (Markham & Buchanan, 2012). Ensuring participant privacy, obtaining informed consent for online data collection, and addressing issues of digital divide and unequal access are among the ethical challenges that researchers face.

Future Directions in Digital Qualitative Research

The future of digital qualitative research holds exciting possibilities and challenges. As technology continues to evolve, researchers are likely to explore:

AI and Natural Language Processing

The integration of artificial intelligence (AI) and natural language processing (NLP) into qualitative research may revolutionize data analysis. AI-driven tools can assist researchers in identifying patterns, sentiments, and themes in textual data, potentially speeding up the coding process (Wang & Kosinski, 2018).

Virtual Reality (VR) and Immersive Research

Virtual reality (VR) offers the potential for immersive qualitative research experiences. Researchers can create virtual environments that simulate real-world contexts, allowing participants to engage in research activities in novel and controlled settings.

Ethical and Privacy Challenges

As digital qualitative research becomes more sophisticated, ethical considerations related to data privacy, informed consent, and the potential misuse of AI and NLP tools will require careful attention (Markham & Buchanan, 2012). Researchers will need to develop robust ethical frameworks to protect both participants and the integrity of the research process.

The digital era has ushered in a new era of possibilities for qualitative research. Advancements in digital qualitative research methods have expanded the scope and reach of qualitative inquiry, offering researchers innovative ways to engage with participants and analyze data. However, these advancements also raise important ethical and methodological considerations that researchers must address as they navigate the evolving landscape of digital qualitative research.

Cross-cultural and Global Qualitative Research

Cross-cultural and global qualitative research has gained increasing prominence in the field of social sciences. This section delves into the recent trends and future directions of cross-cultural and global qualitative research, highlighting its significance and the challenges it presents. Scholarly insights, citations, and references are provided to support the discussion.

The Globalization of Qualitative Research

The advent of globalization and the interconnectedness of societies have given rise to the need for cross-cultural and global qualitative research. Researchers across disciplines recognize the importance of studying human behavior, experiences, and phenomena within diverse cultural contexts (Denzin & Lincoln, 2008). This trend has opened doors to fresh perspectives and innovative approaches in qualitative research.

Global Phenomena and Local Realities

Cross-cultural and global qualitative research allows scholars to explore global phenomena while acknowledging the nuances of local realities. Whether studying migration, health disparities, or environmental issues, researchers aim to capture the complexity of human experiences across different cultural and geographical settings (Wolcott, 2005).

Cultural Sensitivity and Reflexivity

In this research paradigm, cultural sensitivity and reflexivity are paramount. Researchers must be attuned to the cultural nuances that shape participants' experiences and interactions. Reflexivity, as advocated by scholars like Lather (1991), encourages researchers to critically examine their own cultural biases and assumptions throughout the research process.

Collaborative and Participatory Research

Collaborative and participatory research approaches have gained traction in cross-cultural studies (Kindon, Pain, & Kesby, 2007). Engaging local communities and stakeholders in the research process fosters a deeper understanding of cultural dynamics and ensures that research outcomes are relevant and actionable.

Emerging Trends in Cross-cultural and Global Qualitative Research

The future of cross-cultural and global qualitative research holds several promising trends:

Digital Ethnography and Virtual Research

Digital ethnography and virtual research methods allow researchers to engage with global participants in digital spaces (Horst & Miller, 2012). Virtual environments enable the study of online communities, social networks, and digital subcultures across borders.

Transnational Perspectives

Transnational research approaches recognize that many contemporary issues transcend national boundaries. Researchers increasingly adopt transnational perspectives to study topics like globalization, migration, and the flow of ideas and culture (Glick Schiller & Salazar, 2013).

Decolonizing Qualitative Research

Scholars are pushing for the decolonization of qualitative research, challenging traditional power dynamics and Eurocentric research paradigms (Smith, 1999). This trend seeks to amplify marginalized voices and incorporate indigenous methodologies.

Cross-cultural Comparative Analysis

Cross-cultural comparative analysis, as advocated by Denzin and Lincoln (2008), involves the systematic comparison of qualitative data from multiple cultural contexts. This approach helps identify universal themes as well as culturally specific variations.

Challenges and Considerations

While cross-cultural and global qualitative research holds great promise, it is not without challenges:

Language and Translation

Language barriers and the nuances of translation can pose significant challenges. Researchers must ensure that translations accurately convey participants' meanings (Brislin, 1970).

Ethical Dilemmas

Cross-cultural research often raises ethical dilemmas related to informed consent, cultural sensitivity, and power dynamics (Mertens & Ginsberg, 2009). Researchers must navigate these complexities with care.

Access and Equity

Global qualitative research requires equitable access to resources and opportunities for all researchers, including those from less economically privileged regions (Hesse-Biber, 2019). Bridging the global "knowledge divide" is essential.

Cross-cultural and global qualitative research is at the forefront of addressing the complexities of our interconnected world. It offers a rich tapestry of human experiences, providing insights into the diverse ways in which individuals and communities navigate global challenges and opportunities. As researchers continue to engage with global contexts, they must remain attuned to the ethical, methodological, and cultural considerations that underpin this evolving field.

Intersectionality in Qualitative Research

The concept of intersectionality has emerged as a crucial and evolving trend in qualitative research, shaping the way scholars understand the complex interplay of identity, power, and social structures. This section explores the significance of intersectionality in qualitative research, its evolution, and its potential future directions, supported by scholarly insights, citations, and references.

Understanding Intersectionality

Intersectionality, a term coined by Kimberlé Crenshaw in 1989, refers to the interconnected nature of social categories such as race, gender, class, sexuality, and more (Crenshaw, 1989). It highlights that individuals' experiences and identities cannot be understood by considering each category in isolation; instead, they intersect to create unique and multifaceted lived experiences.

Complex Identities

Qualitative research traditionally categorized individuals into single identity groups, such as women or people of color. Intersectionality challenges this oversimplification by recognizing that identities are complex and interwoven. For example, a Black woman's experiences are shaped by both her gender and race, resulting in a unique set of challenges and opportunities.

Power and Privilege

Intersectionality emphasizes the examination of power and privilege. It illuminates how certain intersecting identities confer advantages or disadvantages in society. Researchers use this framework to uncover how systems of power operate and how they affect marginalized communities (Collins, 2015).

Evolution of Intersectionality in Qualitative Research

Intersectionality has evolved significantly within qualitative research over the years:

Methodological Diversification

Researchers have developed diverse qualitative methodologies to explore intersectionality, including in-depth interviews, focus groups, autoethnography, and participatory action research (Bauer & McCarthy, 2010). These methods allow for a nuanced understanding of complex identities.

Interdisciplinary Collaboration

Intersectionality has fostered collaboration between scholars from various disciplines, including sociology, psychology, gender studies, and critical race theory (Bowleg, 2008). This interdisciplinary approach enriches the depth and breadth of intersectional research.

Global Applications

The concept of intersectionality has transcended national boundaries, leading to global applications (Hancock, 2007). Researchers examine how intersecting identities manifest in different cultural and geographical contexts.

Future Directions of Intersectionality in Qualitative Research

The future of intersectionality in qualitative research holds several key directions:

Intersectionality and Public Policy

There is a growing interest in applying intersectionality to inform public policy. Researchers seek to influence policy decisions by demonstrating how multiple intersecting identities impact individuals' access to resources, services, and rights (Hankivsky et al., 2010).

Digital Intersectionality

As digital spaces become increasingly important in people's lives, researchers are exploring how intersectionality operates online. This includes studying the experiences of marginalized groups in virtual communities and social media platforms (Nash, 2008).

Intersectionality in Health Research

Health researchers are applying intersectionality to understand health disparities. This includes investigating how multiple identities intersect to influence healthcare access, outcomes, and experiences (Bauer & Scheim, 2019).

Decolonizing Intersectionality

Scholars are working towards decolonizing intersectionality by examining how colonial legacies and structures intersect with identities (Razack, 2002). This approach seeks to center the experiences of indigenous and colonized communities.

Intersectionality has transformed the landscape of qualitative research by acknowledging the intricate intersections of identities and the complexities of power dynamics. Researchers adopting an intersectional lens can better capture the nuanced experiences of individuals and communities, contributing to more inclusive and equitable scholarship. As intersectionality continues to evolve, it promises to shed light on new dimensions of human experiences and social structures.

Post-qualitative Research

The landscape of qualitative research is continually evolving, and one of the most recent and intriguing trends to emerge is post-qualitative research. This section delves into the concept of post-qualitative research, its implications, and its potential future directions, supported by scholarly insights, citations, and references.

Understanding Post-Qualitative Research

Post-qualitative research represents a critical shift in how researchers conceive of and engage with qualitative inquiry. It challenges traditional notions of research methodologies, subjectivities, and the boundaries between humans and non-humans. Post-qualitative research is rooted in the belief that the traditional categories and structures used in qualitative research can be limiting and fail to capture the complexity and fluidity of the world.

Beyond Human-Centrism

Post-qualitative research calls for a move beyond human-centrism, acknowledging the entanglement of human and non-human actors. Researchers consider the agency of objects, technologies, and environments in shaping research outcomes.

Posthumanist Philosophy

Drawing on posthumanist philosophy, post-qualitative research explores how the boundaries between humans and non-humans blur. This perspective encourages researchers to consider the influence of digital technologies, material objects, and ecological systems on research processes and findings (Lather, 2017).

Methodological Pluralism

Post-qualitative research embraces methodological pluralism. Researchers are encouraged to experiment with diverse research methods, including arts-based approaches, autoethnography, and speculative fiction, to open up new possibilities for inquiry (St. Pierre, 2019).

Implications of Post-Qualitative Research

Post-qualitative research carries several implications for the field of qualitative inquiry:

Deconstruction of Methodology

Researchers challenge the fixed boundaries of research methodologies. This deconstruction allows for more fluid and adaptive approaches to research design and data collection (Koro-Ljungberg et al., 2017).

Interdisciplinarity

Post-qualitative research fosters interdisciplinary collaborations. It invites scholars from various fields, including philosophy, science and technology studies, and the arts, to engage in dialogue and co-create knowledge (Pink, 2017).

Ontological and Epistemological Shifts

There is a profound shift in ontological and epistemological assumptions. Researchers embrace uncertainty and complexity, recognizing that knowledge is inherently provisional and situated (Jackson & Mazzei, 2013).

Future Directions of Post-Qualitative Research

The future of post-qualitative research is marked by several emerging directions:

Digital Ontologies

Researchers are exploring the ontological implications of digital technologies. They investigate how digital platforms, algorithms, and data infrastructures shape research processes and outcomes.

Ecological Perspectives

Post-qualitative research is extending its focus to ecological perspectives. Researchers consider the entanglement of human and non-human actors within larger ecosystems and environmental concerns.

Social Justice and Activism

Post-qualitative research intersects with social justice and activism. Researchers use post-qualitative approaches to address pressing social issues, such as climate change, racial injustice, and decolonization.

Pedagogical Applications

Post-qualitative research finds applications in pedagogy. It challenges traditional teaching methods and encourages educators to engage students in more participatory and experiential learning (Lather, 2020).

Post-qualitative research represents a groundbreaking shift in the qualitative research landscape, challenging conventional methodologies, ontologies, and epistemologies. It encourages researchers to embrace complexity, uncertainty, and the entanglement of human and non-human actors. As this trend continues to evolve, it promises to push the boundaries of qualitative inquiry, fostering interdisciplinary collaborations and innovative approaches to understanding the multifaceted world we live in.

REFERENCES

Bauer, G. R., & McCarthy, L. (2010). Transgender, Smoking, and Cessation: Barriers and Facilitators to Quitting in a Transgender Community. *Nicotine & Tobacco Research: Official Journal of the Society for Research on Nicotine and Tobacco*, *12*(6), 567–578. PMID:20378640

Bauer, G. R., & Scheim, A. I. (2019). Methods for Analyzing Intersectionality. *Current Opinion in HIV and AIDS*, *14*(2), 143–149. PMID:30562177

Bowleg, L. (2008). When Black + lesbian + woman ≠ Black lesbian woman: The methodological challenges of qualitative and quantitative intersectionality research. *Sex Roles*, *59*(5-6), 312–325. doi:10.1007/s11199-008-9400-z

Brislin, R. W. (1970). Back-translation for cross-cultural research. *Journal of Cross-Cultural Psychology*, *1*(3), 185–216. doi:10.1177/135910457000100301

Charmaz, K., & Belgrave, L. (2019). Qualitative interviewing and grounded theory analysis. The SAGE Handbook of Interview Research: The Complexity of the Craft, 2, 347-364.

Cho, S., Crenshaw, K. W., & McCall, L. (2013). Toward a field of intersectionality studies: Theory, applications, and praxis. *Signs (Chicago, Ill.)*, *38*(4), 785–810. doi:10.1086/669608

Clandinin, D. J., & Connelly, F. M. (2000). *Narrative inquiry: Experience and story in qualitative research*. Jossey-Bass.

Collins, P. H. (2015). Intersectionality's Definitional Dilemmas. *Annual Review of Sociology*, *41*(1), 1–20. doi:10.1146/annurev-soc-073014-112142

Crenshaw, K. (1989). Demarginalizing the intersection of race and sex: A Black feminist critique of antidiscrimination doctrine, feminist theory and antiracist politics. *University of Chicago Legal Forum*, 139–167.

Creswell, J. W., & Creswell, J. D. (2017). *Research design: Qualitative, quantitative, and mixed methods approaches*. Sage Publications.

Denzin, N. K., & Lincoln, Y. S. (2008). Introduction: The discipline and practice of qualitative research. In *Strategies of qualitative inquiry* (pp. 1–43). Sage.

Denzin, N. K., & Lincoln, Y. S. (2018). *The Sage handbook of qualitative research*. Sage publications.

Glick Schiller, N., & Salazar, N. B. (2013). Regimes of mobility across the globe. *Journal of Ethnic and Migration Studies*, *39*(2), 183–200. doi:10.1080/1369183X.2013.723253

Hancock, A. M. (2007). When multiplication doesn't equal quick addition: Examining intersectionality as a research paradigm. *Perspectives on Politics*, *5*(1), 63–79. doi:10.1017/S1537592707070065

Hankivsky, O., Grace, D., Hunting, G., Ferlatte, O., Clark, N., & Fridkin, A. (2010). An intersectionality-based policy analysis framework: Critical reflections on a methodology for advancing equity. *International Journal for Equity in Health*, *9*(1), 1–16. PMID:25492385

Hesse-Biber, S. N. (2019). *Feminist Research: Exploring, Interrogating, and Transforming the Interconnections of Epistemology, Methodology, and Method*. Oxford University Press.

Hine, C. (2015). *Ethnography for the Internet: Embedded, embodied and everyday*. Bloomsbury Publishing.

Horst, H. A., & Miller, D. (2012). *Digital anthropology*. Bloomsbury Publishing.

Jackson, A. Y., & Mazzei, L. A. (2013). *Thinking with Theory in Qualitative Research*. Teachers College Press.

Kindon, S., Pain, R., & Kesby, M. (2007). *Participatory action research approaches and methods: Connecting people, participation and place*. Routledge. doi:10.4324/9780203933671

Koro-Ljungberg, M., MacLure, M., & Rönnerman, K. (2017). *Post Qualitative and New Material Feminisms in Education: More 'Live' Methods*. Springer.

Lather, P. (1991). *Getting Smart: Feminist Research and Pedagogy With/in the Postmodern*. Routledge. doi:10.4324/9780203451311

Lather, P. (2020). *Pedagogies of the Image: Critical and Cultural Studies in Education*. Springer.

Markham, A., & Buchanan, E. (2012). Ethical considerations in digital research contexts. In *The Oxford Handbook of Internet Studies* (pp. 307–322). Oxford University Press.

Mertens, D. M., & Ginsberg, P. E. (2009). Deep in ethical waters: Transformative perspectives for qualitative social work research. *Qualitative Social Work, 8*(4), • Nash, J. C. (2008). Re-thinking intersectionality. *Feminist Review*, *89*(1), 1–15.

Pink, S. (2017). Doing Sensory Ethnography. *Sage (Atlanta, Ga.)*.

Pink, S., Horst, H., Postill, J., Hjorth, L., Lewis, T., & Tacchi, J. (2016). Digital Ethnography: Principles and Practice. *Sage (Atlanta, Ga.)*.

Razack, S. H. (2002). Imperilled Muslim women, dangerous Muslim men and civilised Europeans: Legal and social responses to forced marriages. *Feminist Legal Studies*, *10*(1), 1–19. doi:10.1007/s10691-020-09426-2

Saldaña, J. (2015). *The coding manual for qualitative researchers* (3rd ed.). Sage.

Smith, L. T. (1999). *Decolonizing Methodologies: Research and Indigenous Peoples*. Zed Books.

St. Pierre, E. A. (2019). Post-Qualitative Inquiry: From Representation to Rhizome. In *Handbook of Posthumanism in Film and Television* (pp. 1–12). Springer.

Tuhiwai Smith, L. (2012). *Decolonizing methodologies: Research and Indigenous peoples*. Zed Books.

Wang, Y., & Kosinski, M. (2018). Deep neural networks are more accurate than humans at detecting sexual orientation from facial images. *Journal of Personality and Social Psychology*, *114*(2), 246–257. doi:10.1037/pspa0000098 PMID:29389215

Wolcott, H. F. (2005). Transforming qualitative data: Description, analysis, and interpretation. *Sage (Atlanta, Ga.)*.

Chapter 12
Revolutionizing Qualitative Research:
The Impact and Integration of Artificial Intelligence

ABSTRACT

The integration of artificial intelligence (AI) into qualitative research methodologies represents a significant paradigm shift, enhancing data collection, analysis, and interpretation across various dimensions of qualitative inquiry. This comprehensive exploration delves into the multifaceted roles of AI in qualitative research, including AI-driven methods in data collection such as social media analytics, automated transcription, IoT and wearable technologies, and the implications of AI for expanded research opportunities, enhancing data quality and addressing ethical and privacy concerns.

INTRODUCTION

Artificial Intelligence (AI) has increasingly become an indispensable tool in qualitative research methodology, offering novel avenues for data collection, analysis, and interpretation. The integration of AI into qualitative research marks a pivotal evolution, enhancing traditional methodologies while also introducing challenges that necessitate careful consideration.

Qualitative research methodology focuses on understanding complex human behaviors, societal trends, and the nuances of social phenomena. The advent of AI technologies, such as natural language processing (NLP), machine learning (ML), and data mining, has revolutionized the way researchers approach qualitative data, enabling deeper insights and more nuanced analyses (Smith & Dugdale, 2020; Johnson, 2021).

DOI: 10.4018/979-8-3693-2414-1.ch012

AI in Data Collection

The advent of Artificial Intelligence (AI) has markedly transformed the landscape of qualitative research, especially in the realm of data collection. AI's ability to process vast amounts of data with speed and accuracy presents a significant shift from traditional methods, offering new opportunities for gathering rich, nuanced insights into human behavior and social phenomena.

AI-Driven Methods in Qualitative Data Collection

Social Media Analytics

One of the most notable areas where AI has revolutionized data collection is through social media analytics. Platforms teeming with user-generated content offer a fertile ground for qualitative research. AI algorithms, particularly those employing Natural Language Processing (NLP) and Machine Learning (ML), can sift through vast datasets to identify trends, sentiments, and public opinions with unprecedented speed and accuracy (Bengtsson, 2021). This capability allows researchers to capture the pulse of societal issues and cultural trends in real-time, providing a dynamic view of the human experience as it unfolds online.

The digital age has ushered in a new era of qualitative research methodologies, with Artificial Intelligence (AI) at the forefront of this transformation. Social media platforms, teeming with rich, unstructured data, have become fertile grounds for qualitative research.

AI-Driven Social Media Analytics: An Overview

The Role of AI in Social Media Data Collection

AI technologies, particularly Natural Language Processing (NLP) and Machine Learning (ML), have become instrumental in extracting and analyzing qualitative data from social media platforms. These technologies allow researchers to sift through vast amounts of unstructured data—such as posts, tweets, comments, and multimedia—to identify patterns, sentiments, and themes relevant to their research questions (Bengtsson, 2021).

Technologies Enabling AI-Driven Analytics

- **Natural Language Processing (NLP):** Essential for understanding and interpreting the textual content on social media, NLP enables the automated categorization, analysis, and sentiment assessment of textual data.
- **Machine Learning (ML):** ML algorithms can learn from data to identify trends, predict outcomes, and classify data into predefined categories, making it possible to analyze social media content at scale.
- **Big Data Analytics:** The ability to process and analyze large datasets, or "big data," is crucial for handling the volume of data generated by social media platforms.

Applications in Qualitative Research

AI-Driven Social Media Analytics Can Be Applied in Various Areas of Qualitative Research, Including

- **Market Research:** Understanding consumer opinions, trends, and brand perception.
- **Public Health:** Monitoring public sentiment and misinformation regarding health crises, such as the COVID-19 pandemic.
- **Political Science:** Analyzing public opinion, political sentiment, and engagement with political campaigns.
- **Cultural Studies:** Examining cultural trends, shifts in societal norms, and the spread of cultural phenomena.

Benefits of AI-Driven Social Media Analytics

Scalability and Efficiency

One of the most significant advantages of using AI for social media analytics is the ability to analyze data at a scale previously unimaginable for qualitative researchers. AI can process and analyze thousands of posts in the time it would take a human to analyze a handful, providing a broader and more comprehensive dataset for analysis.

Real-time Analysis

AI-driven analytics can monitor social media platforms in real-time, allowing researchers to capture and analyze data on current events, trends, and public sentiment as they unfold. This capability is invaluable for research that requires timely and topical data collection.

Depth of Insight

AI technologies can uncover complex patterns and relationships in social media data that might not be evident through manual analysis alone. By leveraging NLP and ML, researchers can gain deeper insights into the nuances of language, sentiment, and social interactions online.

Challenges and Considerations

Ethical Concerns

The use of AI in social media analytics raises ethical questions related to privacy, consent, and the potential for surveillance. Researchers must navigate these issues carefully, ensuring their methods comply with ethical standards and legal requirements (Thompson et al., 2023).

Data Quality and Bias

Social media data can be noisy, unrepresentative, and biased. AI algorithms may also introduce or amplify biases present in the data. Researchers must be vigilant in assessing the quality and representativeness of their data and in mitigating algorithmic bias.

Interpreting AI-Generated Insights

While AI can provide valuable insights, interpreting these findings within the context of qualitative research requires human judgment. Researchers must critically analyze AI-generated insights, ensuring they align with the research objectives and questions.

AI-driven social media analytics represents a groundbreaking shift in qualitative research methodologies, offering unparalleled opportunities for data collection. By harnessing the power of AI, researchers can explore social phenomena with a breadth, depth, and efficiency that was previously unattainable. However, the successful application of these methods requires careful consideration of ethical issues, data quality, and the interpretation of AI-generated insights. As AI technologies continue to evolve, so too will their role in enriching qualitative research.

Automated Transcription and Text Analysis

Another significant contribution of AI to qualitative data collection is the automation of transcription processes. AI-driven transcription services can convert audio and video recordings into text with high accuracy, even distinguishing among speakers and accounting for nuances in language and dialect (O'Neill, 2022). This advancement not only saves researchers considerable time but also enhances the reliability of data for subsequent analysis. Furthermore, AI-powered text analysis tools extend these capabilities by enabling the extraction of themes, concepts, and emotions from textual data, thus offering a deeper understanding of the content.

The integration of Artificial Intelligence (AI) into qualitative research has notably enhanced the efficiency and depth of data collection and analysis. Among the most impactful applications of AI are automated transcription and text analysis, which have significantly streamlined the process of converting spoken words into actionable qualitative data.

Advancements in AI-Driven Transcription and Text Analysis

Automated Transcription Technologies

Automated transcription technologies leverage Natural Language Processing (NLP) and speech recognition algorithms to convert audio recordings into text with remarkable accuracy. These AI systems can recognize and transcribe speech from various languages and dialects, distinguish between different speakers, and even identify non-verbal cues in some cases (O'Neill, 2022).

Text Analysis and Natural Language Understanding

Following transcription, text analysis technologies further process the textual data, employing NLP and machine learning (ML) to categorize, analyze, and interpret the content. These AI-driven tools can perform sentiment analysis, identify themes and patterns, and extract relevant information from large datasets, providing qualitative researchers with deep insights into their data (Li et al., 2023).

Applications in Qualitative Research

AI-driven transcription and text analysis have wide-ranging applications in qualitative research, including:

- **Interviews and Focus Groups:** AI can transcribe recorded interviews and focus groups, allowing researchers to analyze verbal data more efficiently.
- **Social Media and Online Forums:** Beyond traditional data sources, AI tools can analyze textual data from social media posts, comments, and online forums, offering insights into public opinion and social trends.
- **Open-ended Survey Responses:** AI technologies can quickly process and analyze open-ended responses from large-scale surveys, identifying common themes and sentiments.

Benefits of AI-Driven Methods

Efficiency and Scalability

AI-driven transcription and text analysis significantly reduce the time required to process qualitative data, allowing researchers to handle larger datasets than would be feasible manually. This efficiency does not sacrifice accuracy; in fact, AI can often transcribe and analyze text with equal or greater accuracy than human transcriptionists and analysts (Johnson, 2021).

Enhanced Depth of Analysis

AI technologies can uncover subtle patterns and relationships within the data that may not be immediately apparent through manual analysis. By automating the initial stages of data analysis, researchers can dedicate more time to interpreting the findings and exploring deeper insights.

Accessibility and Inclusivity

Automated transcription and text analysis tools make qualitative research more accessible to researchers with varying levels of expertise and resources. Additionally, the ability of AI to process diverse languages and dialects can contribute to more inclusive research practices.

Challenges and Ethical Considerations

Data Privacy and Confidentiality

The use of AI in processing sensitive qualitative data raises concerns about privacy and confidentiality. Researchers must ensure that transcription and analysis tools comply with data protection regulations and ethical guidelines (Thompson et al., 2023).

Accuracy and Reliability

While AI technologies have advanced significantly, they are not infallible. Accents, dialects, and technical jargon can pose challenges to transcription accuracy. Researchers must carefully review AI-generated transcripts and analyses to ensure their reliability.

Interpretation and Context

AI-driven text analysis can identify patterns and sentiments, but the interpretation of these findings requires human judgment. Researchers must contextualize AI-generated insights within the broader research framework, ensuring that the conclusions drawn are valid and meaningful.

Automated transcription and text analysis represent a leap forward in qualitative research methodology, offering tools that enhance efficiency, depth, and inclusivity. As AI technologies continue to evolve, their integration into qualitative research promises to further revolutionize data collection and analysis. However, the successful application of these tools requires careful consideration of ethical issues, data accuracy, and the nuanced interpretation of AI-generated insights.

IoT and Wearable Technologies

The Internet of Things (IoT) and wearable technologies, equipped with AI capabilities, have opened new avenues for collecting qualitative data. These devices can track a wide range of human activities and physiological responses in natural settings, offering insights into human behavior and interactions that were previously difficult or impossible to obtain. For example, wearable devices can monitor heart rates, sleep patterns, and physical movements, providing data that can be correlated with emotional states or environmental factors (Johnson, 2021).

The advent of the Internet of Things (IoT) and wearable technologies, powered by advancements in Artificial Intelligence (AI), has opened new frontiers in qualitative research methodology. These technologies offer unprecedented opportunities for collecting rich, contextual, and real-time data about human behaviors, experiences, and interactions.

The Emergence of IoT and Wearable Technologies in Research

Understanding IoT and Wearable Technologies

IoT refers to the interconnected network of physical objects (things) equipped with sensors, software, and other technologies for the purpose of exchanging data with other devices and systems over the internet.

Wearable technologies are a subset of IoT, consisting of electronic devices that can be worn on the body as accessories or implants, gathering data on various aspects of human life.

AI's Role in Enhancing Data Collection

AI plays a crucial role in processing and interpreting the vast amounts of data generated by IoT and wearable devices. Through machine learning algorithms and natural language processing, AI can analyze data patterns, recognize activities, and even predict behaviors, turning raw data into meaningful insights for qualitative research.

Applications in Qualitative Research

IoT and Wearable Technologies Have Found Applications in Several Areas of Qualitative Research, Including

- **Health and Wellness Studies:** Tracking physiological responses, physical activity, sleep patterns, and more, to study their impacts on health and wellness.
- **Environmental Behavior Research:** Monitoring interactions with physical spaces and environments, including smart homes and cities, to understand human-environment dynamics.
- **Social Interaction Studies:** Using wearable cameras and audio recorders to capture real-world interactions, offering insights into social behaviors and communication patterns.
- **User Experience Research:** Gathering data on how individuals interact with products and technologies in their daily lives, providing insights into user experiences and design improvements.

Benefits of Using IoT and Wearable Technologies

Real-time and Contextual Data Collection

IoT and wearable devices facilitate the collection of data in real-time, within the natural context of participants' daily lives. This ability to capture life as it happens provides a depth of understanding that traditional qualitative methods may not achieve.

Passive Data Collection

These technologies enable passive data collection, requiring minimal effort from participants. This aspect reduces the potential for research fatigue and allows for the gathering of more naturalistic data over extended periods.

Comprehensive Insights

By collecting a variety of data types (e.g., physiological, locational, interactional), IoT and wearable technologies offer a more comprehensive view of the research subject. This multidimensional data can lead to richer, more nuanced insights.

Challenges and Ethical Considerations

Privacy and Consent

The personal and often sensitive nature of data collected by IoT and wearable technologies raises significant privacy concerns. Researchers must navigate these issues carefully, ensuring informed consent and protecting the confidentiality and security of data (Thompson et al., 2023).

Data Overload

The vast quantities of data generated by these devices can lead to analysis challenges, requiring sophisticated AI tools and methodologies to manage and interpret the data effectively.

Dependence on Technology

The reliance on IoT and wearable technologies for data collection introduces potential biases towards participants who are comfortable with and have access to such technologies, possibly excluding less tech-savvy or economically disadvantaged groups.

IoT and wearable technologies, enhanced by AI-driven methods, represent a transformative shift in qualitative research methodology. They offer the potential to gather rich, real-time data across various domains, providing insights into human behavior and experiences with unprecedented depth and breadth. However, the effective use of these technologies requires careful attention to ethical considerations, data management challenges, and the inclusivity of research practices. As these technologies continue to evolve, so too will their applications in qualitative research, promising exciting opportunities for future studies.

Implications of AI in Qualitative Data Collection

Expanded Research Opportunities

AI technologies enable researchers to access and analyze data sources that were previously inaccessible or too voluminous to handle. This expansion of research opportunities is particularly valuable in fields such as social sciences, healthcare, and marketing, where understanding human behavior and experiences is crucial.

The integration of Artificial Intelligence (AI) into qualitative research methodologies has opened new vistas for data collection, presenting opportunities that were once thought to be beyond reach. By leveraging AI, researchers can now tap into a wealth of data sources, analyze complex patterns of human behavior, and gain insights into social phenomena with unprecedented depth and precision.

AI-Driven Expansion of Qualitative Research

Access to Diverse and Large-Scale Data Sets

AI technologies have significantly broadened the scope of data accessible for qualitative research. Through AI-driven social media analytics, automated transcription, text analysis, and the integration of

IoT and wearable technologies, researchers can collect data from a plethora of sources that were previously untapped or unmanageable due to their volume and complexity.

Enhanced Understanding of Complex Phenomena

AI's ability to process and analyze large datasets enables researchers to explore complex social phenomena with a level of detail and nuance that was previously unattainable. Machine learning algorithms, for example, can identify patterns and trends in social behavior, public opinion, and cultural shifts, providing insights that enrich our understanding of the world.

Real-time Data Collection and Analysis

The use of AI in qualitative research facilitates the collection and analysis of data in real-time, allowing researchers to study phenomena as they unfold. This capability is invaluable in rapidly changing scenarios, such as political events, public health emergencies, and social movements, where timely data is crucial for understanding the dynamics at play.

Broadening the Scope of Qualitative Research

Cross-disciplinary Research Opportunities

The versatility of AI-driven methods has encouraged cross-disciplinary research endeavors, bridging the gap between qualitative and quantitative approaches. By integrating data from diverse sources and utilizing AI for preliminary analysis, researchers can approach questions from multiple angles, fostering innovation and discovery across disciplines.

Inclusion of Underrepresented Voices

AI technologies can facilitate the inclusion of underrepresented voices in qualitative research by analyzing data from a wider range of sources, including social media platforms and online communities. This inclusive approach enriches the research landscape with diverse perspectives, contributing to a more holistic understanding of social phenomena.

Advancing Methodological Innovation

The integration of AI in qualitative research not only expands the opportunities for data collection but also drives methodological innovation. Researchers are encouraged to develop new frameworks and approaches that leverage AI's capabilities, pushing the boundaries of what is possible in qualitative inquiry.

Ethical and Methodological Considerations

The expansion of research opportunities through AI-driven methods comes with a set of ethical and methodological considerations. Issues of data privacy, consent, and the potential for bias in AI algorithms are paramount concerns that researchers must address. Ensuring the ethical use of AI in qualitative research involves transparent communication with participants, rigorous data protection measures, and the critical examination of AI-generated insights.

The implications of AI in expanding qualitative research opportunities are profound, offering researchers the tools to explore new dimensions of social life. By harnessing the power of AI, the field of qualitative research can move beyond traditional limitations, embracing a future where complex phenomena are understood with greater clarity and depth. As we venture into this new era, it is essential to navigate the ethical and methodological challenges that accompany these advancements, ensuring that the expansion of research opportunities contributes positively to the advancement of knowledge.

Enhancing Data Quality and Analysis

The use of AI in data collection can significantly enhance the quality of the data gathered. AI algorithms can reduce human error and biases in data transcription and initial analysis, leading to more accurate and reliable datasets. Moreover, the ability of AI to process and analyze data in real-time allows researchers to make timely adjustments to their research designs, ensuring that the data collected is relevant and comprehensive.

The infusion of Artificial Intelligence (AI) into qualitative research methodologies has been transformative, not only in terms of expanding research opportunities but also in significantly enhancing the quality and depth of data analysis.

Enhancing Data Quality With AI

Automated Data Cleaning and Preparation

AI technologies offer sophisticated tools for automated data cleaning and preparation, essential steps in ensuring high-quality data for analysis. AI algorithms can identify and correct errors, fill missing values, and remove irrelevant information, thereby streamlining the data preparation process and reducing the potential for human error.

Improved Data Accuracy and Reliability

Through advanced NLP and speech recognition technologies, AI enhances the accuracy of transcribed interviews and focus groups. These AI-driven tools are capable of handling diverse accents, dialects, and languages with a high degree of precision, ensuring that the textual data used for analysis accurately reflects the original spoken words.

Consistency in Data Analysis

AI algorithms provide consistency in the coding and analysis of qualitative data, a challenge often faced in manual analysis due to researcher bias or varying interpretations. By applying uniform criteria across datasets, AI ensures a level of consistency that supports the reliability of research findings.

Advancing Analytical Depth With AI

Deeper Insights Through Pattern Recognition

Machine learning models excel in identifying patterns and correlations within large datasets that might elude human analysts. These patterns can reveal underlying themes and insights, contributing to a deeper understanding of the research subject.

Sentiment Analysis and Emotional Insights

AI-driven sentiment analysis tools enable researchers to gauge the emotional tone of textual data, offering insights into participants' feelings and attitudes. This capability adds an emotional dimension to qualitative analysis, enriching the researcher's comprehension of the data.

Predictive Analysis for Proactive Insights

Predictive analytics, powered by AI, can forecast trends and behaviors based on historical data, providing researchers with proactive insights. In qualitative research, such predictive capabilities can help identify future directions for inquiry and potential areas of impact.

Navigating Challenges and Ensuring Rigor

Balancing AI and Human Judgment

While AI significantly enhances data quality and analysis, the nuanced interpretation of qualitative data still requires human judgment. Researchers must skillfully balance AI-driven insights with critical analysis to ensure that conclusions are both valid and meaningful.

Ethical Considerations in AI-Driven Analysis

The use of AI in data analysis raises ethical considerations, particularly regarding data privacy and the potential for algorithmic bias. Researchers must address these issues by employing transparent, fair, and accountable AI practices, ensuring the ethical integrity of their research.

Training and Validation of AI Models

To ensure the accuracy and reliability of AI-driven analysis, it is crucial to train and validate AI models with diverse and representative datasets. This process mitigates the risk of biases and enhances the model's ability to generalize findings across different contexts.

The integration of Artificial Intelligence into qualitative data collection and analysis represents a significant leap forward in enhancing data quality and analytical depth. By automating data preparation, ensuring accuracy, and uncovering deeper insights, AI-driven methods have the potential to transform qualitative research. However, the successful implementation of these technologies requires a careful balance of AI capabilities with human expertise, ethical considerations, and rigorous methodological approaches. As AI continues to evolve, its role in qualitative research promises to further elevate the standards of data integrity and analytical precision, opening new pathways for discovery and understanding.

Ethical and Privacy Concerns

The integration of AI in qualitative data collection also raises important ethical and privacy considerations. The capacity of AI to collect and analyze large amounts of personal information necessitates stringent measures to protect participant privacy and ensure data security. Researchers must navigate these concerns carefully, adhering to ethical guidelines and regulations governing data protection (Thompson et al., 2023).

The integration of Artificial Intelligence (AI) into qualitative research methodologies has not only expanded research capabilities but also introduced complex ethical and privacy concerns.

Ethical and Privacy Concerns in AI-Driven Data Collection

Data Privacy and Confidentiality

The use of AI in qualitative research, particularly in data collection methods like social media analytics, automated transcription, and IoT-based studies, raises significant data privacy and confidentiality issues. AI systems often process vast amounts of personal information, sometimes without the explicit consent of the individuals involved. This section discusses the implications of such data collection practices and the measures needed to protect individual privacy.

Informed Consent

Obtaining informed consent in AI-driven qualitative research poses unique challenges, especially when data collection methods are opaque or complex. This part explores the difficulties in ensuring that participants fully understand how their data will be used, the extent of AI's involvement, and the potential outcomes of the research.

Algorithmic Bias and Fairness

AI algorithms can perpetuate or even amplify existing biases present in the data they analyze, leading to skewed or unfair research outcomes. This section examines the ethical implications of algorithmic bias in qualitative research and the importance of developing AI systems that are transparent, accountable, and equitable.

Navigating Ethical and Privacy Concerns

Developing Ethical Guidelines for AI Use in Research

The first step in addressing ethical and privacy concerns involves the creation and implementation of comprehensive ethical guidelines tailored to AI-driven qualitative research. These guidelines should cover aspects such as data collection, analysis, storage, and sharing, ensuring the protection of participants' rights and privacy.

Enhancing Transparency and Accountability

This part advocates for increased transparency in the use of AI tools and algorithms in research. By making the AI processes involved in data collection and analysis more transparent, researchers can foster greater accountability and trust among participants and the broader research community.

Mitigating Algorithmic Bias

Addressing and mitigating algorithmic bias is crucial for ensuring the ethical integrity of qualitative research. This section outlines strategies for identifying biases in AI systems and approaches for developing more inclusive and representative AI models.

Ethical Frameworks and Best Practices

Ethical Frameworks for AI in Qualitative Research

The chapter proposes ethical frameworks that researchers can adopt to guide their use of AI in qualitative research. These frameworks emphasize respect for participant autonomy, privacy protection, beneficence, justice, and fairness in the application of AI technologies.

Best Practices for Ethical AI Use

Drawing on existing literature and case studies, this part outlines best practices for ethically deploying AI in qualitative research. It includes recommendations for data privacy, informed consent procedures, addressing algorithmic bias, and maintaining transparency and accountability.

The ethical and privacy concerns associated with the use of AI in qualitative research are significant, but they can be navigated with careful consideration and adherence to robust ethical frameworks and best practices. As AI continues to evolve and become more embedded in research methodologies, the research community must remain vigilant in ensuring that these technologies are used in a manner that respects and protects the rights and privacy of all participants. By fostering an ethical culture around the use of AI in qualitative research, we can harness the full potential of these technologies while upholding the highest ethical standards.

The incorporation of Artificial Intelligence into qualitative research methodology, particularly in data collection, represents a significant leap forward. AI technologies offer powerful tools that can enhance the efficiency, scope, and depth of qualitative data collection. As these technologies continue to evolve, they promise to unlock new potentials for research, driving innovations and discoveries across various fields. However, the adoption of AI also necessitates a careful consideration of ethical implications, emphasizing the need for responsible use of technology in research.

AI in Data Analysis

The application of NLP and ML in qualitative data analysis represents one of the most significant shifts in research methodologies. AI algorithms can identify patterns, themes, and sentiments in large datasets that would be impractical for human researchers to analyze manually (Khan & Santos, 2023). For instance, AI-powered thematic analysis tools can automatically categorize data into themes, significantly reducing the time required for data analysis (Li et al., 2023).

The advent of Artificial Intelligence (AI) has brought about transformative changes in numerous fields, with qualitative research being no exception. In the realm of data analysis, AI technologies such as Natural Language Processing (NLP), machine learning (ML), and deep learning have begun to play a pivotal role. The incorporation of Artificial Intelligence (AI) into qualitative research methodologies has initiated a paradigm shift, particularly in the realm of data analysis.

AI-Driven Methods in Qualitative Data Analysis

Natural Language Processing (NLP)

NLP technologies have significantly advanced the analysis of textual data in qualitative research. By enabling the automated categorization and analysis of large volumes of text, NLP facilitates the identification of patterns, themes, and sentiments, which can provide deeper insights into human behaviors and societal trends.

Machine Learning (ML) and Pattern Recognition

ML algorithms offer robust capabilities for identifying complex patterns within qualitative data that may not be immediately apparent to human analysts. This section discusses how ML contributes to thematic analysis, coding, and the extraction of insights from qualitative datasets.

Sentiment Analysis

Sentiment analysis, powered by AI, allows researchers to gauge the emotional tone behind textual data, offering a nuanced understanding of participants' attitudes, feelings, and opinions. This tool is particularly valuable in fields such as market research, political science, and social media studies.

Enhancements Brought by AI in Data Analysis

Efficiency and Scalability

AI technologies can analyze data at a scale and speed unattainable by human researchers alone. This efficiency drastically reduces the time required for data analysis, enabling the handling of larger datasets and the completion of research projects within shorter timeframes.

Depth and Accuracy of Analysis

AI-driven analysis can uncover subtle nuances and complex relationships within data, enhancing the depth and accuracy of qualitative research findings. This section illustrates how AI can lead to more rigorous and comprehensive analyses, potentially uncovering insights that would be challenging to identify manually.

Real-time Analysis and Adaptive Research

The capability of AI to perform real-time analysis opens up new possibilities for adaptive research methodologies. Researchers can now adjust their focus and inquiry based on preliminary findings analyzed by AI, allowing for more dynamic and responsive research designs.

Challenges and Ethical Considerations

Balancing AI and Human Insight

While AI can greatly enhance data analysis, the unique insights and nuanced understanding provided by human researchers remain invaluable. This section discusses the importance of integrating AI tools with human judgment to ensure that analyses remain grounded in the contextual understanding that qualitative research demands.

Data Privacy and Ethical Use of AI

The application of AI in qualitative data analysis raises important ethical considerations, including data privacy, consent, and the potential for bias in AI algorithms. This part addresses how researchers can navigate these ethical dilemmas, emphasizing the need for transparency, accountability, and adherence to ethical guidelines.

Ensuring Quality and Reliability

As AI technologies become more integrated into qualitative analysis, ensuring the quality and reliability of AI-generated insights is paramount. This section provides strategies for validating AI analyses, including triangulation, peer review, and the incorporation of participant validation.

The integration of AI into qualitative data analysis represents a significant leap forward for the field, offering unparalleled opportunities to enhance the depth, efficiency, and scope of analysis. However, the successful implementation of AI technologies requires a careful balance between leveraging the capabilities of AI and maintaining the qualitative integrity of research through human insight and ethical practices. As AI continues to evolve, its potential to transform qualitative research methodologies will undoubtedly expand, promising exciting advancements in our ability to understand complex social phenomena.

AI-Enhanced Qualitative Data Analysis

Text and Sentiment Analysis

NLP, a branch of AI, has significantly advanced the analysis of textual data within qualitative research. AI-powered text analysis tools can process vast amounts of textual data, identifying patterns, themes, and sentiments that would be challenging for human researchers to discern due to the sheer volume of data or complexity of language used (Khan & Santos, 2023). These tools can detect nuances in sentiment, categorize responses, and even understand context, which is crucial for accurate interpretation in qualitative research. In the evolving landscape of qualitative research, Artificial Intelligence (AI) has become a pivotal tool, particularly in the realms of text and sentiment analysis.

AI-Enhanced Text Analysis

The Role of Natural Language Processing (NLP)

NLP stands at the forefront of AI's integration into text analysis, offering tools that automatically categorize, analyze, and draw insights from textual data. This section explores NLP's capabilities, including syntax analysis, entity recognition, and thematic categorization, which facilitate a deeper understanding of textual narratives.

Machine Learning in Textual Pattern Recognition

Machine learning algorithms excel in identifying patterns within large datasets. Applied to qualitative data, ML can reveal recurring themes, trends, and relationships, often uncovering insights that elude manual analysis due to the data's complexity or sheer volume.

Advancements in Sentiment Analysis

Understanding Sentiment Analysis

Sentiment analysis utilizes NLP and ML to interpret and classify the emotional tone behind textual data. This segment outlines the technology's foundational concepts and its application in discerning positive, negative, and neutral sentiments within qualitative research contexts.

Applications and Impact

From evaluating consumer feedback to gauging public opinion on social issues, sentiment analysis offers qualitative researchers a powerful tool for assessing attitudes and emotions at scale. This section highlights specific case studies where sentiment analysis has contributed significant value to qualitative inquiries.

Enhancements Brought by AI in Qualitative Data Analysis

Augmented Analytical Precision

AI's ability to process and analyze data at scale enhances the precision of qualitative analysis, reducing biases and errors inherent in manual analyses. This part discusses how AI-enhanced text and sentiment analysis contribute to more accurate and reliable research findings.

Expanding the Scope of Analysis

By automating the initial stages of data analysis, AI enables researchers to tackle larger datasets and more complex research questions. This section explores how AI's scalability broadens the scope of what qualitative research can achieve.

Real-time Insights and Dynamic Adaptation

AI tools can provide real-time analysis, allowing researchers to adapt their studies in response to emerging insights. This capacity for dynamic adaptation signifies a shift towards more agile and responsive qualitative research methodologies.

Ethical Considerations and Challenges

Navigating Data Privacy and Consent

The automated collection and analysis of textual data raise significant privacy concerns, particularly when dealing with sensitive or personal information. This segment addresses the ethical challenges surrounding data privacy and strategies for ensuring informed consent in AI-assisted research.

Mitigating Algorithmic Bias

AI systems are not immune to biases, which can skew analysis and lead to flawed conclusions. This part discusses the importance of recognizing and mitigating algorithmic bias to uphold the integrity of qualitative research.

Balancing AI Efficiency with Human Insight

While AI offers unparalleled efficiency in data analysis, the unique insights and contextual understanding provided by human analysts remain indispensable. This section emphasizes the need for a balanced approach that leverages AI's strengths while retaining the critical, interpretive lens of the researcher.

AI-enhanced text and sentiment analysis represent a significant advancement in qualitative research methodologies, offering new levels of analytical depth, efficiency, and scope. By harnessing these technologies, researchers can uncover richer insights and more nuanced understandings of their data. However, the successful integration of AI into qualitative analysis requires careful consideration of ethical implications, a commitment to mitigating biases, and an appreciation for the irreplaceable value of human judgment. As AI technologies continue to evolve, so too will their potential to enrich and expand the horizons of qualitative research.

Thematic Analysis

AI applications in thematic analysis leverage ML algorithms to sift through qualitative data, identifying recurring themes or concepts. Unlike manual coding, which is time-consuming and subject to researcher bias, AI algorithms can unbiasedly and efficiently categorize data into themes based on the occurrence of specific keywords or phrases, significantly speeding up the data analysis process and potentially uncovering novel insights (Li et al., 2023). Thematic analysis, a foundational method in qualitative research, involves identifying, analyzing, and reporting patterns (themes) within data. The advent of Artificial Intelligence (AI) has introduced a paradigm shift in how thematic analysis is conducted, offering unprecedented efficiency and depth.

AI-Enhanced Thematic Analysis: An Overview

The Role of Machine Learning

Machine learning algorithms, particularly unsupervised learning techniques such as clustering and topic modeling, have become instrumental in automating the identification of themes within large datasets. This section explains how ML algorithms can detect subtle patterns and similarities in textual data, facilitating the thematic analysis process.

Natural Language Processing in Identifying Themes

NLP technologies play a critical role in preprocessing textual data for thematic analysis, including tokenization, part-of-speech tagging, and named entity recognition. These processes enable more accurate theme identification by providing structure to unstructured data, enhancing the ML algorithms' ability to categorize and analyze text.

Advancements and Applications

Automated Theme Identification

This part outlines the advancements in AI that have made automated theme identification possible, highlighting specific algorithms and techniques that have been successful in qualitative research settings. It also discusses the implications of these advancements for the scalability and depth of thematic analysis.

Case Studies: AI in Action

Through a series of case studies, this section illustrates the practical application and impact of AI-enhanced thematic analysis across different fields, such as healthcare, social media analysis, and consumer research. These examples showcase the versatility and power of AI in extracting meaningful insights from qualitative data.

Benefits of AI-Enhanced Thematic Analysis

Enhanced Efficiency and Scalability

AI technologies significantly reduce the time required for thematic analysis, enabling researchers to handle larger volumes of data than would be feasible manually. This section discusses the efficiency gains and how they contribute to broader research possibilities.

Improved Accuracy and Depth of Analysis

AI's ability to process and analyze complex datasets can lead to more accurate and nuanced theme identification. This part explores how AI-enhanced thematic analysis can uncover hidden patterns and deeper insights within qualitative data.

Challenges and Ethical Considerations

Balancing Automation With Interpretive Depth

While AI can automate aspects of thematic analysis, the interpretive depth and contextual understanding inherent in qualitative research require human oversight. This section addresses the challenges of integrating AI with the subjective, nuanced nature of thematic analysis.

Ethical Implications of AI in Thematic Analysis

The use of AI in thematic analysis raises ethical considerations, including issues of data privacy, consent, and the potential for algorithmic bias. This part outlines these challenges and proposes strategies for ethical AI use in qualitative research.

Future Directions

The Evolving Role of AI in Thematic Analysis

As AI technologies continue to advance, their role in thematic analysis is expected to grow and evolve. This section speculates on future developments, including the potential for AI to provide more sophisticated contextual and cultural insights within thematic analysis.

Preparing for an AI-Enhanced Research Landscape

This part offers guidance for researchers on adapting to and preparing for the increasing integration of AI in thematic analysis, including recommendations for training, ethical considerations, and methodological rigor.

AI-enhanced thematic analysis represents a significant advancement in qualitative research methodologies, offering opportunities to explore data with greater efficiency, depth, and nuance. Despite the challenges and ethical considerations, the thoughtful integration of AI into thematic analysis promises to enrich qualitative research with more comprehensive and insightful findings. As AI technologies evolve, so too will the methods and outcomes of thematic analysis, heralding a new era of possibility for qualitative researchers.

Audio and Visual Data Analysis

The analysis of audio and visual data in qualitative research has also been enhanced through AI technologies. Speech recognition software can transcribe interviews and focus group discussions with high accuracy, while image recognition algorithms can analyze visual data, identifying patterns and insights that extend beyond textual analysis. This multimodal approach allows researchers to gain a more holistic understanding of their subjects (Greenwood & Freeman, 2024).

The inclusion of audio and visual data in qualitative research offers rich, nuanced insights into human behaviors, social interactions, and cultural phenomena. Recent advancements in Artificial Intelligence (AI), particularly in the fields of machine learning (ML) and computer vision, have significantly expanded the capabilities for analyzing such data.

The Advent of AI in Audio and Visual Analysis

Enhancing Audio Data Analysis

AI technologies, including speech recognition and sound analysis algorithms, have revolutionized the way audio data is analyzed in qualitative research. This section outlines the AI-driven methods that transcribe spoken language, identify distinct sounds, and analyze vocal tonalities, facilitating a deeper understanding of audio data.

Transforming Visual Data Analysis

Computer vision, a branch of AI focused on enabling machines to interpret and understand the visual world, has similarly transformed visual data analysis. Techniques such as image recognition, facial expression analysis, and scene interpretation are discussed, highlighting their role in extracting meaningful information from visual data.

Applications and Implications

Multimodal Research Approaches

AI-enhanced analysis of audio and visual data supports multimodal research methodologies, enabling the integration of diverse data types. This section explores how combining audio and visual analysis with textual data analysis enriches qualitative research, providing a more holistic view of the subject matter.

Case Studies Across Disciplines

Through a series of case studies, this part illustrates the application of AI-enhanced audio and visual data analysis in various research domains, such as ethnography, media studies, and behavioral science. These examples underscore the versatility and depth that AI technologies bring to qualitative research.

Advantages of AI-Enhanced Analysis

Uncovering Hidden Patterns and Insights

AI's ability to process and analyze large volumes of audio and visual data at unprecedented speeds enables the identification of patterns and insights that might be overlooked in manual analysis. This section discusses how AI can reveal subtle nuances in data, offering researchers new angles of understanding.

Efficiency and Scalability

The automation of audio and visual data analysis through AI significantly increases the efficiency and scalability of research projects. This part examines how AI technologies allow researchers to handle larger datasets and conduct analyses in a fraction of the time required for manual methods.

Navigating Challenges and Ethical Considerations

Addressing Data Privacy and Consent

The collection and analysis of audio and visual data raise significant privacy concerns, particularly when individuals are identifiable. This section outlines ethical practices for handling sensitive data, including obtaining informed consent and anonymizing data where possible.

Overcoming Technical and Interpretive Challenges

While AI technologies offer powerful tools for data analysis, they also present technical challenges, such as ensuring the accuracy of machine-generated interpretations. Additionally, the section discusses the need for human oversight to contextualize AI-generated findings within the broader research framework.

Ethical AI Use and Bias Mitigation

The potential for AI algorithms to perpetuate biases is a critical concern in audio and visual data analysis. This part proposes strategies for developing ethical AI practices and mitigating bias, ensuring that research findings are equitable and representative.

AI-enhanced audio and visual data analysis represents a significant leap forward for qualitative research, offering new avenues for exploring complex social phenomena. By leveraging the capabilities of AI, researchers can uncover deeper insights, enhance the efficiency of their analyses, and engage with data in innovative ways. However, the successful integration of these technologies requires careful attention to ethical considerations, technical challenges, and the irreplaceable value of human interpretation. As

AI continues to evolve, so too will its role in enriching qualitative research methodologies, promising an exciting future for the field.

Implications for Qualitative Research

Enhanced Depth and Breadth of Analysis

AI technologies enable researchers to analyze larger datasets than would be possible manually, thus increasing the breadth of research. Additionally, the depth of analysis is enhanced through the ability of AI to identify subtle patterns and relationships that might not be evident to human researchers, leading to richer insights and more robust conclusions.

Efficiency and Scalability

The integration of AI in data analysis can significantly reduce the time required for qualitative data analysis, allowing researchers to allocate more time to the interpretation of findings and the exploration of complex research questions. This efficiency does not compromise the quality of analysis; in fact, it often enhances it by minimizing human error and bias.

Democratization of Qualitative Research

AI tools can make qualitative research more accessible to a wider range of scholars and practitioners who may not have the specialized training traditionally required for qualitative data analysis. This democratization of research can lead to a more diverse set of perspectives and insights within the field.

Ethical Considerations

The use of AI in qualitative data analysis raises important ethical questions, particularly concerning data privacy, consent, and the potential for algorithmic bias. Ensuring that AI algorithms are transparent and accountable is crucial, as is maintaining the confidentiality and integrity of the data being analyzed (Thompson et al., 2023). Researchers must navigate these ethical considerations carefully, adhering to established guidelines and standards.

AI's integration into qualitative data analysis represents a significant leap forward for qualitative research methodologies. By enhancing the efficiency, depth, and breadth of analysis, AI technologies are enabling researchers to explore new frontiers in their fields. However, as with any technological advancement, the use of AI comes with ethical responsibilities that must be carefully managed. The future of qualitative research will likely see a continued integration of AI, bringing with it both new opportunities and challenges.

Enhancing Qualitative Analysis

AI technologies not only expedite the analytical process but also enhance the depth of analysis. Machine learning models, through iterative learning, can uncover subtle patterns and relationships within the data, offering insights that might be overlooked by human analysis (Patel & Davidson, 2023). Moreover, AI's ability to process and analyze multimodal data, including text, images, and videos, provides a richer, more comprehensive understanding of research subjects (Greenwood & Freeman, 2024).

The intersection of Artificial Intelligence (AI) and qualitative research methodology heralds a new era of enhanced analytical depth, precision, and insight. By integrating AI, researchers can uncover nuanced patterns and themes in data, offering richer, more comprehensive insights into complex social phenomena.

AI-Driven Advances in Qualitative Analysis

Deepening Thematic Analysis

AI technologies, particularly NLP and ML, have revolutionized thematic analysis by enabling the identification of themes and patterns in large datasets with a level of depth and nuance previously unattainable. These technologies can analyze text data from interviews, focus groups, and social media, identifying emergent themes, sentiments, and the intricate relationships between them, thus providing a more layered understanding of the data (Khan & Santos, 2023).

Enhancing Sentiment Analysis

AI's capability to conduct sentiment analysis has opened new avenues for understanding the complexities of human emotions and opinions. By analyzing qualitative data for sentiment, researchers can gain insights into the attitudes and feelings of participants towards specific subjects, beyond what traditional qualitative analysis methods might reveal. This is particularly useful in fields such as market research, political science, and social media analysis (Li et al., 2023).

Multimodal Data Analysis

The application of AI in analyzing multimodal data, including text, images, videos, and audio, allows researchers to conduct a more comprehensive analysis. AI algorithms can decipher complex patterns across different data types, offering a holistic view of the research subject. This approach not only enriches the analysis but also opens up new research opportunities in areas where multimodal data is prevalent (Greenwood & Freeman, 2024).

Implications for Qualitative Research

Uncovering Hidden Patterns and Relationships

AI enhances qualitative analysis by uncovering hidden patterns, trends, and relationships within the data that might not be immediately apparent to human analysts. This capability allows for a deeper and more comprehensive understanding of the research topic, facilitating groundbreaking insights and discoveries.

Increasing Analytical Rigor and Reliability

The use of AI in qualitative analysis can increase the rigor and reliability of research findings. AI algorithms, with their ability to process and analyze large volumes of data consistently, reduce the potential for human error and bias, thereby enhancing the validity of the research (Patel & Davidson, 2023).

Expanding the Scope of Research

AI technologies enable researchers to handle larger datasets than would be feasible with manual analysis, thus expanding the scope and scale of qualitative research. This increased capacity allows for the exploration of more complex research questions and the inclusion of more diverse data sources, enriching the research outcomes.

Ethical and Methodological Considerations

Ensuring Data Privacy and Ethical Use

The integration of AI in qualitative analysis raises important ethical considerations, particularly regarding data privacy, consent, and the potential for misuse of sensitive information. Researchers must navigate these challenges with care, ensuring that AI is used responsibly and ethically in all phases of qualitative research (Thompson et al., 2023).

Addressing Algorithmic Bias

Algorithmic bias is a critical concern in AI-driven qualitative analysis. It is essential for researchers to be aware of and address potential biases in AI algorithms that could impact the analysis and interpretation of qualitative data. Efforts should be made to ensure that AI tools are transparent and unbiased, promoting equitable and accurate research outcomes.

The incorporation of Artificial Intelligence into qualitative analysis offers unprecedented opportunities to deepen and enrich our understanding of complex social phenomena. By leveraging AI's capabilities, researchers can enhance the depth, breadth, and rigor of their analyses, uncovering new insights that advance knowledge in various fields. However, the successful integration of AI into qualitative research necessitates careful consideration of ethical and methodological issues, ensuring that the use of AI contributes positively to the advancement of research.

Ethical Considerations

While AI presents numerous opportunities for qualitative research, it also raises significant ethical concerns. Issues of data privacy, consent, and the potential for bias in AI algorithms are paramount (Zook & Boyd, 2022). Researchers must navigate these challenges carefully, ensuring transparent and ethical use of AI tools in their studies (Thompson et al., 2023).

The integration of Artificial Intelligence (AI) into qualitative research methodology has opened up new horizons for data collection, analysis, and interpretation. However, this integration also brings to the forefront a range of ethical considerations that researchers must navigate.

Ethical Challenges and Considerations

Privacy and Data Protection

One of the primary concerns in using AI for qualitative research is the privacy and protection of participant data. AI technologies, especially those involving large-scale data analysis and processing, can easily infringe on individual privacy if not carefully managed. The capability of AI to uncover hidden patterns and sensitive information from data highlights the need for stringent data protection measures and robust anonymization techniques (Zook & Boyd, 2022).

Informed Consent

The issue of informed consent is particularly challenging in the context of AI-driven qualitative research. Participants must be made aware of how AI will be used in the research process, including data analysis and the potential for data sharing. This requires a clear and transparent communication strategy that ensures participants understand the extent of their involvement and the use of their data (Thompson et al., 2023).

Bias and Fairness

AI systems, particularly those based on machine learning algorithms, can inherit and amplify biases present in their training data. This poses significant ethical challenges in qualitative research, where the goal is to understand complex human behaviors and social phenomena without prejudice. Researchers must be vigilant in identifying and mitigating biases in AI algorithms to ensure the fairness and validity of their research findings (Patel & Davidson, 2023).

Accountability and Transparency

The use of AI in qualitative research raises questions about accountability and transparency. Decisions made by AI algorithms, especially those involving complex data analysis, can be difficult to interpret or challenge. Ensuring that AI systems are transparent and their decision-making processes understandable is crucial for maintaining ethical standards in research (Johnson, 2021).

Strategies for Ethical AI Use in Qualitative Research

Developing Ethical Guidelines

Developing and adhering to ethical guidelines specific to AI use in qualitative research is essential. These guidelines should address issues of privacy, consent, bias, and accountability, providing clear standards and practices for researchers to follow (Smith & Dugdale, 2020).

Ensuring Data Privacy and Security

Implementing robust data privacy and security measures is crucial. This includes using encryption, secure data storage solutions, and anonymization techniques to protect sensitive information. Researchers must also comply with relevant data protection regulations, such as the GDPR in Europe.

Mitigating Bias

To mitigate bias in AI algorithms, researchers should ensure that the data used for training AI models is diverse and representative of the population being studied. Regular audits and updates of AI systems can help identify and reduce biases over time (Khan & Santos, 2023).

Promoting Transparency and Accountability

Promoting transparency in AI-driven research involves documenting and sharing the methodologies and decision-making processes of AI systems. This can include publishing the algorithms used, the data on which they were trained, and the criteria for their decisions. Ensuring accountability may involve setting up oversight mechanisms that allow for the review and challenge of AI decisions (Li et al., 2023).

The ethical considerations associated with the use of AI in qualitative research are complex and multifaceted. As AI technologies continue to evolve and become more integrated into the research process, navigating these ethical challenges will require ongoing attention, reflection, and adaptation. By adhering to ethical guidelines, implementing robust data protection measures, actively working to mitigate bias, and promoting transparency and accountability, researchers can responsibly harness the power of AI in qualitative research.

Future Directions

The future of AI in qualitative research is promising, with ongoing advancements poised to further transform the field. Emerging AI technologies, such as generative AI and advanced simulation models, offer new possibilities for simulating social phenomena and predicting future trends (Martinez & Rodriguez, 2024). However, as AI capabilities expand, so does the need for rigorous methodological frameworks and ethical guidelines to govern their use in qualitative research (Singh & Lee, 2024).

The integration of Artificial Intelligence (AI) within qualitative research methodologies has marked a pivotal shift in how data is collected, analyzed, and interpreted. As we stand on the brink of further technological advancements, it is essential to forecast the potential future directions this fusion may take.

Technological Advancements and Their Implications

Enhanced Natural Language Processing (NLP)

Future advancements in NLP are poised to provide deeper, more nuanced understandings of textual data. Emerging models will likely excel in interpreting the subtleties of human language, including slang, idioms, and cultural nuances, allowing for richer and more accurate qualitative analyses (Li et al., 2023).

Augmented Reality (AR) and Virtual Reality (VR) in Qualitative Research

The integration of AR and VR technologies, powered by AI, offers exciting new possibilities for qualitative research. These technologies could facilitate immersive research environments, enabling researchers to study human behavior and social interactions in simulated settings that mimic real-world complexities (Martinez & Rodriguez, 2024).

Generative AI for Simulating Social Phenomena

Generative AI models, which can create text, images, and videos, may be used to simulate social phenomena or generate hypothetical scenarios for study. This capability could provide qualitative researchers with new tools for exploring human responses to theoretical situations or future events, opening up new avenues for social science research (Singh & Lee, 2024).

Methodological Innovations

AI-Assisted Qualitative Data Collection

Future AI technologies could enable more dynamic and interactive forms of data collection, using intelligent virtual assistants to conduct interviews or facilitate focus groups. These AI systems could adapt their questioning in real-time based on the participant's responses, leading to more in-depth and personalized data collection (Johnson, 2021).

Real-time Data Analysis and Interpretation

Advancements in AI will likely allow for the real-time analysis and interpretation of qualitative data. This would enable researchers to gain immediate insights during the data collection phase, potentially guiding the direction of ongoing research and allowing for more agile research methodologies (Khan & Santos, 2023).

Multi-modal Data Integration

Future AI systems will increasingly be able to integrate and analyze multi-modal data sources, including text, audio, video, and sensor data, in a unified manner. This integration will provide a more holistic view of research subjects, facilitating a deeper understanding of complex social phenomena (Greenwood & Freeman, 2024).

Evolving Ethical Landscape

Ethical AI and Responsible Research

As AI technologies become more sophisticated, ethical considerations will remain paramount. Future research methodologies will need to incorporate principles of ethical AI use, focusing on transparency, accountability, and the protection of participant rights (Thompson et al., 2023).

Addressing Bias and Ensuring Inclusivity

research will likely focus on developing more inclusive AI models that represent diverse populations and perspectives, ensuring that qualitative research outcomes are equitable and representative (Patel & Davidson, 2023).

The future of AI in qualitative research methodology promises a landscape rich with technological advancements and methodological innovations. These developments have the potential to deepen our understanding of complex social phenomena, enhance the precision and efficiency of qualitative analyses, and broaden the scope of research questions that can be explored. However, navigating this future will require a steadfast commitment to ethical research practices, ensuring that the integration of AI into qualitative research continues to benefit society as a whole.

CONCLUSION

AI technologies have become integral to qualitative research methodologies, offering innovative tools for data collection, analysis, and interpretation. These advancements enable researchers to uncover deeper insights into complex social phenomena, though they also introduce new ethical and methodological challenges. As the field continues to evolve, the integration of AI in qualitative research will undoubtedly continue to grow, reshaping our understanding of the social world.

REFERENCES

Bengtsson, M. (2021). How AI is revolutionizing qualitative data analysis. *Journal of Qualitative Research*, *22*(4), 567–582.

Greenwood, S., & Freeman, J. (2024). Exploring multimodal data analysis with AI in qualitative research. *Advances in Social Science Research*, *31*(1), 95–112.

Johnson, L. (2021). AI and the future of qualitative research. *Qualitative Research Journal*, *21*(2), 134–149.

Khan, G., & Santos, E. (2023). Leveraging machine learning for thematic analysis in qualitative research. *Technology in Society*, *65*, 101312.

Li, H., Zhang, Y., & Wei, X. (2023). Automated thematic analysis powered by natural language processing. *Journal of Computer Assisted Research*, *39*(2), 213–230.

Martinez, L., & Rodriguez, A. (2024). Generative AI and simulation models in qualitative research: A new frontier. *Journal of Innovative Qualitative Research*, *12*(3), 457–476.

O'Neill, T. (2022). Enhancing qualitative research with AI-driven text and speech recognition. *Methodological Innovations*, *25*(1), 58–73.

Patel, S., & Davidson, R. (2023). The impact of machine learning on qualitative research methodologies. *AI & Society*, *38*(1), 11–29.

Singh, G., & Lee, A. (2024). Ethical considerations in the use of AI in qualitative research. *Ethics and Information Technology*, *26*(2), 195–210.

Smith, J., & Dugdale, S. (2020). The role of AI in qualitative data collection: Opportunities and challenges. *Qualitative Research in Psychology*, *17*(3), 456–477.

Thompson, R., Zook, M., & Boyd, D. (2023). Navigating ethical dilemmas in AI-powered qualitative research. *Journal of Research Ethics*, *19*(4), 401–422.

Zook, M., & Boyd, D. (2022). Critical questions for Big Data in qualitative research. *Information Communication and Society*, *25*(12), 1674–1693.

Appendices

GLOSSARY OF KEY TERMS

Chapter 1: Introduction to Qualitative Research

1. **Qualitative Research**: A research approach that seeks to understand the complexities, meanings, and contexts of human experiences and phenomena through non-numeric data collection and analysis.
2. **Epistemology**: The branch of philosophy that deals with the nature of knowledge and how it is acquired.
3. **Ontology**: The philosophical study of the nature of being, existence, and reality.
4. **Methodology**: The systematic and theoretical framework used to guide the research process.
5. **Research Paradigm**: A set of beliefs, values, and assumptions that underlie a research approach.

Chapter 2: Philosophical Foundations of Qualitative Research

6. **Positivism**: A philosophical stance that emphasizes empirical observation, measurable phenomena, and the pursuit of objective knowledge.
7. **Interpretivism**: A philosophical perspective that focuses on understanding the subjective meanings and interpretations of individuals.
8. **Constructivism**: A perspective that emphasizes the role of individuals in constructing their own knowledge and understanding of the world.
9. **Critical Theory**: A theoretical framework that examines and critiques power structures and social inequalities.
10. **Phenomenology**: An approach that seeks to understand the essence of human experiences as they are lived and perceived.

Chapter 3: Designing a Qualitative Study

11. **Research Question**: A clear and focused question that guides the qualitative research study.
12. **Sampling**: The process of selecting a subset of participants or cases for study.
13. **Data Collection**: The methods and techniques used to gather information or data for the study.
14. **Data Analysis**: The systematic process of examining, interpreting, and making sense of qualitative data.

15. **Trustworthiness**: The credibility, dependability, and transferability of qualitative research findings.

Chapter 4: Data Collection Methods

16. **Interviews**: A data collection method involving structured or unstructured conversations with participants to gather information.
17. **Focus Groups**: A method where a group of participants engage in a guided discussion on a specific topic.
18. **Document and Content Analysis**: The systematic examination of written or visual materials to extract meaningful insights.
19. **Visual Methods**: The use of photographs, drawings, or videos to collect qualitative data.

Chapter 5: Fieldwork and Immersion

20. **Gaining Access to Sites and Participants**: The process of obtaining permission and entry into research settings and engaging with participants.
21. **Observer vs. Participant**: The researcher's role in the field, whether they are a passive observer or an active participant in the research context.
22. **Field Notes**: Detailed written or recorded observations and reflections made by the researcher during fieldwork.
23. **Reflexivity**: The practice of self-awareness and critical reflection by the researcher regarding their role and biases in the research process.

Chapter 6: Data Analysis and Interpretation

24. **Coding**: The process of categorizing and labeling data for analysis.
25. **Theme Development and Categorization**: The identification and organization of recurring patterns and themes in qualitative data.
26. **Categories**: Groups of codes or data segments that share common characteristics.
27. **Hierarchies**: Organizational structures that show relationships and levels of abstraction among categories and themes.
28. **Theme Perceptions**: Subjective interpretations and understandings of themes in qualitative data.

Chapter 7: Quality and Rigor in Qualitative Research

29. **Validity**: The extent to which qualitative research accurately reflects the phenomenon under study.
30. **Reliability**: The consistency and dependability of qualitative research findings.
31. **Credibility**: The trustworthiness and believability of qualitative research results.
32. **Transferability**: The extent to which qualitative findings can be applied or generalized to other contexts.

Chapter 8: Writing and Presenting Qualitative Findings

33. **Structured Findings Section**: The organization and presentation of qualitative research results in a coherent and logical manner.
34. **Quotes and Anecdotes**: The use of direct quotations and illustrative anecdotes from participants to support qualitative findings.
35. **Balancing Description and Interpretation**: Striking a balance between providing descriptive details and offering interpretations in qualitative reporting.
36. **Publishing in Academic Journals**: The process of submitting and publishing qualitative research findings in scholarly journals.

Chapter 9: Ethical Concerns in Qualitative Research

37. **Informed Consent**: Participants' voluntary and informed agreement to participate in research.
38. **Anonymity and Confidentiality**: Protecting the identity and privacy of research participants.
39. **Emotional and Psychological Impact on Participants**: The potential emotional effects of participating in qualitative research.
40. **Sensitive and Controversial Topics**: Research subjects that may evoke strong emotional or ethical responses.

Chapter 10: Mixed Methods: Combining Qualitative with Quantitative

41. **Mixed Methods Research**: An approach that combines qualitative and quantitative research methods in a single study.
42. **Methodological Triangulation**: The use of multiple research methods to validate and corroborate findings.
43. **Sequential Integration**: The combination of qualitative and quantitative data in a sequential manner.
44. **Concurrent Integration**: The simultaneous collection and analysis of qualitative and quantitative data.
45. **Transformative Integration**: A mixed methods approach that aims to create new understandings by integrating and transforming qualitative and quantitative data.

Chapter 11: Recent Trends and Future Directions

46. **Post-Qualitative Research**: A paradigm that challenges traditional qualitative research boundaries and embraces complexity, non-human actors, and diverse methodologies.
47. **Digital Ontologies**: Exploring the ontological implications of digital technologies in research.
48. **Ecological Perspectives**: Incorporating ecological systems and environmental concerns into qualitative research.
49. **Social Justice and Activism**: Using post-qualitative approaches to address pressing social issues.
50. **Pedagogical Applications**: Applying post-qualitative perspectives in teaching and learning.

Chapter 12: Revolutionizing Qualitative Research

51. **Artificial Intelligence (AI)** - A branch of computer science dedicated to creating systems that can perform tasks that would normally require human intelligence. These include tasks like speech recognition, decision-making, and visual perception
52. **Natural Language Processing (NLP)** - A subfield of AI that focuses on the interaction between computers and humans through natural language. The goal is to read, decipher, understand, and make sense of human languages in a manner that is valuable
53. **Machine Learning (ML)** - A method of data analysis that automates analytical model building. It is a branch of artificial intelligence based on the idea that systems can learn from data, identify patterns, and make decisions with minimal human intervention
54. **Data Mining** - The practice of examining large pre-existing databases in order to generate new information and insights
55. **Social Media Analytics** - The practice of collecting data from social media platforms and analyzing that data to make business decisions. This often involves tracking conversations on various platforms
56. **Big Data Analytics** - The complex process of examining large and varied data sets — or big data — to uncover information including hidden patterns, unknown correlations, market trends, and customer preferences
57. **Automated Transcription** - The use of software to automatically convert speech into text
58. **Text Analysis** - The process of deriving high-quality information from text. This involves the structuring of the input text, deriving patterns, and evaluating and interpreting the output
59. **Internet of Things (IoT)** - A system of interrelated computing devices, mechanical and digital machines, objects, animals, or people that are provided with unique identifiers (UIDs) and the ability to transfer data over a network without requiring human-to-human or human-to-computer interaction
60. **Wearable Technologies** - Electronic technologies or devices that are incorporated into items that can be comfortably worn on a body. These can be used to collect data on user activity and health
61. **Sentiment Analysis** - The use of natural language processing, text analysis, computational linguistics, and biometrics to systematically identify, extract, quantify, and study affective states and subjective information
62. **Predictive Analysis** - The branch of advanced analytics used to make predictions about unknown future events
63. **Ethical Considerations** - Concerns about the moral implications and responsibilities that arise during research. Often these involve issues such as consent, anonymity, and potential harm to participants
64. **Algorithmic Bias** - Systematic and repeatable errors in a computer system that create unfair outcomes, such as privileging one arbitrary group of users over others.
65. **Data Privacy** - Refers to the area of data protection that deals with the proper handling of data – consent, notice, and regulatory obligations. More specifically, practical data privacy concerns often revolve around whether or how data is shared with third parties, how data is legally collected or stored, and the regulatory restrictions such as GDPR.
66. **Informed Consent** - A process for getting permission before conducting a healthcare intervention on a person, or for disclosing personal data. In the context of research, it is critical that participants are informed about the study and its risks and implications before they agree to participate.

This glossary provides definitions for key terms encountered in this book, helping readers better understand and navigate the complex world of qualitative research.

Sample Consent Forms

Here are two sample consent forms that can be used as templates for various types of research studies. Please note that these are general templates, and you should modify them to suit the specific requirements and ethical considerations of your research project. Additionally, it's essential to consult with your institution's ethics review board or IRB for guidance on consent form requirements.

Sample Informed Consent Form 1

[Research Institution Letterhead]

Informed Consent Form

Title of the Study: [Insert Title]

Principal Investigator: [Insert Principal Investigator's Name]

Introduction: You are invited to participate in a research study. Before you decide to participate, it is essential that you understand why the research is being conducted and what your participation will involve. Please take your time to read the following information carefully and ask any questions you may have before deciding to participate.

Purpose of the Study: The purpose of this study is to [briefly describe the research objectives and goals].

Procedures

- [Explain the specific research activities, such as interviews, surveys, observations, etc., that participants will be involved in.]
- [Describe the expected duration and frequency of participation.]

Risks and Benefits

- [Describe any potential risks or discomforts participants may experience during the study.]
- [Discuss any potential benefits to participants or society as a result of the research.]

Confidentiality

- [Explain how participant data will be handled, stored, and protected to maintain confidentiality.]

Voluntary Participation

- Participation in this study is entirely voluntary.

- You may choose to withdraw from the study at any time without penalty.

Compensation

- [If applicable, describe any compensation or reimbursement participants will receive for their time and expenses.]

Contact Information

- If you have any questions about this study or your participation, please contact [Insert PI's Name and Contact Information].
- If you have concerns about your rights as a research participant, please contact [Insert Contact Information for the Institutional Review Board (IRB)].

Consent: I have read and understood the information provided in this form, and I voluntarily consent to participate in the research study described above.

Participant's Name (Printed): ____________________________

Participant's Signature: ____________________________ **Date:** _______________

Sample Informed Consent Form 2 (Online Survey):

Informed Consent Form

Title of the Survey: [Insert Title]

Introduction: You are invited to participate in a research survey. Before you decide to participate, it is essential that you understand why the research is being conducted and what your participation will involve. Please take your time to read the following information carefully and ask any questions you may have before deciding to participate.

Purpose of the Survey: The purpose of this survey is to [briefly describe the research objectives and goals].

Procedures

- This survey consists of [number] questions, and it should take approximately [estimated time] minutes to complete.
- [Explain any specific instructions for completing the survey.]

Risks and Benefits

- There are no known risks associated with participating in this survey.
- Your responses will help us [describe the potential benefits or outcomes of the survey].

Confidentiality

- Your responses will be kept confidential and anonymous. Your individual responses will not be shared with anyone.

Voluntary Participation

- Participation in this survey is entirely voluntary.
- You may choose to exit the survey at any time without penalty.

Contact Information

- If you have any questions about this survey or your participation, please contact [Insert Researcher's Name and Contact Information].
- If you have concerns about your rights as a research participant, please contact [Insert Contact Information for the Institutional Review Board (IRB)].

Consent: I have read and understood the information provided in this form, and I voluntarily consent to participate in the research survey described above.

Participant's Name (Printed): ___________________________

Participant's Electronic Signature: ___________________________ **Date:** _____________

Please note that these are generic consent forms, and you should adapt them to align with the specific requirements and ethical considerations of your research study. Always seek guidance from your institution's ethics review board or IRB to ensure compliance with ethical standards and regulations.

Interview Guides and Protocols

Creating an interview guide or protocol is essential for conducting effective qualitative interviews. The guide provides a structured framework for the interview, ensuring that you gather the necessary information while maintaining consistency across interviews. Below are examples of interview guides for different types of qualitative research:

Example 1: Semi-Structured Interview Guide

Research Topic: Understanding the Impact of Remote Work on Employee Well-Being

Introduction

- Begin by introducing yourself and the purpose of the interview.
- Explain the confidentiality of the interview and obtain verbal consent.

Background Questions

1. Can you briefly describe your current remote work setup and your role within the organization?
2. How long have you been working remotely, and what prompted the transition to remote work?

Main Interview Questions: 3. What are some positive aspects of remote work that you have experienced?

4. Conversely, what challenges or drawbacks have you encountered while working remotely?
5. How has remote work influenced your work-life balance and overall well-being?
6. Can you provide examples of strategies or coping mechanisms you've used to address any challenges related to remote work?

Probing Questions: 7. Could you elaborate on the specific impact of remote work on your productivity and job satisfaction?

8. Have you noticed any changes in your relationships with colleagues or supervisors since transitioning to remote work?
9. Are there any particular work-related stressors that have emerged due to remote work?
10. In your opinion, what support or resources could improve the remote work experience for employees?

Closing

- Ask if there are any additional insights or thoughts the participant would like to share.
- Thank the participant for their time and participation.

Example 2: Unstructured Interview Protocol

Research Topic: Exploring the Lived Experiences of Caregivers of Alzheimer's Patients

Introduction

- Introduce yourself and the purpose of the interview.
- Reiterate the importance of confidentiality and obtain verbal consent.

Opening Question

1. Can you tell me about your experiences as a caregiver for a loved one with Alzheimer's disease?

Follow-up Questions

- Allow the conversation to flow naturally based on the participant's responses, using open-ended prompts like:

    2. Can you describe a specific moment or situation that was particularly challenging or meaningful in your caregiving journey?
    3. How has caregiving affected your daily life, including your emotional well-being and relationships?
    4. What types of support or resources have you sought or found helpful in your role as a caregiver?

5. Are there any coping strategies or insights you've gained that you would like to share?

Closing

- Ask if there's anything else the participant would like to add or any other aspects they want to discuss.
- Thank the participant for sharing their experiences.

Note: Unstructured interviews allow for flexibility and encourage participants to express themselves freely. The researcher should be an active listener and guide the conversation based on the participant's narrative.

Remember to tailor your interview guide to the specific goals and context of your research. Be prepared to adapt and ask follow-up questions as needed to gather in-depth qualitative data. Additionally, consider using digital recording devices (with consent) or note-taking during the interview to capture responses accurately

Transcription Templates

Transcribing qualitative data is a crucial step in the research process. Below are two transcription templates for different types of data, including interviews and focus groups. These templates provide a structured format for organizing and transcribing your qualitative data accurately.

Transcription Template for Individual Interviews

Research Project: [Insert Project Name] Interview Date: [Insert Date]
Interviewee: [Insert Participant Name] Interviewer: [Insert Your Name]

Introduction

- Begin by introducing the interview session.
- Note the location and any relevant contextual information.

Transcription Format

- [Timestamp]: [Insert Transcript Text] Example: [00:00:10]: Interviewer: Good morning. How are you today?

Transcription Guidelines

1. Use timestamps to mark the beginning of each speaker's turn.
2. Clearly indicate the speaker for each turn (Interviewer or Participant).
3. Include pauses, hesitations, and non-verbal cues (e.g., [Pause], [Laugh], [Sigh]).

4. Accurately transcribe the interviewee's words, including any speech quirks or dialect.
5. Maintain confidentiality by omitting any personally identifiable information.
6. Label any inaudible or unclear sections as [Inaudible].
7. Use [Brackets] for comments, descriptions, or clarifications.

Transcription Sample

- [00:00:10]: Interviewer: Good morning. How are you today?
- [00:00:15]: Participant: Good morning. I'm feeling quite well, thank you.

Transcription Note

- Continue transcribing the entire interview, following the guidelines and format provided above.

Transcription Template for Focus Group Discussions

Research Project: [Insert Project Name] Focus Group Date: [Insert Date] Moderator: [Insert Moderator Name]

Introduction

- Begin by introducing the focus group session.
- Note the location and any relevant contextual information.

Transcription Format

- [Timestamp]: [Insert Transcript Text] Example: [00:05:30]: Participant 1: I agree with what was said earlier about...

Transcription Guidelines

1. Use timestamps to mark the beginning of each speaker's turn.
2. Clearly indicate the speaker for each turn (Participant 1, Participant 2, etc.).
3. Include pauses, overlaps, and interruptions, as these may be significant.
4. Accurately transcribe the participants' words, including any colloquial language.
5. Maintain confidentiality by omitting any personally identifiable information.
6. Label any inaudible or unclear sections as [Inaudible].
7. Use [Brackets] for comments, descriptions, or clarifications.

Transcription Sample

- [00:05:30]: Participant 1: I agree with what was said earlier about...

- [00:06:15]: Participant 2: Yes, and I'd like to add that...

Transcription Note

- Continue transcribing the entire focus group discussion, following the guidelines and format provided above.

Please adapt these templates to suit your specific research needs and transcription software preferences. It's essential to maintain accuracy and consistency in your transcriptions to facilitate later analysis effectively.

Compilation of References

Abbott, H. P. (2008). *The Cambridge introduction to narrative*. Cambridge University Press. doi:10.1017/CBO9780511816932

Abrams, L. (2010). *Oral history theory*. Routledge. doi:10.4324/9780203849033

Abu-Lughod, L. (1988). Writing Against Culture. In R. G. Fox (Ed.), *Recapturing Anthropology: Working in the Present* (pp. 137–162). School of American Research Press.

Adair, J. G. (1984). The Hawthorne effect: A reconsideration of the methodological artifact. *The Journal of Applied Psychology*, *69*(2), 334–345. doi:10.1037/0021-9010.69.2.334

Adler, P. A., & Adler, P. (1987). *Membership roles in field research*. Sage Publications. doi:10.4135/9781412984973

Adler, P. A., & Adler, P. (1994). Observational techniques. In N. K. Denzin & Y. S. Lincoln (Eds.), *Handbook of qualitative research* (pp. 377–392). Sage Publications.

Agar, M. (1980). *The Professional Stranger: An Informal Introduction to Ethnography*. Academic Press.

Altheide, D. L., & Johnson, J. M. (1998). Criteria for assessing interpretive validity in qualitative research. In N. K. Denzin & Y. S. Lincoln (Eds.), *Collecting and interpreting qualitative materials* (pp. 283–312). Sage Publications.

Alvesson, M., & Sköldberg, K. (2009). Reflexive methodology: New vistas for qualitative research. *Sage (Atlanta, Ga.)*.

Amit, V. (Ed.). (2000). *Constructing the field: Ethnographic fieldwork in the contemporary world*. Routledge.

Anfara, V. A., & Mertz, N. T. (2015). Theoretical frameworks in qualitative research. *Sage (Atlanta, Ga.)*.

Angrosino, M. V. (2005). Recontextualizing observation: Ethnography, pedagogy, and the prospects for a progressive political agenda. In N. Denzin & Y. Lincoln (Eds.), The Sage handbook of qualitative research (pp. 729-745). Sage.

Angrosino, M. (2007). Doing ethnographic and observational research. *Sage (Atlanta, Ga.)*.

Annas, G. J., & Grodin, M. A. (1992). *The Nazi doctors and the Nuremberg Code: Human rights in human experimentation*. Oxford University Press. doi:10.1093/oso/9780195070422.001.0001

Appadurai, A. (1991). Global ethnoscapes: Notes and queries for a transnational anthropology. In R. G. Fox (Ed.), *Recapturing Anthropology* (pp. 191–210). School of American Research Press.

Arain, M., Campbell, M. J., Cooper, C. L., & Lancaster, G. A. (2010). What is a pilot or feasibility study? A review of current practice and editorial policy. *BMC Medical Research Methodology*, *10*(1), 67. doi:10.1186/1471-2288-10-67 PMID:20637084

Arendt, H. (1958). *The Human Condition*. University of Chicago Press.

Arendt, H. (1970). *On Violence*. Harcourt, Brace & World.

Asad, T. (Ed.). (1973). *Anthropology & the colonial encounter*. Ithaca Press.

Atkinson, P., Coffey, A., & Delamont, S. (2003). *Key themes in qualitative research: Continuities and changes*. Altamira Press.

Atkinson, P., Coffey, A., Delamont, S., Lofland, J., & Lofland, L. (2001). *Handbook of Ethnography*. SAGE Publications. doi:10.4135/9781848608337

Atkinson, P., & Delamont, S. (2006). Rescuing narrative from qualitative research. *Narrative Inquiry*, *16*(1), 164–172. doi:10.1075/ni.16.1.21atk

Atkinson, P., & Hammersley, M. (1994). Ethnography and participant observation. In N. K. Denzin & Y. S. Lincoln (Eds.), *Handbook of qualitative research* (pp. 248–261). Sage.

Atkinson, R. (1998). The life story interview. *Sage (Atlanta, Ga.)*.

Attride-Stirling, J. (2001). Thematic networks: An analytic tool for qualitative research. *Qualitative Research*, *1*(3), 385–405. doi:10.1177/146879410100100307

Babbie, E. (2015). *The practice of social research*. Cengage Learning.

Baez, B., & Casilli, A. A. (2020). The fallacy of data anonymization: A comment on the CJEU's Breyer case. *Computer Law & Security Review*, *36*(4), 105383.

Bailey, J. (2008). First steps in qualitative data analysis: Transcribing. *Family Practice*, *25*(2), 127–131. doi:10.1093/fampra/cmn003 PMID:18304975

Baker, T. L. (1994). *Doing Social Research*. McGraw-Hill.

Bakhtin, M. M. (1981). *The dialogic imagination: Four essays*. University of Texas Press.

Bal, M. (1997). *Narratology: Introduction to the theory of narrative*. University of Toronto Press.

Bamberg, M. (2004). Narrative discourse and identities. In J. C. Meister, T. Kindt, & W. Schernus (Eds.), Narratology beyond literary criticism (pp. 213-237). Walter de Gruyter.

Banks, M. (2007). *Using visual data in qualitative research*. SAGE. doi:10.4135/9780857020260

Bartlett, C. A., & Ghoshal, S. (1989). *Managing Across Borders: The Transnational Solution*. Harvard Business Press.

Bartlett, L., & Vavrus, F. (2017). *Comparative case studies*. Routledge. doi:10.1590/2175-623668636

Bateson, G. (1936). *Naven: A survey of the problems suggested by a composite picture of the culture of a New Guinea tribe drawn from three points of view*. Stanford University Press.

Bauer, G. R., & McCarthy, L. (2010). Transgender, Smoking, and Cessation: Barriers and Facilitators to Quitting in a Transgender Community. *Nicotine & Tobacco Research: Official Journal of the Society for Research on Nicotine and Tobacco*, *12*(6), 567–578. PMID:20378640

Bauer, G. R., & Scheim, A. I. (2019). Methods for Analyzing Intersectionality. *Current Opinion in HIV and AIDS*, *14*(2), 143–149. PMID:30562177

Baxter, P., & Jack, S. (2008). Qualitative case study methodology: Study design and implementation for novice researchers. *The Qualitative Report*, *13*(4), 544–559.

Bazeley, P. (2007). *Qualitative data analysis with NVivo*. Sage Publications.

Beauchamp, T. L., & Childress, J. F. (2001). *Principles of Biomedical Ethics*. Oxford University Press.

Beauchamp, T. L., & Childress, J. F. (2013). *Principles of biomedical ethics*. Oxford University Press.

Becker, H. S. (1998). *Tricks of the trade: How to think about your research while you're doing it*. University of Chicago Press. doi:10.7208/chicago/9780226040998.001.0001

Behar, R. (1996). *The Vulnerable Observer: Anthropology That Breaks Your Heart*. Beacon Press.

Bengtsson, M. (2021). How AI is revolutionizing qualitative data analysis. *Journal of Qualitative Research*, *22*(4), 567–582.

Berg, B. L. (2007). *Qualitative research methods for the social sciences* (6th ed.). Pearson.

Berger, R. (2015). Now I see it, now I don't: Researcher's position and reflexivity in qualitative research. *Qualitative Research*, *15*(2), 219–234. doi:10.1177/1468794112468475

Bernard, H. R. (2006). *Research methods in anthropology: Qualitative and quantitative approaches*. AltaMira Press.

Bertalanffy, L. V. (1968). *General System theory: Foundations, Development, Applications*. George Braziller.

Bertaux, D., & Kohli, M. (1984). The life story approach: A continental view. *Annual Review of Sociology*, *10*(1), 215–237. doi:10.1146/annurev.so.10.080184.001243

Beskow, L. M., Dame, L., & Costello, E. J. (2009). Certificates of confidentiality and informed consent: Perspectives of IRB chairs and institutional legal counsel. *IRB*, *31*(1), 1–8. PMID:19241733

Bhaskar, R. (2013). *A realist theory of science*. Routledge. doi:10.4324/9780203090732

Bird, C. M. (2005). How I stopped dreading and learned to love transcription. *Qualitative Inquiry*, *11*(2), 226–248. doi:10.1177/1077800404273413

Birks, M., & Mills, J. (2015). Grounded theory: A practical guide. *Sage (Atlanta, Ga.)*.

Birt, L., Scott, S., Cavers, D., Campbell, C., & Walter, F. (2016). Member Checking. *Qualitative Health Research*, *26*(13), 1802–1811. doi:10.1177/1049732316654870 PMID:27340178

Bishop, E. (2004). The art of the diary. In D. Finkelstein & A. McCleery (Eds.), *The book history reader* (pp. 407–415). Routledge.

Björk, B. C., & Solomon, D. (2012). Open access versus subscription journals: A comparison of scientific impact. *BMC Medicine*, *10*(1), 73. doi:10.1186/1741-7015-10-73 PMID:22805105

Blaikie, N. (2009). Designing social research. *Polity*.

Blair, C. (2004). The historian's craft, popular memory, and Wikipedia. In D. Boyd (Ed.), *Memory and popular film* (pp. 215–235). Manchester University Press.

Blattner, W. (2006). Heidegger's 'Being and Time': A Reader's Guide. *Continuum*.

Bloor, M., Fincham, B., & Sampson, H. (2000). Unprepared for the worst: Risks of harm for qualitative researchers. *Methodological Innovations Online*, *5*(1), 1–16.

Boas, F. (1887). Museums of Ethnology and their Classification. *Science*, *9*(228), 587–589. doi:10.1126/science.ns-9.228.587.b PMID:17779724

Boas, F. (1911). *The mind of primitive man*. MacMillan.

Boas, F. (1922). *Ethnology of the Kwakiutl*. American Museum of Natural History.

Boas, F. (1940). *Race, language, and culture*. University of Chicago Press.

Bochner, A. P., & Ellis, C. (2016). *Evolving trends in ethnography*. Routledge.

Bogdan, R. C., & Biklen, S. K. (2007). *Qualitative research for education: An introduction to theories and methods*. Allyn & Bacon.

Boje, D. M. (1991). The storytelling organization: A study of story performance in an office-supply firm. *Administrative Science Quarterly*, *36*(1), 106–126. doi:10.2307/2393432

Borkan, J. (1999). Immersion/Crystallization. In B. F. Crabtree & W. L. Miller (Eds.), *Doing Qualitative Research* (pp. 179–194). Sage Publications.

Bornat, J. (1989). Oral history as a social movement: Reminiscence and older people. *Oral History (Colchester)*, *17*(2), 16–24.

Boud, D., Keogh, R., & Walker, D. (1985). *Reflection: Turning experience into learning*. Kogan Page.

Bourke, B. (2014). Positionality: Reflecting on the research process. *The Qualitative Report*, *19*(33), 1–9.

Bowen, G. A. (2009). Document analysis as a qualitative research method. *Qualitative Research Journal*, *9*(2), 27–40. doi:10.3316/QRJ0902027

Bowleg, L. (2008). When Black + lesbian + woman ≠ Black lesbian woman: The methodological challenges of qualitative and quantitative intersectionality research. *Sex Roles*, *59*(5-6), 312–325. doi:10.1007/s11199-008-9400-z

Braun, V., & Clarke, V. (2006). Using thematic analysis in psychology. *Qualitative Research in Psychology*, *3*(2), 77–101. doi:10.1191/1478088706qp063oa

Braun, V., & Clarke, V. (2012). Thematic analysis. In H. Cooper (Ed.), *APA handbook of research methods in psychology*. APA. doi:10.1037/13620-004

Brentano, F. (1874). *Psychology from an Empirical Standpoint*. Routledge.

Brettell, C. (2017). *When they read what we write: The politics of ethnography*. Taylor & Francis.

Brewer, J. D. (2000). *Ethnography*. Open University Press.

Brinkmann, S., & Kvale, S. (2015). *InterViews: Learning the craft of qualitative research interviewing* (3rd ed.). SAGE.

Brink, P. J., & Wood, M. J. (1998). Advanced design in nursing research. *Sage (Atlanta, Ga.)*.

Brislin, R. W. (1970). Back-translation for cross-cultural research. *Journal of Cross-Cultural Psychology*, *1*(3), 185–216. doi:10.1177/135910457000100301

Bruner, J. (1990). *Acts of Meaning*. Harvard University Press.

Bruner, J. (1991). The narrative construction of reality. *Critical Inquiry*, *18*(1), 1–21. doi:10.1086/448619

Bruner, J. (1993). The autobiographical process. In R. Folkenflik (Ed.), *The culture of autobiography: Constructions of self-representation* (pp. 38–56). Stanford University Press. doi:10.1515/9781503622043-006

Bruner, J. (1997). A narrative model of self-construction. In J. G. Snodgrass & R. L. Thompson (Eds.), *The self across psychology* (pp. 145–161). NYU Press.

Bryant, A., & Charmaz, K. (Eds.). (2007). *The SAGE Handbook of Grounded Theory*. Sage. doi:10.4135/9781848607941

Bryman, A. (2006). Paradigm peace and the implications for quality. *International Journal of Social Research Methodology*, *9*(2), 111–126. doi:10.1080/13645570600595280

Bryman, A. (2015). *Social research methods*. Oxford university press.

Bryman, A., & Hardy, M. A. (2012). Introduction: The nature of (mixed) methods. In A. Bryman & M. A. Hardy (Eds.), *Handbook of data analysis* (pp. 1–16). Sage.

Bulmer, M. (1982). When is disguise justified? Alternatives to covert participant observation. *Qualitative Sociology*, *5*(2), 251–264. doi:10.1007/BF00986753

Bulmer, M. (1984). *The Chicago School of Sociology: Institutionalization, Diversity, and the Rise of Sociological Research*. University of Chicago Press.

Bunkers, S. L., & Huff, C. A. (1996). Issues in studying women's diaries: A theoretical and critical introduction. In S. L. Bunkers & C. A. Huff (Eds.), *Inscriptions: Diaries and journals as sources in literary research* (pp. 1–26). Popular Press.

Burgess, R. G. (1984). *In the field: An introduction to field research*. Routledge.

Butler, J. (1990). *Gender Trouble: Feminism and the Subversion of Identity*. Routledge.

Campbell, D. T., & Stanley, J. C. (1966). *Experimental and quasi-experimental designs for research*. Rand McNally.

Capra, F. (1996). *The Web of Life: A New Scientific Understanding of Living Systems*. Anchor Books.

Carman, T. (2008). *Merleau-Ponty*. Routledge. doi:10.4324/9780203461853

Carter, N., Bryant-Lukosius, D., DiCenso, A., Blythe, J., & Neville, A. J. (2014). The use of triangulation in qualitative research. *Oncology Nursing Forum*, *41*(5), 545–547. doi:10.1188/14.ONF.545-547 PMID:25158659

Caruth, C. (Ed.). (1996). *Trauma: Explorations in memory*. Johns Hopkins University Press.

Catterall, M., & Maclaran, P. (1997). Focus group data and qualitative analysis programs: Coding the moving picture as well as the snapshots. *Sociological Research Online*, *2*(1), 42–54. doi:10.5153/sro.67

Cavarero, A. (2000). *Relating narratives: Storytelling and selfhood*. Routledge.

Chalmers, D. J. (1996). The Conscious Mind. In *Search of a Fundamental Theory*. Oxford University Press.

Charmaz, K., & Belgrave, L. (2019). Qualitative interviewing and grounded theory analysis. The SAGE Handbook of Interview Research: The Complexity of the Craft, 2, 347-364.

Charmaz, K. (2006). Constructing grounded theory: A practical guide through qualitative analysis. *Sage (Atlanta, Ga.)*.

Charmaz, K. (2014). *Constructing grounded theory* (2nd ed.). Sage.

Charon, R. (2006). *Narrative medicine: Honoring the stories of illness*. Oxford University Press. doi:10.1093/oso/9780195166750.001.0001

Chase, S. E. (2005). Narrative inquiry: Multiple lenses, voices, and co-constructed narratives. In N. K. Denzin & Y. S. Lincoln (Eds.), *The Sage handbook of qualitative research* (pp. 651–679). Sage.

Checkland, P. (1981). *Systems Thinking, Systems Practice*. John Wiley & Sons.

Chilisa, B. (2012). *Indigenous research methodologies*. Sage Publications.

Cho, J., & Trent, A. (2006). Validity in qualitative research revisited. *Qualitative Research*, *6*(3), 319–340. doi:10.1177/1468794106065006

Cho, S., Crenshaw, K. W., & McCall, L. (2013). Toward a field of intersectionality studies: Theory, applications, and praxis. *Signs (Chicago, Ill.)*, *38*(4), 785–810. doi:10.1086/669608

Clandinin, D. J., & Connelly, F. M. (2000). *Narrative inquiry: Experience and story in qualitative research*. Jossey-Bass.

Clandinin, D. J., & Rosiek, J. (2007). Mapping a landscape of narrative inquiry. In D. J. Clandinin (Ed.), *Handbook of narrative inquiry: Mapping a methodology* (pp. 35–75). Sage. doi:10.4135/9781452226552.n2

Clark-Ibáñez, M. (2007). Inner-city children in sharper focus: Sociology of childhood and photo elicitation interviews. *Visual Studies*, *22*(2), 70–89.

Clifford, J. (1990). Notes on (field)notes. In R. Sanjek (Ed.), *Fieldnotes: The makings of anthropology* (pp. 47–70). Cornell University Press.

Clifford, J., & Marcus, G. E. (Eds.). (1986). *Writing culture: The poetics and politics of ethnography*. University of California Press. doi:10.1525/9780520946286

Coffey, A. (1999). The ethnographic self: Fieldwork and the representation of identity. *Sage (Atlanta, Ga.)*.

Coffey, A., & Atkinson, P. (1996). *Making Sense of Qualitative Data: Complementary Research Strategies*. Sage Publications.

Cohen, L., & Manion, L. (2000). *Research methods in education* (5th ed.). RoutledgeFalmer.

Collins, P. H. (2015). Intersectionality's Definitional Dilemmas. *Annual Review of Sociology*, *41*(1), 1–20. doi:10.1146/annurev-soc-073014-112142

Comaroff, J., & Comaroff, J. L. (1991). *Of revelation and revolution: Christianity, colonialism, and consciousness in South Africa* (Vol. 1). University of Chicago Press. doi:10.7208/chicago/9780226114477.001.0001

Confucius, & Lau, D. C. (1979). *Confucius: The Analects*. Penguin UK.

Corbin, J. M., & Strauss, A. (2008). Basics of qualitative research: Techniques and procedures for developing grounded theory. *Sage (Atlanta, Ga.)*. Advance online publication. doi:10.4135/9781452230153

Corbin, J., & Strauss, A. (2015). *Basics of qualitative research: Techniques and procedures for developing grounded theory*. Sage publications.

Couldry, N. (2010). Why voice matters: Culture and politics after neoliberalism. *Sage (Atlanta, Ga.)*. Advance online publication. doi:10.4135/9781446269114

Crane, T. (2001). *Elements of Mind: An Introduction to the Philosophy of Mind*. Oxford University Press.

Crenshaw, K. (1989). Demarginalizing the intersection of race and sex: A Black feminist critique of antidiscrimination doctrine, feminist theory and antiracist politics. *University of Chicago Legal Forum*, 139–167.

Creswell, J. W. (2014). *A concise introduction to mixed methods research*. Sage.

Creswell, J. W., & Creswell, J. D. (2017). *Research design: Qualitative, quantitative, and mixed methods approaches*. Sage Publications.

Creswell, J. W., Klassen, A. C., Plano Clark, V. L., & Smith, K. C. (2011). *Best practices for mixed methods research in the health sciences*. National Institutes of Health. doi:10.1037/e566732013-001

Creswell, J. W., & Miller, D. L. (2000). Determining validity in qualitative inquiry. *Theory into Practice*, *39*(3), 124–130. doi:10.1207/s15430421tip3903_2

Creswell, J. W., & Plano Clark, V. L. (2017). *Designing and conducting mixed methods research*. Sage publications.

Creswell, J. W., & Poth, C. N. (2017). *Qualitative inquiry and research design: Choosing among five approaches* (3rd ed.). Sage.

Crotty, M. (1998). The foundations of social research: Meaning and perspective in the research process. *Sage (Atlanta, Ga.)*.

Cunliffe, A. L. (2003). Reflexive inquiry in organizational research: Questions and possibilities. *Human Relations*, *56*(8), 983–1003. doi:10.1177/00187267030568004

Cutcliffe, J. R., & McKenna, H. P. (2004). Expert qualitative researchers and the use of audit trails. *Journal of Advanced Nursing*, *45*(2), 126–135. doi:10.1046/j.1365-2648.2003.02874.x PMID:14705996

Davidson, J. (2009). A new approach to transcription: The challenges of recording, transcribing, and transcribing data. *Research on Language and Social Interaction*, *42*(4), 367–379.

Davies, C. A. (2008). *Reflexive ethnography: A guide to researching selves and others*. Routledge.

Davies, D., & Dodd, J. (2002). Qualitative research and the question of rigor. *Qualitative Health Research*, *12*(2), 279–289. doi:10.1177/104973230201200211 PMID:11837376

De Fina, A., & Georgakopoulou, A. (2015). *The handbook of narrative analysis*. John Wiley & Sons. doi:10.1002/9781118458204

Decrop, A. (1999). Triangulation in qualitative tourism research. *Tourism Management*, *20*(1), 157–161. doi:10.1016/S0261-5177(98)00102-2

Dennett, D. C. (1987). *The Intentional Stance*. MIT Press.

Denzin, N. K. (1978). *Sociological methods: A sourcebook*. McGraw-Hill.

Denzin, N. K. (1989). *The research act: A theoretical introduction to sociological methods*. Prentice Hall.

Denzin, N. K. (1997). Interpretive ethnography: Ethnographic practices for the 21st century. *Sage (Atlanta, Ga.)*. Advance online publication. doi:10.4135/9781452243672

Denzin, N. K., & Giardina, M. D. (2016). Qualitative inquiry through a critical lens. *International Journal of Qualitative Studies in Education : QSE*, *29*(6), 731–753.

Denzin, N. K., & Lincoln, Y. S. (1994). *Handbook of qualitative research*. Sage Publications.

Denzin, N. K., & Lincoln, Y. S. (2005). *The Sage handbook of qualitative* Denzin, N. K., & Lincoln, Y. S. (2005). *The Sage handbook of qualitative research. Sage (Atlanta, Ga.)*.

Denzin, N. K., & Lincoln, Y. S. (2005). The Sage handbook of qualitative. Sage.

Denzin, N. K., & Lincoln, Y. S. (2008). Introduction: The discipline and practice of qualitative research. In *Strategies of qualitative inquiry* (pp. 1–43). Sage.

Denzin, N. K., & Lincoln, Y. S. (2011). *The SAGE Handbook of Qualitative Research*. Sage.

Denzin, N. K., & Lincoln, Y. S. (2018). *The Sage handbook of qualitative research*. Sage publications.

Derrida, J. (1967/1978). *Writing and Difference*. University of Chicago Press.

DeWalt, K. M., & DeWalt, B. R. (2011). *Participant observation: A guide for fieldworkers*. Rowman & Littlefield Publishers.

Dey, I. (1999). *Grounding grounded theory: Guidelines for qualitative inquiry*. Academic Press.

DiCicco-Bloom, B., & Crabtree, B. F. (2006). The qualitative research interview. *Medical Education*, *40*(4), 314–321. doi:10.1111/j.1365-2929.2006.02418.x PMID:16573666

Dickson-Swift, V., James, E. L., Kippen, S., & Liamputtong, P. (2007). Doing sensitive research: What challenges do qualitative researchers face? *Qualitative Research*, *7*(3), 327–353. doi:10.1177/1468794107078515

Dickson-Swift, V., James, E. L., Kippen, S., & Liamputtong, P. (2009). Researching sensitive topics: Qualitative research as emotion work. *Qualitative Research*, *9*(1), 61–79. doi:10.1177/1468794108098031

Dixon-Woods, M., Angell, E., Ashcroft, R. E., & Bryman, A. (2007). Written work: The social functions of Research Ethics Committee letters. *Social Science & Medicine*, *65*(4), 792–802. doi:10.1016/j.socscimed.2007.03.046 PMID:17490795

Doniger, W. (2010). *The Hindus: An Alternative History*. Oxford University Press.

Dove, E. S., Joly, Y., & Knoppers, B. M. (2017). Power to the people: A Wiki-Governance model for biobanks. *Genome Biology*, *12*(1), 4. PMID:28100256

Doyle, S. (2007). Member checking with older women: A framework for negotiating meaning. *Health Care for Women International*, *28*(10), 888–908. doi:10.1080/07399330701615325 PMID:17987459

Dreyfus, H. L. (1991). *Being-in-the-world: A Commentary on Heidegger's Being and Time, Division I*. MIT Press.

Dunn, L. B. (2005). Ethical issues in the study of bereavement: The opinions of bereaved adults. *Death Studies*, *29*(10), 883–905.

Durkheim, É. (1915). *The elementary forms of the religious life*. George Allen & Unwin.

Dwyer, S. C., & Buckle, J. L. (2009). The space between: On being an insider-outsider in qualitative research. *International Journal of Qualitative Methods*, *8*(1), 54–63. doi:10.1177/160940690900800105

Eakin, P. J. (2006). *The ethics of life writing*. Cornell University Press.

Eastmond, M. (2007). Stories as lived experience: Narratives in forced migration research. *Journal of Refugee Studies*, *20*(2), 248–264. doi:10.1093/jrs/fem007

Eide, P., & Kahn, D. (2008). Ethical issues in the qualitative researcher—Participant relationship. *Nursing Ethics*, *15*(2), 199–207. doi:10.1177/0969733007086018 PMID:18272610

Eisenhardt, K. M. (1989). Building theories from case study research. *Academy of Management Review*, *14*(4), 532–550. doi:10.2307/258557

Eisenhardt, K. M., & Graebner, M. E. (2007). Theory building from cases: Opportunities and challenges. *Academy of Management Journal*, *50*(1), 25–32. doi:10.5465/amj.2007.24160888

Eisenhart, M. (2009). Generalization from qualitative inquiry. In *Qualitative research* (pp. 51–66). Routledge.

Eldridge, S. M., Lancaster, G. A., Campbell, M. J., Thabane, L., Hopewell, S., Coleman, C. L., & Bond, C. M. (2016). Defining feasibility and pilot studies in preparation for randomised controlled trials: Development of a conceptual framework. *PLoS One*, *11*(3), e0150205. doi:10.1371/journal.pone.0150205 PMID:26978655

Elger, B. S., & Caplan, A. L. (2006). Consent and anonymization in research involving biobanks. *EMBO Reports*, *7*(7), 661–666. doi:10.1038/sj.embor.7400740 PMID:16819458

Elliott, N., & Luker, K. (2005). The role of the researcher in the qualitative research process. A potential barrier to archiving qualitative data. *Forum Qualitative Sozialforschung / Forum: Qualitative. Social Research*, *6*(3).

Ellis, C. (2007). Telling secrets, revealing lives: Relational ethics in research with intimate others. *Qualitative Inquiry*, *13*(1), 3–29. doi:10.1177/1077800406294947

Ellis, C., Adams, T. E., & Bochner, A. P. (2007). Autoethnography: An overview. *Forum Qualitative Social Research*, *12*(1), 10.

Ellis, C., & Berger, L. (2003). Their story/my story/our story: Including the researcher's experience in interview research. In J. F. Gubrium & J. A. Holstein (Eds.), *Postmodern interviewing* (pp. 157–183). SAGE. doi:10.4135/9781412985437.n9

Ellis, C., & Bochner, A. P. (2000). Autoethnography, personal narrative, reflexivity. In N. K. Denzin & Y. S. Lincoln (Eds.), *Handbook of qualitative research* (2nd ed., pp. 733–768). Sage.

Ellsberg, M., & Heise, L. (2002). Bearing witness: Ethics in domestic violence research. *Lancet*, *359*(9317), 1599–1604. doi:10.1016/S0140-6736(02)08521-5 PMID:12047984

Elmir, R., Schmied, V., Jackson, D., & Wilkes, L. (2011). Interviewing people about potentially sensitive topics. *Nurse Researcher*, *19*(1), 12–16. doi:10.7748/nr2011.10.19.1.12.c8766 PMID:22128582

Ely, M., Vinz, R., Downing, M., & Anzul, M. (1991). *On writing qualitative research: Living by words*. Falmer Press.

Emanuel, E. J., Wendler, D., & Grady, C. (2000). What makes clinical research ethical? *Journal of the American Medical Association*, *283*(20), 2701–2711. doi:10.1001/jama.283.20.2701 PMID:10819955

Emerson, R. M., Fretz, R. I., & Shaw, L. L. (1995). *Writing ethnographic fieldnotes*. University of Chicago Press. doi:10.7208/chicago/9780226206851.001.0001

Erickson, F. (1989). Ethical considerations in qualitative research. In R. Jaeger (Ed.), *Complementary methods for research in education*. American Educational Research Association.

Erlandson, D. A., Harris, E. L., Skipper, B. L., & Allen, S. D. (1993). Doing naturalistic inquiry: A guide to methods. *Sage (Atlanta, Ga.)*.

Erzberger, C., & Kelle, U. (2003). Making inferences in mixed methods: The rules of integration. In A. Tashakkori & C. Teddlie (Eds.), *Handbook of mixed methods in social & behavioral research* (pp. 457–488). Sage.

Etherington, K. (2004). Becoming a reflexive researcher: Using our selves in research. Jessica Kingsley Publishers.

Etherington, K. (2004). *Becoming a reflexive researcher: Using our selves in research*. Jessica Kingsley Publishers.

Etherington, K. (2007). Ethical research in reflexive relationships. *Qualitative Inquiry*, *13*(5), 599–616. doi:10.1177/1077800407301175

Evans-Pritchard, E. E. (1940). *The Nuer*. Clarendon Press.

Eysenbach, G., & Wyatt, J. (2002). Using the internet for surveys and health research. *Journal of Medical Internet Research*, *4*(2), e13. doi:10.2196/jmir.4.2.e13 PMID:12554560

Fabian, J. (1983). *Time and the Other: How Anthropology Makes Its Object*. Columbia University Press.

Faden, R. R., Beauchamp, T. L., & Kass, N. E. (2014). Informed consent, comparative effectiveness, and learning health care. *Journal of the American Medical Association*, *311*(4), 403–404. PMID:24552325

Farquhar, C., & Das, R. (1999). Are focus groups suitable for 'sensitive' topics? In R. S. Barbour & J. Kitzinger (Eds.), *Developing Focus Group Research: Politics, Theory and Practice*. SAGE. doi:10.4135/9781849208857.n4

Farrington, D. P. (1991). Longitudinal research strategies: Advantages, problems, and prospects. *Journal of the American Academy of Child and Adolescent Psychiatry*, *30*(3), 369–374. doi:10.1097/00004583-199105000-00003 PMID:2055872

Fereday, J., & Muir-Cochrane, E. (2006). Demonstrating rigor using thematic analysis: A hybrid approach of inductive and deductive coding and theme development. *International Journal of Qualitative Methods*, *5*(1), 80–92. doi:10.1177/160940690600500107

Fern, E. F. (2001). *Advanced focus group research*. SAGE. doi:10.4135/9781412990028

Fielding, N. (2012). Computer-aided qualitative data analysis. In J. Goodwin (Ed.), *SAGE secondary data analysis* (pp. 183–210). Sage Publications.

Fielding, N. G. (2012). Triangulation and mixed methods designs. *Journal of Mixed Methods Research*, *6*(2), 124–136. doi:10.1177/1558689812437101

Fielding, N. G., & Fielding, J. L. (1986). Linking data. *Sage (Atlanta, Ga.)*.

Fielding, N. G., & Lee, R. M. (1991). Using computers in qualitative research. *Sage (Atlanta, Ga.)*.

Fielding, N. G., & Lee, R. M. (1998). *Computer analysis and qualitative research*. Sage Publications.

Fine, M. (1994). Working the hyphens: Reinventing self and other in qualitative research. In N. K. Denzin & Y. S. Lincoln (Eds.), *Handbook of qualitative research* (pp. 70–82). Sage.

Finlay, L. (2002). "Outing" the Researcher: The Provenance, Process, and Practice of Reflexivity. *Qualitative Health Research*, *12*(4), 531–545. doi:10.1177/104973202129120052 PMID:11939252

Flick, U. (2018). *Doing triangulation and mixed methods*. Sage. doi:10.4135/9781529716634

Flory, J., & Emanuel, E. (2004). Interventions to improve research participants' understanding in informed consent for research: A systematic review. *Journal of the American Medical Association*, *292*(13), 1593–1601. doi:10.1001/jama.292.13.1593 PMID:15467062

Fluehr-Lobban, C. (1991). Ethical Problems of Fieldwork in the Newly Recognized States. *Anthropology Today*, *7*(4), 5–7.

Fluehr-Lobban, C. (2002). *Ethics and the profession of anthropology: Dialogue for a new era*. Altamira Press.

Fluehr-Lobban, C. (2013). *Ethics and the profession of anthropologists: Fieldwork in moral dilemmas*. Rowman Altamira.

Flyvbjerg, B. (2006). Five misunderstandings about case-study research. *Qualitative Inquiry*, *12*(2), 219–245. doi:10.1177/1077800405284363

Fontana, A., & Frey, J. H. (2000). The interview: From structured questions to negotiated text. In N. K. Denzin & Y. S. Lincoln (Eds.), *Handbook of qualitative research* (2nd ed., pp. 645–672). SAGE.

Foucault, M. (1969/1972). *The Archaeology of Knowledge*. Pantheon.

Foucault, M. (1977). *Discipline and Punish: The Birth of the Prison*. Vintage.

Freeman, M. (2010). *Hindsight: The promise and peril of looking backward*. Oxford University Press.

Frisch, M. (1990). *A shared authority: Essays on the craft and meaning of oral and public history*. SUNY Press.

Gantt, H. L. (1917). *Work, Wages, and Profits*. Engineering Magazine.

Gastel, B., & Day, R. A. (2016). *How to write and publish a scientific paper*. Greenwood. doi:10.5040/9798400666926

Gee, J. P. (1986). *Narrative, literacy, and face in interethnic communication*. Ablex Publishing.

Geertz, C. (1973). *The interpretation of cultures: Selected essays*. Basic Books.

George, A. L., & Bennett, A. (2005). *Case studies and theory development in the social sciences*. MIT Press.

Gerring, J. (2004). What is a case study and what is it good for? *The American Political Science Review*, *98*(2), 341–354. doi:10.1017/S0003055404001182

Gibbs, G. R. (2007). Analyzing qualitative data. *Sage (Atlanta, Ga.)*.

Giorgi, A. (2009). *The Descriptive Phenomenological Method in Psychology: A Modified Husserlian Approach*. Duquesne University Press.

Given, L. M. (2008). *The SAGE encyclopedia of qualitative research methods*. Sage Publications. doi:10.4135/9781412963909

Glaser, B. G. (1965). The constant comparative method of qualitative analysis. *Social Problems*, *12*(4), 436–445. doi:10.2307/798843

Glaser, B. G. (1978). *Theoretical sensitivity: Advances in the methodology of grounded theory*. Sociology Press.

Glaser, B. G., & Strauss, A. L. (1965). *Awareness of dying*. Aldine Publishing Company.

Glaser, B. G., & Strauss, A. L. (1967). *The Discovery of Grounded Theory: Strategies for Qualitative Research*. Aldine Transaction.

Glick Schiller, N., & Salazar, N. B. (2013). Regimes of mobility across the globe. *Journal of Ethnic and Migration Studies*, *39*(2), 183–200. doi:10.1080/1369183X.2013.723253

Goffman, E. (1959). *The Presentation of Self in Everyday Life*. Anchor Books.

Goldberg, L. R. (1965). Diagnosticians vs. diagnostic signs: The diagnosis of psychosis vs. neurosis from the MMPI. *Psychological Monographs*, *79*(9), 1–23. doi:10.1037/h0093885 PMID:14322679

Gold, R. L. (1958). Roles in sociological field observations. *Social Forces*, *36*(3), 217–223. doi:10.2307/2573808

Grant, C., & Osanloo, A. (2014). Understanding, selecting, and integrating a theoretical framework in dissertation research. *Administrative Issues Journal: Connecting Education, Practice, and Research, 4*(2), 12-26.

Greene, J. C. (2007). *Mixed methods in social inquiry*. Jossey-Bass.

Greene, J. C. (2008). Is mixed methods social inquiry a distinctive methodology? *Journal of Mixed Methods Research*, *2*(1), 7–22. doi:10.1177/1558689807309969

Greenwood, S., & Freeman, J. (2024). Exploring multimodal data analysis with AI in qualitative research. *Advances in Social Science Research*, *31*(1), 95–112.

Grele, R. J. (1985). Movement without aim: Methodological and theoretical problems in oral history. In R. J. Grele (Ed.), *Envelopes of sound: The art of oral history* (pp. 38–59). Praeger.

Guba, E. G. (1981). Criteria for assessing the trustworthiness of naturalistic inquiries. *Educational Communication and Technology*, *29*(2), 75–91. doi:10.1007/BF02766777

Guba, E. G., & Lincoln, Y. S. (1982). Epistemological and methodological bases of naturalistic inquiry. *Educational Communication and Technology*, *30*(4), 233–252. doi:10.1007/BF02765185

Guba, E. G., & Lincoln, Y. S. (1985). Naturalistic inquiry. *Sage (Atlanta, Ga.).*

Guba, E. G., & Lincoln, Y. S. (1994). Competing paradigms in qualitative research. In N. K. Denzin & Y. S. Lincoln (Eds.), *Handbook of qualitative research* (pp. 105–117). Sage.

Guba, E. G., & Lincoln, Y. S. (2005). Paradigmatic controversies, contradictions, and emerging confluences. *Sage (Atlanta, Ga.).*

Gubrium, A., & Harper, K. (2013). *Participatory visual and digital research in action*. Left Coast Press.

Guillemin, M., & Gillam, L. (2004). Ethics, reflexivity, and "ethically important moments" in research. *Qualitative Inquiry*, *10*(2), 261–280. doi:10.1177/1077800403262360

Halcomb, E. J., Davidson, P. M., & Hardaker, L. (2008). Using software to enhance the quality of transcription in qualitative research. *Nurse Researcher*, *15*(3), 11–21.

Hall, B., & Howard, K. (2008). A synergistic approach: Conducting mixed methods research with typological and systemic design considerations. *Journal of Mixed Methods Research*, *2*(3), 248–269. doi:10.1177/1558689808314622

Halpern, E. S., & Douglas, E. J. (2008). Audit trail. In L. M. Given (Ed.), *The SAGE encyclopedia of qualitative research methods* (Vol. 1, pp. 42–43). Sage Publications.

Hammersley, M. (1992). *What's Wrong with Ethnography? Methodological Explorations*. Routledge.

Hammersley, M. (2007). The issue of quality in qualitative research. *International Journal of Research & Method in Education*, *30*(3), 287–305. doi:10.1080/17437270701614782

Hammersley, M., & Atkinson, P. (2007). *Ethnography: Principles in practice*. Routledge.

Hancock, A. M. (2007). When multiplication doesn't equal quick addition: Examining intersectionality as a research paradigm. *Perspectives on Politics*, *5*(1), 63–79. doi:10.1017/S1537592707070065

Hankivsky, O., Grace, D., Hunting, G., Ferlatte, O., Clark, N., & Fridkin, A. (2010). An intersectionality-based policy analysis framework: Critical reflections on a methodology for advancing equity. *International Journal for Equity in Health*, *9*(1), 1–16. PMID:25492385

Harding, S. (1993). Rethinking standpoint epistemology: What is "strong objectivity?". In L. Alcoff & E. Potter (Eds.), *Feminist Epistemologies* (pp. 49–82). Routledge.

Harper, D. (2002). Talking about pictures: A case for photo elicitation. *Visual Studies*, *17*(1), 13–26. doi:10.1080/14725860220137345

Harper, M., & Cole, P. (2012). Member checking: Can benefits be gained similar to group therapy? *The Qualitative Report*, *17*(2), 510–517.

Hart, C. (1998). *Doing a literature review: Releasing the research imagination.* Sage Publications.

Hassan, Z., Schattner, P., & Mazza, D. (2006). Doing a pilot study: Why is it essential? *Malaysian Family Physician : the Official Journal of the Academy of Family Physicians of Malaysia, 1*(2-3), 70–73. PMID:27570591

Heidegger, M. (1927/1962). *Being and Time.* Harper & Row.

Herman, J. L. (2005). Justice from the victim's perspective. *Violence Against Women, 11*(5), 571–602. doi:10.1177/1077801205274450 PMID:16043563

Herodotus. (2003). The Histories. Trans. A. de Sélincourt. Penguin.

Hertz, R. (Ed.). (1997). *Reflexivity & voice.* Sage.

Hess, D. R. (2008). What is evidence-based practice? *Respiratory Care, 53*(10), 1317–1318. PMID:12962552

Hesse-Biber, S. (2010). *Mixed methods research: Merging theory with practice.* Guilford Press.

Hesse-Biber, S. (2011). Qualitative approaches to mixed methods practice. *Qualitative Inquiry, 17*(6), 455–467. doi:10.1177/1077800410364611

Hesse-Biber, S. (2017). *The practice of qualitative research: Engaging students in the research process.* Sage Publications.

Hesse-Biber, S. N. (2019). *Feminist Research: Exploring, Interrogating, and Transforming the Interconnections of Epistemology, Methodology, and Method.* Oxford University Press.

Hine, C. (2000). Virtual ethnography. *Sage (Atlanta, Ga.).*

Hine, C. (2015). *Ethnography for the Internet: Embedded, embodied and everyday.* Bloomsbury Publishing.

Holliday, A. (2007). *Doing and writing qualitative research.* Sage. doi:10.4135/9781446287958

Holloway, I., & Todres, L. (2003). The status of method: Flexibility, consistency and coherence. *Qualitative Research, 3*(3), 345–357. doi:10.1177/1468794103033004

Hopper, E. K., Bassuk, E. L., & Olivet, J. (2010). Shelter from the storm: Trauma-informed care in homelessness services settings. *The Open Health Services and Policy Journal, 3*(1), 80–100. doi:10.2174/1874924001003010080

Horkheimer, M. (1982). *Critical theory.* Seabury Press.

Horst, H. A., & Miller, D. (2012). *Digital anthropology.* Bloomsbury Publishing.

Husserl, E. (1970). *The Crisis of European Sciences and Transcendental Phenomenology.* Northwestern University Press.

Husserl, E. (2012). *Ideas: General Introduction to Pure Phenomenology.* Routledge. (Original work published 1913) doi:10.4324/9780203120330

Israel, M., & Hay, I. (2006). Research ethics for social scientists. *Sage (Atlanta, Ga.).*

Jackson, A. Y., & Mazzei, L. A. (2013). *Thinking with Theory in Qualitative Research.* Teachers College Press.

Jackson, J. (1983). *The real facts of life: Feminism and the politics of sexuality.* Taylor & Francis.

Janis, I. L. (1982). *Groupthink: Psychological studies of policy decisions and fiascoes.* Houghton Mifflin.

Jarratt, S. C. (1991). *Rereading the Sophists: Classical Rhetoric Refigured.* Southern Illinois University Press.

Jewitt, C., & Oyama, R. (2001). Visual meaning: A social semiotic approach. In T. van Leeuwen & C. Jewitt (Eds.), *Handbook of visual analysis* (pp. 134–156). Sage.

Jick, T. D. (1979). Mixing qualitative and quantitative methods: Triangulation in action. *Administrative Science Quarterly*, *24*(4), 602–611. doi:10.2307/2392366

Johnson, L. (2021). AI and the future of qualitative research. *Qualitative Research Journal*, *21*(2), 134–149.

Johnson, R. B., & Christensen, L. B. (2012). *Educational research: Quantitative, qualitative, and mixed approaches* (4th ed.). Sage Publications.

Johnson, R. B., & Onwuegbuzie, A. J. (2004). Mixed methods research: A research paradigm whose time has come. *Educational Researcher*, *33*(7), 14–26. doi:10.3102/0013189X033007014

Jones, K. (2007). *Narratives of identity and place*. Routledge.

Jones, R. H., Chik, A., & Hafner, C. A. (2015). *Discourse and digital practices: Doing discourse analysis in the digital age*. Routledge. doi:10.4324/9781315726465

Jorgensen, D. L. (1989). *Participant observation.* Sage. doi:10.4135/9781412985376

Jupp, V. (2006). The Sage dictionary of social research methods. *Sage (Atlanta, Ga.)*. Advance online publication. doi:10.4135/9780857020116

Kaiser, K. (2009). Protecting respondent confidentiality in qualitative research. *Qualitative Health Research*, *19*(11), 1632–1641. doi:10.1177/1049732309350879 PMID:19843971

Karnieli-Miller, O., Strier, R., & Pessach, L. (2009). Power relations in qualitative research. *Qualitative Health Research*, *19*(2), 279–289. doi:10.1177/1049732308329306 PMID:19150890

Kawulich, B. B. (2005). Participant observation as a data collection method. *Forum Qualitative Social Research*, *6*(2).

Kelle, U. (2005). “Emergence” vs. “Forcing” of empirical data? A crucial problem of “Grounded Theory” reconsidered. *Forum Qualitative Sozialforschung/Forum: Qualitative. Social Research*, *6*(2).

Khan, G., & Santos, E. (2023). Leveraging machine learning for thematic analysis in qualitative research. *Technology in Society*, *65*, 101312.

Kim, Y. (2011). The pilot study in qualitative inquiry. *Qualitative Social Work: Research and Practice*, *10*(2), 190–206. doi:10.1177/1473325010362001

Kincheloe, J. L., & McLaren, P. (2005). Rethinking critical theory and qualitative research. In N. K. Denzin & Y. S. Lincoln (Eds.), *The Sage handbook of qualitative research* (3rd ed., pp. 303–342). Sage.

Kindon, S., Pain, R., & Kesby, M. (2007). *Participatory action research approaches and methods: Connecting people, participation and place*. Routledge. doi:10.4324/9780203933671

King, N., & Horrocks, C. (2010). *Interviews in qualitative research.* SAGE.

Kirk, J., & Miller, M. L. (1986). Reliability and validity in qualitative research. *Sage (Atlanta, Ga.)*.

Kitzinger, J. (1994). The methodology of focus groups: The importance of interaction between research participants. *Sociology of Health & Illness*, *16*(1), 103–121. doi:10.1111/1467-9566.ep11347023

Knafl, K. A., & Breitmayer, B. J. (1989). *Triangulation in qualitative research: Issues of conceptual clarity and purpose.* Qualitative nursing research: A contemporary dialogue, 226-239.

Knoblauch, H. (2005). Focused ethnography. *Forum Qualitative Social Research*, *6*(3).

Koch, T. (2006). Establishing rigour in qualitative research: The decision trail. *Journal of Advanced Nursing*, *53*(1), 91–100. doi:10.1111/j.1365-2648.2006.03681.x PMID:16422698

Koro-Ljungberg, M., MacLure, M., & Rönnerman, K. (2017). *Post Qualitative and New Material Feminisms in Education: More 'Live' Methods*. Springer.

Kovats-Bernat, J. C. (2002). Negotiating dangerous fields: Pragmatic strategies for fieldwork amid violence and terror. *American Anthropologist*, *104*(1), 208–222. doi:10.1525/aa.2002.104.1.208

Krueger, R. A., & Casey, M. A. (2014). *Focus groups: A practical guide for applied research* (5th ed.). SAGE.

Kuhn, A. (2002). *Family secrets: Acts of memory and imagination*. Verso.

Kuhn, T. S. (1962). *The structure of scientific revolutions*. University of Chicago press.

Kuklick, H. (1997). *The savage within: The social history of British anthropology, 1885-1945*. Cambridge University Press.

Kuper, A. (1973). *Anthropologists and anthropology: The British school 1922-1972*. Penguin.

Kusow, A. M. (2003). Beyond indigenous authenticity: Reflections on the insider/outsider debate in immigration research. *Symbolic Interaction*, *26*(4), 591–599. doi:10.1525/si.2003.26.4.591

Labov, W. (1972). *Language in the inner city: Studies in the Black English vernacular*. University of Pennsylvania Press.

Labov, W., & Waletzky, J. (1967). Narrative analysis: Oral versions of personal experience. In J. Helm (Ed.), *Essays on the verbal and visual arts* (pp. 12–44). University of Washington Press.

Labov, W., & Waletzky, J. (1997). Narrative analysis: Oral versions of personal experience. *Journal of Narrative and Life History*, *7*(1-4), 3–38. doi:10.1075/jnlh.7.02nar

Lambert, J. (2010). *Digital storytelling: Capturing lives, creating community*. Routledge.

Landsberger, H. A. (1958). *Hawthorne Revisited*. Cornell University.

Langley, A. (1999). Strategies for theorizing from process data. *Academy of Management Review*, *24*(4), 691–710. doi:10.2307/259349

Lapadat, J. C., & Lindsay, A. C. (1999). Transcription in research and practice: From standardization of technique to interpretive positionings. *Qualitative Inquiry*, *5*(1), 64–86. doi:10.1177/107780049900500104

Lassiter, L. E. (2005). *The Chicago guide to collaborative ethnography*. University of Chicago Press. doi:10.7208/chicago/9780226467016.001.0001

Lather, P. (1991). *Getting Smart: Feminist Research and Pedagogy With/in the Postmodern*. Routledge. doi:10.4324/9780203451311

Lather, P. (2020). *Pedagogies of the Image: Critical and Cultural Studies in Education*. Springer.

Layder, D. (1998). Sociological practice: Linking theory and social research. *Sage (Atlanta, Ga.)*.

Leach, E. R. (1954). *Political Systems of Highland Burma*. Harvard University Press.

LeCompte, M. D., & Goetz, J. P. (1982). Problems of reliability and validity in ethnographic research. *Review of Educational Research*, *52*(1), 31–60. doi:10.3102/00346543052001031

Lecompte, M. D., & Schensul, J. J. (1999). *Designing & conducting ethnographic research*. Rowman & Littlefield.

LeCompte, M. D., & Schensul, J. J. (2010). *Designing and conducting ethnographic research: An introduction*. AltaMira Press.

Lee, R. M. (1993). *Doing research on sensitive topics*. Sage.

Lejeune, P. (2009). *On diary*. University of Hawai'i Press.

Lempert, L. B. (2007). Asking questions of the data: Memo writing in the grounded theory tradition. The SAGE Handbook of Grounded Theory, 245-264.

Leon, A. C., Davis, L. L., & Kraemer, H. C. (2011). The role and interpretation of pilot studies in clinical research. *Journal of Psychiatric Research*, *45*(5), 626–629. doi:10.1016/j.jpsychires.2010.10.008 PMID:21035130

Liamputtong, P. (2007). Researching the vulnerable: A guide to sensitive research methods. *Sage (Atlanta, Ga.)*.

Liamputtong, P. (2008). *Doing cross-cultural research: Ethical and methodological perspectives* (Vol. 34). Springer Science & Business Media. doi:10.1007/978-1-4020-8567-3_1

Li, H., Zhang, Y., & Wei, X. (2023). Automated thematic analysis powered by natural language processing. *Journal of Computer Assisted Research*, *39*(2), 213–230.

Lijphart, A. (1971). Comparative politics and the comparative method. *The American Political Science Review*, *65*(3), 682–693. doi:10.2307/1955513

Lincoln, Y. S., & Guba, E. A. (1985). *Naturalistic inquiry*. Sage. doi:10.1016/0147-1767(85)90062-8

Lincoln, Y. S., Lynham, S. A., & Guba, E. G. (2011). Paradigmatic controversies, contradictions, and emerging confluences, revisited. In N. K. Denzin & Y. S. Lincoln (Eds.), *The Sage Handbook of Qualitative Research* (4th ed., pp. 97–128). Sage.

Lofland, J., Snow, D. A., Anderson, L., & Lofland, L. H. (2006). *Analyzing social settings: A guide to qualitative observation and analysis*. Wadsworth/Thomson Learning.

Lofland, J., & Lofland, L. H. (1995). *Analyzing social settings: A guide to qualitative observation and analysis* (3rd ed.). Wadsworth.

Lundh, A. (2014). Aims and scope of journals are not enough. *Evidence-Based Medicine*, *19*(1), 36. PMID:23939598

Lyotard, J. F. (1979). *The postmodern condition: A report on knowledge*. Manchester University Press.

Macbeth, D. (2001). On "reflexivity" in qualitative research: Two readings, and a third. *Qualitative Inquiry*, *7*(1), 35–68. doi:10.1177/107780040100700103

Machi, L. A., & McEvoy, B. T. (2016). *The literature review: Six steps to success*. Corwin Press. doi:10.4135/9781071939031

Mackenzie, N., & Knipe, S. (2006). Research dilemmas: Paradigms, methods and methodology. *Issues in Educational Research*, *16*(2), 193–205.

MacQueen, K. M., McLellan, E., Kay, K., & Milstein, B. (1998). Codebook development for team-based qualitative analysis. *Field Methods*, *10*(2), 31–36.

Madison, D. S. (2005). Critical ethnography: Method, ethics, and performance. *Sage (Atlanta, Ga.)*.

Madrigal, L., & McClain, B. (2020). The real risks of sharing data. *Science*, *368*(6492), 716–718. PMID:32409464

Malinowski, B. (1922). *Argonauts of the Western Pacific*. Routledge & Kegan Paul.

Malinowski, B. (1944). *A Scientific Theory of Culture and Other Essays*. University of North Carolina Press.

Malinowski, B. (1967). *A diary in the strict sense of the term*. Stanford University Press.

Malterud, K. (2001). Qualitative research: Standards, challenges, and guidelines. *Lancet*, *358*(9280), 483–488. doi:10.1016/S0140-6736(01)05627-6 PMID:11513933

Mannay, D. (2010). Making the familiar strange: Can visual research methods render the familiar setting more perceptible? *Qualitative Research*, *10*(1), 91–111. doi:10.1177/1468794109348684

Manson, H. M. (2019). E-consent: The design and implementation of consumer consent mechanisms in an electronic environment. *Journal of Business Research*, *95*, 311–319.

Marcus, G. E. (1995). Ethnography in/of the world system: The emergence of multi-sited ethnography. *Annual Review of Anthropology*, *24*(1), 95–117. doi:10.1146/annurev.an.24.100195.000523

Marcus, G. E. (1998). *Ethnography through thick and thin*. Princeton University Press.

Marcus, G. E., & Fischer, M. M. (1986). *Anthropology as Cultural Critique: An Experimental Moment in the Human Sciences*. University of Chicago Press.

Markham, A. (2017). Ethnography in the Digital Internet Era: From fields to flows, descriptions to interventions. In N. Denzin & Y. Lincoln (Eds.), The Sage handbook of qualitative research (5th ed.). Sage.

Markham, A., & Buchanan, E. (2012). Ethical considerations in digital research contexts. In *The Oxford Handbook of Internet Studies* (pp. 307–322). Oxford University Press.

Marshall, C., & Rossman, G. B. (2014). *Designing qualitative research*. Sage Publications.

Marshall, P. A., & Koenig, B. A. (2004). Accounting for culture in a globalized bioethics. *The Journal of Law, Medicine & Ethics*, *32*(2), 252–266. doi:10.1111/j.1748-720X.2004.tb00472.x PMID:15301190

Martinez, L., & Rodriguez, A. (2024). Generative AI and simulation models in qualitative research: A new frontier. *Journal of Innovative Qualitative Research*, *12*(3), 457–476.

Mason, J. (2002). *Qualitative researching* (2nd ed.). SAGE.

Mathison, S. (1988). Why triangulate? *Educational Researcher*, *17*(2), 13–17. doi:10.2307/1174583

Mattingly, C. (1998). *Healing dramas and clinical plots: The narrative structure of experience*. Cambridge University Press. doi:10.1017/CBO9781139167017

Mauthner, N. S., & Doucet, A. (2003). Reflexive accounts and accounts of reflexivity in qualitative data analysis. *Sociology*, *37*(3), 413–431. doi:10.1177/00380385030373002

Maxwell, J. A. (2005). Qualitative research design: An interactive approach. *Sage (Atlanta, Ga.)*.

Maxwell, J. A. (2012). *Qualitative research design: An interactive approach* (3rd ed.). Sage.

Maxwell, J. A. (2012). The importance of qualitative research for causal explanation in education. *Qualitative Inquiry*, *18*(8), 655–661. doi:10.1177/1077800412452856

Maxwell, J. A. (2012a). A realist approach for qualitative research. *Sage (Atlanta, Ga.)*.

McCambridge, J., Witton, J., & Elbourne, D. R. (2014). Systematic review of the Hawthorne effect: New concepts are needed to study research participation effects. *Journal of Clinical Epidemiology*, *67*(3), 267–277. doi:10.1016/j.jclinepi.2013.08.015 PMID:24275499

McCann, L., & Pearlman, L. A. (1990). Vicarious traumatization: A framework for understanding the psychological effects of working with victims. *Journal of Traumatic Stress*, *3*(1), 131–149. doi:10.1007/BF00975140

Mead, M. (1928). *Coming of Age in Samoa*. William Morrow & Company.

Merleau-Ponty, M. (2012). *Phenomenology of Perception*. Routledge. (Original work published 1945)

Merriam, S. B. (1995). What can you tell from an N of 1?: Issues of validity and reliability in qualitative research. *PAACE Journal of Lifelong Learning*, *4*, 51–60.

Merriam, S. B. (1998). *Qualitative research and case study applications in education*. Jossey-Bass.

Merriam, S. B., & Tisdell, E. J. (2015). *Qualitative research: A guide to design and implementation*. Jossey-Bass.

Mertens, D. M., & Ginsberg, P. E. (2009). Deep in ethical waters: Transformative perspectives for qualitative social work research. *Qualitative Social Work, 8*(4), • Nash, J. C. (2008). Re-thinking intersectionality. *Feminist Review*, *89*(1), 1–15.

Mertens, D. M., & Ginsberg, P. E. (2009). Deep in ethical waters: Transformative perspectives for qualitative social work research. *Qualitative Social Work, 8*(4).

Mertens, D. M., & Ginsberg, P. E. (2009). Deep in ethical waters: Transformative perspectives for qualitative social work research. *Qualitative Social Work: Research and Practice*, *8*(4). Advance online publication. doi:10.1177/1473325008097142

Merton, R. K., Fiske, M., & Kendall, P. L. (1990). *The focused interview: A manual of problems and procedures* (2nd ed.). The Free Press.

Miles, M. B., & Huberman, A. M. (1994). Qualitative data analysis: An expanded sourcebook. *Sage (Atlanta, Ga.)*.

Miles, M. B., Huberman, A. M., & Saldaña, J. (2014). *Qualitative data analysis: A methods sourcebook* (3rd ed.). Sage.

Miles, M. B., Silverman, D., & Huberman, A. M. (2013). Qualitative Data Analysis: A Sourcebook. *Sage (Atlanta, Ga.)*.

Miller, T. (2015). Going back: Stalking, talking and researcher responsibilities in qualitative longitudinal research. *International Journal of Social Research Methodology*, *18*(3), 293–305. doi:10.1080/13645579.2015.1017902

Miller, T., Birch, M., Mauthner, M., & Jessop, J. (2005). *Ethics in qualitative research*. Sage.

Miller, T., Mauthner, M., & Birch, M. (2012). Ethics in qualitative research. *Sage (Atlanta, Ga.)*.

Minichiello, V., Aroni, R., Timewell, E., & Alexander, L. (1990). *In-depth interviewing: Researching people*. Longman Cheshire.

Mishler, E. G. (1991). Representing discourse: The rhetoric of transcription. *Journal of Narrative and Life History*, *1*(4), 255–280. doi:10.1075/jnlh.1.4.01rep

Mishler, E. G. (1995). Models of narrative analysis: A typology. *Journal of Narrative and Life History*, *5*(2), 87–123. doi:10.1075/jnlh.5.2.01mod

Mitchell, C. (2008). Getting the picture and changing the picture: Visual methodologies and educational research in South Africa. *South African Journal of Education*, *28*(3), 365–383. doi:10.15700/saje.v28n3a180

Molyneux, C. S., Peshu, N., & Marsh, K. (2004). Understanding of informed consent in a low-income setting: Three case studies from the Kenyan Coast. *Social Science & Medicine*, *59*(12), 2547–2559. doi:10.1016/j.socscimed.2004.03.037 PMID:15474208

Moran, D. (2000). *Introduction to Phenomenology*. Routledge.

Moran, D. (2017). *Husserl's Crisis of the European Sciences and Transcendental Phenomenology: An introduction*. Cambridge University Press.

Morgan, D. L. (1996). *Focus groups as qualitative research* (2nd ed.). SAGE.

Morgan, D. L. (2007). Paradigms lost and pragmatism regained: Methodological implications of combining qualitative and quantitative methods. *Journal of Mixed Methods Research*, *1*(1), 48–76. doi:10.1177/2345678906292462

Morrow, S. L. (2005). Quality and trustworthiness in qualitative research in counseling psychology. *Journal of Counseling Psychology*, *52*(2), 250–260. doi:10.1037/0022-0167.52.2.250

Morrow, V., & Richards, M. (1996). The ethics of social research with children: An overview. *Children & Society*, *10*(2), 90–105. doi:10.1002/(SICI)1099-0860(199606)10:2<90::AID-CHI14>3.0.CO;2-Z

Morse, J. M. (1991). Approaches to qualitative-quantitative methodological triangulation. *Nursing Research*, *40*(2), 120–123. doi:10.1097/00006199-199103000-00014 PMID:2003072

Morse, J. M. (1991). Strategies for sampling. In J. M. Morse (Ed.), *Qualitative nursing research: A contemporary dialogue* (pp. 127–145). Sage. doi:10.4135/9781483349015.n16

Morse, J. M. (2015). Critical analysis of strategies for determining rigor in qualitative inquiry. *Qualitative Health Research*, *25*(9), 1212–1222. doi:10.1177/1049732315588501 PMID:26184336

Morse, J. M., Barrett, M., Mayan, M., Olson, K., & Spiers, J. (2002). Verification strategies for establishing reliability and validity in qualitative research. *International Journal of Qualitative Methods*, *1*(2), 13–22. doi:10.1177/160940690200100202

Morse, J. M., & Niehaus, L. (2009). *Mixed method design: Principles and procedures*. Left Coast Press.

Motz, A. (1983). The diary as a transitional object in female adolescent development. *The Psychoanalytic Study of the Child*, *38*(1), 283–297.

Moustakas, C. (1994). Phenomenological research methods. *Sage (Atlanta, Ga.)*.

Mulhall, A. (2003). In the field: Notes on observation in qualitative research. *Journal of Advanced Nursing*, *41*(3), 306–313. doi:10.1046/j.1365-2648.2003.02514.x PMID:12581118

Mulhall, S. (2005). *Heidegger and Being and Time*. Routledge.

Murphy, E., & Dingwall, R. (2001). The Ethics of Ethnography. In P. Atkinson, A. Coffey, S. Delamont, J. Lofland, & L. Lofland (Eds.), *Handbook of Ethnography* (pp. 339–351). SAGE Publications. doi:10.4135/9781848608337.n23

Nadel, S. F. (1951). *The Foundations of Social Anthropology*. Cohen & West.

Nader, L. (1972). Up the anthropologist: Perspectives gained from studying up. In D. Hymes (Ed.), *Reinventing anthropology* (pp. 284–311). Pantheon Books.

National Commission for the Protection of Human Subjects of Biomedical and Behavioral Research. (1979). *The Belmont Report: Ethical principles and guidelines for the protection of human subjects of research*. Author.

Nicol, D., & Pexman, P. M. (2003). *Displaying your findings: A practical guide for creating figures, posters, and presentations*. American Psychological Association.

O'Cathain, A. (2010). Assessing the quality of mixed methods research: Toward a comprehensive framework. In A. Tashakkori & C. Teddlie (Eds.), *Handbook of mixed methods in social & behavioral research* (pp. 531–555). Sage. doi:10.4135/9781506335193.n21

O'Cathain, A., Murphy, E., & Nicholl, J. (2007). Why, and how, mixed methods research is undertaken in health services research in England: A mixed methods study. *BMC Health Services Research*, *7*(1), 1–11. doi:10.1186/1472-6963-7-85 PMID:17570838

O'Cathain, A., Murphy, E., & Nicholl, J. (2010). Three techniques for integrating data in mixed methods studies. *BMJ (Clinical Research Ed.)*, *341*(sep17 1), c4587. doi:10.1136/bmj.c4587 PMID:20851841

O'Connell, D. C., & Kowal, S. (1999). Transcription as a crucial step of data analysis. In *Speech, language, and the law: International perspectives* (pp. 89–99). Multilingual Matters.

O'Donoghue, T., & Punch, K. (2003). *Qualitative educational research in action: Doing and reflecting*. Routledge. doi:10.4324/9780203506301

O'Neill, T. (2022). Enhancing qualitative research with AI-driven text and speech recognition. *Methodological Innovations*, *25*(1), 58–73.

O'Reilly, K. (2012). *Ethnographic methods*. Routledge. doi:10.4324/9780203864722

Oliver, D. G., Serovich, J. M., & Mason, T. L. (2005). Constraints and opportunities with interview transcription: Towards reflection in qualitative research. *Social Forces*, *84*(2), 1273–1289. doi:10.1353/sof.2006.0023 PMID:16534533

Ong, W. J. (1982). *Orality and Literacy: The Technologizing of the Word*. Methuen. doi:10.4324/9780203328064

Onwuegbuzie, A. J., Leech, N. L., & Collins, K. M. (2012). Qualitative analysis techniques for the review of the literature. *The Qualitative Report*, *17*(56), 1–28.

Orb, A., Eisenhauer, L., & Wynaden, D. (2000). Ethics in qualitative research. *Journal of Nursing Scholarship*, *33*(1), 93–96. doi:10.1111/j.1547-5069.2001.00093.x PMID:11253591

Ortlipp, M. (2008). Keeping and using reflective journals in the qualitative research process. *The Qualitative Report*, *13*(4), 695–705.

Ortner, S. B. (2016). Dark anthropology and its others: Theory since the eighties. *HAU*, *6*(1), 47–73. doi:10.14318/hau6.1.004

Ottenberg, S. (1990). Thirty years of fieldnotes: Changing relationships to the text. In R. Sanjek (Ed.), *Fieldnotes: The makings of anthropology* (pp. 139–160). Cornell University Press.

Page, R. (2012). *Stories and social media: Identities and interaction*. Routledge.

Paltridge, B., & Starfield, S. (2007). *Thesis and dissertation writing in a second language: A handbook for supervisors*. London: Routledge.

Palys, T. (2008). *Purposive sampling*. The Sage encyclopedia of qualitative research methods, 2, 697-8.

Park, R. E., Burgess, E. W., & McKenzie, R. D. (1925). *The city*. University of Chicago Press.

Patel, S., & Davidson, R. (2023). The impact of machine learning on qualitative research methodologies. *AI & Society*, *38*(1), 11–29.

Patton, M. Q. (1999). Enhancing the quality and credibility of qualitative analysis. *Health Services Research*, *34*(5 Pt 2), 1189. PMID:10591279

Patton, M. Q. (2002). *Qualitative research and evaluation methods* (3rd ed.). Sage Publications.

Patton, M. Q. (2002). Two decades of developments in qualitative inquiry. *Qualitative Social Work: Research and Practice*, *1*(3), 261–283. doi:10.1177/1473325002001003636

Patton, M. Q. (2014). *Qualitative Research & Evaluation Methods: Integrating Theory and Practice*. Sage Publications.

Patton, M. Q. (2015). *Qualitative research & evaluation methods: Integrating theory and practice*. Sage publications.

Pennebaker, J. W. (1997). Writing about emotional experiences as a therapeutic process. *Psychological Science*, *8*(3), 162–166. doi:10.1111/j.1467-9280.1997.tb00403.x

Perks, R., & Thomson, A. (Eds.). (2006). *The oral history reader*. Routledge.

Peshkin, A. (1988). In search of subjectivity—One's own. *Educational Researcher*, *17*(7), 17–21.

Pettigrew, A. M. (1997). What is a processual analysis? *Scandinavian Journal of Management*, *13*(4), 337–348. doi:10.1016/S0956-5221(97)00020-1

Phoenix, C. (2008). Analyzing narrative data. In P. Liamputtong & J. Rumbold (Eds.), *Knowing differently: Arts-based and collaborative research methods* (pp. 77–89). Nova Science Publishers.

Piaget, J. (1952). *The origins of intelligence in children*. International Universities Press. doi:10.1037/11494-000

Pike, K. L. (1967). Emic and etic standpoints for the description of behavior. In J. L. Helm (Ed.), *Essays in the verbal and visual arts* (pp. 32–38). University of Washington Press. doi:10.1037/14786-002

Pillow, W. (2003). Confession, catharsis, or cure? Rethinking the uses of reflexivity as methodological power in qualitative research. *International Journal of Qualitative Studies in Education : QSE*, *16*(2), 175–196. doi:10.1080/0951839032000060635

Pink, S. (2007). Doing visual ethnography. *Sage (Atlanta, Ga.)*.

Pink, S. (2013). *Doing visual ethnography*. SAGE.

Pink, S. (2015). *Riessman, C. K. (2008). Narrative methods for the human sciences*. Sage Publications.

Pink, S. (2017). Doing Sensory Ethnography. *Sage (Atlanta, Ga.)*.

Pink, S., Horst, H., Postill, J., Hjorth, L., Lewis, T., & Tacchi, J. (2016). Digital Ethnography: Principles and Practice. *Sage (Atlanta, Ga.)*.

Plano Clark, V. L., & Creswell, J. W. (2008). *The mixed methods reader*. Sage.

Plano Clark, V. L., & Ivankova, N. V. (2016). *Mixed methods research: A guide to the field*. Sage Publications. doi:10.4135/9781483398341

Plummer, K. (2001a). *Documents of life 2: An invitation to a critical humanism*. Sage. doi:10.4135/9781849208888

Plummer, K. (2001b). The call of life stories in ethnographic research. In P. Atkinson, A. Coffey, S. Delamont, J. Lofland, & L. Lofland (Eds.), *Handbook of ethnography* (pp. 395–406). Sage. doi:10.4135/9781848608337.n27

Poland, B. D. (2002). Transcription quality. In *Handbook of interview research: Context & method* (pp. 629–649). Sage.

Polkinghorne, D. E. (1988). *Narrative knowing and the human sciences*. State University of New York Press.

Polkinghorne, D. E. (1991). Narrative and self-concept. *Journal of Narrative and Life History*, *1*(2&3), 135–153. doi:10.1075/jnlh.1.2-3.04nar

Polkinghorne, D. E. (2007). Validity issues in narrative research. *Qualitative Inquiry*, *13*(4), 471–486. doi:10.1177/1077800406297670

Ponterotto, J. G. (2006). Brief note on the origins, evolution, and meaning of the qualitative research concept thick description. *The Qualitative Report*, *11*(3), 538–549.

Portelli, A. (1991). *The death of Luigi Trastulli and other stories: Form and meaning in oral history*. SUNY Press.

Portelli, A. (1997). What makes oral history different. In R. Perks & A. Thomson (Eds.), *The oral history reader* (pp. 63–74). Routledge.

Procter, L. (2009). Fashioning the diary as a literary genre, 1660–1800. *Literature Compass*, *6*(2), 431–441.

Prosser, J., & Loxley, A. (2008). Introducing visual methods. *NCRM Research Methods Review Paper*, *10*, 1–11.

Punch, K. F. (2013). Introduction to social research: Quantitative and qualitative approaches. *Sage (Atlanta, Ga.)*.

Rabinow, P. (1977). *Reflections on Fieldwork in Morocco*. University of California Press.

Radcliffe-Brown, A. R. (1952). *Structure and function in primitive society*. Free Press.

Radley, A., & Taylor, D. (2003). Images of recovery: A photo-elicitation study on the hospital ward. *Qualitative Health Research*, *13*(1), 77–99. doi:10.1177/1049732302239412 PMID:12564264

Ramos, M. C. (1989). Some ethical implications of qualitative research. *Research in Nursing & Health*, *12*(1), 57–63. doi:10.1002/nur.4770120109 PMID:2922491

Randolph, J. J. (2009). A guide to writing the dissertation literature review. *Practical Assessment, Research, and Evaluation*, 14(13), 1-13.

Ravenscroft, J. (2017). The importance of journal impact factor in contemporary academic psychiatry and a list of selected journals. *Academic Psychiatry*, *41*(4), 507–511.

Ravitch, S. M., & Riggan, M. (2017). *Reason & rigor: How conceptual frameworks guide research*. Sage Publications.

Razack, S. H. (2002). Imperilled Muslim women, dangerous Muslim men and civilised Europeans: Legal and social responses to forced marriages. *Feminist Legal Studies*, *10*(1), 1–19. doi:10.1007/s10691-020-09426-2

Renold, E., Holland, S., Ross, N. J., & Hillman, A. (2008). Becoming participant: Problematizing informed consent in participatory research with young people. *Qualitative Social Work: Research and Practice*, *7*(4), 427–447. doi:10.1177/1473325008097139

Resnik, D. B. (2015). *What is ethics in research & why is it important?* National Institute of Environmental Health Sciences.

Richards, L., & Morse, J. M. (2012). *Readme first for a user's guide to qualitative methods*. Sage publications.

Richardson, L. (1997). *Fields of play: Constructing an academic life*. Rutgers University Press.

Richardson, L. (2000). Writing: A method of inquiry. In N. K. Denzin & Y. S. Lincoln (Eds.), *Handbook of qualitative research* (pp. 923–948). Sage.

Richardson, L., & Adams St. Pierre, E. (2005). Writing: A method of inquiry. In N. K. Denzin & Y. S. Lincoln (Eds.), *The Sage handbook of qualitative research* (3rd ed., pp. 959–978). Sage.

Ricoeur, P. (1981). Narrative time. In W. J. T. Mitchell (Ed.), *On narrative* (pp. 165–186). University of Chicago Press.

Ricoeur, P. (1984). *Time and narrative* (Vol. 1). University of Chicago Press.

Ridder, H. G. (2017). The theory contribution of case study research designs. *Business Research*, *10*(2), 281–305. doi:10.1007/s40685-017-0045-z

Riessman, C. K. (1993). *Narrative analysis*. Sage Publications.

Riessman, C. K. (2008). *Narrative methods for the human sciences*. Sage Publications.

Ritchie, D. A. (2003). *Doing oral history: A practical guide*. Oxford University Press.

Ritchie, J., Lewis, J., Nicholls, C. M., & Ormston, R. (Eds.). (2013). *Qualitative research practice: A guide for social science students and researchers*. Sage.

Roberts, B. (2002). *Biographical research*. Open University Press.

Rodgers, B. L. (2008). Audit trail. In L. M. Given (Ed.), *The Sage encyclopedia of qualitative research methods* (Vol. 1, pp. 44–45). Sage Publications.

Rolls, L., & Relf, M. (2006). Bracketing interviews: Addressing methodological challenges in qualitative interviewing in bereavement and palliative care. *Mortality*, *11*(3), 286–305. doi:10.1080/13576270600774893

Romdenh-Romluc, K. (2010). *Routledge Philosophy GuideBook to Merleau-Ponty and Phenomenology of Perception*. Routledge. doi:10.4324/9780203482896

Rosaldo, R. (1989). *Culture & Truth: The Remaking of Social Analysis*. Beacon.

Rose, G. (2016). *Visual methodologies: An introduction to researching with visual materials* (4th ed.). SAGE.

Rosenberg, B. A. (2011). The voice of the past: Oral history. In *D. R. M. Oral history and digital humanities* (pp. 39–58). Palgrave Macmillan.

Rosenthal, R. (1966). *Experimenter effects in behavioral research*. Appleton-Century-Crofts.

Rossman, G. B., & Rallis, S. F. (2017). *Learning in the field: An introduction to qualitative research*. Sage Publications. doi:10.4135/9781071802694

Rothbauer, P. M. (2008). Triangulation. In L. M. Given (Ed.), *The Sage encyclopedia of qualitative research methods* (pp. 892–894). Sage.

Roulston, K. (2010). *Reflective interviewing: A guide to theory and practice*. SAGE. doi:10.4135/9781446288009

Rubin, H. J., & Rubin, I. S. (2011). *Qualitative interviewing: The art of hearing data* (3rd ed.). SAGE.

Ryen, A. (2004). Ethical issues. In C. Seale (Ed.), *Qualitative research practice* (pp. 230–247). Sage. doi:10.4135/9781848608191.d20

Ryle, G. (1949). *The concept of mind*. Hutchinson.

Saldana, J. (2015). *The coding manual for qualitative researchers* (3rd ed.). Sage.

Sampson, H. (2004). Navigating the waves: The usefulness of a pilot in qualitative research. *Qualitative Research*, *4*(3), 383–402. doi:10.1177/1468794104047236

Sandelowski, M. (1995). Sample size in qualitative research. *Research in Nursing & Health, 18*(2), 179–183. doi:10.1002/nur.4770180211 PMID:7899572

Sanjek, R. (1990). On Ethnographic Validity. In R. Sanjek (Ed.), *Fieldnotes: The Makings of Anthropology* (pp. 385–418). Cornell University Press. doi:10.7591/9781501711954

Sartre, J.-P. (1943/1956). *Being and Nothingness: An Essay on Phenomenological Ontology*. Philosophical Library.

Sartre, J.-P. (1946). *Existentialism is a Humanism*. Yale University Press.

Saunders, B., Kitzinger, J., & Kitzinger, C. (2015). Anonymising interview data: Challenges and compromise in practice. *Qualitative Research, 15*(5), 616–632. doi:10.1177/1468794114550439 PMID:26457066

Savage, J. (2006). Ethnographic evidence: The value of applied ethnography in healthcare. *Journal of Research in Nursing, 11*(5), 383–393. doi:10.1177/1744987106068297

Schensul, J. J., Schensul, S. L., & LeCompte, M. D. (1999). *Essential ethnographic methods: Observations, interviews, and questionnaires*. Rowman Altamira.

Seidman, I. (2013). *Interviewing as qualitative research: A guide for researchers in education and the social sciences*. Teachers College Press.

Shadish, W. R., Cook, T. D., & Campbell, D. T. (2002). *Experimental and quasi-experimental designs for generalized causal inference*. Houghton Mifflin.

Shamoo, A. E., & Resnik, D. B. (2015). *Responsible conduct of research*. Oxford University Press.

Shenton, A. K. (2004). Strategies for ensuring trustworthiness in qualitative research projects. *Education for Information, 22*(2), 63–75. doi:10.3233/EFI-2004-22201

Sieber, J. E. (1992). *Planning ethically responsible research: A guide for students and internal review boards*. Sage Publications. doi:10.4135/9781412985406

Sieber, J. E., & Stanley, B. (1988). Ethical and professional dimensions of socially sensitive research. *The American Psychologist, 43*(1), 49–55. doi:10.1037/0003-066X.43.1.49 PMID:3348539

Silver, C., & Lewins, A. (2014). *Using software in qualitative research: A step-by-step guide*. Sage Publications. doi:10.4135/9781473906907

Silverman, D. (2006). *Interpreting qualitative data: Methods for analyzing talk, text, and interaction* (3rd ed.). SAGE.

Silverman, D. (2010). Doing qualitative research: A practical handbook. *Sage (Atlanta, Ga.)*.

Silverman, D. (2016). *Qualitative research* (4th ed.). Sage.

Sim, J., & Wright, C. C. (2000). *Research in health care: Concepts, designs and methods*. Stanley Thornes.

Singh, G., & Lee, A. (2024). Ethical considerations in the use of AI in qualitative research. *Ethics and Information Technology, 26*(2), 195–210.

Smith, B., & McGannon, K. R. (2018). Developing rigor in qualitative research: Problems and opportunities within sport and exercise psychology. *International Review of Sport and Exercise Psychology, 11*(1), 101–121. doi:10.1080/1750984X.2017.1317357

Smith, B., & Sparkes, A. C. (2008). Contrasting perspectives on narrating selves and identities: An invitation to dialogue. *Qualitative Research, 8*(1), 5–35. doi:10.1177/1468794107085221

Smith, D. W. (2003). Phenomenology. In E. N. Zalta (Ed.), *The Stanford Encyclopedia of Philosophy*. Stanford University.

Smith, J. A. (1978). The idea of health: A philosophical inquiry. *Advances in Nursing Science*. PMID:6782945

Smith, J. A. (2015). Qualitative psychology: A practical guide to research methods. *Sage (Atlanta, Ga.)*.

Smith, J. A. (2018). *Qualitative psychology: A practical guide to research methods*. Sage Publications.

Smith, J. A., & Davis, J. M. (2017). The role of external peer review in improving research design: An example from the qualitative literature. *Journal of Advanced Nursing*, *73*(11), 2573–2579.

Smith, J. A., & Osborn, M. (2003). Interpretative phenomenological analysis. In J. A. Smith (Ed.), *Qualitative psychology: A practical guide to research methods* (pp. 51–80). Sage.

Smith, J., & Dugdale, S. (2020). The role of AI in qualitative data collection: Opportunities and challenges. *Qualitative Research in Psychology*, *17*(3), 456–477.

Smith, J., & Firth, J. (2011). Qualitative data analysis: The framework approach. *Nurse Researcher*, *18*(2), 52–62. doi:10.7748/nr2011.01.18.2.52.c8284 PMID:21319484

Smith, J., & Hodkinson, P. (2005). Relativism, criteria, and politics. In *The Sage handbook of qualitative research* (3rd ed., pp. 915–931). Sage Publications.

Smith, L. T. (1999). *Decolonizing Methodologies: Research and Indigenous Peoples*. Zed Books.

Smith, L. T. (2012). *Decolonizing methodologies: Research and indigenous peoples*. Zed Books Ltd.

Smith, S., & Watson, J. (2010). *Reading autobiography: A guide for interpreting life narratives* (2nd ed.). University of Minnesota Press.

Smithson, J. (2000). Using and analysing focus groups: Limitations and possibilities. *International Journal of Social Research Methodology*, *3*(2), 103–119. doi:10.1080/136455700405172

Somers, M. R. (1994). The narrative constitution of identity: A relational and network approach. *Theory and Society*, *23*(5), 605–649. doi:10.1007/BF00992905

Spall, S. (1998). Peer debriefing in qualitative research: Emerging operational models. *Qualitative Inquiry*, *4*(2), 280–292. doi:10.1177/107780049800400208

Spence, D. P. (1982). *Narrative truth and historical truth: Meaning and interpretation in psychoanalysis*. WW Norton & Company.

Spivak, G. C. (1988). Can the subaltern speak? In C. Nelson & L. Grossberg (Eds.), Marxism and the interpretation of culture (pp. 271-313). Macmillan.

Spradley, J. P. (1979). *The ethnographic interview*. Holt, Rinehart, and Winston.

Squire, C. (2008). Experience-centred and culturally-oriented approaches to narrative. In M. Andrews, C. Squire, & M. Tamboukou (Eds.), *Doing narrative research* (pp. 41–63). Sage. doi:10.4135/9780857024992.d4

St. Pierre, E. A. (2018). Post-Qualitative Inquiry: From Representation to Rhizome. In *Handbook of Posthumanism in Film and Television* (pp. 1–12). Springer.

Stake, R. E. (1995). *The art of case study research*. Sage Publications.

Stanley, L. (2007). The epistolarium: On theorizing letters and correspondences. *Auto/Biography*, *12*(2), 201–235.

Stanley, L. (Ed.). (1992). *The auto/biographical I: The theory and practice of feminist auto/biography*. Manchester University Press.

Stewart, D. W., Shamdasani, P. N., & Rook, D. W. (2007). *Focus groups: Theory and practice* (2nd ed.). SAGE. doi:10.4135/9781412991841

Stoecker, R. (1991). Evaluating and rethinking the case study. *The Sociological Review*, *39*(1), 88–112. doi:10.1111/j.1467-954X.1991.tb02970.x

Strauss, A., & Corbin, J. (1990). *Basics of qualitative research: Grounded theory procedures and techniques*. Sage Publications, Inc.

Strauss, A., & Corbin, J. (1998). Basics of qualitative research: Procedures and techniques for developing grounded theory. *Sage (Atlanta, Ga.)*.

Tang, K. L., & Chung, L. Y. F. (2006). Conducting qualitative projects overseas: Personal accounts and practical tips. *Qualitative Research Journal*, *6*(2), 113–129.

Tashakkori, A., & Teddlie, C. (2010). *Handbook of mixed methods in social & behavioral research* (2nd ed.). Sage. doi:10.4135/9781506335193

Teddlie, C., & Tashakkori, A. (2009). Foundations of mixed methods research: Integrating quantitative and qualitative approaches in the social and behavioral sciences. *Sage (Atlanta, Ga.)*.

Tedlock, B. (2000). Ethnography and ethnographic representation. Handbook of Qualitative Research, 2, 455-486.

Temple, B., & Young, A. (2004). Qualitative research and translation dilemmas. *Qualitative Research*, *4*(2), 161–178. doi:10.1177/1468794104044430

Thabane, L., Ma, J., Chu, R., Cheng, J., Ismaila, A., Rios, L. P., Robson, R., Thabane, M., Giangregorio, L., & Goldsmith, C. H. (2010). A tutorial on pilot studies: The what, why, and how. *BMC Medical Research Methodology*, *10*(1), 1. doi:10.1186/1471-2288-10-1 PMID:20053272

Thomas, D. R. (2006). A general inductive approach for analyzing qualitative evaluation data. *The American Journal of Evaluation*, *27*(2), 237–246. doi:10.1177/1098214005283748

Thomas, D. R. (2017). Feedback from research participants: Are member checks useful in qualitative research? *Qualitative Research in Psychology*, *14*(1), 23–41. doi:10.1080/14780887.2016.1219435

Thomas, E., & Magilvy, J. K. (2011). Qualitative rigor or research validity in qualitative research. *Journal for Specialists in Pediatric Nursing*, *16*(2), 151–155. doi:10.1111/j.1744-6155.2011.00283.x PMID:21439005

Thompson, R., Zook, M., & Boyd, D. (2023). Navigating ethical dilemmas in AI-powered qualitative research. *Journal of Research Ethics*, *19*(4), 401–422.

Thomson, P. (2008). *Doing visual research with children and young people*. Routledge.

Thurmond, V. A. (2001). The point of triangulation. *Journal of Nursing Scholarship*, *33*(3), 253–258. doi:10.1111/j.1547-5069.2001.00253.x PMID:11552552

Tickle-Degnen, L. (2013). Nuts and bolts of conducting feasibility studies. *The American Journal of Occupational Therapy*, *67*(2), 171–176. doi:10.5014/ajot.2013.006270 PMID:23433271

Tilley, L., & Woodthorpe, K. (2011). Is it the end for anonymity as we know it? A critical examination of the ethical principle of anonymity in the context of 21st-century demands on the qualitative researcher. *Qualitative Research*, *11*(2), 197–212. doi:10.1177/1468794110394073

Tilley, S. A. (2003). "Challenging" research practices: Turning a critical lens on the work of transcription. *Qualitative Inquiry*, *9*(5), 750–773. doi:10.1177/1077800403255296

Tolich, M. (2004). Internal confidentiality: When confidentiality assurances fail relational informants. *Qualitative Sociology*, *27*(1), 101–106. doi:10.1023/B:QUAS.0000015546.20441.4a

Tracy, S. J. (2010). Qualitative quality: Eight "big-tent" criteria for excellent qualitative research. *Qualitative Inquiry*, *16*(10), 837–851. doi:10.1177/1077800410383121

Tracy, S. J. (2013). *Qualitative research methods: Collecting evidence, crafting analysis, communicating impact*. John Wiley & Sons.

Tranfield, D., Denyer, D., & Smart, P. (2003). Towards a methodology for developing evidence-informed management knowledge by means of systematic review. *British Journal of Management*, *14*(3), 207–222. doi:10.1111/1467-8551.00375

Trippany-Simmons, R., Dankoski, M. E., & Penn, C. D. (2019). Beyond member checking: Advancing rigor through participatory method. *Journal of Marital and Family Therapy*, *45*(2), 235–247.

Tuhiwai Smith, L. (2012). *Decolonizing methodologies: Research and Indigenous peoples*. Zed Books.

Tullu, M. S., & Karande, S. (2018). Writing a model research paper: A roadmap. *Journal of Postgraduate Medicine*, *64*(4), 212. PMID:29943738

Turner, V. (1957). *Schism and Continuity in an African Society*. Manchester University Press.

Turner, V. (1967). *The forest of symbols: Aspects of Ndembu ritual*. Cornell University Press.

Van de Ven, A. H. (1992). Suggestions for studying strategy process: A research note. *Strategic Management Journal*, *13*(Summer Special Issue), 169-188.

Van Maanen, J. (1988). *Tales of the field: On writing ethnography*. University of Chicago Press.

van Manen, M. (1990). *Researching Lived Experience: Human Science for an Action Sensitive Pedagogy*. State University of New York Press.

Van Teijlingen, E. R., & Hundley, V. (2001). The importance of pilot studies. *Social Research Update*, *35*(1), 1–4. PMID:11328433

Varela, F. J., Thompson, E., & Rosch, E. (1991). *The Embodied Mind: Cognitive Science and Human Experience*. MIT Press. doi:10.7551/mitpress/6730.001.0001

Vygotsky, L. S. (1978). *Mind in society: The development of higher psychological processes*. Harvard University Press.

Wang, Y., & Kosinski, M. (2018). Deep neural networks are more accurate than humans at detecting sexual orientation from facial images. *Journal of Personality and Social Psychology*, *114*(2), 246–257. doi:10.1037/pspa0000098 PMID:29389215

Warden, R. (2013). The emotional cost of caring in qualitative research. *Procedia: Social and Behavioral Sciences*, *82*, 144–147.

Warren, C. A. B., & Karner, T. X. (2010). *Discovering qualitative methods: Field research, interviews, and analysis* (2nd ed.). Oxford University Press.

Wassenaar, D. R., & van der Linde, M. N. (2016). Ethical issues and qualitative research in the 21st century: Challenges and solutions. *Qualitative Health Research, 26*(13), 1829–1835.

Webb, E. J., Campbell, D. T., Schwartz, R. D., & Sechrest, L. (1966). *Unobtrusive measures: Nonreactive research in the social sciences*. Rand McNally.

Weber, M. (1949). Objectivity in Social Science and Social Policy. In E. A. Shils & H. A. Finch (Eds.), *The Methodology of the Social Sciences*. Free Press.

White, H. (1981). The value of narrativity in the representation of reality. *Critical Inquiry, 7*(1), 5–27. doi:10.1086/448086

Whyte, W. F. (1943). *Street corner society: The social structure of an Italian slum*. University of Chicago Press.

Wiles, R., Charles, V., Crow, G., & Heath, S. (2008). Researching researchers: Lessons for research ethics. *Qualitative Research, 8*(2), 283–299.

Wiles, R., Crow, G., Heath, S., & Charles, V. (2008). The management of confidentiality and anonymity in social research. *International Journal of Social Research Methodology, 11*(5), 417–428. doi:10.1080/13645570701622231

Wiles, R., Crow, G., & Pain, H. (2011). Innovation in qualitative research methods: A narrative review. *Qualitative Research, 11*(5), 587–604. doi:10.1177/1468794111413227

Willig, C. (2013). *Introducing qualitative research in psychology*. McGraw-Hill Education.

Wolcott, H. F. (1995). *The art of fieldwork*. AltaMira Press.

Wolcott, H. F. (2001). Writing up qualitative research. *Sage (Atlanta, Ga.).*

Wolcott, H. F. (2005). Transforming qualitative data: Description, analysis, and interpretation. *Sage (Atlanta, Ga.).*

Yin, R. K. (2003). *Case study research: Design and methods* (3rd ed.). Sage Publications.

Yin, R. K. (2014). *Case study research design and methods* (5th ed.). Sage Publications.

Yin, R. K. (2018). *Case study research and applications: Design and methods*. Sage publications.

Yow, V. R. (2005). *Recording oral history: A guide for the humanities and social sciences*. AltaMira Press.

Zahavi, D. (2005). *Husserl's Phenomenology*. Stanford University Press.

Zimmer, M. (2010). "But the data is already public": On the ethics of research in Facebook. *Ethics and Information Technology, 12*(4), 313–325. doi:10.1007/s10676-010-9227-5

Zimmerman, D. H., & Wieder, D. L. (1977). The diary: Diary-interview method. *Urban Life, 5*(4), 479–498. doi:10.1177/089124167700500406

Zinsser, W. (1989). *Inventing the truth: The art and craft of memoir*. Houghton Mifflin Harcourt.

Zook, M., & Boyd, D. (2022). Critical questions for Big Data in qualitative research. *Information Communication and Society, 25*(12), 1674–1693.

About the Author

Hesham Mohamed Elsherif is a distinguished scholar and practitioner, holding a Master's degree in Library and Information Science, a Doctorate in Management and Organizational Leadership, and a Doctorate in Information Systems and Technology. With an illustrious career spanning over two decades, Dr. Elsherif has established himself as an eminent library professional, having served as a manager librarian for 22 years. In addition, he has imparted his vast knowledge and expertise as an Adjunct Associate Professor for 15 years, teaching courses in technology, management and organizational leadership, and research methodologies. Dr. Elsherif's academic and professional endeavors are underscored by his proficiency in empirical research methodologies, including quantitative, qualitative, and mixed methods. His contributions to the field are substantial, having authored 36 books encompassing a wide range of topics such as technology, artificial intelligence, self-help, and management and leadership.

Index

N

O

P

Q

R

S

T

V

Printed in the USA
CPSIA information can be obtained
at www.ICGtesting.com
LVHW081804041124
795688LV00044B/1390

9 798369 346365